C0-BOF-815

OXFORD MONOGRAPHS ON GEOLOGY AND GEOPHYSICS NO. 19

OXFORD MONOGRAPHS ON GEOLOGY AND GEOPHYSICS

1. DeVerle P. Harris: *Mineral resources appraisal: mineral endowment, resources, and potential supply; concepts, methods and cases*
2. J. J. Veevers (ed.): *Phanerozoic earth history of Australia*
3. Yang Zunyi, Wang Hogzhen and Cheng Yuqi (eds.): *The geology of China*
4. Lin-gun Liu and William A. Bassett: *Elements, oxides, and silicates: high pressure phases with implications for the earth's interior*
5. Antoni Hoffman and Matthew H. Nitecki (eds.): *Problematic fossil taxa*
6. S. Mahmood Naqvi and John J. W. Rogers: *Precambrian geology of India*
7. Chih-Pei Chang and T. N. Krishnamurti (eds.): *Monsoon meteorology*
8. Zvi-Ben-Avraham (ed.): *The evolution of the Pacific Ocean margins*
9. Ian McDougall and T. Mark Harrison: *Geochronology and thermochronology by the* $^{40}Ar/^{39}Ar$ *method*
10. Walter C. Sweet: *The Conodonta: morphology, taxonomy, paleoecology, and evolutionary history of a long-extinct animal phylum*
11. H. J. Melosh: *Impact cratering: a geologic process*
12. J. W. Cowie and M. D. Brasier (eds.): *The Precambrian-Cambrian boundary*
13. C. S. Hutchinson: *Geological evolution of southeast Asia*
14. Anthony J. Naldrett: *Magmatic sulfide deposits*
15. D. R. Prothero and R. M. Schoch (eds.): *The evolution of perissodactyls*
16. M. Menzies (ed.): *Continental mantle*
17. R. J. Tingey (ed.): *Geology of the Antarctic*
18. Thomas J. Crowley and Gerald R. North: *Paleoclimatology*
19. Gregory J. Retallack: *Miocene paleosols and ape habitats of Pakistan and Kenya*

MIOCENE PALEOSOLS AND APE HABITATS OF PAKISTAN AND KENYA

GREGORY J. RETALLACK
University of Oregon

New York Oxford
OXFORD UNIVERSITY PRESS CLARENDON PRESS
1991

Dedicated to Nicholas, Jeremy and the dreams of youth

Oxford University Press

Oxford New York Toronto
Delhi Bombay Calcutta Madras Karachi
Petaling Jaya Singapore Hong Kong Tokyo
Nairobi Dar es Salaam Cape Town
Melbourne Auckland

and associated companies in
Berlin Ibadan

Published by Oxford University Press, Inc.,
200 Madison Avenue, New York, New York 10016

Oxford is a registered trademark of Oxford University Press

Library of Congress Cataloging-in-Publication Data
Miocene paleosols and ape habitats of Pakistan and Kenya/ Gregory J. Retallack
346 p. 26 cm. - (Oxford monographs on geology and geophysics; no. 19)
Includes bibliographical references (p. 298) and index (p. 323).
ISBN 0-19-506002-4
1. Paleopedology — Pakistan. 2. Paleopedology — Kenya.
3. Apes, fossil. 4. Paleontology — Miocene.
5. Paleontology — Pakistan. 6. Paleontology — Kenya.
I. Title. II. Series.
QE473.R472 1991 90-25410
560′.45 — dc20

Printing 9 8 7 6 5 4 3 2 1

Printed in the United States of America
on acid-free paper

PREFACE

One of the purest joys offered by paleopedological studies is the opportunity to travel, not only in space, but also in time. For me, raised and educated in Australia and now enjoying a second home in the Pacific Northwest of the United States, there are hardly two more exotic or different locales then Kenya and Pakistan. One is a stark rural desert of the sort caricatured in bible stories from Sunday school. The other is home to those splendid mammalian faunas so prevalent still in children's books. Both now are bending to the ways of western civilization, but much can still be found of the old India of Rudyard Kipling and E. H. Forrester, or of the untamed Kenya of Karen Blixen, Beryl Markham and Ernest Hemingway. As different as they seem now, both had diverse wildlife during Miocene time when apelike ancestors of humans were widespread through the Old World tropics. Most of what we know about our apelike ancestors is based on collections of fossil apes from the Himalayan foothills of Pakistan and India and from the volcanic hills of southwestern Kenya. The work reported here has focused on these important fossil sites. It is an attempt to travel back in time to the days well before human evolution, using as evidence for ancestral habitats the ancient soils on which these creatures lived and died. A paleosol, or fossil soil, is not only a distinctive bed in a sedimentary or volcanic sequence. It is also a trace fossil of an ancient ecosystem in the very place where it once flourished. An attempt at reconstructing the habitats of these Miocene apes is made in a final section of this book, but much also has been learned of the detailed appearance and composition of paleosols, and how paleosols are recognized and altered after burial. This work has been a grand adventure and education. May similar research continue in the same spirit.

Eugene, Oregon G.J. Retallack
January 1991

Acknowledgments

This research was made possible by a variety of federal and private grants of funds. Initial fieldwork in Pakistan in 1981 was done with the help and direction of Kay Behrensmeyer and John Barry, as part of a joint expedition of the Geological Survey of Pakistan and Yale University directed by David Pilbeam and S. Ibrahim Shah. This expedition was funded by Smithsonian Foreign Currency grant 80-254100. Thanks are due to my Pakistani driver Mulik Ghulam Raza. Pakistani colleagues Muhammad Asif and Khalid Ayub Sheikh offered varied local assistance, including an introduction to Hindi movies. Thanks also are due to other expedition members, Andrew Hill, Lisa Tauxe, David Archibald, Larry Flynn and Will Downs.

Initial fieldwork in Kenya in 1984 was supported by a grant from the Wenner-Gren Foundation for Anthropological Research. This and a subsequent trip in 1987 depended on official permission from Ms C. A. Mwango of the Office of the President of Kenya, and sponsorship of Mr. R. E. Leakey of the National Museums of Kenya. Martin Pickford's advice and the generous hospitality of Mr. Norman Brooks of Homa Lime Company were invaluable on this first trip. Also helpful were caretakers at the National Monuments, Wambara Musomba, Joseph Odongo, and Joseph Tuda at Fort Ternan and Elias Suchen and Matias Daya at Songhor. On a second trip to Kenya in 1987, Erick Bestland, Daniel Dugas, and Glenn Thackray helped in various ways while pursuing their own thesis studies. Ann Rockette Ndege and Peter Okombo made camping on Rusinga Island quite civilized. Laboratory work in Eugene, Oregon, and the second trip to Kenya were funded by National Science Foundation grants EAR8206183 and EAR8503232.

Meave Leakey helped with curatorial arrangements. Jessica Bondy, Catharine Clifton, Sandra Newman, Janet Puckett, Connie Scolla and Lu Willis prepared numerous excellent thin sections from crumbling claystone samples. Sterling Cook, Robert Goodfellow, Grant Smith and Barbara Tegge helped with point counting the thin sections. Rebecca Kimbel and Michael Shaffer aided with scanning electron microscopy. John Baham, Sterling Cook, Mark Humphrey and Andy Ungerer helped with x-ray diffractometry. Chemical analyses were performed by Christine McBirney, Colin Cool, Don Horneck and Tony Irving. The basic data of this book could not have been obtained so readily or pleasantly without their help.

For discussion and guidance during the long period of reflection on the meaning of these data I thank Peter Andrews, Erick Bestland, Peter Birkeland, Greg Botha, Thomas Bown, Stan Buol, Daniel Dugas, Leon Follmer, Richard Hay, Mary Kraus, Rob Lander, S.N. Rajaguru, Pat Shipman, Glenn Thackray, and John Wolff. They made it easy and fun, as did Darby Dyar, who introduced me to the joys of word processing with Word Perfect© on the Garnet computer net she established for the Department of Geological Sciences of the University of Oregon.

I am also grateful to the following individuals and organizations for permission to redraw copyright material (figure numbers in parentheses): © 1961 figs 46 and 87 from *Woody plants of Ghana* by F. R. Irvine, Oxford University Press (6.2); © 1986 figs 3 and 10 from *Anacardiaceae* in *Flora of tropical East Africa*, by J. O. Kokwaro and Royal Botanical Gardens (Kew), A. A. Balkema (6.2); © 1987 reconstruction of *Rangwapithecus* from *The field guide to early man* by D. Lambert and The Diagram Group (6.3); © 1988 reconstructions of *Proconsul, Limnopithecus* and *Dendropithecus* from *Primate adaptation and evolution* by John Fleagle, Academic Press (6.3); © 1986 reconstructions of *Hyaenodon, Chalicotherium* and *Sivapithecus* from *Mammal evolution* by R. J. G. Savage and M. R. Long, British Museum of Natural History (6.3, 6.10); © 1958 and 1972, figs 201, and also 9 and 10, respectively, from *Connaraceae*, and from *Rhamnaceae* in *Flora of tropical East Africa*, by J. Hutchinson, J. M. Dalziel, and R. W. J. Keay, and by M. C. Johnston, Crown Agents for Oversea Governments (6.6); © 1960 fig. 8 from *Meliaceae* in *Die naturliche Pflanzenfamilien* by H. Harms, Duncker & Humblot (6.9); © 1976 figs 27, 56, 65 from *Forest flora of the Bombay Presidency and Sind* by W. A. Talbot, Today's and Tomorrow's Publishers (6.9); © 1986 reconstruction of hipparionine horse by G. Paul from *Earth and life through time* by S. M. Stanley, W. H. Freeman and company (6.10); © 1942 neg. #281080 of *Choerolophodon corrugatus* courtesy Department of Library Services, American Museum of Natural History (6.10).

Last but not least, thanks to Diane, Nicholas and Jeremy Retallack for adapting so gracefully to the unreasonable demands of a book such as this.

CONTENTS

1. Introduction, 3

2. Features of the fossil soils, 9

2.1 Traces of life, 9
2.1.1 Root traces, 9
2.1.2 Plant remains, 13
2.1.3 Invertebrate fossils, 15
2.1.4 Burrows and other trace fossils, 15
2.1.5 Fossil bones, 16
2.2 Soil horizons, 17
2.2.1 Clayey surface horizons, 17
2.2.2 Subsurface clayey horizons, 18
2.2.3 Subsurface calcareous nodules and stringers, 20
2.2.4 Subsurface horizons rich in iron-manganese, 26
2.2.5 Chemical depth functions, 27
2.2.6 Paleokarst, 27
2.3 Soil structure, 34
2.3.1 Cracking and veining, 34
2.3.2 Claystone grains, 36
2.3.3 Destruction of mineral grains, 37
2.3.4 Microfabric of clayey matrix, 38
2.3.5 Clay minerals, 39
2.4 Recognizing paleosols, 41

3. Alteration of paleosols after burial, 43

3.1 Loss of organic matter, 43
3.2 Burial gleization, 43
3.3 Burial reddening, 45
3.4 Calcite cementation, 46
3.5 Zeolitization, 46
3.6 Compaction, 47
3.7 Thermal maturation of organic matter, 49
3.8 Illitization, 49
3.9 Feldspathization, 51
3.10 Recrystallization, 51
3.11 Peering through the veil of alteration, 51

4. Early and middle Miocene paleosols in southwestern Kenya, 53

4.1 Geological setting, 55
4.1.1 Geology of the area around Koru, 59
4.1.2 Geology of the area around Songhor National Monument, 68
4.1.3 Geology of the area around Fort Ternan National Monument, 72

4.2 Description and interpretation of the paleosols, 83
4.2.1 Mobaw Series, 85
4.2.2 Kwar Series, 91
4.2.3 Kiewo Series, 95
4.2.4 Choka Series, 100
4.2.5 Tut Series, 107
4.2.6 Buru Series, 111
4.2.7 Lwanda Series, 113
4.2.8 Rairo Series, 118
4.2.9 Rabuor Series, 121
4.2.10 Dhero Series, 125
4.2.11 Chogo Series, 130
4.2.12 Onuria Series, 140
4.3 Summary and prospects for Kenyan Miocene paleosols, 145

5. Late Miocene paleosols in northern Pakistan, 153

5.1 Geological Setting, 154
5.1.1 Geology of the area around Khaur, 158
5.2 Description and interpretation of the paleosols, 172
5.2.1 Sarang Series, 172
5.2.2 Bhura Series, 177
5.2.3 Khakistari Series, 183
5.2.4 Sonita Series, 187
5.2.5 Lal Series, 192
5.2.6 Pila Series, 196
5.2.7 Pandu Series, 202
5.2.8 Kala Series, 206
5.2.9 Naranji Series, 211
5.3 Summary and prospects for Miocene paleosols in Indo-Pakistan, 217

6. Reconstructed life and landscapes of the Old World Tropics, 227

6.1 Early Miocene in Kenya, 228
6.2 Middle Miocene in Kenya, 235
6.3 Late Miocene in Pakistan, 241
6.4 What, if anything, is characteristic of tropical paleosols? 249
6.5 Expansion of grasslands in the Old World tropics, 252
6.6 Initial steps in the evolution of human ancestors from apes, 256

Appendix 1. Individual named Miocene paleosols, 261
Appendix 2. Munsell colors of Miocene paleosols, 262
Appendix 3. Textures and calcareousness of Miocene paleosols, 274
Appendix 4. Mineral composition of Miocene paleosols, 278
Appendix 5. Major element chemical analyses and density of paleosols, 282
Appendix 6. Trace element chemical analyses of Miocene paleosols, 285
Appendix 7. Molecular weathering ratios of Miocene paleosols, 287
Appendix 8. Early Miocene (20 Ma) fossils from Koru and Songhor, Kenya, 290
Appendix 9. Middle Miocene (14 Ma) fossils from Fort Ternan, Kenya, 293
Appendix 10. Late Miocene (8.5-7.8 Ma) fossils from Khaur, Pakistan, 296
References, 298
Index, 323

1. INTRODUCTION

For if such holy song
Enwrap our fancy long
Time will run back, and fetch the age of gold
And speckled vanity
Will sicken soon and die.
John Milton (1629)

The idea of a long-distant time of human harmony with nature in a world unsullied by human folly is a persistent theme of Western culture. The Golden Age for classical Greek and Roman poets, playwrights, and philosophers was a time when humans lived in a land fit for gods, compared with the dross and practical Bronze and Iron Ages. In Judaeo-Christian tradition, the Garden of Eden was an oasis of natural abundance, in which humans were created, unlike the deforested and dry Mediterranean lands of historic experience. These twin themes of human origins and of landscapes untrammeled by the ravages of civilization are of scientific as well as philosophical interest. They are the main motivations for this study of Miocene paleosols of the Old World tropics.

Buried soils of the distant geological past can be considered evidence of soil-forming conditions during their formation, by analogy with environmental factors influential in modern soil formation (Retallack, 1990a). For example, a dry climate can be inferred from the presence of a shallow horizon of calcareous nodules in a paleosol (Arkley, 1963; de Wit, 1978; Jenny, 1941; Sehgal, Sys & Bhumbla, 1968). Some of the factors important in soil formation, such as vegetation, also have been thought important to human origins.

A large group of theories on human origins implicate the change from forest and woodland to open and wooded grassland. These grassland vegetation types often are imprecisely called "savanna" (M. M. Cole, 1986). Here the term *wooded grassland* is used for grassy vegetation with 10 to 40 percent cover by trees (White, 1983). In such open vegetation, the transition from quadrupedal apes to erect human ancestors has been envisaged as promoted by martial needs to throw projectiles during big-game hunting, or to see potential predators at a distance (Ardrey, 1976; Dart, 1957). Also relying on the advent of open vegetation, an opportunistic set of theories stress the value of erect stance in covering large distances in order to locate carcasses to scavenge (Shipman, 1986b; Szalay, 1975), or in freeing the hands to manipulate smaller fruits and seeds than available in woodlands and forests (Jolly, 1970). Other theories are independent of vegetation, seeking to explain the transition to erect stance as an adaptation for provisioning a family in a home base (Lovejoy, 1981). Thus a careful analysis of the likely ancient vegetation of paleosols containing fossils important to human ancestry may constrain speculation on this fundamental problem of human origins.

The special value of paleosols in this enterprise stems not only from their independence from other paleontological lines of evidence, but from the simple fact that fossil soils remain in the place of their formation. They cannot be transported and remain recognizable, like a fossil bone for example. Particular profile forms and associated trace fossils, such as root traces, reflect particular paleoenvironmental conditions, including ancient climate and vegetation (Fig. 1.1). As shown by many

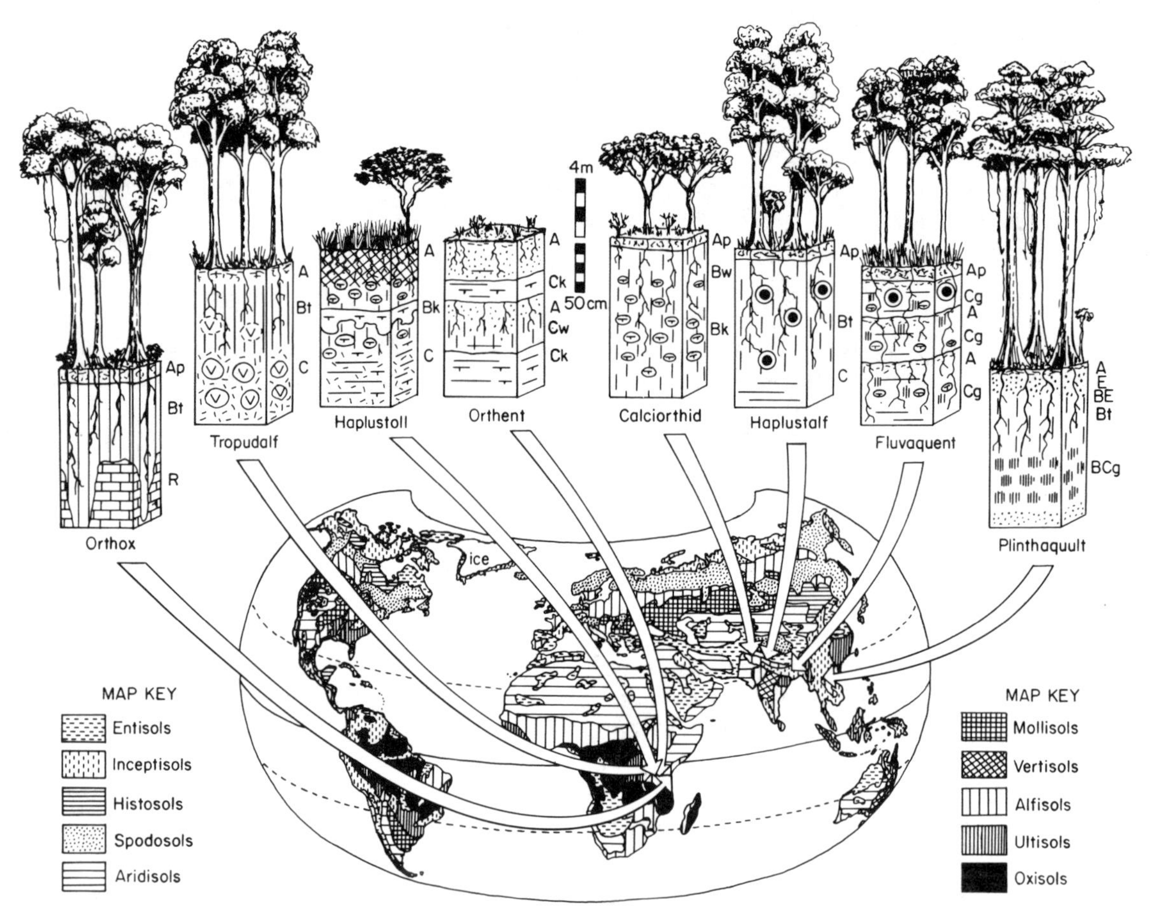

Fig. 1.1. Vegetation and climate is reflected in the profile form of soils, as shown by profiles selected for comparison with the paleosols described in this book. Key shown is for map (after Buol, Hole, & Mc Cracken, 1980). For key to lithological symbols in profiles see Figs 4.12 and 5.17. Soil profiles (from left) include profile 8 of Muchena & Sombroek (1981) on coral (Orthox), Sossok Series of Thorp *et al.* (1960) on nephelinite agglomerate (Tropudalf), profile 44 of de Wit (1978) on carbonatite-nephelinite tuff (Haplustoll), Watuni profile of Anderson (1963) on carbonatite-nephelinite tuff (Orthent), profile 1 of Sehgal, Sys, & Bhumbla (1968) on alluvium (Calciorthid), profile 2 of same authors (Haplustalf), Kanagarh Series of Kooistra (1982) on alluvium (Fluvaquent), and profile near Chang Mai of Food and Agriculture Organization (1979) on alluvium (Plinthaquult). Estimated natural vegetation is shown even though some of the profiles are now plowed (Ap horizon).

detailed surveys (Agarwal & Mukherji, 1951; Andrews, Groves & Horne, 1975; Bogdan, 1958; Botha, 1985; Courty & Féderoff, 1985; Gertenbach, 1983; Harrop, 1960; Touber, Van der Pouw & Van Engelen, 1982), soils reflect local mosaics of vegetation and other environmental conditions. Ancient mosaics of vegetation have proven very difficult to map from the study of fossil bones, shells, and plants, but mosaics of vegetation figure prominently in some theories of human evolution (Foley, 1987; Kortlandt, 1983). Miocene paleosols at sites well known for fossil mammals have much to offer in resolving paleoenvironmental heterogeneity and environmental preferences of extinct creatures.

A baseline study of landscapes unscarred by the effects of civilization is

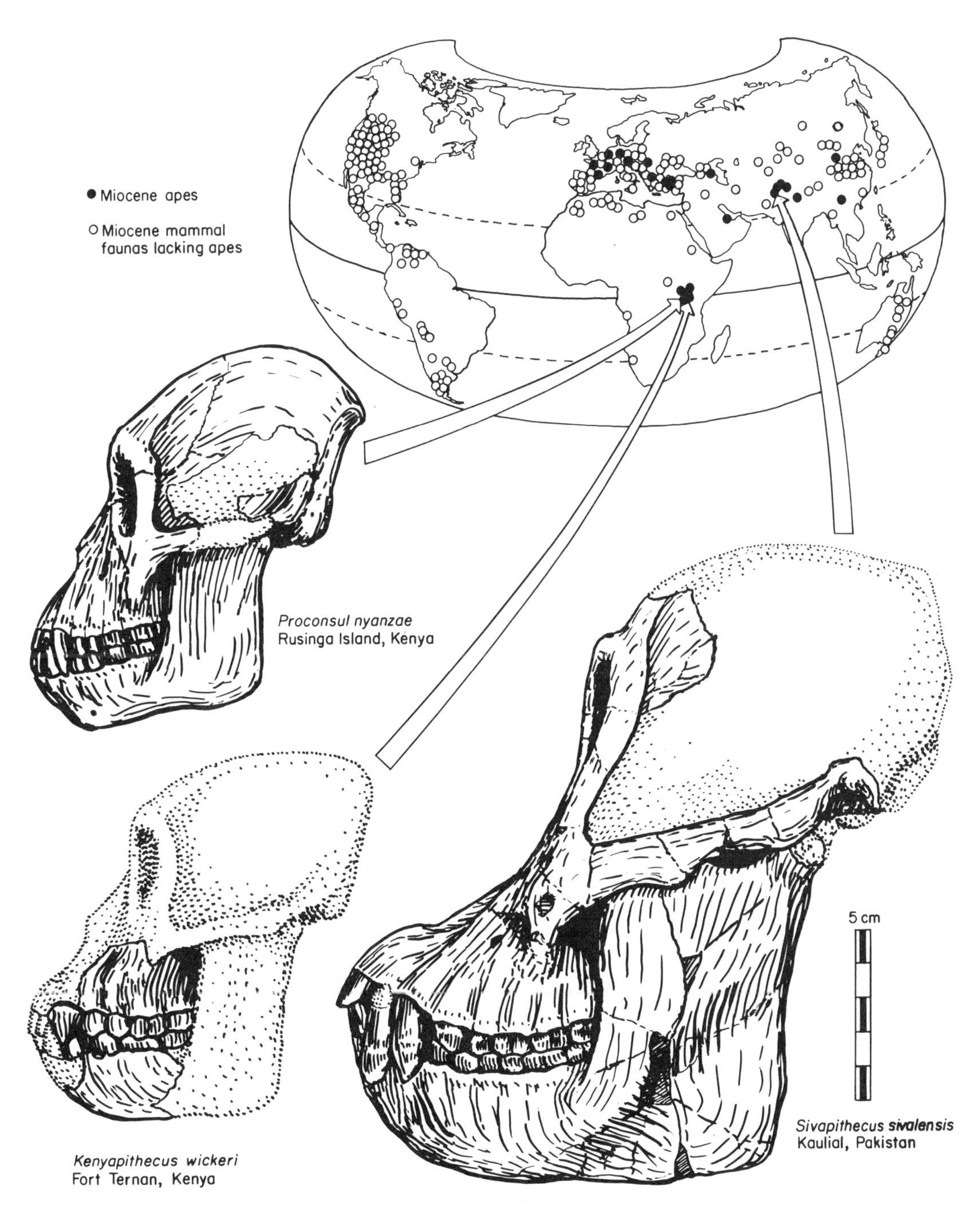

Fig. 1.2. The global distribution of Miocene fossil mammalian faunas without apes (white dots) and with apes (black dots), with restorations of selected fossil ape skulls. Fossil sites compiled from data of Bernor (1983), Savage & Russell (1983), Munthe *et al.* (1983); ape skulls from A. C. Walker & Andrews (1973), A. C. Walker *et al.* (1983), Ward & Pilbeam (1983), Fleagle & Kay (1983), Pilbeam (1984).

Fig. 1.3. Several well known fossil sites for Miocene apes in paleosols near Songhor, Koru and Fort Ternan, are scattered around the carbonatite-nephelinite volcanic center of Tinderet. The deeply eroded basanite cone of 5 million years ago is visible here on the skyline, viewed from near Muhoroni sugar factory, southwestern Kenya.

an important control for understanding of surficial processes such as soil formation. It is no longer an exaggeration to claim that no part of the world has escaped human influence. The urban jungles of New York and Athens, and the tailored agricultural landscapes of England and Illinois, are only the most obvious cases of such influence. Humans have from time immemorial altered the landscape by firing woodland to encourage fodder for game, to aid in hunting, to reduce insect pests, or to combat enemies. The use of fire by human ancestors is well established back at least 500,000 years and may extend back as far as 1.4 million years (Gowlett *et al.*, 1981; Brain *et al.*, 1988). Even in national parks, fire management and other policies have had significant effects on soils and ecosystems (Romme & Despain, 1989). How soils and landscapes form is more or less veiled by human influence. Miocene paleosols can be regarded as evidence of soil formation well before human impact.

Information on how soils form also has cultural biases. From the Russian origins of soil science, usually traced back to the monograph on grassland soils by Dokuchaev (1883), soils of northern, temperate, glaciated lands have had a privileged position in the development of soil science theory and terminology (Birkeland, 1984). Much fine research on tropical soils and landscapes is now redressing this imbalance. The global soil maps of the Food and Agriculture Organization (1974, 1977a, 1977b, 1979) of the United Nations Educational, Scientific, and Cultural Organization (UNESCO) are outstanding summaries. Intertropical land areas are greater than those affected by glaciation, and paleoclimatic studies ind-

Fig. 1.4. The most productive source of Miocene fossil apes in northern Pakistan is in badlands of clayey and sandy alluvium of ancient Himalayan outwash some 8 million years old, seen here near the village of Thaganwali Dhok, northern Pakistan.

icate that ice ages like the one in which we find ourselves are atypical of the long sweep of geological history (Frakes, 1979). Thus a study of Miocene tropical paleosols dating back to before human modification of the landscape may have value for constructing general models of soil formation.

These twin problems of human origins and tropical soil formation without human influence are more than can be adequately addressed by a single study such as this, not because suitable paleosols are rare, but because there are so many from which to choose. Miocene paleosols in sedimentary sequences containing fossil apes and human ancestors are widespread in the Old World, from France to China and south into Africa at least as far as Kenya and Tanzania (Fig. 1.2). There are many thousands of paleosols in the highly fossiliferous volcaniclastic sequences of southwestern Kenya, and most of the fossil bones and snails in this region come from paleosols (Bishop & Whyte, 1962; Pickford, 1986b; Pickford & Andrews, 1981). Similarly, paleosols are abundant in the extraordinarily thick (6 km according to Wadia, 1975) Miocene alluvial outwash from the Himalayan ranges: the Siwalik Group of northern Pakistan and India (Behrensmeyer, 1987; Behrensmeyer & Tauxe, 1982; G. D. Johnson *et al.*, 1981). Miocene paleosols in association with fossil apes are known in other parts of the world also (Badgley *et al.*, 1988; Bestland, 1990a), but the thick sequences around Kenyan carbonatite-nephelinite volcanoes (Fig. 1.3) and quartzofeldspathic alluvium of the Potwar Plateau of northern Pakistan (Fig. 1.4) were chosen here for several reasons beyond the fact that they are rich in paleosols.

Sequences of paleosols associated with alkaline volcanic centers (Retallack & Dilcher, 1988) and with quartzofeldspathic outwash of major mountain ranges (Retallack, 1985) as old as Paleozoic are known elsewhere. A study of Miocene examples serves as a near modern analog for the interpretation of much more ancient paleosols, for which soil forming factors such as vegetation may have been very different.

In addition, alkaline volcanic rocks like the Kenyan examples and calcareous paleosols like the Pakistani examples commonly preserve fossil bone unusually well (Pickford, 1986b; Retallack, 1984a). Some of the Miocene paleosols in Kenya and Pakistan have proven to be extraordinarily fossiliferous, with fossil apes such as *Proconsul*, *Kenyapithecus*, and *Sivapithecus*. These creatures have been considered significant to human ancestry (L. S. B. Leakey, 1962; Pickford & Andrews, 1981; Pilbeam, 1982; Shipman *et al.*, 1981). Although *Sivapithecus* from Pakistan was once considered a plausible distant ancestor of humans (Lewin, 1987), it is now thought to represent a side branch to our evolutionary tree and to be instead, ancestral to Asiatic apes such as the living orang-utan, *Pongo pygmaeus* (Andrews & Cronin, 1982). *Proconsul*, *Kenyapithecus*, and other Kenyan fossil apes are more likely ancestors of chimpanzees, gorillas, and australopithecine ancestors of humans (Andrews, 1986). Because of their unusually great paleoanthropological interest, this study has focused on paleosols containing these fossil apes, together with associated paleosols of their ancient landscapes.

This book is thus a detailed account of a few paleosols selected from among many because of their pedological and paleoanthropological interest. It remains for others to demonstrate how representative these paleosols are of Miocene tropical pedogenesis or of the habitats of human evolution from apes. This work is merely an outline of a new way of approaching these enduring problems of natural history through the study of paleosols.

2. FEATURES OF THE FOSSIL SOILS

Paleosols have been recognized in Miocene rocks of East Africa and Indo-Pakistan for many years (Bishop & Whyte, 1962; G. D. Johnson, 1977; Pilgrim, 1913). These studies have been careful to indicate the reasons why a particular bed was interpreted as a paleosol rather than a graded bed, a volcanic mudflow, a mudcracked surface, a hydrothermally altered horizon or a zone of intrastratal alteration. These concerns remain as important as ever. This chapter outlines the various features used to recognize paleosols within their enclosing rocks. Interpretation of these features is included where appropriate. Modification of the paleosols and their features after burial are discussed in chapter 3. Names for different kinds of paleosols, used in this chapter merely as convenient labels, are defined in chapters 4 and 5.

2.1 Traces of life

To many soil scientists (Buol, Hole and McCracken, 1980), it is life that distinguishes soils from sediments and rocks. Traces of land life, such as root traces and burrows, are the best possible evidence for the existence of paleosols. This is not to say that paleosols or soils must show traces of life. Because of chemical and other conditions prevailing during formation and burial of a soil, very little of their original life is preserved (Retallack, 1984a). Few Precambrian paleosols or Antarctic soils show much trace of life, because little life was present (Retallack, 1990a). In the case of Miocene paleosols from Kenya and Pakistan, much research has been directed toward their rich fossil record of mammals (Barry *et al.*, 1985; Pickford, 1986a; Pickford & Andrews, 1981; Shipman *et al.*, 1981). These paleosols also contain a variety of fossil root traces, burrows, and other useful paleontological indicators of their nature.

2.1.1 Root traces

Both burrows and root traces are abundant and diverse in paleosols of Pakistan and Kenya. In some cases they are difficult to distinguish because root traces show varying degrees of encrustation with calcium carbonate or because burrows have been invaded by root traces. In general, root traces are irregularly shaped, branching and tapering downward, whereas burrows are parallel sided, branching systematically, if at all.

Calcareous rhizoconcretions are especially common in the Miocene paleosols from Pakistan (Fig. 2.1). In some cases they are so encrusted with carbonate as to appear like nodules (Fig. 2.2). In thin section, their central tube, where the root once lived, is filled with sparry calcite or siltstone. Their encrustation, or *neocalcan* in the terminology of Brewer (1976), is micritic, with concentric zones of iron staining. The inner contact of the micrite with the calcite crystal tube or silt-filled tube is sharp. The outer boundary of the micrite is gradational into the paleosol matrix, as in calcareous nodules in the same paleosols. The calcite crystal tubes may be a cement precipitated from shallow groundwater in the hole left by the decaying root shortly after burial. This sparry calcite has all the appearance of a cavity fill, and lacks the regularity of structure seen in plant-mediated calcite crystals (Jaillard, 1987) or the irregular crystal boundaries of recrystallized carbonate (Bathurst, 1975). The siltstone filling the root traces has the character of silty coatings found in cavities in seasonally flooded soils of Bangladesh ("gleyans"

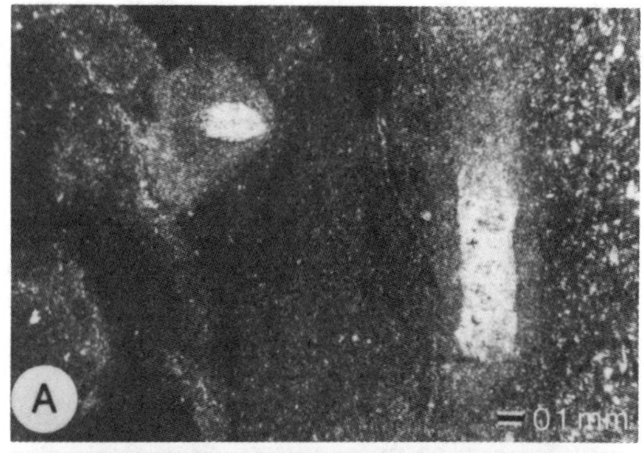

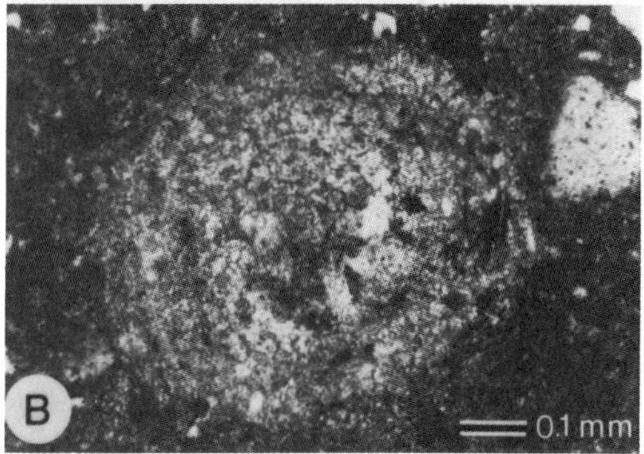

Fig. 2.1. Photomicrographs of calcareous rhizoconcretions: A, slender calcite crystal tubes after fine roots, with micritic neocalcans, from C horizon of type Sarang clay paleosol: B, cross-section of concentrically banded micrite around root tube filled with calcite and silt, from A horizon of Bhura clay ferric-concretionary variant paleosol: C, calcite crystal tubes intermingled with ferruginized margins and micritic neocalcans, from 2Bk horizon of Pila clay nodular-variant paleosol. All specimens from lower measured section in Dhok Pathan Formation, west of Kaulial village, Pakistan. Smithsonian specimens 353769, 353822 and 353791 respectively. Scale bars 0.1 mm.

of Brammer, 1971). Generally similar rhizoconcretions are widespread in calcareous dune sands in humid climates and in noncalcareous soils of dry climates (Cohen, 1982; Semeniuk & Meagher, 1971). The small size, limited length, prominent internal lamination, and replacive micritic fabric in these Miocene rhizoconcretions are characteristic of soils in dry climates (Kooistra, 1982; Sehgal & Stoops, 1972; Wieder & Yaalon, 1982).

An unusual combination of roots and burrows was seen in the type Pandu clay paleosol in Pakistan: calcareous, parallel-sided tubes filled with pellets, but distinctly more green-gray than their matrix and penetrated by irregular tubes filled with silt and sparry calcite (Fig. 2.2, far right). The tubes can be identified as the trace fossil genus *Edaphichnium*, thought to be burrows filled with fecal pellets by earthworms (Bown & Kraus, 1983). The penetrating irregular tubes with their drab-colored halo are most like calcareous rhizoconcretions, as can be seen especially well from the fine lateral branches of the root traces protruding from the sides of the burrow. One could argue that these were encrusted mycorrhizal roots, with the pelletoidal surface representing fungally infested rootlets, but such dense and parallel-sided mycorrhizae are not known in modern roots (Richards, 1987; Sen, 1980). A simpler explanation is that these burrows, with their loose fecal fill, were favored sites of root penetration, and then became a nucleus for the formation of calcareous rhizoconcretions.

Many of the root traces in these Miocene paleosols show drab haloes: diffuse zones of green-gray to blue-gray carbonate or clay extending outward from the sharp inner boundary of the root hole (Figs 2.3 and 2.4). Such drab-haloed root traces are common in paleosols (Retallack, 1983, 1990a). Their absence in some of these Miocene paleosols provides additional evidence on how drab haloes form around root traces. Drab haloes were not seen around root traces in paleosols at

Koru or Fort Ternan in Kenya (Mobaw, Kwar, Lwanda, Rabuor, Rairo, Chogo, Dhero, or Onuria Series), although paleosols at Songhor in Kenya (Kiewo, Choka, and Tut Series) contained them, and so did all the paleosols near Kaulial village in Pakistan (Sarang, Bhura, Sonita, Khakistari, Lal, Pila, Pandu, Kala, and Naranji Series). At Fort Ternan and Koru, the calcitic or clayey root traces are flanked by brown or red clay skins (*argillans* in the terminology of Brewer, 1976), black zones of iron-manganese stain (*mangans*), or clayey matrix. In contrast, paleosols at Songhor in Kenya and in Pakistan have prominent drab haloes, with widths proportional to the diameter of the central root trace filled with calcite or clay. Some of the paleosols (such as Pila and Naranji Series) have markedly finer and more copiously branched drab-haloed root traces (Fig. 2.3) than seen in other paleosols (Lal and Sarang Series: Fig. 2.4). Even paleosols with grayish-yellow (Pandu Series) or grayish-pink (Khakistari Series) overall color contain root traces with haloes that are even more drab in color. Within some of the drab-haloed root traces (for example in Khakistari Series), there are preserved indistinct carbonaceous remains of the root itself (Fig. 2.4). Calcareous rhizoconcretions and burrows invaded by them (as in Pandu Series) also are drab in color compared with their soil matrix (Fig. 2.2). In all of the paleosols in which drab-haloed root traces are found, the root traces penetrate deeply into the profile. They commonly occur in paleosols with calcareous nodules, although paleosols with abundant relict bedding (Buru and Sarang Series) have drab-haloed root traces without calcareous nodules and the paleosols from Fort Ternan have calcareous nodules but no drab haloes on their root traces. In one kind of paleosol (Kala Series of Pakistan), drab-haloed root traces are found along with both iron-manganese and calcareous nodules.

These observations of the drab haloes as diffuse areas of discoloration around root

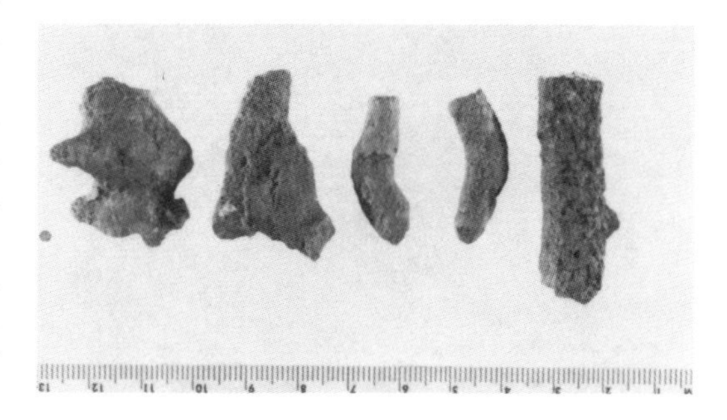

Fig. 2.2. Calcareous rhizoconcretions and earthworm (?) burrow (to right), also invaded by calcareous rhizoconcretion, from Bk horizon of type Pandu clay paleosol, in lower measured section, Dhok Pathan Formation, west of Kaulial village, Pakistan. Smithsonian specimens 353884, 353882A, 353882B, 353-883A, 353883B, from left to right. Scale in mm.

Fig. 2.3. Drab-haloed calcareous rhizoconcretion, of light gray (5Y7/1) micrite and greenish-gray (5GY-6/1) claystone in matrix mottled red (2.5YR4/6) and dark brown (7.5YR4/4), BA horizon of type Naranji clay paleosol, upper measured section of Dhok Pathan Formation, east of Kaulial village, Pakistan. Smithsonian specimen 353831. Scale in mm.

Fig. 2.4. Drab-haloed root trace of greenish-gray (5GY6/1), showing remnant black (5BG2.5/1) fibrous organic matter of the root and reddish-brown (5YR3/4) matrix, from A horizon of unnamed Sonita Series paleosol, at 9.7 m in lower measured section of Dhok Pathan Formation, west of Kaulial village, Pakistan. Smithsonian specimen 353786. Scale in mm.

traces in paleosols with calcareous nodules, deeply penetrating root traces, and other evidence of formerly good drainage, are indications against their origin as holes filled with drab-colored material from overlying layers (*krotovinas* in soil terminology), as areas unaffected by burial reddening of the paleosols, or as features of originally waterlogged soils. The way in which the haloes extend out from stout parts of the root traces, as well as from fine rootlets, is incompatible with the idea that the drab haloes represent an ancient rhizosphere depleted of iron and other nutrients. The active rhizosphere is at the tips of the rootlets where there are abundant root hairs, not on the old main roots (Russell, 1977). Thus the most likely of the various possible origins for drab haloes around root traces (also discussed by Retallack, 1990a), is as products of anaerobic decay during early burial of organic matter of the root. Such an explanation is also compatible with the absence of drab haloes from root traces at Fort Ternan and Koru, where paleosols formed in colluvium of a volcanic foot-slope and of a paleokarst on carbonatite welded-tuff, respectively, rather than the river floodplains likely at Songhor (Pickford & Andrews, 1981) and in Pakistan (Behrensmeyer & Tauxe, 1982). In well-drained soils away from alluvial bottomlands, buried organic matter of root traces may have decomposed aerobically, thus leaving iron-bearing minerals in their oxidized state. In alluvial lowlands, in contrast, buried soils would be at or below the water table, where anaerobic decay of buried organic matter would chemically reduce iron-bearing minerals (Robert & Berthelin, 1986). Because dead roots in these well-drained lowland soils would already have decayed aerobically before burial, the drab-haloed root traces represent the last crop of vegetation and humus before burial. Drab-haloed root traces can be a guide to the nature of former vegetation.

Several different patterns of root traces were observed during excavation of these paleosols. Patterning and ecology of modern roots now is increasingly well documented (some major works and a summary are given by M. B. Jackson, 1986), even in East Africa (Glover, 1950, 1951a, 1951b, 1952; Kerfoot, 1963) and Indo-Pakistan (Bhimaya & Kaul, 1965; S. B. Deshpande & Sharma, 1975; Sen, 1980; Tanwar & Sen, 1980). All the fossil root traces seen in these Miocene paleosols were deeply penetrating, and most of them were near vertical, with the exception of root traces deflected by boulders of volcanic lava (in Lwanda and Chogo Series) or by hardpans of carbonatite-nephelinite tuff (in Kiewo Series). None of the paleosols were seen with the tabular root systems found in permanently waterlogged soils (Jenik, 1978) or low-nutrient soils of tropical rain forest (Sanford, 1987). Among the Miocene paleosols, three general groups of root branching and density could be distinguished: (1) sparsely branched, stout (more than 3 mm in diameter) root traces (as in Kiewo, Choka, Tut, Buru, Sarang, Khakistari, Sonita, Bhura, and Lal Series); (2) copiously branched, stout root traces (Mobaw, Kwar, and Naranji Series); and (3) copiously branched, fine (2 mm or less), dense (at least one root every centimeter) root traces, with rare stout root traces (Rairo, Rabuor, Chogo, Onuria, Pila, and Kala Series). Like modern root systems (Bowen, 1983; Kerfoot, 1963; Sen, 1980), these groups may represent, respectively, (1) forest or woodland of dry, well drained clayey soils, (2) forest or woodland of moist, well-drained sandy soils, and (3) wooded grassland or grassy woodland.

2.1.2 Plant remains

The studied Miocene paleosols were selected largely because of their significance for mammalian fossils, but they also contain rare and diverse plant fossils (Appendices 8-10) in addition to the root traces already discussed. Both the nature and style of preservation of these remains are compatible with interpretation of these rocks as sequences of fossil soils.

All the paleosols examined in the Miocene quartzofeldspathic alluvium of Pakistan had deeply penetrating root traces, and most of them were red and highly oxidized. As would be expected from general models of fossil preservation in such sequences of paleosols (Retallack, 1984a), no plant fossils were seen other than root traces and charcoal. Most of the plant material probably decayed aerobically, but charcoal is somewhat resistant to these natural processes of decomposition. Charcoal was seen in several paleosols (principally in the type Kala clay, type Pandu clay, and an unnamed Lal paleosol at 25.3 m in Fig. 5.9). It is less than 3 mm in size and easy to overlook. The charcoal specks are cubic to rectangular in shape, are sharply broken at the ends, and have a fibrous surface texture, quite different from coalified wood chips (Harris, 1981). Under the scanning electron microscope, the charcoal specks show uncrushed cell cavities and fused middle lamellae between adjacent cell walls (Fig. 2.5), as is diagnostic of charcoal. They do not show well-preserved details of wood structure, such as bordered pits, that are found in some fossil charcoal (Cope & Chaloner, 1985). This may be related to the degree of microbial degradation in the original soil, to the intensity of combustion in these well-drained paleosols compared with seasonally waterlogged paleosols containing well-preserved charcoal, or to differences in the ratio of carbon dioxide to oxygen in the atmosphere during the Miocene compared with other times (J. M. Robinson, 1989). Charcoal is found in Quaternary paleosols formed under a wide range of pH and Eh (Retallack, 1984a), but probably is best preserved under acidic and anoxic conditions like other fossil plant materials.

Fig. 2.5. Scanning electron micrograph showing fused middle lamellae between vessel and tracheid walls of charcoalified dicot wood, from A horizon of unnamed Lal Series paleosol, at 25.2 m in lower measured section of Dhok Pathan Formation, west of Kaulial village, Pakistan. Smithsonian specimen 353787. Scale bar 10 μm.

Fossil pollen, leaf impressions and silicified wood are known from the Siwalik Group in Pakistan and India (Awasthi, 1982; Lakhanpal & Guleria, 1986; Y. K. Mathur, 1984; Prakash and Prasad, 1984; M. Prasad & Prakash, 1984; Tripathi & Tiwari, 1983; Vishnu Mittre, 1984), mainly from drab colored swamp and river deposits. The plant fossils show a bias toward wet and mesophytic vegetation types, whereas these paleosols represent vegetation of seasonally wet and well-drained sites in which plant remains were seldom preserved.

The Kenyan paleosols examined also are brown or red and have deeply penetrating root traces, but contrary to expectation (Retallack, 1984a), plant fossils in some of them are abundant and well preserved. At Koru, thin white siltstones overlying the uppermost paleosol studied (Kwar silty clay paleosol) contain indistinct impressions of dicot leaves, like those found elsewhere in the Walkers Limestone (Pickford, 1986a). At Songhor, some white silty beds overlying paleosols (unnamed Kiewo Series below measured section) contain casts of small branching twigs and some beds (Calcified Tuff Member) bear impressions of dicot leaves (Pickford & Andrews, 1981). At Fort Ternan, casts of twigs are common in sandstones, casts of logs (Fig. 2.6) in laharlike breccia beds, and fossil stumps in the surface of several paleosols. Fossil pollen, fragmentary leaves, thorny twigs, and stones and pods of dicot fruits also have been found in the main fossiliferous paleosols at Fort Ternan (Bonnefille, 1984; Shipman, 1977; see Appendix 9). Above one paleosol (type Onuria clay) at Fort Ternan, sandstones contain abundant fossil grasses, uniformly bent obliquely away from the north (vector mean azimuth 351 degrees magnetic; Dugas, 1989). Some of these fossil grasses showed patches of well-preserved cuticle and phytoliths, which allowed their identification as three species of panicoid grasses and two species of chloridoids. A few fragments of dicot

Fig. 2.6. Natural cast of a small tree trunk, showing bark impression and multiple boles, and filled with calcite and clay. Found loose in excavation at Fort Ternan, probably from type Dhero clay paleosol, where similar remains are found in place. Kenyan National Museum specimen FT-F22653. Scale bar graduated in millimeters.

cuticles also were found (Dugas, 1989). Despite local preservation of cellular detail, these plant remains are in most cases fragmentary and dominated by woody parts of plants, as would be expected in leaf litters of paleosols. Thin leaf cuticles and pollen were preserved in some cases, but most of the fruits and wood fragments were natural casts filled with sparry calcite or clay that preserved none of their cellular detail.

This style of preservation of plant remains is similar to that documented from Miocene rocks of Rusinga Island, Kenya (Chesters, 1957; Collinson, 1983). The fossil grass bed at Fort Ternan also is similar to others in Miocene rocks of East Africa; at Bukwa in Uganda (A. C. Hamilton, 1968), and at Ombo and Majiwa in Kenya (Pickford, 1984, 1986a). The preservation of plant remains on and around these formerly well-drained paleosols probably is due to the hard-setting behavior and high alkalinity of carbonatite-nephelinite ashes (Hay, 1978, 1986; Pickford, 1986b). They can be regarded as a natural Portland cement (Barker & Nixon, 1983). Comparable modern volcanic ashes from the active volcano Oldoinyo Lengai, Tanzania, have a pH in excess of 9 and are white with salts. These salts are rapidly leached by rain as the ash becomes cemented hard with calcite (Hay, 1978, 1989). These ashes are an excellent chemical preservative, and can be voluminous enough to bury plant remains away from large animals, but not from the aerobic decomposing microbes that destroy cellular structure. Fossil plant remains are occasionally preserved in volcanic rocks of various composition, provided they are sufficiently thick (Chaney & Sanborn, 1933; B. F. Jacobs and Kabuye, 1987; Leys, 1983; Lockwood & Lipman, 1980).

Unlike most of the plant remains found in these paleosols, the stones of hackberries (*Celtis*, found in both Chogo and Kwar Series paleosols and on Rusinga Island, Kenya: Chesters, 1957; Shipman, 1977), may have been more than just organic matter. Hackberry stones are commonly mineralized on the tree with both calcium carbonate and silica (Yanovsky, Nelson & Kingsbury, 1932). Such fruits are thus preserved as hard parts, similar to snail shells and mammal bones, rather than as casts, compressions or permineralizations of organic matter.

2.1.3 Invertebrate fossils

As expected from general models of fossil preservation in paleosols (Retallack, 1984a), fossil snail shells were found in many of the calcareous paleosols. Only nonaquatic snails were seen in the Kenyan paleosols, where snails are extraordinarily abundant and diverse (Appendices 8 and 9: Pickford, 1983, 1984, 1985, 1986a; Wenz & Zilch, 1960). Similar fossil land snail assemblages are well known from Miocene rocks of Rusinga Island (Verdcourt, 1963; A. C. Walker & Pickford, 1983), where some collections include aquatic snails and clams. The fossil snails were found both complete and broken, without preferred orientation. All were discolored to some extent by their paleosol matrix, and some were marked by root traces. They look like assemblages that accumulated in the paleosols through attritional mortality.

Fossil snails were much less common in Miocene paleosols in Pakistan, and most of them were found in drab paleosols (Khakistari and Pandu Series) interpreted to have formed in seasonally flooded lowlands. A few fragments of shell were found, but most of the remains were thick opercula of the ampullariid pond snail *Pila prisca* (Prashad, 1924; as emended by Wenz, 1938). This is an aquatic pulmonate of seasonally flooded soils. They aestivate in a closed shell during the dry season (Hawkins, 1986). Although opercula are locally common, snail shells of any kind are virtually absent from these paleosols. The Pakistani paleosols are less developed and less calcareous than the Kenyan paleosols. Presumably snail shells were dissolved by rain, which even when unpolluted has a pH of 5.6 (Prinn & Fegley, 1987), or by acidic soil solutions that were less effectively buffered in these paleosols compared with the highly calcareous volcaniclastic Kenyan paleosols.

2.1.4 Burrows and other trace fossils

A variety of burrows and other invertebrate trace fossils were found in many of the paleosols (Appendices 8 to 10). Especially noteworthy in the Pakistani paleosols are pinnately branched burrow systems (Fig. 2.7) and concentrically zoned spherical casts, similar to galleries and fungus gardens of termites (Bown, 1982; Sands, 1987). One of the Kenyan paleosols has abundant ellipsoidal shaped internal casts (Fig. 2.8) like those of the larval cells of dung beetles (Retallack, 1984b). The abundance of burrows in these paleosols is an aid to identifying the tops of the paleosols, which can be taken as the planes that truncate the burrows. The variety of forms found are quite consistent with what is known about soil invertebrates (Retallack, 1990a), as opposed to trace fossil assemblages of lakes or rivers.

The makers of these burrows were not commonly preserved, as would be expected in such well-drained and oxidized paleosols (Retallack, 1984a), because of aerobic decay that destroyed most of the plant biomass in these paleosols. Natural casts of a fossil beetle and whip scorpion have been found at Fort Ternan (Shipman, 1977). Similarly preserved calcite casts of spiders, insects, and caterpillars have been found in Miocene carbonatite-nephelinite ashes on Rusinga Island, Kenya (L. S. B. Leakey, 1952; Paulian, 1976; Wilson & Taylor, 1964). As for the plant remains already discussed, the fossil invertebrates were preserved in leaf litters overwhelmed by highly alkaline and early cemented ash (Hay, 1986).

Fossil millipedes (Pickford, 1986a; Shipman, 1981) and isopods (Morris, 1979) also are found in the Kenyan paleosols, but these are exuviae preserved as hard parts, a preservational style more like that of snails than of calcite casts of insects. Some modern millipedes have unusually durable exoskeletons, fortified by as much as 55 weight percent calcium carbonate and phosphate (Neville, 1975). Both round-backed, or burrowing, and flat-backed, or litter-splitting, millipedes were found.

2.5.1 Fossil bones

Northern Pakistan and southwestern Kenya are well known for mammal fossils (Barry *et al.*, 1985; Pickford, 1986a), but few of these fossils make good museum displays. Few articulated skeletons or complete skulls have been found among tens of thousands of specimens. Most of the mammalian fossils in Kenya (Andrews & Van Couvering, 1975; Pickford & Andrews, 1981; Shipman *et al.*, 1981) and in Pakistan (Badgley, 1982, 1986) are fragmentary, cracked, broken, and discolored.

Fig. 2.7. Pinnately branching, pellet-filled burrows, similar to galleries of termites, from C horizon type Sarang clay, in lower measured section of Dhok Pathan Formation, west of Kaulial village, Pakistan. Smithsonian specimen 353764. Pen for scale.

Detailed studies of bones in two of the paleosols (both Chogo Series) at Fort Ternan, Kenya (Shipman, 1977, 1981; Shipman *et al.*, 1981) showed that the bones had no particular orientation, indicating little or no sorting by water, but signs of surface weathering and of gnawing, as would be expected during the natural accumulation of dead animals in a soil. Paleosols at Koru and Songhor, Kenya, also contain fossil bone in a similar state of preservation (Pickford, 1986b; Pickford & Andrews, 1981), although some nearby Kenyan Miocene fossil sites, such

Fig. 2.8. Natural cast of hollow structure similar to dung-lined brood cells of a dung beetle, from type Choka clay at Songhor Kenya. Kenyan National Museum specimen SO-F22652. Scale graduated in mm.

as Meswa Bridge, appear to be fluvial accumulations of bones. All the Pakistani bones that I collected also had the earmarks of attritional assemblages on land surfaces. Some of the localities with abundant bones in paleosols in Pakistan have been excavated and show some indications that they were carnivore or scavenger accumulations (Badgley, 1982, 1986). In Pakistan also there are assemblages of waterworn and consistently oriented bone scrap in fluvial paleochannel sandstones and conglomerates (Badgley & Behrensmeyer, 1980). In Pakistan, bone assemblages of paleochannels and of paleosols are quite distinct in their preservational style.

Fossil bones in Miocene rocks of both Kenya and Pakistan are largely those of land animals, even in paleochannels. In Pakistan there also are bones of fish, turtle, gavial, crocodile, and otter (Barry, Behrensmeyer, & Monaghan, 1980; Barry, Lindsay, & Jacobs, 1982; Roe, 1987,1988). Turtle and otter also have been found at Songhor in Kenya (Pickford & Andrews, 1981; Schmidt-Kittler, 1987). At Koru and Fort Ternan in Kenya, however, no unequivocal aquatic fossils have been found in more than 11,000 specimens (Pickford, 1985, 1986a). Although bones can be scattered widely through landscapes, the dominance of non-aquatic fossils also is consistent with interpreting the enclosing beds as paleosols.

Excavations at Fort Ternan have yielded two notable exceptions to the general rule of fragmentary preservation of fossil vertebrates in these Miocene rocks. In one area (Nzuve trench or northwest corner) of one of the fossiliferous paleosols (Chogo clay eroded phase) were found two fully articulated juvenile skeletons of mongooses, *Kanuites* sp. (Savage & Long, 1986; Shipman, 1977, 1981; Shipman *et al.*, 1981). In the area of the excavation where they were found are several large calcareous nodulelike structures that appear to be burrows filled with calcareous sandstone less weathered than the surrounding soil matrix. This small group of carnivores may have been buried within a den.

A second exceptional find at Fort Ternan is an articulated head and shoulders of a fossil chameleon, *Chameleo intermedius*, partly enveloped with a natural cast of the skin (Hillenius, 1978). This specimen was found loose in the bottom of the quarry, but judging from the matrix, it came from the silty ash and laharlike breccias forming the high wall of the quarry. Comparable soft-part preservation has been seen in other Kenyan Miocene localities, such as Rusinga Island, where there was found a lizard skull with a natural cast of its tongue (Estes, 1962). These may be additional cases of the preservation of soft parts by the early cementation of highly alkaline tuff, as for the fossil insects and spiders from Miocene rocks of southwestern Kenya (Barker & Nixon, 1983; Hay, 1986).

2.2 Soil horizons

A second general category of features diagnostic of paleosols, as opposed to other geological phenomena, is the layering of alteration down from the ancient land surface. Soil horizons usually are truncated abruptly at the land surface but show gradational contacts downward into their parent material (Fig. 2.9). Sedimentary beds, on the other hand, are in general sharply bound and often more numerous and thin. These general differences were used in conjunction with other features, such as root traces and soil structure to recognize paleosols in these Miocene sedimentary sequences.

2.2.1 Clayey surface horizons

In some of the paleosols (especially Rairo, Chogo, and Onuria Series at Fort Ternan and Kala Series in Pakistan), the most clayey part of the profile is the uppermost

Fig. 2.9. Modern soil (Hapludoll upper left) and two Chogo clay paleosols (Haplustolls) in the main excavation at Fort Ternan, Kenya, showing sharp upper contacts and gradational boundaries of calcareous nodular horizons. Hammer gives scale.

20 cm or so (Fig. 2.9). Was this clayey layer formed by weathering of a more silty parent material? Or was it a separate clayey bed of the parent material? These alternatives are not mutually exclusive: some component of each is likely in a soil, but no weathering in place would be detectable if it were only a sedimentary bed. For these paleosols, the contribution of weathering is most obvious in thin sections, where etched and deeply weathered minerals are common (see Fig 2.30, discussed in a later section). Neither in thin section nor in outcrop is there much trace of bedding, of the clastic original texture of coarse-grained parent materials or of the sharp basal contact of sedimentary beds. Instead, most of the paleosols are riddled with root traces and a variety of structures like those found in soils. In some paleosols (Kala Series and Chogo clay ferruginized-nodule variant), the lower contact of the clayey horizon is a gradational and laterally irregular change downward into material with clear relict sedimentary bedding. In other cases (type Chogo clay and Onuria Series), a distinct but irregular lower contact coincides with a calcareous hardpan that has obscured a former lithological contact. These clayey surface horizons thus appear to be former A horizons of soils.

Although surface clayey horizons are unlikely to be merely a fine-grained upper part of a graded bed, some sedimentary additions to them are likely. Chemical analyses, particularly of materials stable in soils such as titania and zirconium, demonstrate that the upper horizon of the type Kala clay in Pakistan had a parent material different from the lower horizon. Soil formation in this case added to an originally fine grained upper bed. In the paleosols from Fort Ternan in Kenya, the persistence of pyroxene and biotite in these clayey horizons is evidence against thorough weathering. Some of these mineral grains were fresh enough to yield radiometric ages that are plausible compared with results from overlying and underlying phonolite lavas (Shipman *et al.*, 1981). Continued input of volcanic ash is likely for these clayey surface horizons, as in similar surface soils in the Serengeti Plains of Tanzania today (Jager, 1982). In all these cases, the rate of influx of new material was slow enough for the material to be incorporated into the soil fabric. Weathering hand in hand with eolian influx is now recognized as common in soils (Brimhall *et al.*, 1988; Muhs *et al.*, 1987).

2.2.2 Subsurface clayey horizons

Many of the paleosols (Lal, Naranji, and Pila Series in Pakistan and Tut and Choka Series at Songhor, Kenya) have a diffuse zone of clay enrichment below the less clayey, truncated top of the profile. These paleosols are all in alluvial sequences, so could these clayey zones represent merely finer-grained beds within their parent material? There is no suggestion of this in the field: no stone lines or relict bedding. These clayey subsurface horizons are well homogenized by root traces, burrows, and fecal pellets and have gradational boundaries into coarser-grained material above and below. Nor were any subtle discontinuities revealed by point counting for mineral composition or chemical analysis of trace elements, as detailed in a later section describing individual profiles. In addition, a pedogenic component of the clay can be recognized in the distinctive mineral composition of ultrafine (less than 0.5 μm) clay (see Fig 2.32). In thin section, pedogenic clay is visible as thin rims to the grains from which it has hydrolyzed (see Fig. 2.30) and as highly birefringent wisps in a less oriented clayey matrix (see Fig. 2.31). Whatever was the original arrangement of these parent materials, the abundant and conspicuous bioturbation in these paleosols has mixed and altered it beyond recognition and has imposed a pedogenic horizonation. This is not to say that there was no clay in the parent materials. Clay was probably at least as abundant in the parent alluvium as it is in associated very weakly developed paleosols (Buru, Dhero, Rairo, and Sarang Series). However, the clay bulge in their depth functions for grain size, that is to say, the excess over a line connecting the amount of clay at the surface and the amount below the paleosol, probably includes at least some pedogenic clay. In this respect, subsurface clayey horizons of these Miocene paleosols are very similar to those in comparable modern soils, such as those of the Serengeti Plains of Tanzania (Jager, 1982) and of the Indo-Gangetic alluvium of northern India (Kooistra, 1982).

Conventionally, the formation of clayey subsurface horizons in soils has been regarded as a process of formation of clay in place by weathering of primary minerals and its washing down into the cracks and root holes in the soil. An alternative view has recently been suggested to accommodate the widespread role of eolian dust in soil formation (Muhs *et al.*, 1987). Soils can be considered very slowly accumulating eolian sediments within the zone of weathering, so that subsurface horizons are more clayey because they have weathered for longer than surface horizons. In addition, fine clays transported by wind easily penetrate soil cracks, and their high surface-to-volume ratio allows more thorough weathering than that found in coarser grains of parent material. Not only are these processes widespread in desert soils (McFadden, 1988), but they also have been proposed to produce lateritic and bauxitic soils under humid forests (Brimhall *et al.*, 1988), in a manner analogous to the accumulation of brown clays in the deep ocean (Clauer & Hoffert, 1985). For these Miocene paleosols, there is in addition to wind, a steady input of sediment from flooding or eruption of volcanic ash. As indicated by Simonson (1976), such slow sedimentary additions to soils should be considered a process of soil formation rather than of sedimentation. The thorough incorporation of this material in the soil fabric is an indication that it is well under the control of the soil system, as opposed to the wider depositional system that occasionally overwhelms soils with thick deposits or destroys soils by eroding them away.

Evidence of continued sediment influx during formation of those Miocene paleosols with subsurface clayey horizons is not difficult to find. Both Tut and Choka paleosols of southwestern Kenya have abundant easily weathered minerals such as pyroxene and calcite globules in their surface horizons. These probably represent additions of carbonatite-nephelinite ash, as found also in surface soils of the Serengeti Plain of Tanzania (de Wit, 1978; Hay, 1978; Jager, 1982). Cracks and burrows in the surface horizon of the Lal clay paleosol of Pakistan are filled with silt-sized grains of easily weathered metamorphic rock fragments not seen in the adjacent clayey soil matrix, as in surface soils of the Indo-Gangetic Plains (Sidhu, 1977). Thus it is likely that the subsurface clayey horizons of these Miocene paleosols were formed partly by weathering in place and partly by the addition of dust, volcanic ash, and flood-borne silt. There are no indications that these diffuse, clayey, subsurface horizons were inherited from a preexisting clayey bed in the parent material of the paleosols.

Fig. 2.10. Calcareous nodules (white) on a modern weathered surface of the Pandu clay-loam paleosol, in lower measured section of Dhok Pathan Formation, west of Kaulial Kas, Pakistan. Hammer gives scale.

2.2.3 Subsurface calcareous nodules and stringers

Many of the Miocene paleosols of Pakistan and Kenya are strongly calcareous with stringers, nodules, and dispersed calcite (Figs 2.9 and 2.10). They preserve a variety of indications that much of this carbonate accumulated during soil formation. For example, calcareous nodules like those in paleosols are found also as clasts in conglomerates of associated fluvial paleochannels in Pakistan (personal observations of the Kamlial Formation on the ridge 300 m east of Khaur Rest House). In thin sections of the Pakistani paleosols there are resorted nodules with broken and sharp edges, as well as nodules with a diffuse replacive outer margin that appears to have formed in place (Fig. 2.11). Within the nodules, partly replaced grains float within a micritic matrix, again as in caliche nodules of soils (Tandon & Narayan, 1981). Carbonate also is intergrown with iron-manganese concretions similar to those found in soils (Fig. 2.12). Fragments of micritic Eocene limestone also are found in the Pakistani paleosols and sediments (Krynine, 1937), but can be distinguished from pedogenic nodules by their relict bedding and remains of large marine foraminifera.

Rock fragments of micritic and sparry calcite and heavily micritized nephelinite, pyroxene, and vermiculite are common in thin sections of the Kenyan paleosols, in which micritic nodules with replacive outer margins also are found. Some of the calcite in the Kenyan paleosols is of carbonatitic volcanic origin. Many of the carbonate clasts have a trachytoidal texture of weakly oriented laths of carbonate (see Fig. 2.13B). These may be clasts of the carbonatite variety called alvikite (Deans & Roberts, 1984). Other clasts are sparry calcite and may be the carbonatite variety sövite. These latter are coarser grained than usual for pedogenic carbonate, and are texturally similar to early burial cements of sparry calcite also found in these rocks. One clue to the volcanic origin of some of these grains is a thin rim of radially oriented fibrous calcite, as found in Quaternary carbonatite tuffs elsewhere in East Africa (Downie & Wilkinson, 1962; Hay 1978, 1989). In places where the calcite rim has flaked off into the paleosol (Fig. 2.13A), the rim can be seen to predate soil formation. Sparry calcite also fills the spaces between expanded books of vermiculite in some of these paleosols. Although this is a form of displacive fabric, it is here regarded as a product of volcanic eruption or hydrothermal alteration of the parent material, because such grains are common in deep layers of the paleosols, where roots and burrows are

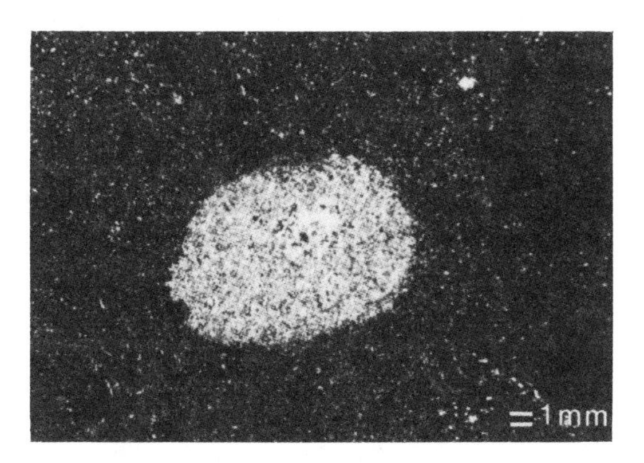

Fig. 2.11. Petrographic thin section under crossed nicols of a micritic nodule, showing indistinct quasisesquans and included clastic grains with partly replaced outer margins, from Bt horizon of type Lal clay paleosol, in measured section of Dhok Pathan Formation, west of Kaulial village, Pakistan. Smithsonian specimen 353780. Scale bar 1 mm.

rare, and because most of them have a ferruginized weathering rind (Fig. 2.13B).

In contrast to these volcanic carbonates, pedogenic carbonate is found in these paleosols in typical nodular form. In thin section, pedogenic carbonate is micritic, with small-scale replacive margins (Fig.2.14). Also found is displacive fabric

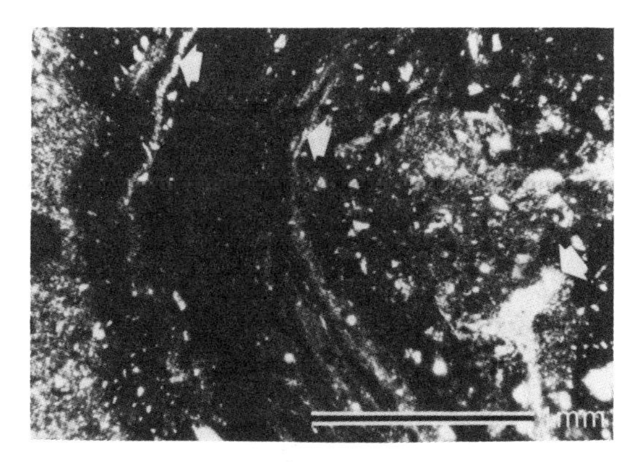

Fig. 2.12. Petrographic thin section under crossed nicols of admixed micritic and sparry calcite (at arrows) as interlayers within, and crystal tubes penetrating, a ferric concretion, from C horizon of type Pila clay paleosol, in lower measured section west of Kaulial village, Pakistan. Smithsonian specimen 353821. Scale bar 1 mm.

(Fig. 2.15) in which carbonate fills and expands the width of cracks, as is common during soil formation, including early subaerial leaching and cementation of carbonatite-nephelinite ashy parent material (Hay, 1978, 1989). A pedogenic generation of carbonate also is indicated by calcareous rhizoconcretions and stump casts in the paleosols, as already discussed. Calcareous nodules in Kenyan paleosols preserve partly embayed but little weathered clasts of nephelinite and fenite, grain types that are deeply weathered in the clayey matrix outside the nodule. This observation and the slight deformation of rhizoconcretions, insect cocoons, and burrows are evidence that the calcareous nodules were not formed following deep burial and compaction of the paleosols.

Also recognizable in calcareous horizons of paleosols from Kenya and Pakistan are variations in microscopic and field character similar to sequences of development in time observed in surface soils in calcareous quartzofeldspathic alluvium of modern Indo-Gangetic plains of northern India (Courty & Fèderoff, 1985; Sehgal & Stoops, 1972; Singh & Lal, 1946) and in carbonatite-nephelinite volcaniclastic sediments of northern Tanzania (Hay, 1978; Hay & Reeder, 1978). The development of calcareous horizons on these two distinct kinds of parent materials (Fig. 2.16) differs in detail from the well-known scheme for development of calcareous horizons in weakly calcareous quartzofeldspathic alluvium of the desert southwest of the United States (Gile, Hawley, & Grossman, 1980; Gile, Peterson, & Grossman, 1966; Machette, 1985), and is detailed in the following paragraphs.

Miocene paleosols of Pakistan, like surface soils on the modern Indo-Gangetic Plains, can be arranged in a chronosequence as follows. Very weakly developed paleosols (Sarang Series) have prominent relict bedding, with fragments of limestone and redeposited soil in sandy layers. There may be some dispersed micr-

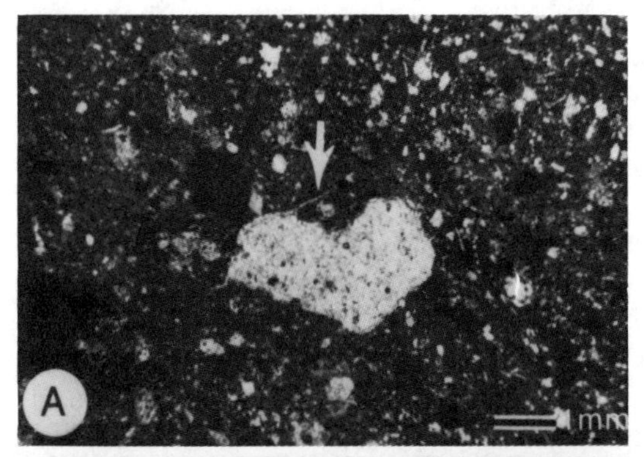

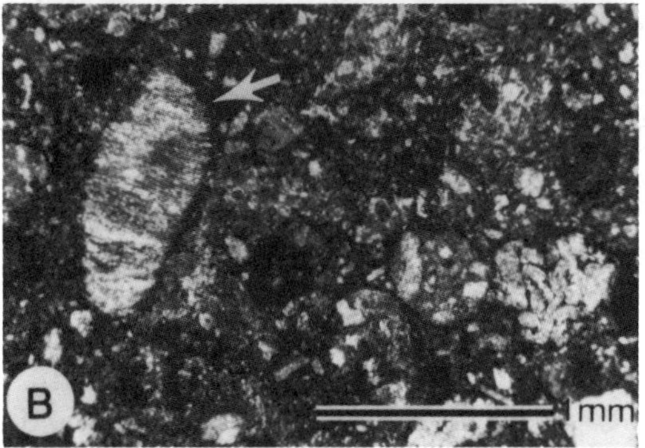

Fig. 2.13. Petrographic thin section under crossed nicols of unusual calcareous volcaniclastic grains from Kenya: A, carbonatite clast, with a thin calcan flaked off in places (at arrow), from Bt horizon of type Tut clay at Songhor (Kenyan National Museum specimen SO-R127): B, large grain of vermiculite (at arrow) expanded by calcite and rimmed by a sesquioxidic weathering rind; also seen are grains of nephelinite (has equant phenocrysts, lower right) and carbonatite (has intergrown laths, also lower right), from A horizon of type Mobaw silty clay loam paleosol near Koru (Kenyan National Museum specimen KO-R84). Scale bars 1 mm.

itic carbonate in clayey layers and a sparry calcite cement in sandy layers. In weakly developed paleosols (Bhura and Sonita Series) there are small (1-2 mm in diameter) nodules of micritic low-magnesium calcite replacing surrounding matrix, as well as calcareous rhizoconcretions already discussed. In moderately developed paleosols (Lal and Pandu Series, and in undescribed paleosols seen in the Chinji Formation 3 km south of Khaur and Kamlial Formation 1 km southeast and 1.5 km south of Khaur), the micritic nodules are larger (2-50 mm in diameter) and more prominent. Their replacive nature can be seen in thin section from gradational and embayed outer margins, and from their etched and matrix-supported included grains (Fig. 2.11). Some of the carbonate also is intergrown with the concentric layers of iron-manganese concretions (Fig. 2.12), as in surface soils of monsoonal India (Sehgal & Stoops, 1972). Strongly developed paleosols (only one seen, undescribed in Kamlial Formation northwest of main trunk road 17 km northeast of Khaur) have nodules that have merged to form a laterally continuous horizon of micritic carbonate.

In general, this interpreted chronosequence is not very different from that proposed for noncalcareous parent materials (Gile, Hawley, & Grossman, 1980; Gile, Peterson, & Grossman, 1966). An important difference for calcareous parent materials is that carbonate is redistributed and focused in the calcareous horizon, rather than being newly formed during weathering or introduced in dust (Wieder & Yaalon, 1982).

A distinctive feature of these Pakistani paleosols and of Indo-Gangetic surface soils (Sehgal, Sys, & Bhumbla, 1968), is the very diffuse spread of nodules in the calcareous horizon (Fig. 2.10). Nodules decline in abundance gradually above and below the horizon, and occur sparingly near the surface of the profiles, so that it is difficult to define an exact top and bottom of the calcareous horizon.

A second distinctive feature of both paleosols and surface soils (Sehgal & Stoops, 1972) is intergrowth of calcareous nodules and rhizoconcretions with iron-manganese concretions (Figs 2.1C and 2.12). Both the diffuse nature of the calcic horizon and its ferric-calcic intergrowths have been attributed to the highly seasonal, tropical, monsoonal climate of northern India (Sehgal & Stoops, 1972; Sehgal, Sys, & Bhumbla, 1968). Alternatively, the spread of carbonate and intergrowth with ferric nodules could, in

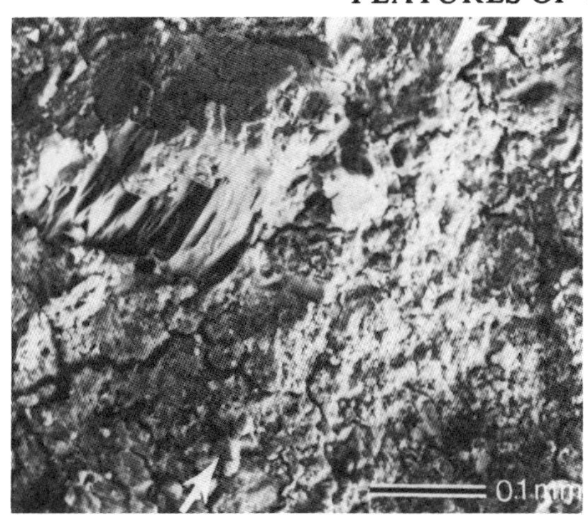

Fig. 2.14. Scanning electron micrograph, of replacive micritic calcite (light and granular, lower right, enhanced by deliberate back scatter) and its irregular contact (between arrows) with a relict clast of nephelinite (cracked and clayey) that includes a prominent corroded grain of pyroxene (upper left), from Bk horizon of Chogo clay ferruginized-nodule variant paleosol, at Fort Ternan, Kenya. Kenyan National Museum specimen FT-R14. Scale bar 0.1 mm.

part, reflect local oxidation in open spaces and carbonate precipitation mediated by fungi in different parts of termite nests (Lee & Wood, 1971). Recognizable traces of termites are common in these paleosols (Fig. 2.7), and termites are most common

Fig. 2.15. Petrographic thin section under crossed nicols of displacive sparry calcite filling old root holes and circumgranular cracks, in Bk horizon of type Chogo clay, Fort Ternan, Kenya. Kenyan National Museum specimen FT-R6. Scale bar 1 mm.

in tropical monsoonal regions (Emerson, 1952). Such diffuse calcareous horizons and ferric-calcic intergrowths have not been reported from calcareous soils of temperate regions such as Israel (Wieder & Yaalon, 1972), or the North American Mojave Desert (Arkley, 1963), Chihuahua Desert (Gile, Hawley, & Grossman, 1980; Gile, Peterson, & Grossman, 1966) or Great Plains (Aandahl, 1982; Jenny, 1941).

Miocene paleosols of carbonatite-nephelinite tuffs of Kenya show calcareous horizons comparable to those of soils forming on similar parent materials around the active volcano Oldoinyo Lengai (Hay, 1989), and near Laetoli (Hay, 1978) and Olduvai Gorge (de Wit, 1978; Hay & Reeder, 1978; Jager, 1982), all in Tanzania. Very weakly developed paleosols (Dhero Series) show clear relict bedding and well preserved accretionary lapilli, scoria fragments, and crystals of easily weathered minerals such as pyroxene. They are cemented with dispersed micrite and riddled with a network of cracks filled with sparry calcite. All of this cement could well be original, because similar textures are seen in carbonatite-melilitite tuff from historic eruptions of Oldoinyo Lengai (Hay, 1989). Salts and zeolites are weathered from these ashes on time scales of less than 600 years, leaving mainly pyroxene, biotite, and calcite (Hay, 1978).

Weakly developed paleosols (Mobaw, Kwar, Kiewo, and Rairo Series) are not so pervasively cemented with calcite. They have clayey layers, with root traces and soil structure, as well as strongly calcareous layers with relict bedding. In thin section, these distinctive calcareous layers with relict bedding are riddled with subhorizontal, anastomosing veins less than 1 mm thick of narrow blades of sparry calcite (Figs 2.17 and 2.18). In places these veins separate the broken halves of crystals or rock fragments. Thus they formed in cavities in loose soil, as is usual for displacive microfabric of calcretes (Braithwaite, 1989; Watts, 1978, 1979). Similar joint fills of bladed calcite are seen

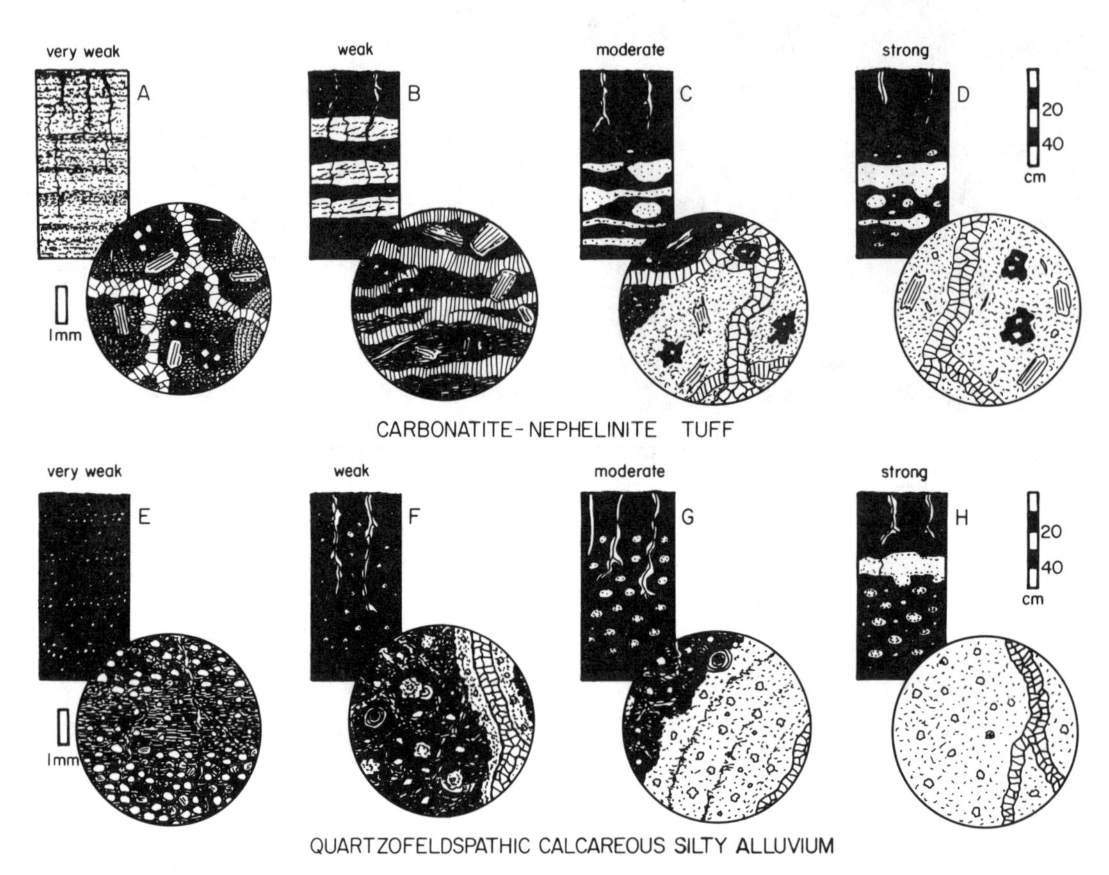

Fig. 2.16. Models for the development of calcareous horizons in soil profiles (rectangles) and thin section (circles) on carbonatite-nephelinite volcaniclastic rocks in a subhumid climate (upper row) and on quartzofeldspathic alluvium in a strongly seasonal climate (lower row). Based on observations of surface soils and Miocene paleosols described in text, and emphasizing minor differences from the well known developmental sequence for quartzofeldspathic alluvium in a semiarid cool temperate climate (Gile, Hawley, & Grossman, 1980; Gile, Peterson, & Grossman, 1966).

in limestone bedrock to modern soils (L. T. West *et al.*, 1988), and this fabric in the paleosols may represent disruption of the initial hardpan phase of soil development on carbonatite-nephelinite ashes. However, I am not aware that such a dense aggregation of displacive veins has been reported from any modern soil or tuff. This distinctive microstructure is here called pervasively displacive and is considered an extreme manifestation of displacive fabrics developed as a result of volume changes during initial subaerial leaching and cementation of carbonatite-nephelinite tuff (Hay, 1978, 1989).

In moderately developed paleosols (Choka, Tut, and Chogo Series) there are calcareous stringers and nodules, but these are massive in their appearance, as is usual for pedogenic nodules. They lack relict bedding and pervasively displacive veining. In thin section, there is much sparry calcite in irregular cracks, old root holes, and burrows, and sometimes filling circum-granular cracks (Fig. 2.15). There also are areas of replacive micrite, in which are floating marginally corroded but otherwise fresh crystals of pyroxene and fragments of nephelinite (Fig. 2.14). Strongly developed paleosols (approached

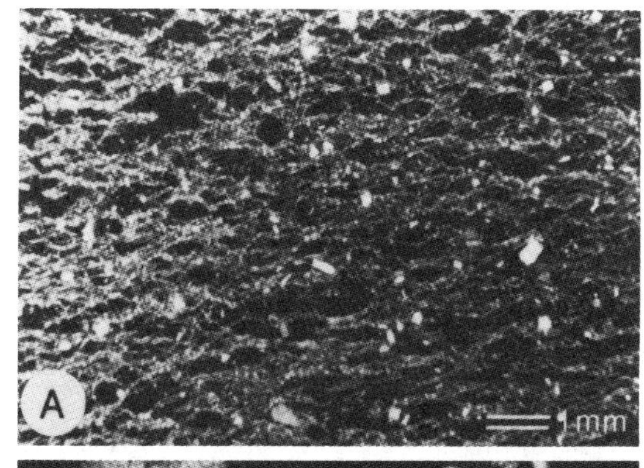

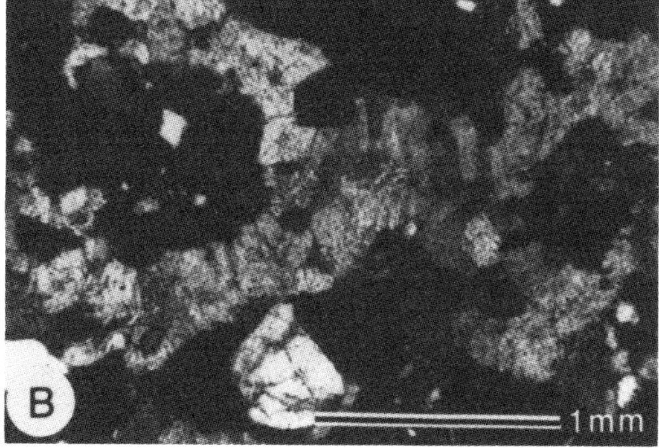

Fig. 2.17. Petrographic thin sections under crossed nicols of pervasively displacive fabric of bladed sparry calcite (light colored) in clayey matrix (dark) with crystals of pyroxene, feldspar and quartz, from C horizon of Kiewo clay single-stringer variant paleosol, Songhor, Kenya. Kenyan National Museum specimens SO-R147 (for A) and SO-R145 (B). Scale bars are 1 mm.

by Onuria Series) have a laterally continuous calcareous horizon, comparable in thin section with that of moderately developed paleosols.

One difference between this reconstructed chronosequence for Miocene paleosols and comparable modern soils of Tanzania is the lack of evidence for lamellar caps to the strongly developed calcic horizons in the paleosols. This may be related to a different geomorphic setting of most of the studied Tanzanian soils on thinner increments of ash more distant from the volcano (Anderson, 1963; de Wit, 1978) than appears to have been the case for the paleosols (Pickford & Andrews, 1981; Shipman *et al.*, 1981). Alternatively, it could reflect a longer time of formation for the Tanzanian soils than for the paleosols (Hay & Reeder, 1978).

Another important difference between modern and ancient soils on this unusual volcanic parent material is the pervasively displacive fabric of the paleosols. A possible explanation, consistent with other geological and paleontological evidence (Andrews & Van Couvering, 1975), is a more humid paleoclimate and thus more thorough leaching of the paleosols (especially Kiewo Series) than for tuffs in the desert and dry grasslands around Oldoinyo Lengai in Tanzania. Pervasively displacive fabric is here considered the result of processes that produce large changes in volume: leaching during early weathering of the tuff and hydrolysis of silicates to clay. These processes may have been more rapid under regimes of weathering more profound than in deserts of the Tanzanian portion of the East African Rift. I know of no comparable feature to pervasively displacive fabric in surface soils, but if this idea is correct, pervasively displacive fabric may yet be discovered in soils developed on carbon-

Fig. 2.18. Scanning electron micrograph of pervasively displacive bladed sparry calcite (light colored due to deliberate back scatter) in clayey matrix, from C horizon of type Kiewo clay paleosol, Songhor, Kenya. Kenyan National Museum specimen SO-R143. Scale bar 1 mm.

atite-nephelinite ash in the Western Rift volcanics near the border between Ruanda and Zaire, where the climate is much wetter than in northern Tanzania.

2.2.4. Subsurface horizons rich in iron-manganese

Many of the paleosols contain small, dark brown to bluish-black, opaque patches of noncrystalline iron-manganese on slickensided argillans, as nodules or concretions in the matrix, or as void fills after root traces. Similar features are common in soils, especially those with clayey texture, free carbonate, or prone to puddling or waterlogging (Brewer, Sleeman, & Foster, 1983; Rahmatullah, Dixon, & Golden, 1990; Sidhu *et al.*, 1977b). They are a minor component of most of the paleosols, which contain low overall amounts of manganese, in many cases depleted from the amount in underlying parent materials (see Figs 2.19 to 2.23). Two paleosols are noteworthy in showing unusual accumulations of manganese.

The type Kala clay paleosol in Pakistan has a zone of small (1-2 mm), dark, iron-manganese nodules and shows overall manganese enrichment below the surface. This dark zone also contains scattered organic fragments of roots and overall enrichment in organic carbon a little higher in the profile (Fig. 2.22). This paleosol does have a clayey upper and sandy lower layer: probably separate beds, judging from their distinct petrographic and chemical composition. It is unlikely that the manganese spotting is an alteration after burial related to this lithological contrast. The zone of enrichment is 30 cm below the contact. In addition the manganese nodules have a diffuse appearance. Also compatible with the idea that the manganese spots were part of the original soil is the preservation of fragments of organic matter, which requires a reducing Eh, unusual for these Pakistani paleosols.

This horizon of the Kala clay paleosol is similar to that found in modern seasonally waterlogged soils (Kanagarh Series of Kooistra, 1982; Gujranwala and Satghara Series of Rahmatullah, Dixon, & Golden, 1990), and may be an early stage in the development of a manganese-cemented horizon (*placic horizon* in the terminology of Soil Survey Staff, 1975).

As a second example, the brownish-yellow clay filling paleokarst fissures into a carbonatite welded-tuff below the Mobaw bouldery-clay paleosol near Koru, Kenya, encloses horizontal and vertical stringers of black, opaque, noncrystalline iron-manganese. Most of these stringers are very thin (2-4 mm). They are internally laminated. The surface of loose examples is pockmarked with depressions from plucked grains of the matrix that were partly enveloped by the iron-manganese cement. Similar grains are included within the stringers. Some of the vertical stringers reach 7 mm thick and extend down as much as 50 cm. In these large stringers, both surfaces have a nodular texture of lumps 6 to 13 mm in diameter and jutting outward 2 to 3 mm from the general level of the surface. The interiors of the large stringers are opaque, with some included mineral and claystone grains.

In their symmetry these large stringers are similar to thick iron-manganese coatings (or *neomangans* in the terminology of Brewer, 1976) that line voids in modern waterlogged soils (such as the Hangram Series of Kooistra, 1982). The iron-manganese stringers could have been original features of a zone of slow drainage in closed karst depressions within a carbonatite welded-tuff. It could also be argued that these mangans are features of modern weathering of the outcrop. However, they are surprisingly well developed for a quarry face exposure, are well below the natural surface, and are covered by several Miocene paleosols lacking such mangans. More likely they are features of Miocene weathering.

2.2.5 Chemical depth functions

Soil horizons, with few exceptions (Retallack, 1988a), are distinct from sedimentary or volcanic beds in showing gradational changes in texture, mineral weathering, color, and other features downward from a sharp erosional plane that represents the ancient land surface. Such gradational depth functions also can be seen in chemical data and are particularly well expressed in moderately developed paleosols (Tut, Choka, Chogo, Lal, and Naranji Series of Figs 2.21 to 2.23). Both the nature and slope of chemical depth functions of the paleosols compare well with those of similar surface soils of Tanzania (de Wit, 1978; Jager, 1982) and Indo-Pakistan (Sidhu, Sehgal, & Randhawa, 1977a; Tomar, 1987). In contrast, very weakly to weakly developed paleosols show little chemical change (as in Kwar Series) or abrupt changes reflecting differences in beds of parent material (Kiewo, Kala, and Pila nodular variant paleosols). In most cases these differences are less marked than those between sedimentary beds.

The down-profile variation in most of these elements is probably a function of soil formation, but some elements such as titanium, zirconium, and gold are relatively unaffected by weathering. Their uniformity in most of the paleosols (with the exception of Kiewo, Chogo clay ferruginized nodule variant, Onuria, Kala, and Khakistari paleosols of Figs 2.19 to 2.23) is an indication of a single parent material.

It could be argued that even gradational depth functions reflect parent material variation, such as fining-upwards sequences deposited with waning flood flow in alluvial sequences (Allen, 1965) or the grain size sorting that may accompany emplacement of pyroclastic flows (Cas & Wright, 1987). If these were the cause, however, it is difficult to understand the divergent behavior of easily weathered oxides (lime and soda) versus resistate oxides (titania), and the other details of the variation so much like modern weathering. Also difficult to understand as parent material effects are bulges in chemical abundance (especially well shown by Choka and Tut Series paleosols).

Another possibility is that the observed chemical depth functions reflect merely porosity-dependent passage of groundwater (Pavich & Obermeier, 1985) or composition-dependent alteration during burial or metamorphism (J. A. Palmer, Phillips, & McCarthy, 1989). These can be serious objections to interpretation of much older and more altered rocks (Retallack, 1989). Most of the Miocene paleosols were clayey and thus low in porosity. Sandy layers within and between paleosols (as in Choka, Tut, Chogo and Naranji paleosols) do not show alteration of a kind distinctly different from that in more clayey parts of the profiles. These paleosols also are highly oxidized, as indicated by a much greater abundance of ferrous than ferric iron, whereas environments of burial alteration are chemically reducing (Thompson, 1972). Nor are there any unusual high-temperature minerals, recrystallization textures, schistosity, or other features of alteration during deep burial or metamorphism. This is not to say that these paleosols were unaltered by burial. Many changes after burial are outlined in the next chapter. However, the chemical depth functions observed are consistent with evidence from root traces, soil horizons and soil structures that these are indeed paleosols.

2.2.6 Paleokarst

In the abandoned lime quarry 2.5 miles north northwest of Koru, and elsewhere around Legetet Hill in Kenya (Pickford, 1984, 1986a), the upper surface of carbonatite tuffs are exceedingly irregular, with a relief of up to 5.8 m (Fig. 2.24). The carbonatites include some that were partly welded following eruption, as well as unwelded tuffs that were cemented by calcite

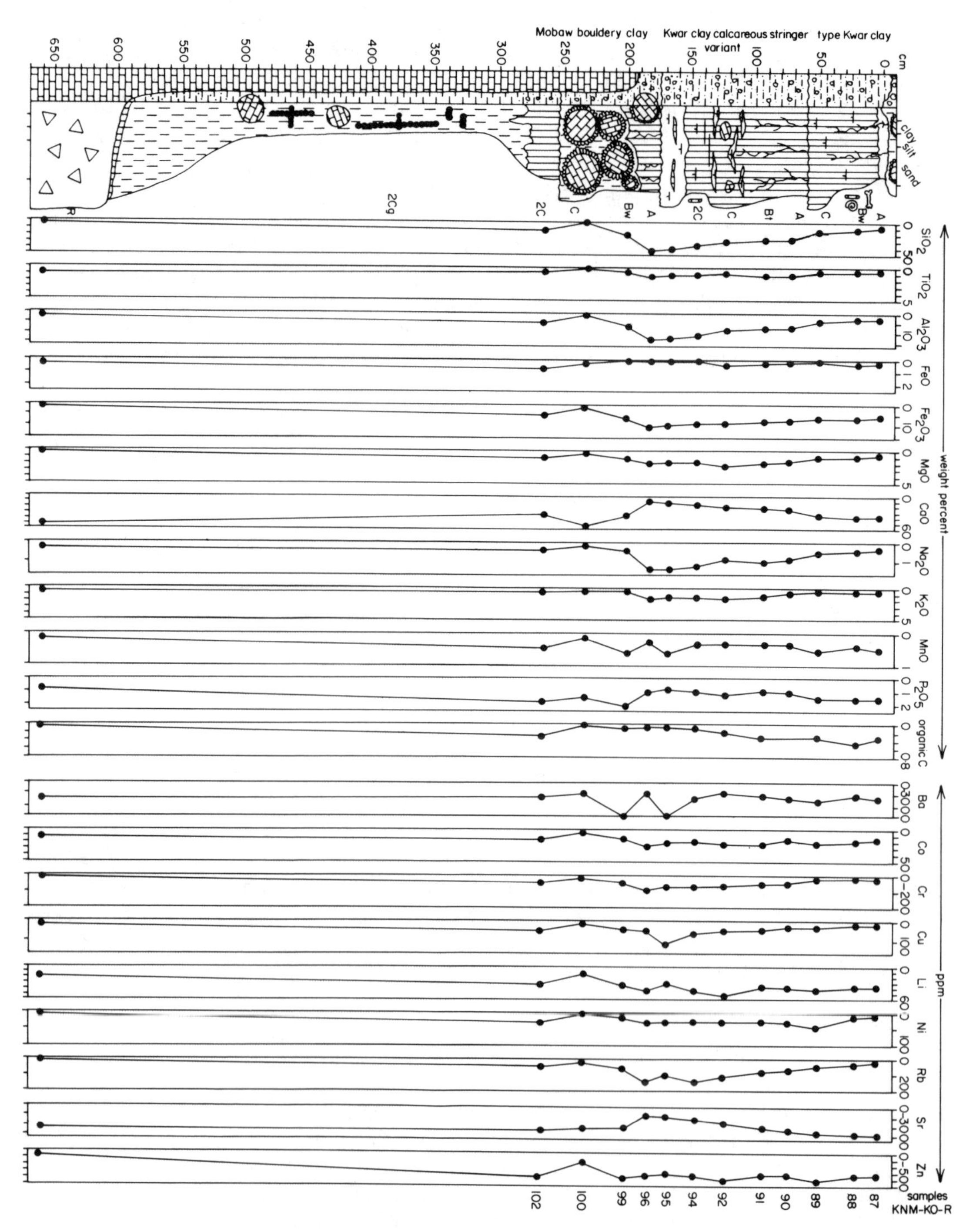

Fig. 2.19. Chemical depth functions of Mobaw and Kwar Series paleosols from a measured section in an abandoned quarry 3 km north of Koru, Kenya. Raw analyses and errors are in Appendices 5 and 6.

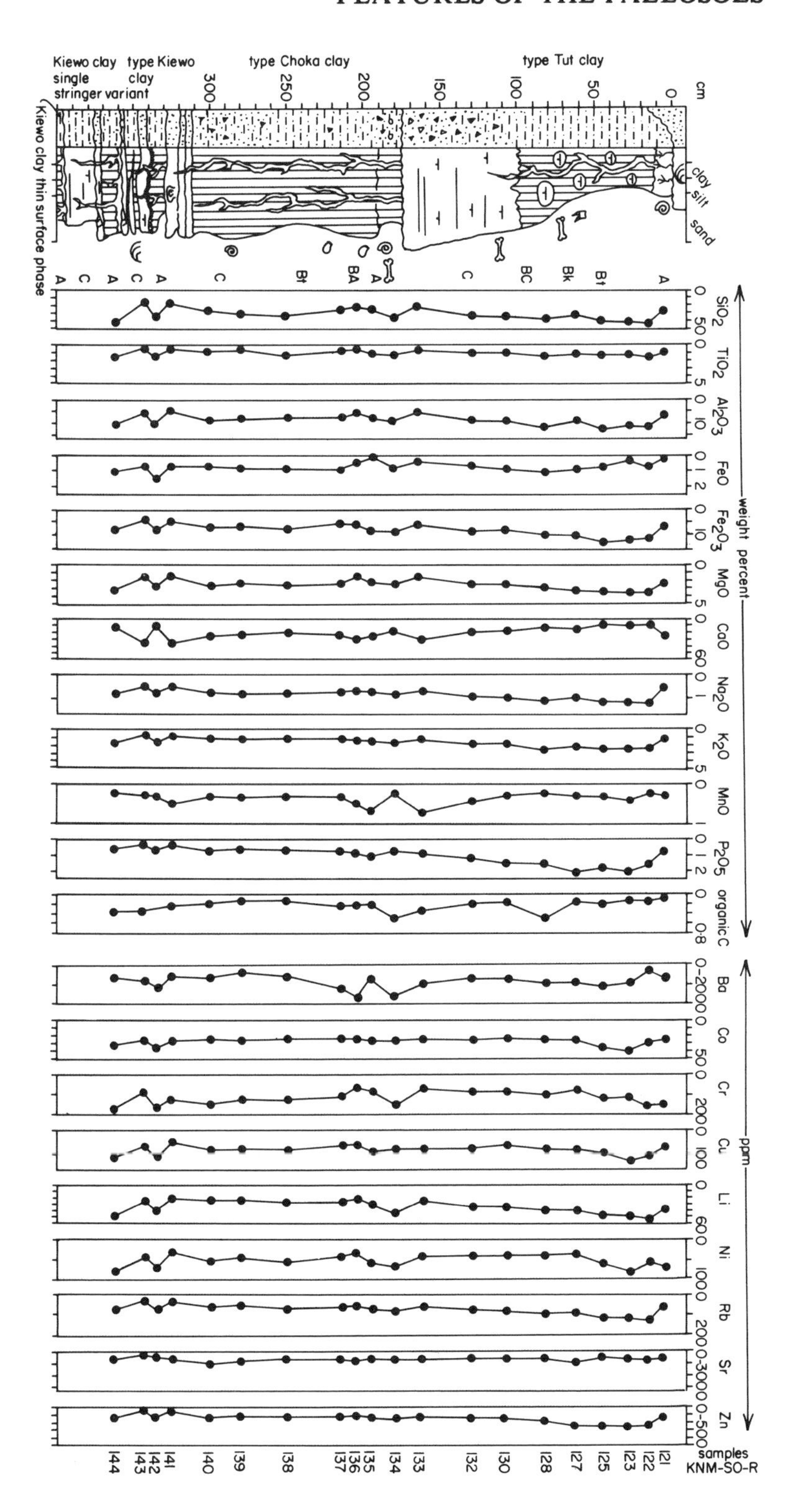

Fig. 2.20. Chemical depth functions of Kiewo, Choka and Tut Series paleosols from measured section in the western exposures of Songhor National Monument, Kenya. Analyses and errors in Appendices 5 and 6.

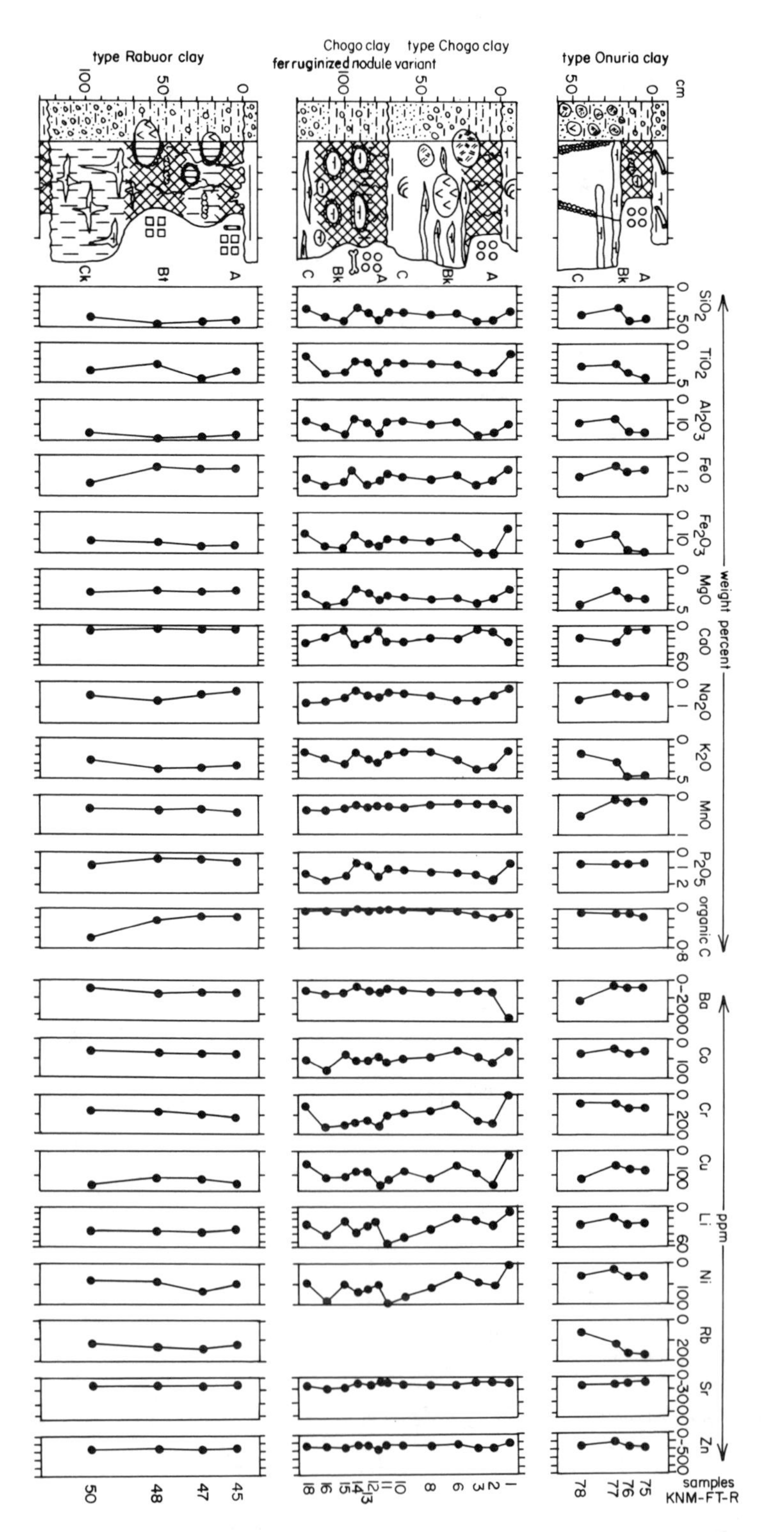

Fig. 2.21. Chemical depth functions of Rabuor, Chogo, and Onuria Series paleosols in the main excavation at Fort Ternan National Monument, Kenya. Raw analyses and errors are in Appendices 5 and 6.

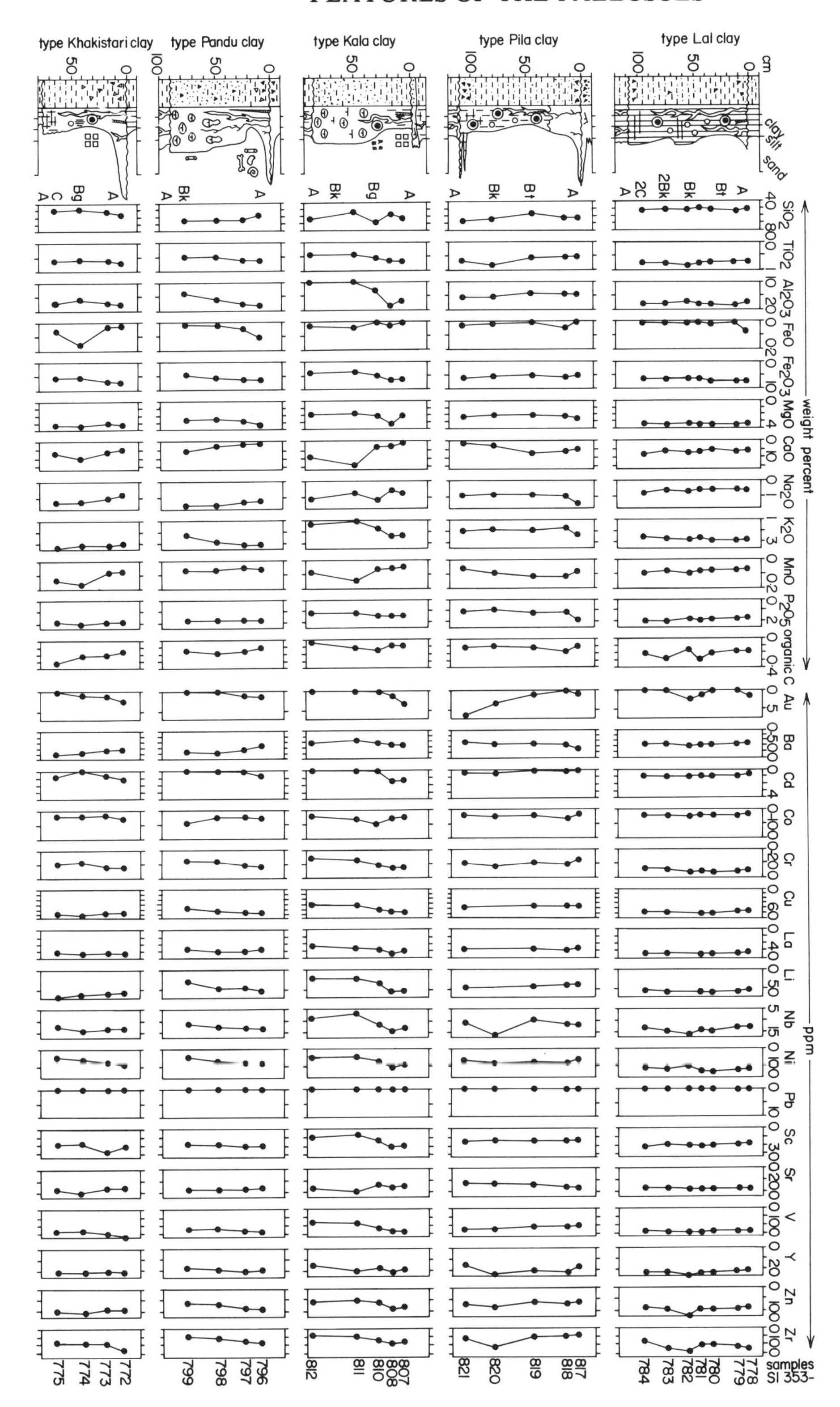

Fig. 2.22. Chemical depth functions of Khakistari, Pandu, Kala, Pila, and Lal Series paleosols in lower measured section of Dhok Pathan Formation, west of Kaulial village, Pakistan. See also Appendices 5 and 6.

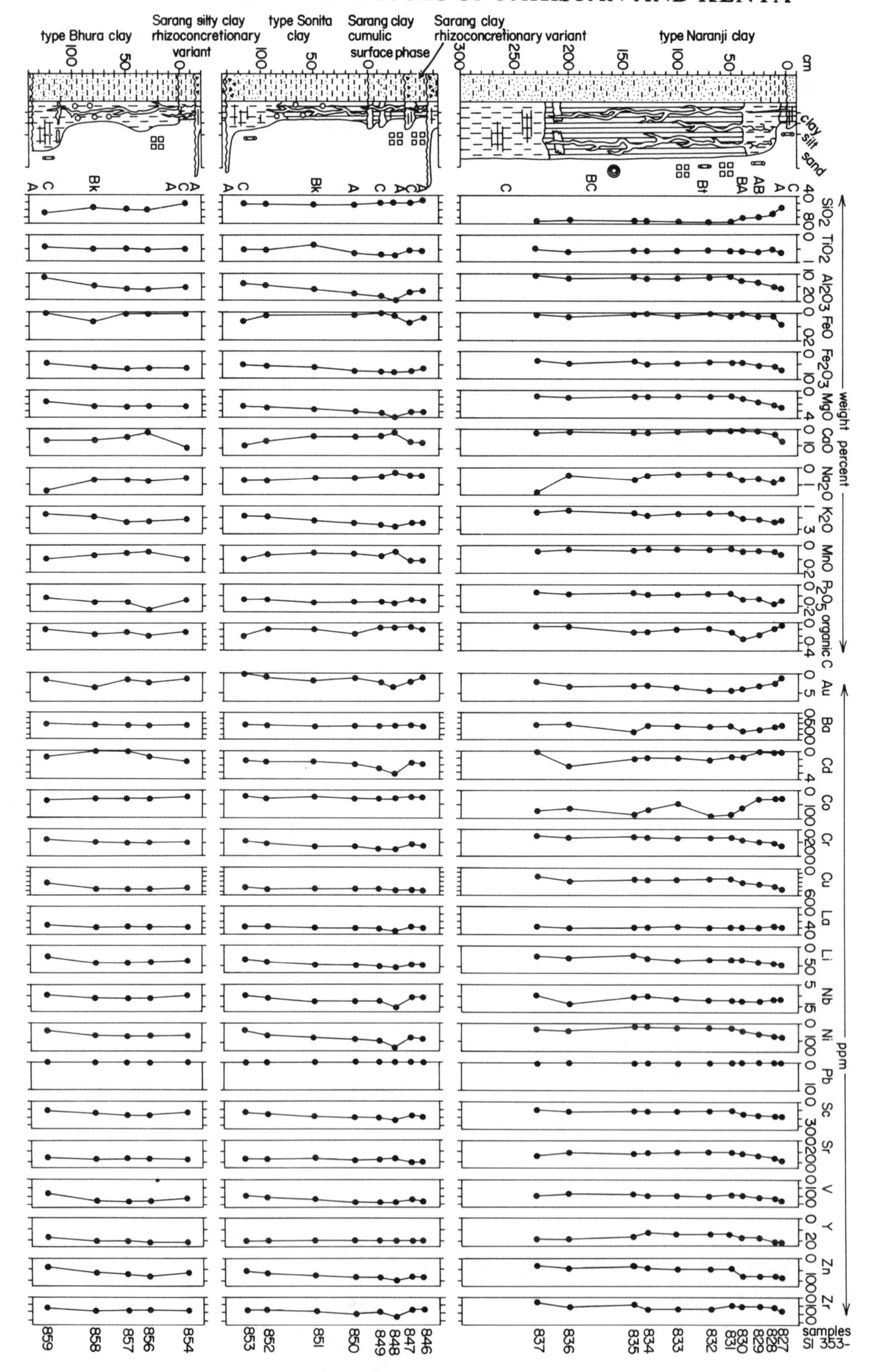

Fig. 2.23. Chemical depth functions of Bhura, Sonita and Naranji Series paleosols in upper measured section of Dhok Pathan Formation east of Kaulial village, Pakistan. See also Appendices 5 and 6.

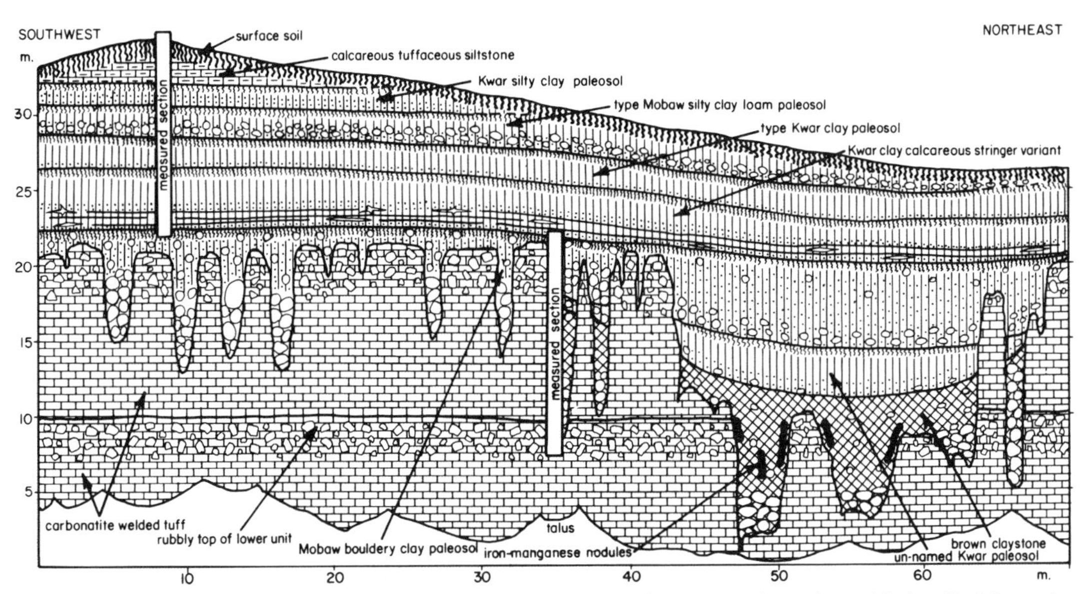

Fig. 2.24. Sketch, drawn from a field grid, of paleosols and paleokarst on carbonatite welded tuff of the early Miocene, Legetet Carbonatite, on the north wall of an abandoned lime quarry, 3 km north of Koru, Kenya.

during subaerial weathering shortly after eruption (Deans & Roberts, 1984). There can be little doubt that they were hard during Miocene time, because large blocks of carbonatite tuff litter the base of the erosional depressions. These blocks and abundant nephelinite lapilli also are evidence that this was an ancient karst landscape, rather than a dissolutional feature that developed beneath a soil mantle (cryptokarst of J. N. Jennings, 1985; or Rindkarren and Deckenkarren of Sweeting, 1973). The red clayey paleosols that now mantle the paleokarst thus developed on volcaniclastic fill of open depressions.

The paleokarst includes deep narrow chasms (grikes, or Kluftkarren), which merge in places to form complex depressions (centripetal drainage pits, or Karrenrohren). No fine surface sculpture was seen on the paleokarst, other than an irregular and lumpy texture most prominent near the tops of carbonatite tuff units, where blocks of carbonatite lava have resisted dissolution a little better that their finer-grained matrix.

A weathering rind seen on both the walls of the paleokarst and on boulders littering the base of the depressions consisted of a thin (generally about 4 mm, but up to 6 cm) zone of dark reddish-brown (5YR3/3) and black (5YR2.5/1) carbonatite. Red colors were seen on exposed surfaces where the adjacent paleosol was also red, and black colors within the dolines near the iron-manganese stringers already discussed. They probably reflect Miocene weathering under oxidizing and reducing conditions, respectively. Blackening of limestone by forest fires is sometimes observed on paleokarsts elsewhere (Shinn & Lidz, 1988). No charcoal was found in these paleosols, so this seems an unlikely explanation in this case.

The top of the karstified carbonatite tuff is still a recognizably planar surface as in limestone pavements of karsts that are geologically young or found in humid temperate climates (Choquette & James, 1987). The density of grikes, and especially the large complex depressions, are indications that this paleokarst could have developed into a more complex form similar to karsts of the humid tropics today (J. N. Jennings, 1985).

The paleoenvironmental significance of this paleokarst could be regarded as uncertain because it is not developed on limestone like well known examples of karst and paleokarst. However karst has been reported on carbonatite near Bonga in Angola, Matongo in Burundi, Tororo in Uganda, and Mbeya in Tanzania (Bloomfield, 1973; Fawley & James, 1955; Mariano, 1989). There also are carbonatites dissected by paleokarst of Jurassic-Cretaceous age on the Chadobets Plateau of Siberia and of late Cenozoic age in northern Ontario, Canada (Bosák *et al.*, 1989). Although these have not been the subject of detailed geomorphological studies, karst on carbonatite has a range of morphology from narrow fissures to broad depressions separating large towers similar to tropical karst in limestone.

Fig. 2.25. Granular-structured A horizon of the type Onuria clay paleosol (dark, below rubber handle of hammer), compared to stratified overlying nephelinitic sandstone, including well preserved fossil grasses in growth position (at arrow), from main excavation at Fort Ternan National Monument, Kenya.

2.3 Soil structure

Beds now known to be paleosols commonly have been described in geological reports as massive, featureless, blocky, jointy, hackly, nodular, or mottled. Such descriptions make the point well that paleosols lack sedimentary, igneous, and metamorphic structures and have distinctive structures of their own (Figs 2.9 and 2.25). For an adequate terminology for these structures one must turn to soil science (Brewer, 1976). Some structures of soils, such as nodules and mottles, are found also in marine sediments and hydrothermally altered rocks (Retallack, 1990a). Other structures are produced by the randomly oriented expansion and contraction of materials under low confining pressures and temperatures that are virtually unique to soil environments. Such structures include peds and cutans visible in the field, and displacive and sepic plasmic fabric seen in thin section. These structures are diagnostic of soils as opposed to other kinds of geological phenomena, and are richly represented in Miocene paleosols from Kenya and Pakistan.

2.3.1 Cracking and veining

Many of the paleosols show complex systems of clay-filled cracks that have disrupted remains of original bedding and grains of their parent material. For example, the type Onuria clay at Fort Ternan, Kenya, has an olive-brown (2.5Y4/4), hackly clay layer overlain by bedded nephelinitic sandstone and devel-

Fig. 2.26. Granular peds defined by cracks and dark clay skins from A horizon of type Chogo clay paleosol, in main excavation at Fort Ternan National Monument, Kenya. Kenyan National Museum specimen FT-R2. Scale graduated in mm.

oped on a massive carbonatite-nephelinite tuff breccia (Fig. 2.25). Hackly clay layers of Chogo paleosols at Fort Ternan are equally distinct from enclosing bedded rocks (Fig. 2.9). The hackly appearance is due to small (6-15 mm in diameter) equant nuclei of massive claystone surrounded by irregular slickensided clayey surfaces (Fig. 2.26). In thin section the slickensided clay can be seen to be highly birefringent in a central plane and less birefringent symmetrically on either side. In soil terminology these are coarse granular peds defined by stress argillans. The stress in this case may in part be due to deformation during burial compaction of the clay skins. These granular peds are not so well developed as those found in some modern grassland soils (Duchafour, 1982), but are still recognizable. Their modest expression can be blamed on volcanic parent materials subject to renewal by airfall tuff, as in similar surface soils of Tanzania (Jager, 1982).

Soil peds at a larger scale (5-10 cm) are common in other paleosols (Rabuor, Pila, Kala, Bhura, Sonita, and Lal Series). These are similar to coarse angular-blocky peds, defined by slickensided ferri-argillans (Fig. 2.27). Also seen (in Rabuor, Rairo, Khakistari, Sonita, Lal, and Kala Series) were narrow (up to 1 cm) near-vertical clastic dikes, filled with sandy material from above (*silans* in the terminology of Brewer, 1976). Some of these paleosols (Khakistari and Sonita Series) had surficial zones of claystone breccia, where loose soil peds have been infiltrated and displaced during burial by more coarse-grained overlying alluvium. These also are evidence of surface cracking of some of the paleosols. In no case however, was this expressed to the extent of tepee structures, conjugate shears or mukkara structure found in seasonally cracking soils such as Vertisols (Allen, 1986a; Paton, 1974).

A crude, very-coarse prismatic structure seen in Chogo paleosols of Kenya is of uncertain origin. This was seen only on dry weathered exposures. It could have been produced by drying of the outcrop after exposure. This structure is not nearly so pronounced or well defined by clay skins as the columnar peds with domed tops that are characteristic of saline and sodic soils (Northcote & Skene, 1972).

Some of these soil structures are no more complex than desiccation cracks, but others show a degree of irregularity that is characteristic of soils. With the possible

Fig. 2.27. Blocky ped defined by slickensided clay skins, and showing a prominent spherical ferric concretion (dark) and irregular drab mottles after root traces (light), from Bt horizon of type Pila clay paleosol, in lower measured section of Dhok Pathan Formation, west of Kaulial village, Pakistan. Smithsonian specimen 353819. Scale graduated in mm.

exception of weakly expressed prismatic peds in Chogo paleosols, these structures do not show the orthogonal and conjugate planar form of joints or faults, which also are present at some outcrops (notably at Songhor in Kenya: Pickford & Andrews, 1981). Some of the soil peds have been ripped up into claystone breccia beds, but these are easy to distinguish from peds in place because of their coarse-grained matrix and lack of associated clay skins. Volcanic or tectonic brecciation also is an unlikely explanation for clayey, complexly shaped and interlocking pedlike features in laterally extensive strata-bound zones lacking any indication of high-temperature mineralization.

2.3.2 Claystone grains

In many of the paleosols examined in thin section there are abundant claystone grains of sand to granule size. These vary considerably in the sharpness of their outer boundary and in the degree to which they are darkened by sesquioxide stain. Only those with sharp outer boundaries and staining or other fabric that set them apart from the matrix were counted as clay clasts during petrographic studies. Claystone nuclei with ragged and diffuse margins, on the other hand, were regarded as paleosol peds of the matrix. The distinct claystone grains appear to be of different origins in Kenyan compared with Pakistani paleosols.

In the Kenyan paleosols, rounded to ellipsoidal claystone grains commonly are ferruginized to a near-opaque reddish-brown color, as seen in thin section under plane light. Some of them include indistinct transparent areas shaped like rectangles, squares, or triangles. These are interpreted as relict or pseudomorphed phenocrysts of nepheline, feldspar, or pyroxene in the highly altered fine-grained groundmass of what was once a granule of nephelinite, phonolite, or carbonatite. Little-weathered examples of these rock fragments also are widespread in these paleosols and their parent materials (Deans & Roberts, 1984). Intermediate stages of granules with moderately fresh cores but a wide ferruginized weathering rind also are found. Not all this weathering necessarily occurred within the paleosol now containing them. A phonolite boulder, some 6 by 14 cm in size, within the type Chogo clay, was much more deeply weathered than other boulders and rock fragments in this paleosol. It may have been transported with the parent material of this paleosol from a pre-existing soil further up the slope of the volcano. Other paleosols such as the type Onuria clay, contain in their surface horizons pyroxene grains much fresher than others in the same horizon. These may have been additions of fresh volcanic ash, as in similar soils on the Serengeti Plains of Tanzania (Jager, 1982). Thus both preweathered and fresh grains were present in the parent materials of some of these paleosols. Weathering of volcanic clasts in place within the paleosols also was widespread, as can be seen from the many grains with gradational and irregular, deeply weathered margins. It could be argued that this alteration in place occurred entirely after burial, and was the work of intrastratal solution or hydrothermal injection. This is unlikely because the most highly altered grains are within the most clayey and least porous parts of the profiles. The freshest grains were found in sandy lower parts of the profiles, with fewer roots, burrows, or other signs of soil formation than higher in the profiles. None of these horizons had extensive veining or mineralization with talc, tremolite, pyrrhotite, or other minerals that would be expected in hydrothermally altered rocks. In contrast, the altered grains were smectitic and ferruginized, as is usual during weathering of volcanic rocks.

A striking feature of many thin sections of the Pakistani paleosols, but especially the type Lal clay, are near-perfect spheres

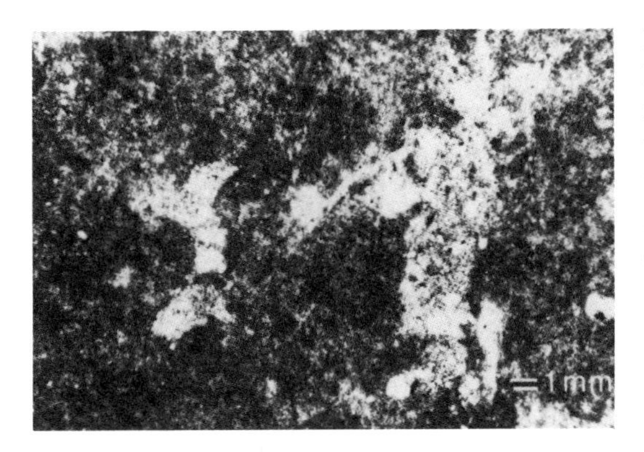

Fig. 2.28. Petrographic thin section under plain light showing spherical micropeds (dark spheres) and likely galleries of termites (silt filled), in A horizon of type Lal clay paleosol, in lower measured section west of Kaulial village, Pakistan. Smithsonian specimen 353778. Bar scale is 1 mm.

of sand to granule size of near-opaque ferruginized claystone, including scattered specks of silt-sized quartz (Figs 2.28 and 2.29). Usually they have abrupt outer boundaries. Some sharply broken rounds also were seen. Thus they do not appear to have formed as nodules. Nor do they look like parts of cavities filled with clay or rolled-up parts of thick clay skins, because they show no internal lamination. Some of these spherical claystone grains may be redeposited parts of soils, but clear examples of redeposited soil clasts also are found and are less ferruginized and more varied in size and shape. These "spherical micropeds" (Stoops, 1983) or "claystone pseudosands" (Mohr & Van Baren, 1954) are widespread in tropical soils. Most of them probably formed as fecal and constructional pellets of termites (Mermut, Arshad & St. Arnaud, 1984). Likely galleries (Fig. 2.7) and calies (similar to fossil examples described by Bown, 1982, and by Sands, 1987) of termites also are common in these Pakistani paleosols. No clear examples of termite mounds were seen, but these may have contributed to claystone breccias found in the uppermost horizons of many of the paleosols (such as the type Lal clay).

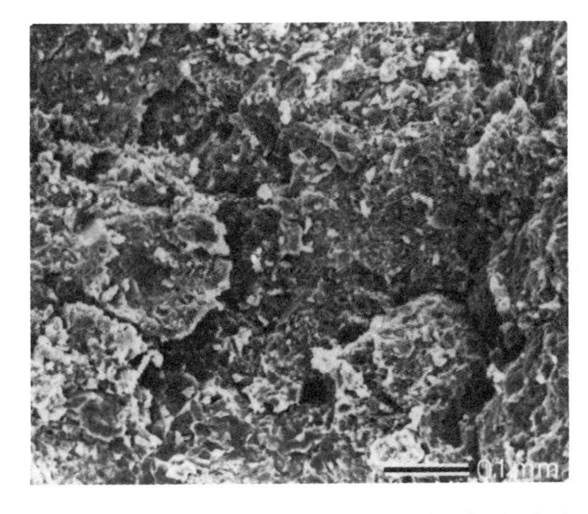

Fig. 2.29. Scanning electron micrograph of spherical micropeds (clay spheres) in gallery (defined by vertical parallel cracks) in micritic structure interpreted as the fungus garden of a termite, from Bt horizon of type Pandu clay paleosol, in lower measured section of Dhok Pathan Formation, west of Kaulial village, Pakistan. Smithsonian specimen 353881. Scale bar 1 mm.

2.3.3 Destruction of mineral grains

An additional line of evidence that paleosols are present in Miocene rocks of Kenya and Pakistan are mineral grains altered by weathering to a delicate skeleton that could not withstand transport. The Onuria clay paleosol, for example, contains large pyroxene grains reduced to irregular islands within the clay to which they have been weathered (Fig. 2.30A). Large biotite grains within the Chogo clay paleosols have missing chunks and sheets locally expanded and crumpled by weathering (Fig. 2.30B). The Pakistani paleosols are finer grained and have fewer mafic minerals, so that examples of mineral weathering are less photogenic. Nevertheless, they include comparable etching, bending, and general degradation of grains of biotite, muscovite, and feldspar. Critical observations that support a weathering origin are the range of examples from nearly fresh to almost completely weather-

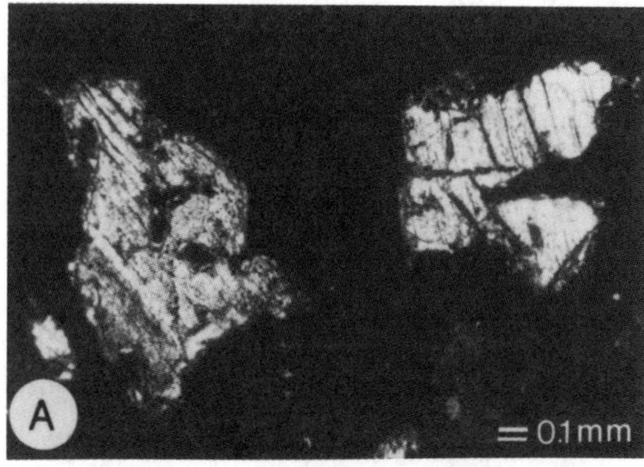

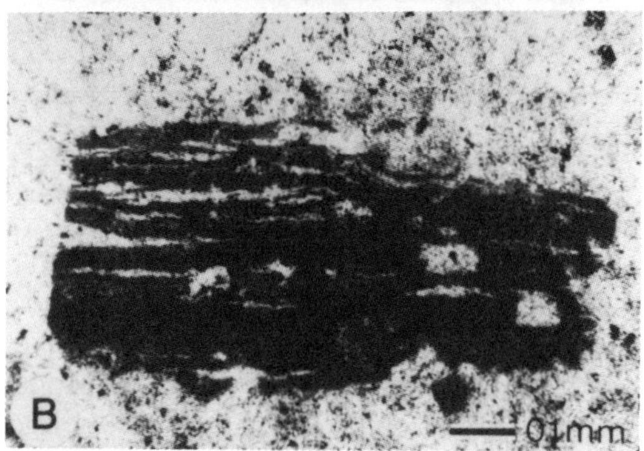

Fig. 2.30. Petrographic thin sections of deeply weathered minerals from paleosols at Fort Ternan National Monument, Kenya: A, deeply etched pyroxene under crossed nicols, from A horizon of type Onuria clay (Kenyan National Museum specimen FT-R76); B, buckled and expanded grain of biotite, under plane light, from A horizon of Chogo clay ferruginized-nodule variant (Kenyan National Museum specimen FT-R14). Scale bars 0.1 mm.

ed away, the delicate extensions of deeply weathered grains that would not withstand transport, and the closely associated clayey weathering products surrounding degraded mineral grains.

These observations rule out a depositional origin of these features but not their formation during deep burial (as has been argued for other comparable cases by T. R. Walker, Ribbe & Honea, 1967). Deep-burial alteration of the grains seems unlikely in this case because altered grains were seen mainly in clayey matrix rather than in sandy or pebbly intervening beds, interpreted as less weathered lower parent material that also would have been more permeable to intrastratal solution. The considerable expansion of some grains, especially of biotite, would not be expected under the confining pressures in deep burial environments. The alteration was also oxidizing, as can be seen from the ferruginized clay skins around the grains, whereas groundwater flowing through rocks as rich in mafic minerals as these Kenyan volcaniclastics would be chemically reducing. Finally there is the chemical conundrum of mineral grains altered by hydrolysis in calcareous paleosols both in Kenya and Pakistan. Soils are geochemically out of equilibrium and variable in both time and space. Weakly acidic rain and soil solutions can attack mineral grains and then evaporate, leaving carbonate in the matrix. Deep groundwaters, in contrast, are more chemically uniform, so that acidic alteration of mineral grains should be accompanied by extensive dissolution of carbonate, which was not seen in the paleosols. In very deep burial environments, the thermal cracking of organic matter to hydrocarbons and carbon dioxide can generate acidic solutions capable of dissolving out a secondary porosity (Schmidt & McDonald, 1979). Vuggy, cross-cutting porosity of this kind was not seen in the paleosols. Nor would it be expected considering their likely depth of burial, discussed in detail in a later chapter (p. 47).

2.3.4 *Microfabric of clayey matrix*

Additional features of the paleosols seen in thin section under crossed polarizers are highly birefringent streaks of oriented clay within the generally flecked clayey matrix. This distinctive soil microfabric can be called "bright clay" (Retallack, 1990a) or "sepic plasmic fabric" (Brewer, 1976). Good examples of this were seen in paleosols from both Kenya and Pakistan (Fig. 2.31). The concentration of highly birefringent clay around mineral grains could have been due in part to original rolling of the grain during deposition in the parent material, or perhaps even rocking of the grain during thin-section preparation, as much as to stresses generated during shrinking and swelling of the original soil. Other streaks of birefringent clay are indications of a highly deviatoric system of local stresses under low temperature and confining pressure; that is to say, in a soil rather than in a depositional or deep burial environment.

Sepic plasmic fabrics are best expressed in well or seasonally drained, clayey, and moderately developed soils (Brewer & Sleeman, 1969; J. F. Collins & Larney, 1983). Their expression is inhibited in weakly developed or waterlogged soils, in soils dominated by amorphous weathering products of volcanic ash, and in soils whose clayey fabric is masked by abundant near-opaque sesquioxides.

These observations on modern soils tally well with observations of the paleosols. In the Kenyan paleosols on volcaniclastic parent materials, sepic plasmic fabric was better displayed in paleosols formed on nephelinitic sandstones derived in part from preexisting soils (Buru, Tut, Rabuor, Chogo, and Onuria Series), compared with paleosols developed on a greater proportion of fresh carbonatite-nephelinite tuff (Mobaw, Kwar, Choka, and Dhero Series). Among the Pakistani quartzofeldspathic paleosols, sepic plasmic fabric is better developed (skelmasepic) in moderately to weakly developed paleosols (Khakistari, Lal, Pila, Naranji, Pandu, and Kala Series), but is only weakly expressed (insepic to skelmosepic) in very weakly to weakly developed paleosols (Sarang, Bhura, and Sonita Series). Those paleosols with gray color, ferric mottles, and other indications of waterlogging (Khakistari, Kala, and Pandu Series) had well-developed sepic plasmic fabrics, perhaps because they were seasonally dry, as indicated also by their deeply penetrating root traces and calcareous nodules. These microfabrics are evidence not only that these were paleosols, but of the kinds of soils they were.

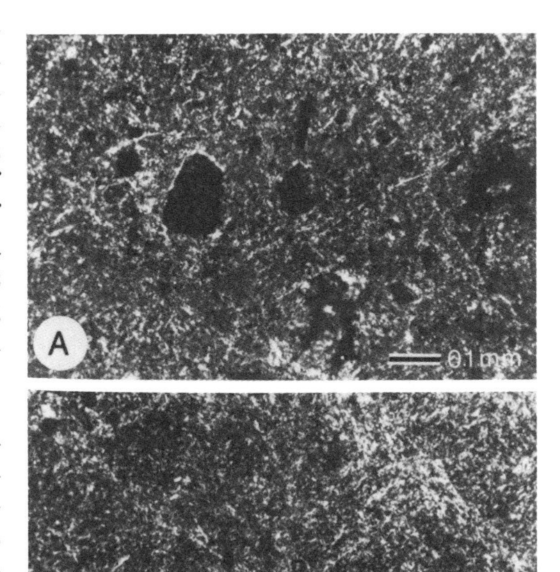

Fig. 2.31. Petrographic thin sections under crossed nicols of sepic plasmic fabric: A, skelmasepic plasmic fabric from A horizon of the type Onuria clay paleosol in the main excavation at Fort Ternan National Monument, Kenya (Kenyan National Museum specimen FT-R76): B, clinobimasepic plasmic fabric from A horizon of type Kala clay paleosol in lower measured section of Dhok Pathan Formation, west of Kaulial village, Pakistan (Smithsonian specimen 353808). Scale bars 0.1 mm.

2.3.5 *Clay minerals*

The nature of clays in paleosols of alluvial and volcaniclastic sequences are uncertain as guides to whether they actually were soils, because clays from windblown dust or flood-borne alluvium originally present in the parent material of a paleosol also are derived from soils of sedimentary source regions. As a generalization, however, base-poor clays (such as kaolinite) reflect

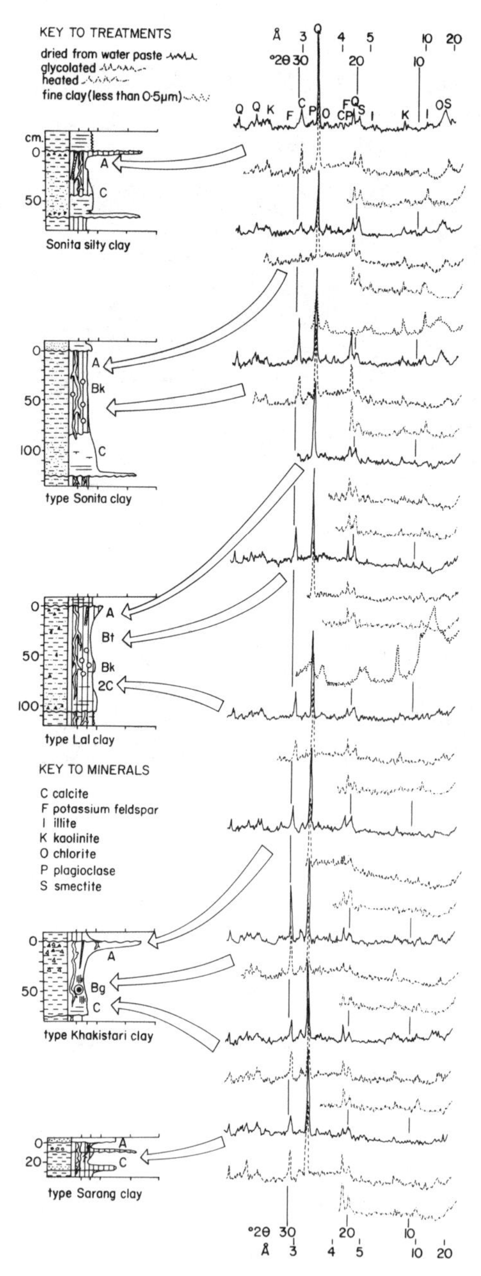

Fig. 2.32. X-ray diffractograms revealing clay mineral composition of Sarang, Khakistari, Lal, and Sonita Series paleosols of Pakistan, using three treatments of crushed whole rock: (1) deposition from distilled water, (2) glycolation for 6 hours at room temperature, and (3) heating to 300-400°C for 2-4 hrs. Lal clay Bt and Sonita clay A also run for fine clay (<0.5 μm) settled from water. All runs from using a Phillips-Norelco 12045 instrument and Cu-Ni radiation, by S. Cook, at Department of Geological Sciences, University of Oregon, Eugene. Rock symbols as for Fig. 5.17.

deeper weathering than base-rich clays (illite, smectite: M. L. Jackson *et al.*, 1948). In addition, soil clays tend to have broader and less distinct peaks on an x-ray diffractogram, reflecting their finer grain size and poor crystallization compared with metamorphic or hydrothermal clays (Townsend & Reed, 1971).

All the paleosols from Pakistan whose clay minerals were studied had a mixture of kaolinite, illite, smectite, and chlorite (Fig. 2.32). This was seen even in weakly developed profiles (Sarang and Khakistari Series), so probably reflects in large part derivation from Himalayan source areas that included soils (for smectites), schists (for illite and chlorite) and even some bauxitic Eocene paleosols (for kaolinite: Cotter, 1933; Wadia, 1928). The degree of weathering of the clays is much more noticeable in very weakly developed paleosols (Sarang profile of Fig. 2.32) associated with ancient drainage of the Himalayan foothills (buff sandstone system of Behrensmeyer & Tauxe, 1982) than in weakly developed paleosols (Khakistari profile of Fig. 2.32) of the axial drainage of a proto-Gangetic River reaching well into the Himalaya (blue-gray sandstone system). There also is a noticeable broadening of the x-ray diffractometer (XRD) peaks in the surface horizons compared with lower horizons in moderately developed paleosols of the proto-Gangetic drainage (Lal and Sonita profiles of Fig. 2.32). This is consistent with weathering in place. This effect is not so marked in the Sonita as in the Lal profile, but is especially striking in the very fine (less than 0.5 μm) clay fraction. These observations are consistent with the neoformation of smectite in lowland soils of weak to moderate development and subhumid to semiarid climate in modern Indo-Gangetic alluvium (Gupta, 1961, 1968; Kanwar, 1959; Pundeer, Sidhu & Hall, 1978; Razzaq & Herbillon, 1979).

In Kenyan paleosols also, crystallinity of clays was poor in surface horizons compared with subsurface horizons (especially in the Chogo, Tut, and Kwar Series paleosols of Fig. 2.33). The dominant clay was a smectite, probably mixed-layer montmorillonite-vermiculite, as in surface soils of East Africa formed on volcanic ash in dry climates (less than 1200 mm mean annual rainfall, Mizota & Chapelle, 1988; Mizota, Kawasaki & Wakatsuki, 1988). There are lesser amounts of illite. Large altered books of biotite (Fig. 2.30B) and of vermiculite (Fig. 2.13B) are probably products of eruption of these carbonatite-nephelinite tuffs, again as elsewhere in East Africa (Hay, 1978). These volcanic phyllosilicates are a minor component, considering the low intensity of their peaks in the XRD traces and the marked expansion of the 15-to 13-Å peak on glycolation. Well-crystallized clay of subsurface horizons of these paleosols may consist, in part, of material transported with colluvial (Chogo and Kwar Series) and alluvial (Tut Series) parent material, or washed down within the profile away from the primary minerals from which it formed. At the surface of the paleosols, in contrast, the clays are poorly crystallized, with broad, low peaks on the XRD traces, as also observed in surface soils on East African volcanic ash (Jager, 1982).

2.4 Recognizing paleosols

Interpretation of many of the beds in Miocene sequences of Pakistan and Kenya as

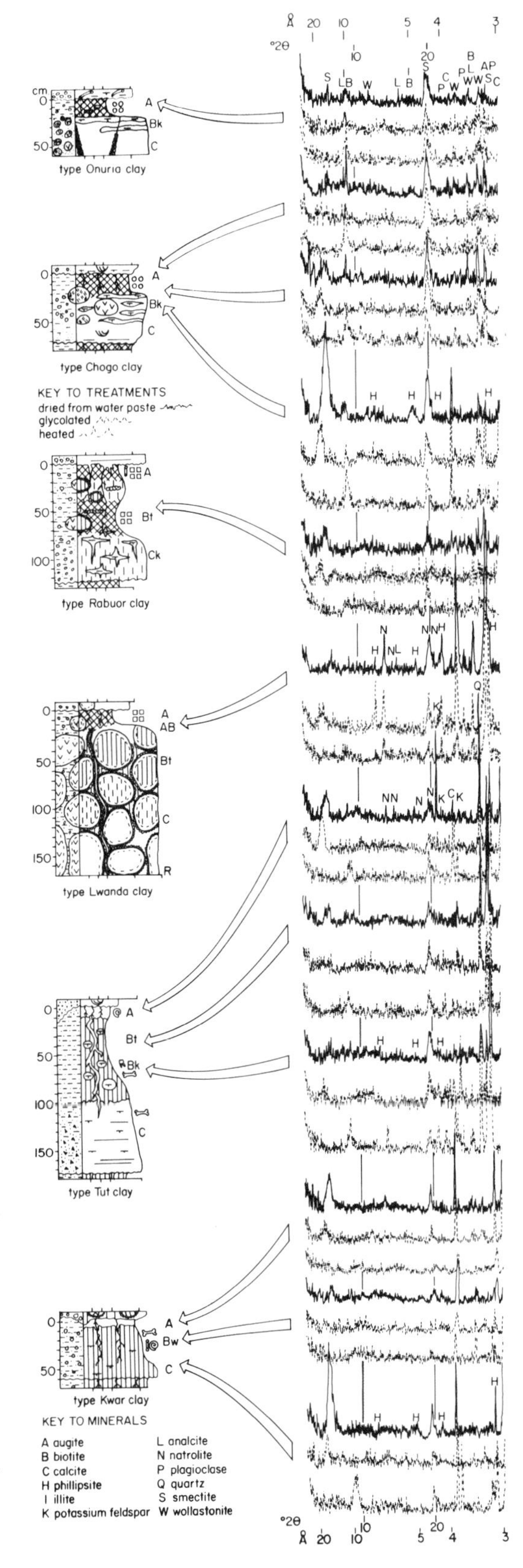

Fig. 2.33 (to right). X-ray diffractograms revealing clay mineral composition of Kwar, Tut, Lwanda, Rabuor, Chogo, and Onuria paleosols, Kenya, using three treatments of crushed whole rock: (1) deposition from distilled water; (2) exposure to ethylene glycol for 1 hour at 60°C; (3) heating to 550°C for 2 hours. Run on a Scintag Pad V instrument using Cu-Kα radiation by J. Baham, M. Humphreys and A. Ungerer, at School of Oceanography, Oregon State University, Corvallis. Rock symbols as for Fig.4.12.

paleosols is well supported by many lines of evidence gathered in both the field and the laboratory. Some of this evidence is merely compatible with a paleosol interpretation, and is not compelling in itself. Laboratory data may appear impressive but are not always convincing. Although a variety of nonpedogenic beds were sampled, laboratory studies naturally focused in more detail on those beds identified in the field as paleosols. Diagnostic field features of paleosols are thus most critical for their recognition in sedimentary and volcaniclastic sequences. For this study, the best indicators of paleosols were root traces, the diffuse contacts, and mineral segregations typical of soil horizons, and the cracking, veining, and other soil structures that obliterated preexisting structures of the parent material.

Paleopedology, like soil science, is fundamentally a field science. Laboratory testing and reference to the wider theory of soil science add a degree of sophistication and conviction to the recognition of paleosols, but field observations remain of primary importance.

3. ALTERATION OF PALEOSOLS AFTER BURIAL

The alteration of paleosols during burial and metamorphism can seriously compromise their recognition and interpretation. The effects of extreme metamorphic alteration, perhaps surprisingly, are often clear from the development of such structures as schistosity or of high-temperature minerals such as garnet. More troublesome in the interpretation of paleosols are modifications during burial that fall short of metamorphism, that is to say, diagenetic changes. Especially problematic are changes that occur soon after burial when the paleosol is still within the reach of surficial processes. These modifications in aqueous solutions, with the aid of microbes and under low temperatures and pressures, are not always distinct from soil formation. Indeed, if diagenesis is defined as alteration of sediments after their deposition, then diagenesis includes both soil formation and alteration after burial. Research on the distinction between these two kinds of diagenesis now is progressing apace (Retallack, 1990a, 1991). The following sections constitute an assessment of the degree to which specific kinds of diagenetic alteration affected the studied Miocene paleosols from Kenya and Pakistan.

3.1 Loss of organic matter

Studies of Quaternary paleosols and equivalent surface soils in central North America have shown that soon after burial, soils lose up to an order of magnitude of organic carbon as determined by the Walkley-Black technique, but they preserve the general trend of their depth function for organic matter (Stevenson, 1969). This loss of organic matter has been found in paleosols similar to well drained soils but not in paleosols similar to peaty, waterlogged soils. The lost organic matter was probably metabolized by aerobic microbial decomposers which were part of the ecosystem that formed soil on the sedimentary increment that buried the paleosol.

All of the Miocene paleosols from Kenya and Pakistan appear to have lost substantial amounts of organic matter. They have very low analytical values of Walkley-Black organic matter (Figs 2.19-2.23). This is surprising in view of the abundant root traces, burrows, and other evidence of life in them. Generally comparable surface soils of the Serengeti Plains of Tanzania have organic carbon contents of 0.01 to 2 weight percent, and in some cases up to 9 weight percent (de Wit, 1978; Jager, 1982). Similar surface soils of the Songhor area of Kenya have organic carbon values of 0 to 5.64 weight percent (Thorp *et al.*, 1960). On the Indo-Gangetic Plains of India and Pakistan organic carbon is 0.03 to 2.68 weight percent (Murthy *et al.*, 1982). Loss of organic matter in the Miocene paleosols from values similar to those in these surface soils may have caused changes in color of the paleosols so that they are now less dirty (higher Munsell chroma). Loss of organic matter also would make the paleosols more prone to badlands weathering in the modern outcrop, compared with the original soils stabilized by organic coatings and roots.

3.2 Burial gleization

Drab haloes around root traces in some of these Kenyan and Pakistani paleosols are best interpreted as products of the anaerobic decay of organic matter buried with the soil, as has already been discussed with respect to root traces in these paleosols. This phenomenon of burial gleization is widespread and well known in paleosols (Retallack, 1976, 1983, 1988a, 1990a), soils (de Villiers, 1965), and sediments (Allen, 1986b). In some paleosols (type Sonita clay, type Bhura clay, and type Naranji clay) there also are drab layers deep within the profile of a similar greenish-gray or bluish-gray color unusual for surface soils. These may also reflect areas reduced in association with anaerobic decay of organic matter buried in the paleosol. More pervasive burial gleization may be responsible for the overall greyish color of Pandu and Khakistari paleosols of Pakistan, but iron-manganese concretions and ferric mottles are evidence that these paleosols were prone to seasonal water-logging as they formed. Root traces in Khakistari paleosols contain preserved organic matter and have a drab halo as well as a diffuse sheath enriched in manganese (Fig. 3.1). This indicates movement of manganese and preservation of organic matter as is found in similar gleyed surface soils of India and Pakistan (Kooistra, 1982). One would expect lowland soils such as these to be more prone to burial gleization than well-oxidized soils of uplands, and there is more burial gley discoloration in paleosols of fluvial sedimentary sequences (in Pakistan and at Songhor in Kenya) than of paleokarst (at Koru) or of volcanic footslopes (at Fort Ternan).

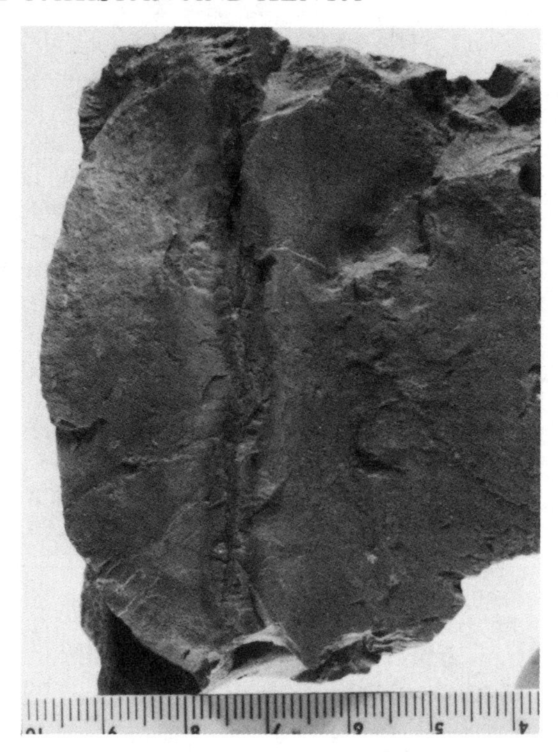

Fig. 3.1. Drab-haloed root trace with original organic matter preserved and a diffuse dark outer mangan attributed to gleization during soil formation, from A horizon of type Khakistari clay paleosol, in lower measured section, west of Kaulial village, Pakistan. Smithsonian specimen 353772. Scale in mm.

A critical question for interpretation of these paleosols is the extent to which burial gleization has altered more than just the redox state of iron and the color of the paleosols within drab mottles and root haloes. Could it be responsible also for the ferric mottles, concretions, and manganese stringers in these paleosols? This applies especially to Mobaw, Khakistari, Pandu, and Kala Series paleosols, which have these structures. A hallmark of drab-haloed root traces attributed to burial gleization is conservation of amounts of iron and manganese despite their chemically reduced state, compared with amounts of these elements in the surrounding paleosol matrix (Retallack, 1976, 1983; G. S. Smith, 1988). This may be interpreted as evidence of chemical reduction in a closed system, unlike losses of large amounts of iron typical of gleyed soils. Pervasive burial gleization of the Pakistani and Kenyan paleosols also is unlikely in view of their negligible analytical values for ferrous iron compared with ferric iron (Figs 2.19-2.23). These paleosols are very highly oxidized, with barely enough reduced iron to account for the pyroxene and other mafic minerals observed in them. A final indication that the iron-manganese concretions of the Pakistani paleosols are original is their co-precipitation with carbonate and penetration by calcareous rhizoconcretions (Fig. 2.12).

There are thus grounds for regarding

the effects of burial gleization as locally restricted to the discolored areas. The bulk of the physical and chemical properties of the paleosols were little affected, even in paleosols originally gleyed. Effects of burial gleization were probably much less in paleosols with only scattered drab-haloed root traces and were negligible in paleosols lacking even these.

3.2 Burial reddening

Another widespread effect of burial of paleosols is the dehydration of ferric hydroxide minerals such as goethite to oxide minerals such as hematite (T. R. Walker, 1967). This change results in reddening of color from yellow or brown to brick red. In the few Quaternary paleosols and comparable surface soils of central North America for which this change has been assessed, the change in color during shallow burial for only a few tens to hundreds of thousands of years amounted to two to three Munsell hue units, from 10YR to 7.5YR or from 7.5YR to 5YR (Ruhe, 1969; Simonson, 1941). No change in hue was noticed in drab-colored (5Y) paleosols compared with similar gleyed surface soils. These changes in color compromise the interpretation of paleosols because dehydration of ferric hydroxides to ferric oxides also occurs with increasing age and climatic temperature during the formation of surface soils (Birkeland, 1984).

Such changes in hue are especially likely for the red paleosols from Pakistan. These show at least three distinct components of magnetization that can be isolated by different temperature treatments: (1) a low temperature (less than 300°C) pigmentary iron stain usually in the general direction of the modern magnetic field; (2) medium temperature (300-650°C) pigmentary hematite magnetized in the direction thought to have prevailed during soil formation; and (3) high temperature (650-685°C) specular hematite, often randomized in its magnetic direction within pedogenically altered zones, but in C horizons and sediments taken to reflect the magnetic field prevailing during deposition of the sediments (Tauxe & Badgley, 1988). In some cases two separate magnetization directions can be recognized at medium temperatures (Tauxe & Badgley, 1984). In all cases the red paleosols contained a more complex mix of magnetization directions than associated red sediments or gray paleosols and sediments. The alignment of all but the low temperature components with the ancient magnetic field, even within times of field reversal temporally constrained to less than a few thousand years, is an indication that the medium temperature components were acquired during soil formation or early after burial (Tauxe & Badgley, 1988). The duration of soil formation indicated by these paleomagnetic studies (thousands of years) is less than that required (hundreds of thousands of years) for the aging of iron hydroxides to hematite in surface soils (Birkeland, 1984), so it is likely that much of the medium temperature component was produced during early burial reddening.

Surface soils as red as the Pakistani paleosols (especially Lal and Sonita Series) are known in India but are found mainly on very old land surfaces of Precambrian shield areas. Modern soils of the Indo-Gangetic Plains are either drab colored when waterlogged, or brown (10YR-7.5YR: Murthy *et al.*, 1982). Although these alluvial soils are siltier and probably are forming within a less vegetated, overpopulated landscape, they are similar in general profile form and soil structures to the paleosols. These brown colors are more reasonable original colors for the paleosols. Their current red hue (7.5YR-5YR) is best attributed to burial dehydration of ferric hydroxides.

A similar degree of reddening during burial may have changed the appearance of the red paleosols in Kenya (Mobaw, Kwar, Kiewo, Choka, Tut, and Buru

Series). Similarly red surface soils are found in the Songhor area today (Thorp *et al.*, 1960), but these are either very ancient exhumed paleosols or have a detrital component from preexisting paleosols on Precambrian gneissic basement, unlike the Miocene paleosols interbedded within volcanic rocks. Unlike these red soils and paleosols, brown colors (10YR) predominate in paleosols at Fort Ternan (Rairo, Rabuor, Chogo, and Onuria Series) and in surface soils of the Serengeti Plains (de Wit, 1978; Jager, 1982). The Fort Ternan paleosols could not have been more drab in color considering their present oxidation state, revealed by their ferric iron/ferrous iron ratios (Appendix 7). Burial reddening of the paleosols at Fort Ternan may have been overwhelmed by the darkening effect of organic matter and volcanic mineral and rock fragments.

3.4 Calcite cementation

Calcite is a common component of these paleosols as clastic grains, as replacive micrite, and as displacive spar (Figs 2.11, 2.12, 2.14-2.17). Calcite is also closely associated with root traces and burrows (Figs 2.1 and 2.2), and in these forms was probably part of the original soils. There also are root holes and intergranular voids filled with a cement of sparry calcite (Figs 2.1 and 3.2). Such cements are found in the lower parts of Chogo, Pandu, and Sarang Series paleosols and associated sandstones in both Kenya and Pakistan. In each case examined, the cements appear to have predated compaction. They preserve uncollapsed and little-distorted root traces and little flattened, though deeply weathered, volcanic and shale rock fragments. This void-filling sparry calcite is probably a cement formed soon after burial of the soils.

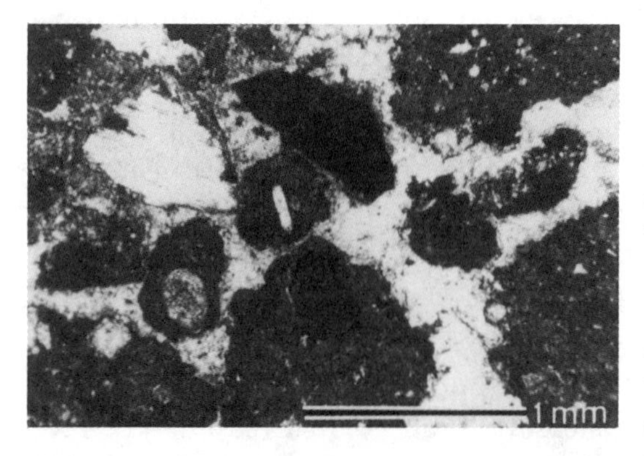

Fig. 3.2. Petrographic thin section under crossed nicols of sparry calcite cement to nephelinitic sandstone overlying the type Chogo clay in the main excavation at Fort Ternan National Monument, Kenya. Kenya National Museum specimen FT-R1. Scale bar 1 mm.

3.5 Zeolitization

No zeolites were detected in thin sections or x-ray diffractograms of samples from Miocene quartzofeldspathic paleosols of Pakistan. Zeolites are a minor component of volcaniclastic Kenyan paleosols, but are more common in associated volcaniclastic parent materials (J. A. Van Couvering for Shipman *et al.*, 1981). No zeolites were seen in contact with unquestionably pedogenic features, such as root traces or burrows. Furthermore, all the Kenyan paleosols analyzed chemically had low soda/potash ratios that declined up the profile (Appendix 7), so that zeolitization during soil formation is unlikely.

Phillipsite needles and small equant crystals of natrolite were observed in deep (C) horizons of several paleosols (Kwar, Tut, Lwanda, and Chogo Series), both in thin sections and in XRD traces (Fig. 2.33). The most spectacular concentrations of phillipsite, natrolite, and some analcite were seen in vesicles of phonolite corestones and clasts (Lwanda Series). Zeolites were occasionally seen to have crystallized before the void-filling calcite cement thought to have predated substantial compaction, as also documented in paleosols on Rusinga Island, Kenya (Thackray, 1989). Very small amounts of natrolite and of analcite were detected by x-ray only in the tuffaceous uppermost horizons of a few paleosols (type Tut and Onuria paleosols, respectively).

From these relationships, as well as the widespread preservation of volcanic lapilli and overall rarity of zeolites, pervasive zeolitization during burial seems unlikely. Instead, zeolites probably were formed during the initial emplacement and weathering of tuffaceous parent materials, including later dustings of volcanic ash. It is likely that zeolites were then progressively destroyed by subsequent soil formation and alteration after burial.

In historically erupted carbonatite-nephelinite tuffs from Oldoinyo Lengai, initial weathering of unstable minerals such as gregoryite and nyerereite produces highly alkaline and saline pores in which salts and zeolites precipitate. Even in the current dry climate of the Tanzanian portion of the East African Rift, these salts and zeolites are destroyed over periods of a few tens to hundreds of years as soils develop and accumulate clay (Hay, 1978, 1989; Hay & O'Neill, 1983). Their disappearance may have been even more rapid in soils of the much wetter paleoclimates envisaged here for these Kenyan paleosols.

3.6 Compaction

Burial of paleosols results in compaction as the void spaces, organisms, and water are crushed by the weight of overburden. The resulting changes in thickness can be of importance for interpreting such factors in soil formation as former rainfall from depth within the profile to the horizon of calcareous nodules (Jenny, 1941). Fortunately, paleosols from Kenya and Pakistan are young enough that their burial history can be reconstructed and compared with general curves for compaction with depth of different kinds of sediments (Baldwin & Butler, 1985).

The paleosols from Pakistan are within the Siwalik Group, for which aggregate thicknesses of 6 km have been reported (Wadia, 1975). Recent research using paleomagnetic stratigraphy to unravel deformation of the Himalayan foothills (Burbank *et al.*, 1988; N. M. Johnson *et al.*, 1988) has shown that this is not a single pile of sediments, but that local depocenters shifted through time. The initiation of thrusting and folding to create the Potwar Plateau, where the paleosols were examined, began about 5 million years ago and was completed 2 million years ago, by which time the uplifted sedimentary sequence had begun to be dissected.

In Kaulial Kas, there is an additional 930 m of Dhok Pathan Formation above the lower measured section of paleosols and 518 m above the upper measured section, but the top of the formation is not exposed in the core of the Soan Syncline that formed with thrusting of the Potwar Plateau. The maximum likely thickness of the Dhok Pathan and overlying Soan Formations in the Kaulial area can be reconstructed by assuming a monotonic thickness decrease from north to south and east to west in these formations from sections where both tops and bottoms of the formations are exposed (using thicknesses given by Fatmi, 1974; Shah, 1977). This gives a combined thickness of both formations near Kaulial of up to 1972 m. Assuming that these formations were mapped in the sense of Cotter (1933), rather than the unnecessarily complex later interpretation of Barry *et al.* (1980), the lower measured section would have been 2018 m deep and the upper one 1637 m deep.

These results are broadly compatible with estimates based on paleomagnetic data (Tauxe & Opdyke, 1982) extrapolating the rate of sediment accumulation from the geological age of the two measured sections (8.5 and 7.7 million years, as readjusted to the time scale of Berggren, Kent & Van Couvering, 1985) to the youngest part of the Dhok Pathan Formation in Kaulial Kas (6.5 million years) and then to the time of initial deformation of the Potwar Plateau (5 million years) or uplift to its modern elevation (2 million years). Extrapolating to 5 million years at constant rates gives cover over the lower section of 1705 m and over the upper section as 1295 m. Extrapolating to 2 million years gives cover of the lower section by 3255 m and the upper section by 2849 m. These last two estimates do not take into account a reduced sedimentation rate likely as deformation of the Potwar Plateau was initiated before its culmination 2 million years ago.

Both these paleomagnetic and geological estimates are compatible with the occasional concavo-convex contacts between quartz grains in thin sections of sandstones within these measured sections, but complete lack of sutured contacts, unlike very deeply buried sandstones found elsewhere (J. M. Taylor, 1950). Of the three lines of evidence, the figures derived from formation thickness are used here: roughly 2000 m for the lower section and 1600 m for the upper section. Using standard compaction curves for marine shale (Baldwin & Butler, 1985), burial by this amount should compact clayey parts of the paleosols to about 60 percent of their former thickness in the upper measured section and 58 percent in the lower section. Using curves for marine sandstone (Baldwin & Butler, 1985), sandy parts of the paleosols may have been compacted by 75 percent in the upper section and 70 percent in the lower section. A similar reconstruction was made of an undescribed clayey paleosol within the Chinji Formation 3 km south of Khaur, Pakistan. This was buried by at least 4300 m and compacted by about 52 percent so that its calcareous horizon now at 54 cm within the profile was probably once closer to 109 cm.

Considering the more abundant cracks and air-filled voids in soils than in marine sediments, these estimates are likely to be conservative. They are compatible with observed deformation of drab-haloed calcareous rhizoconcretions and with the vertical impersistence of clastic dikes in these paleosols.

Compaction also was significant for paleosols in volcaniclastic sequences in Kenya. Depths of burial can be calculated by reconstructing the former extent of the ancient volcanoes around which they formed as similar in relative height above the plains (about 3000 m) to the well-studied Miocene carbonatite-nephelinite volcano Kisingiri, in southwestern Kenya (Le Bas, 1977). Interpolating from this height centered on the Pliocene vent of Tinderet volcano out to the most distal preserved lava flows gives estimates of overburden of 625 m for paleosols at Fort Ternan National Monument, 700 m for Songhor National Monument, and 600 m for the abandoned lime quarry 3 km northwest of Koru. Using standard curves for shales (Baldwin & Butler, 1985), clayey parts of paleosols at Fort Ternan were compacted to 74 percent of their former thickness, those at Songhor to 72 percent, and those at Koru to 75 percent. Standard curves for sandstones yield estimated compaction of 87 percent for sandy parts of paleosols at Fort Ternan, 85 percent for Songhor, and 88 percent for Koru. Similarly, two paleosols in the Miocene Hiwegi Formation on Rusinga Island were probably buried by as much as 2111 m, so that their clayey surface horizons were compacted to 63 percent, bringing their calcareous horizons from original depths of 96 and 25 cm to the observed depth of 55 and 16 cm, respectively.

Compaction of originally calcareous parts of some Kenyan paleosols would have been resisted by an early burial cement of sparry calcite especially evident in the coarse-grained lower portion of many of these paleosols. Some calcite-cemented fossil burrows and pupal cells in the paleosols (Fig. 2.8) are little flattened. Other burrows have been deformed either during burial or by collapse in the original soil, so that the implications of these observations for compaction are uncertain.

3.7 Thermal maturation of organic matter

Coalification and cracking of hydrocarbons at depth can produce acidic reducing brines capable of considerable chemical alteration and the development of secondary porosity in deeply buried sedimentary rocks (Schmidt & McDonald, 1979). This is unlikely to have affected these paleosols at the depths of burial already outlined. Using a typical geothermal gradient of 25°C/km, gives burial temperatures of about 41°C for paleosols at Fort Ternan, 40°C for Koru, and 43°C for Songhor in Kenya; and 65°C for the upper measured section west of Kaulial village and 75°C for the lower section east of Kaulial village in Pakistan. Even at an extremely highly geothermal gradient of 65°C/km, possible for volcaniclastic Kenyan paleosols but very unlikely for Pakistani paleosols, burial temperatures would not have been high by geological standards: 66°C for Fort Ternan, 64°C for Koru, 71°C for Songhor, 129°C for the upper Kaulial section, and 155°C for the lower Kaulial section. Only temperatures and depths of burial for the Pakistani paleosols approach conditions needed for the generation of oil, gas, or coal of a higher grade than lignite (Hunt, 1979; Stach *et al.*, 1975; Tissot & Welte, 1978).

These theoretical predictions of the likely state of organic matter in these paleosols agree with direct observations of what little organic matter remains. Compressed remains of the woody portions of roots seen in some of the paleosols in the lower measured section west of Kaulial village, Pakistan (Fig. 3.1: also in Kala, Pandu, and Lal Series) were dark brown and fibrous in thin section, rather than opaque and broken into coal cleat. In the Kenyan paleosols, original plant organic matter is rarely preserved. In some paleosols (Chogo, Dhero, and Kiewo Series) natural casts of stumps, twigs, and grasses showed clear outlines of bark and veins, unfractured by coal cleat. Fossil grass remains (associated with the type Onuria clay) have cuticular remains showing cell outlines, stomates and phytoliths (Dugas, 1989; Retallack, 1990a; Retallack, Dugas & Bestland, 1990), that are not discernably coalified or graphitized. None of the paleosols in Kenya or Pakistan were seen in thin section to have the inflated and invasive vugs with multiple generations of cement that are characteristic of secondary porosity developed during deep burial. Thus significant alteration of these paleosols due to thermal maturation of organic matter is considered unlikely.

3.8 Illitization

Another potential alteration of clayey paleosols is destruction of smectite and growth of illite by transfer of potassium in groundwater from dissolution of potassium-bearing minerals such as microcline and muscovite (Eberl *et al.*, 1990; Weaver, 1989). For marine shales, this mineralogical transformation is common when burial depths reach 1200 to 2300 m and burial temperatures are from 55° to 100°C, but the transformation slows dramatically once easily mobilized pore water is lost (J. P. Morton, 1985). Theoretically, however, the transformation of smectite to illite could occur at much lower temperatures over very long periods of geological time (Bethke & Altaner, 1986). Indeed, illitization of smectite may occur to a limited extent during shrink-swell behavior of swelling clay soils, such as Vertisols (D. Robinson & Wright, 1987).

There were traces of potash feldspar in almost all the Miocene paleosols studied, and thus a potential for illitization of their clays. The Kenyan paleosols were less deeply buried and at lower burial temperatures than is usual for illitization, but the Pakistani paleosols may have experienced burial conditions conducive to illitization for as long as a million years at some time between their burial about 8 million years ago and the beginnings of

TABLE 3.1. Indices of illite "crystallinity" in Miocene paleosols of Kenya and Pakistan

Paleosol Name	Horizon	Specimen Number	Weaver index	Kubler index	Weber index
Sonita silty clay	A	SI353840	1.3	0.23	105
type Sonita clay	A	SI353850	1.1	0.68	166
	Bk	SI353851	2.4	0.79	247
mean and error for Kaulial west			1.6 ±0.6	0.56 ±0.2	173 ±58
type Lal clay	A	SI353778	1.4	0.93	137
	Bt	SI353780	1.6	0.38	136
	2Bk	SI353783	2.2	0.35	106
type Khakistari clay	A	SI353772	1.9	0.64	164
	A	SI353773	2.1	0.56	329
	C	SI353775	2.0	0.38	129
type Sarang clay	C	SI353768	2.0	0.57	300
mean and error for Kaulial east			1.9 ±0.3	0.54 ±0.2	186 ±83
mean and error for Pakistan			1.8 ±0.4	0.55 ±0.2	182 ±77
type Onuria clay	A	FT-R76	1.2	1.07	-
type Chogo clay	A	FT-R2	1.7	0.41	175
	A	FT-R3	1.1	0.46	-
	Bk	FT-R6	1.1	0.93	-
type Rabuor clay	Bt	FT-R48	1.2	0.39	-
type Lwanda clay	A	FT-R58	1.6	0.17	-
mean and error for Fort Ternan			1.3 ±0.2	0.57 ±0.3	-
type Tut clay	A	SO-R121	1.2	0.52	371
	Bt	SO-R125	0.8	0.65	382
	Bk	SO-R128	1.5	0.41	152
type Kwar clay	A	KO-R87	1.8	0.39	-
	Bw	KO-R88	1.1	0.27	-
	C	KO-R89	1.3	0.37	-
mean and error for Songhor-Koru			1.3 ±0.3	0.44 ±0.1	302 ±106
mean and error for Kenya			1.3 ±0.3	0.50 ±0.3	270 ±107

Note: All XRD traces were from powders of whole rock dissolved in distilled water, mounted on glass slides, and run on instruments specified for Figs 2.32 and 2.33. Methods of calculation of the indices are outlined by Frey (1987). The Weber index could not be calculated for Kenyan samples lacking quartz. The transition to metamorphic illite occurs when values become greater than 2.3 (Weaver index), less than 0.42 (Kubler index), and less than 181 (Weber index).

erosion of the sedimentary pile 5 million years ago (Burbank *et al.*, 1988; N. M. Johnson *et al.*, 1988). Several of the Pakistani paleosols show a slight surficial increase in potash, along with an up-section decline in soda (Figs 2.19-2.23), unlike comparable surface soils (Tomar, 1987). This is much less marked than in profoundly illitized and sericitized paleosols (Feakes & Retallack, 1988; Retallack, 1986c). Taken together with the much greater abundance of smectite than illite in these paleosols (Fig. 2.32), this is an indication of discernable but slight illitization of the Pakistani paleosols.

Another way of approaching the problem is to estimate the crystallinity of illite present from x-ray diffractometer traces (Frey, 1987). As can be seen from the raw traces (Figs 2.32 and 2.33) and a variety of standard crystallinity indices (Table 3.1), illite in the paleosols is poorly crystallized with respect to that in their parent materials, which in some cases include highly illitized schists and volcanogenic micas. The variation in crystallinity of illite in these paleosols is much as would be expected in weakly to moderately developed soils on such parent materials. Metamorphic illites, in contrast, have sharper 10Å peaks, with higher values for the Weaver index and lower values for the Kubler and Weber indices. Nevertheless, these paleosols do not have as poorly crystallized illites as can be found in soils, and slight illitization is compatible with these data.

3.9 Feldspathization

Authigenic feldspars form in a variety of rocks during deep burial (Duffin *et al.*, 1989). Enrichment in feldspar in a process of potassium enrichment called fenitization is commonly associated with the emplacement of carbonatite intrusions (Le Bas, 1977).

None of the feldspars seen in thin sections of the paleosols showed the abrupt, euhedral crystal faces or cut across preexisting textures as is usual for authigenic crystals. Nor was feldspar abundant or associated with the veining and other injection features of fenites. In contrast, the feldspars seen were etched or coated in ferruginized clay, as is usual in soils. Fenitization can extend for a few hundred meters around carbonatite intrusions (Le Bas, 1977), but geological mapping on this fine scale (Figs 4.6-4.8) failed to reveal such intrusions anywhere near the studied paleosols. Alteration of the paleosols by feldspathization is thus very unlikely.

3.10 Recrystallization

Many of the Miocene paleosols described here are strongly calcareous, and as is well known (Bathurst, 1975), calcite is prone to recrystallization to coarse grain size during burial. In thin section there is no textural evidence of recrystallization of calcite in these paleosols. Most of the calcite is micritic. Sparry calcite was found in many profiles (see Figs 2.1, 2.15, 2.17, and 2.18), but in all cases appears to be filling intergranular voids or old root holes. None of the equiangular crystal junctions, cross-cutting relationships, or other features of neomorphic spar (Folk, 1965) were observed. The array of carbonate crystal forms observed in the paleosols was comparable with that seen in surface soils (Courty & Féderoff, 1985; Sehgal & Stoops, 1972; Wieder & Yaalon, 1985).

Two exceptions to this deserve comment. First, no needle fiber calcite was seen in the paleosols, despite its occurrence in comparable modern soils (Kooistra, 1982) and its persistence in some paleosols as ancient as early Carboniferous (V. P. Wright, 1984). Such slender crystals are found arranged randomly and loosely in soil voids. They would be prone to breakage and dissolution during compaction, as well as recrystallization during burial.

Second, the bladed sparry calcite filling abundant veins in very weakly developed paleosols on carbonatite ash (Kiewo Series) are without precise modern analog, as has already been discussed. The abundance of this sparry calcite in some thin sections is similar to recrystallization fabrics, but the narrow crystal form perpendicular to the walls of the veins (Figs 2.17 and 2.18) is quite unlike recrystallization textures.

3.11 Peering through the veil of burial alteration

From this assessment of the burial alteration of Miocene paleosols of Kenya and

Pakistan, these paleosols can be considered little altered by recrystallization, feldspathization, zeolitization or thermal maturation of organic matter. Their burial depths (600-2000 m) resulted in compaction that was significant enough (74 to 52 percent of original thickness) to make it very difficult to follow clastic dikes within these paleosols, but compaction is not always evident from root traces or burrows fortified by calcite cement. The Pakistani paleosols may have been somewhat illitized during deep burial, but illitization is less likely for the Kenyan profiles.

The most profound changes to these paleosols resulted not from deep burial, but from a variety of alterations that occurred soon after burial and near the surface, in part aided by groundwater flow and microbial activity. These changes include depletion of organic matter to as much as a tenth of its original abundance, chemical reduction of oxidized iron in minerals near buried organic matter, and dehydration of ferric hydroxides to oxides. These changes significantly affected the color of the paleosols, altering them to purer color (lower chroma, perhaps by one or two Munsell units), to blue-gray hue in formerly organic parts of the profiles (to Munsell 5Y to 5G from original brownish-gray 10YR to 5Y) and to red hue in weakly organic parts of formerly well-drained paleosols (to Munsell 5YR to 7.5YR from original 10YR to 7.5YR). Some of the profiles once dark brown over orange brown (Tut, Choka, Lal, Sonita Series) are now red with blue-gray surface horizons and root mottles. Other paleosols once gray (Pandu, Khakistari Series) now have a distinctive bluish-gray cast. Other paleosols (Rairo, Rabuor, Chogo, Onuria Series) now brown (Munsell 5YR to 10YR) do not seem to have been altered greatly in color, and can be matched with similar surface soils. Like these changes in color and local redox state, calcite cementation of coarse-grained parts of the profiles also appears to have occurred soon after burial.

It is perhaps surprising that changes during burial that are most significant for interpretation of these paleosols occurred early after burial. There was little discernable effect of recrystallization and other deep burial alterations, which now are well understood and are very distinct from soil formation (Retallack, 1990a, 1991). The great array of alterations during burial that could compromise interpretation of paleosols can be intimidating. Although they limit interpretation of these particular Miocene paleosols in some ways, many features of ancient soil formation remain or can be reconstructed.

4. EARLY AND MIDDLE MIOCENE PALEOSOLS IN SOUTH-WESTERN KENYA

Paleosols have been recognized for some time at Miocene fossil mammal sites in southwestern Kenya (Fig. 4.1; Bishop & Whyte, 1962), but only recently has their abundance in these rocks been appreciated (Pickford, 1986b; Retallack, 1986b). A part of the problem in recognizing paleosols in this sequence was the traditional view, influenced by nearby Lake Victoria, that many of the limestones in the sequence were lake beds (Gregory, 1921; Kent, 1944). Many of these same limestones are now recognized as pedogenically altered subaerial carbonatite tuffs (Bishop, 1968; Deans & Roberts, 1984). The eruption from volcanoes of material with a chemical composition close to that of limestone was not widely believed possible until the historic eruption of the Tanzanian volcano Oldoinyo Lengai (Dawson, 1962; Dawson, Garson, & Roberts, 1987; Hay, 1989), and experimental petrological studies of the $CaO-CO_2-H_2O$ chemical system (Wyllie & Tuttle, 1960). Paleoenvironmental reassessment of Miocene sediments of Kenya employing advances in sedimentary facies models, as well as these volcanological discoveries, have shown that very few of these sediments, such as the Kulu Beds of Rusinga Island (Drake *et al.*, 1988; J. A. H. Van Couvering, 1982), were deposited in lakes. Most of the Miocene fossiliferous strata of southwestern Kenya and adjacent Uganda accumulated subaerially as tuffs, mudflows, and fluvial deposits around carbonatite-nephelinite stratovolcanoes and cinder cones (Pickford, 1986b).

Not only are paleosols abundant in volcaniclastic sediments of this region but they also are unusually fossiliferous with mammal bones, teeth, and snail shells. Martin Pickford (1986b) has estimated that at least 75 percent of the tens of thousands of fossils from western Kenya have come from paleosols, as opposed to lake beds, river paleochannels, or pyroclastic beds. As in other parts of the world (Retallack, 1984a), the abundance of these calcareous and phosphatic remains increases with the calcareousness and degree of development of the paleosols (Pickford, 1986b). In other words, bones and shells are preserved best in alkaline paleosols, and it takes time for them to accumulate to levels that are impressively fossiliferous. The alkalinity of these Kenyan paleosols, and their precursor soils, owes much to their carbonatite-nephelinite parent material (Hay, 1986). These chemically unusual volcanic ashes also have been responsible for some remarkable cases of preservation of fossil insects, spiders, and lizard skin and tongues (Estes, 1962; L. S. B. Leakey, 1952; Paulian, 1976; Wilson & Taylor, 1964) not usually found in sequences of well-drained paleosols (Retallack, 1984a). Thus detailed study of paleosols in southwestern Kenya can provide information on preservational biases in the assemblages of abundant fossils that many of them contain.

Paleosols also can be helpful in reconstructing the environmental context of Miocene mammalian evolution. Initial geological sections and collections of early Miocene ruminants from Koru, Songhor, and Mfangano and Rusinga Islands (Whitworth, 1953, 1958) were interpreted as evidence for vegetation much like that around Lake Victoria today: a mosaic of wooded grassland, woodland, and local thicket forest. Additional hints of grasslands were found in the nature of fossil dung beetles (Paulian, 1976), lizards (Estes, 1962), and some fossil snails (Verdcourt, 1963). However, most of the fossil snails, as well as fossil fruits and seeds (Chesters, 1957) and mammalian faunas (Andrews & Evans, 1979; Andrews,

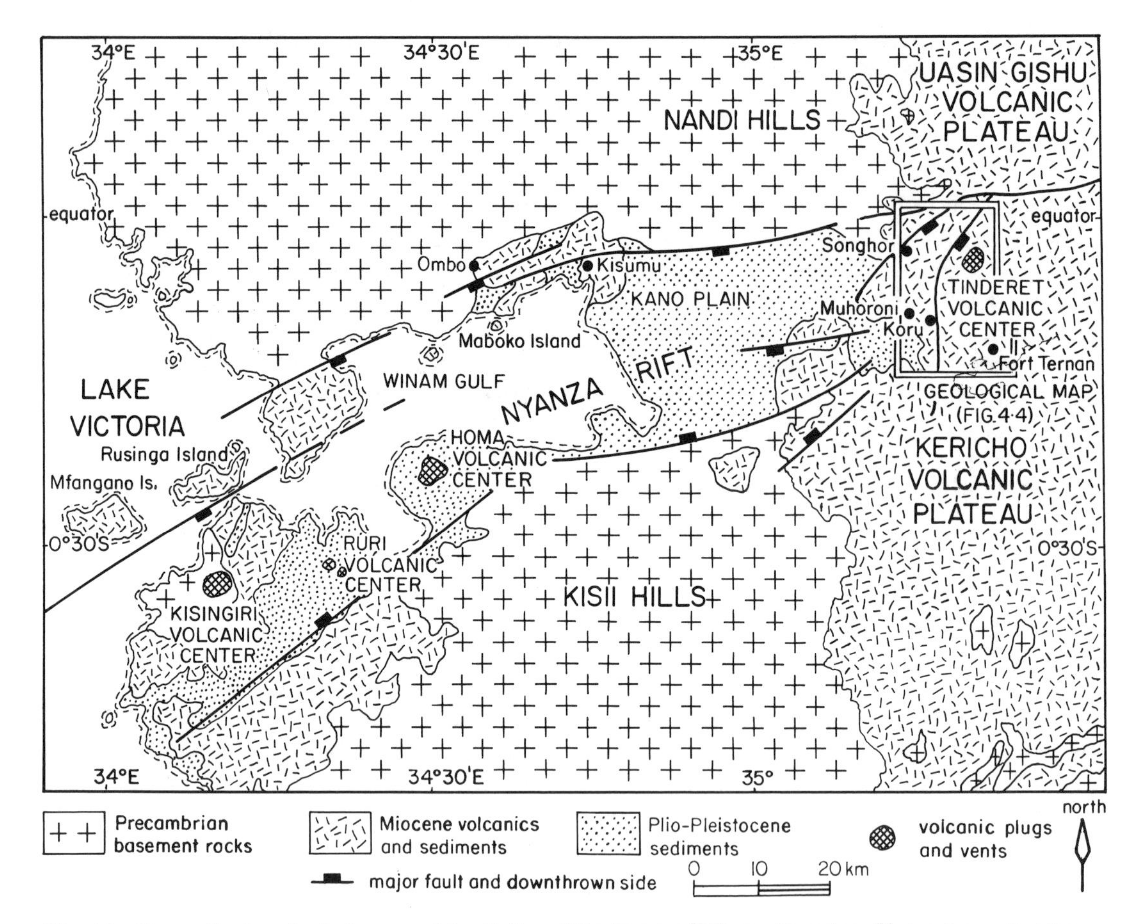

Fig. 4.1. Generalized geological map and fossil sites of the Nyanza Rift, southwestern Kenya (compiled from Pickford 1982, 1984, 1986a).

Lord, & Evans, 1979; Evans, Van Couvering, & Andrews, 1981) have been taken to indicate forested conditions. There are some indications from excavation of mammalian fossils for a mosaic of vegetation on a scale of only hundreds of meters (Andrews & Van Couvering, 1975), and there still are some advocates of open vegetation for these faunas (Kortlandt, 1983). Such interpretations have implications for the paleoecology of early ape fossils known from these deposits. These have been reconstructed (by Pickford, 1983, using evidence from sedimentary facies) as having lived in floodplain dry forests (*Dendropithecus macinnesi*, *Proconsul africanus*, *P. nyanzae*), dry upland forest (*P. major*, *Kalepithecus songhorensis*), and wet upland forests (*Limnopithecus legetet*, *Micropithecus clarki*). These interpretations can now be reassessed from the evidence of paleosols.

Differences of opinion have been especially pronounced concerning the former vegetation at the middle Miocene fossil site near Fort Ternan. The degree of hypsodonty and the microscopic wear pattern of fossil antelope teeth have been taken as indications of wooded grassland (Shipman, 1986a; Shipman *et al.*, 1981), as have recently discovered fossil grasses (Dugas, 1989; Retallack, Dugas, & Bestland, 1990). Fossil pollen grains from the site were mostly from grasses, with a conspicuous component of Afromontane trees (Bonnefille, 1984). However, an overall ecological assessment of the large fossil mammalian fauna from this site was taken as evidence of woodland or bushland (Andrews & Evans, 1979; Andrews, Lord, & Evans, 1979; Evans, Van Couvering, & Andrews, 1981; J. A. H. Van Couvering, 1980). The fossil antelope have a limb structure like, and taxonomic affinities with, living antelope of woodland, rather than open vegetation (Gentry, 1970; Kappelman, 1991; Thomas, 1984). Fossil fruits and seeds (Shipman, 1977) and snails (Pickford, 1985, 1987) also are mainly forms from woodlands. Considering these various views of former vegetation at Fort Ternan, did the fossil ape *Kenyapithecus wickeri* live there in forest, woodland, or wooded grassland?

Here are additional interpretations to be tested against the fossil record of paleosols. Two basic features of paleosols make them especially suited to this task. They can be considered evidence of vegetation and other aspects of paleoenvironment independent of fossils. Paleosols also are by definition in place, and so reflect paleoenvironmental mosaics, unlike transportable fossils.

4.1 Geological setting

The Nyanza Rift, formerly called the Kavirondo Rift, has played an important role in our understanding of the tectonic evolution of the East African Rift Valleys, including the nearby Gregory Rift Valley (Pickford, 1982). It was in the Nyanza Rift that evidence and arguments were first developed concerning compressional versus tensional origins of the African rifts (Kent, 1944; Shackleton, 1951). Detailed mapping since that time, swept along by plate tectonic theory, has now firmly established a tensional origin as an early phase in what may result in the opening of new ocean basins (B. H. Baker, 1986). The western portion of the Nyanza Rift is submerged under Winam Gulf of Lake Victoria (Fig. 4.1). From there its marginal fault scarps can be traced eastward for 130 km, where they fade out unconnected with the Gregory Rift (Pickford, 1982). Basement structures where the Nyanza Rift approaches the Gregory Rift are difficult to study because they are buried under the large central volcanoes of Tinderet, Londiani, and Timboroa.

The general geological history of both the Nyanza and Gregory Rifts can be divided into three phases (Fig. 4.2). From 15 to 25 million years ago the great rift valleys had not yet formed. This deeply weathered region of Precambrian basement rocks formed a domed inland plateau, which probably included the watershed between the Congo Basin and coastal Kenya (Andrews & Van Couvering, 1975). Basement rocks to the west included a belt of greenstones and banded iron formations of the Nyanzian System some 3000 million years old, unconformably overlain by sandstones and conglomerates of the Kavirondian System some 2600 million years old (Cahen & Snelling, 1984). Both of these suites of rocks were highly deformed and intruded by granitic rocks, which were locally overlain by sandstones, shales, and dolomites of the Bukoban System, some 1221 to 546 million years old. The Bukoban rocks were deposited on a stable continental platform to the west of rocks some 930 to 444 million years old, now complexly deformed in the Mozambique Belt. This extensive belt of gneiss and schist may have formed during continent-to-continent collision, which created a late Precambrian mountain range comparable in scale with the present European Alps (Vail, 1983).

By early Miocene time these ancient mountains had been eroded deeply and the region was covered by thick lateritic soils (McFarlane, 1976). Basement structure retained a role in the placement of Miocene faults and volcanoes. Early Miocene strike ridges and fault scarps were aligned north-south with the grain of the Precambrian rocks (Bishop & Trendall, 1967). There also were several carbonatite-

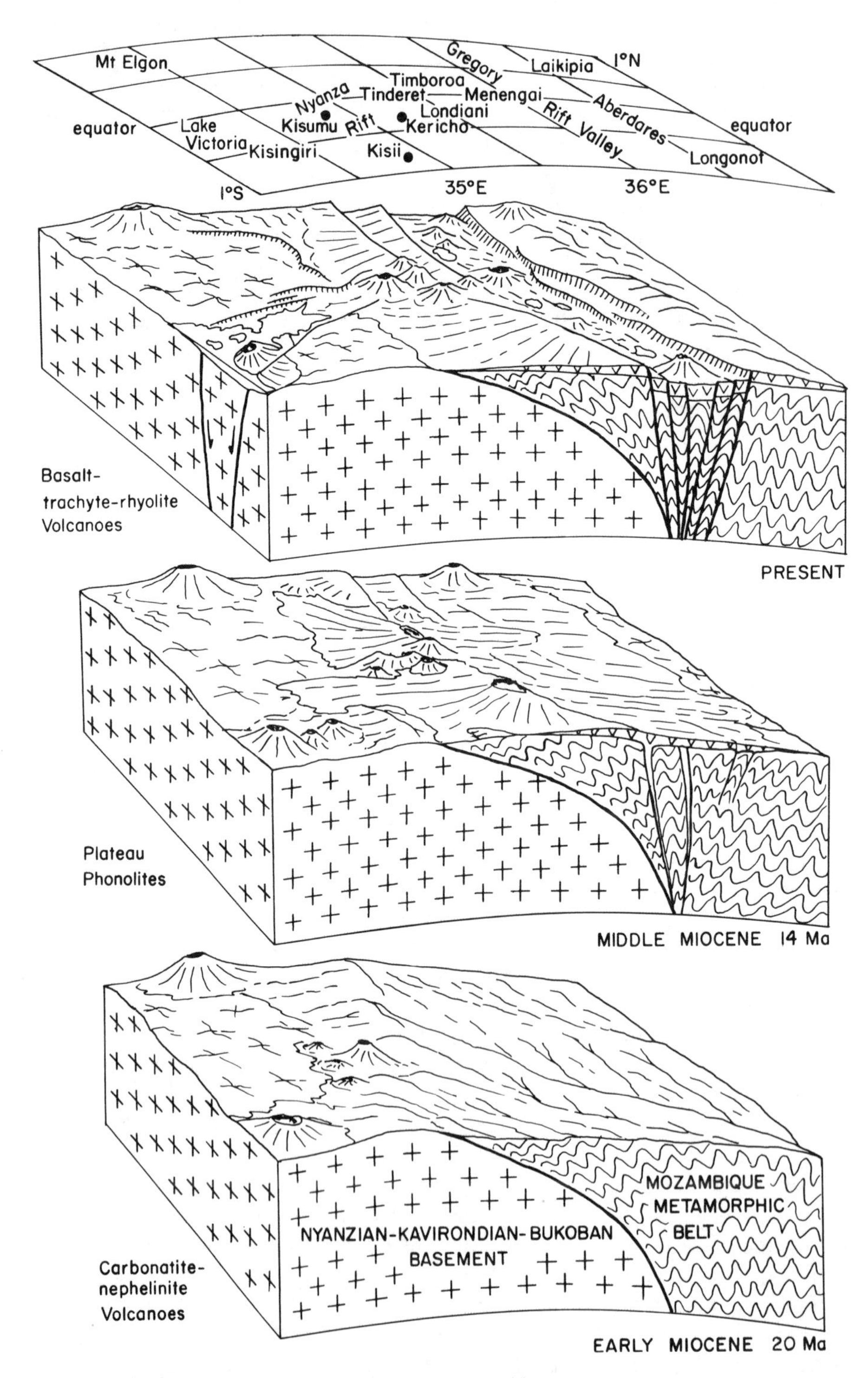

Fig. 4.2. An interpretative geological history of the Nyanza and Gregory Rifts of Kenya (based on data of B. H. Baker, 1986, 1987; Pickford, 1982).

nephelinite stratovolcanoes, now represented by eroded remnants: Mt Elgon, the hills of Kisingiri and the foothills of Tinderet (Le Bas, 1977). While active, these were similar to the modern Tanzanian volcanoes Oldoinyo Lengai and Kerimasi (Dawson, 1962; Hay, 1989; Mariano & Roeder, 1983). The large volcanoes also were associated with numerous small early Miocene intrusions of carbonatite (Deans & Roberts, 1984; King, Le Bas, & Sutherland, 1972; Le Bas, 1977), some of which have associated tuffs that may have formed small cinder cones like those now seen around Arusha and Basotu in Tanzania (Dawson, 1962; Downie & Wilkinson, 1962). Paleomagnetic studies of the Turkana Lavas, some 17 million years old in northwestern Kenya (Reilly *et al.*, 1976), have shown that during the early Miocene, Kenya was about 5 degrees of latitude south of its present equatorial position.

During middle Miocene time, about 15 to 12 million years ago there were massive outpourings of plateau-forming phonolites. These covered some 10,000 km^2 of the central Kenyan dome. These lavas are thought to have erupted from fissures or from coalescing shield volcanoes in a north-south faulted zone at the site of the future Gregory Rift (Lippard, 1973a, 1973b). The flood lavas onlapped the still-active carbonatite-nephelinite stratovolcanoes at Tinderet and elsewhere. Paleomagnetic studies of middle Miocene lavas in and around the Nyanza Rift (Reilly *et al.*, 1976) indicate that it was about 4 degrees south of the equator 12 million years ago.

During late Miocene time, some 5-11 million years ago, both structural and volcanic styles changed. Progressively less feldspathoidal lavas are found in the chain of geologically younger volcanoes eastward from Tinderet, to Timboroa, Londiani, Kapkut, Kilombe and Menengai (B. H. Baker, 1987). Phonolites and nephelinites 8 to 9 million years old around Timboroa are succeeded by 6-7-million-year-old trachytes of Kapkut, 5-6-million-year-old basanites around Tinderet and Londiani, and trachytes 3 million years old on Londiani, 2 million years old on Kilombe, and 330,000 years old in the giant caldera of Menengai (W. B. Jones & Lippard, 1979). Some 9 to 6 million years ago the western boundary fault of the Gregory Rift created that structure as a half graben, and the Nyanza Rift also was initiated as a shallow simple graben (Pickford, 1982). By 4 million years ago, major faults developed on the eastern side of the Gregory Rift, making it also a full graben. Faulting of the Nyanza Rift has continued since that time (Pickford, 1982), but not to the same extent as in the Gregory Rift, which has an axial zone of active faults and volcanoes (B. H. Baker, 1986).

Although the broad outlines of the geological history of southwestern Kenya are becoming well established, for any particular area the geological sequence is complicated by local faulting and volcanism. The present work focused on three small areas in the complex sequence around Tinderet volcanic center: near Koru, Songhor, and Fort Ternan (Figs 4.3 and 4.4). Unraveling of the geological succession in this area has been greatly aided by radiometric dating. Nevertheless, stratigraphic mapping around Tinderet has been of a reconnaissance nature. In attempting to compile stratigraphic nomenclature for this area (Table 4.1), I have taken the type localities to be those outcrops that figured prominently in discussions by the author of the names, which in most cases were proposed informally. In only two cases have names been corrected from their original source: the Grey Sandstone Member rather than tuff (herein) and Polymict (rather than Polygenetic) Agglomerate (following Pickford, 1984, 1986a). An attempt also was made to characterize lavas and intrusions petrographically (Table 4.2) within the small areas mapped. Each of these small areas in which paleosols were

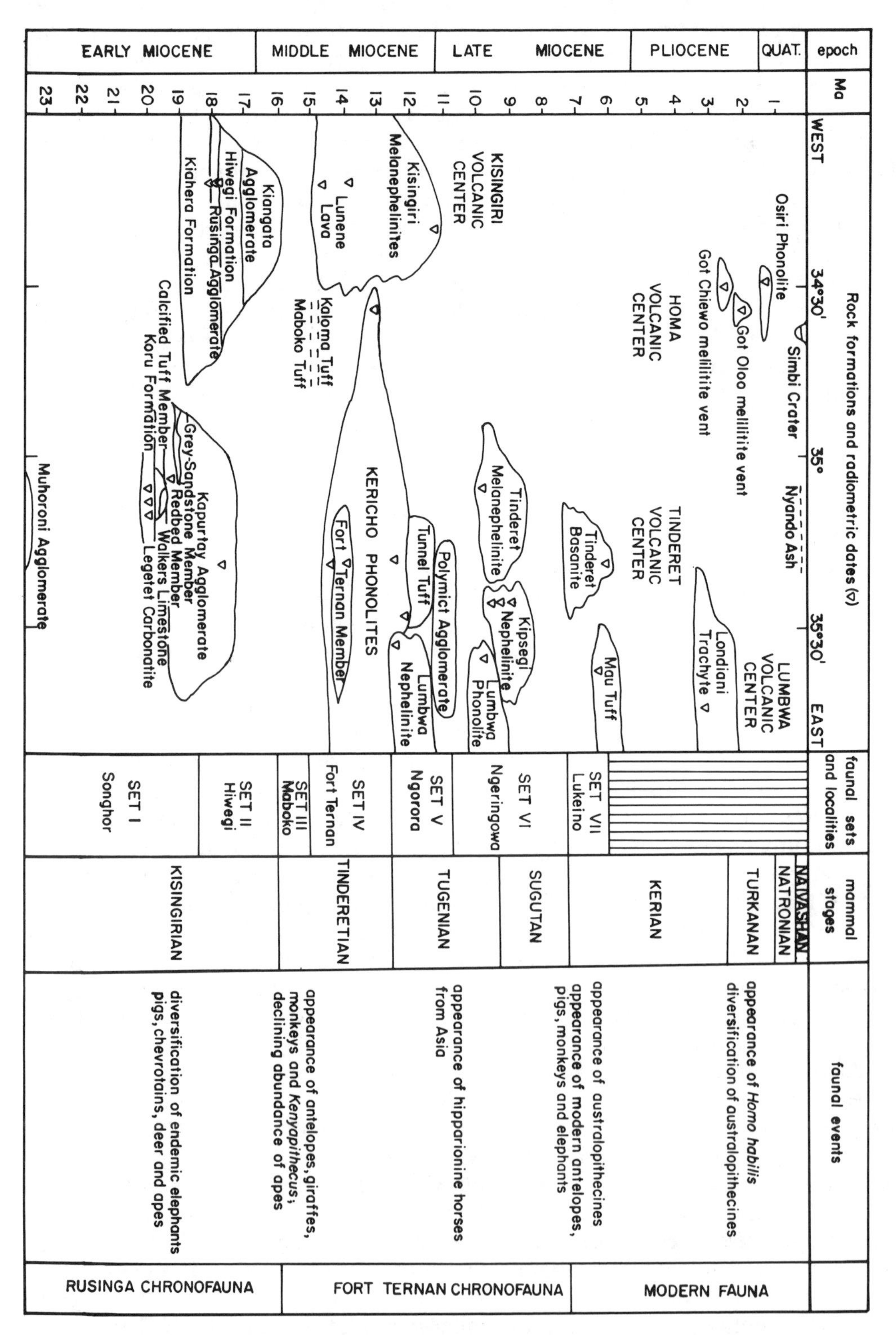

Fig. 4.3. A stratigraphic framework for southwestern Kenya (compiled from data of Berggren, Kent, & Van Couvering, 1985; Drake *et al.*, 1988; Pickford, 1981, 1982; J. A. H. Van Couvering, 1980).

studied will now be considered in order of geological age.

4.1.1 Geology of the area around Koru

Koru is a village nestled in the southwestern footslopes of Tinderet volcano, and a station on the main railway between Nairobi and Kisumu (Fig. 4.4). The red volcanic soils around Koru (Ferralsols or Orthox: Sombroek, Braun, & Van Der Pouw, 1982) are extensively planted with sugarcane. Vegetation before European settlement was a mosaic of lowland forest with secondary grassland in a transitional zone between Afromontane forest of Tinderet Peak and wooded grassland of the Kano Plains to the west (Trapnell *et al.*, 1969; White, 1983). Koru enjoys both high rainfall (1247 mm annually, with 20 mm in January and 158 mm in July) and temperature (mean annual 21°C: Thorp *et al.*, 1960). Koru has long been known for its limestone quarries and their Miocene mammalian fossils (Kent, 1944; Shackleton, 1951; Wayland, 1928). The combined exposures of the working and abandoned quarries of Homa Lime Company proved a natural focus for geological investigations, which are hampered by the soils, crops, and forest elsewhere. Paleosols from near Koru studied here were in an abandoned quarry 0.5 km south of the Brooks' residential compound east of Legetet Hill (Figs 4.5 and 4.6).

4.1.1.1 Turoka System. This metamorphic basement to the Miocene sequence exposed along the railway west of Koru consists of gneisses of the Mozambique Belt. These gneisses have been radiometrically dated elsewhere in Kenya as about 440 to 930 million years old, or late Precambrian to Cambrian (Cahen & Snelling, 1984). Near Koru the gneisses are coarse grained, light gray, crystalline rocks of foliated microcline, quartz, biotite, and muscovite (Binge, 1962), similar to gneisses to the west and north in the Nandi Hills and near Muhoroni to the west and south (Sanders, 1965). These highly deformed rocks are about the same geological age as little-altered basalts, quartzite, and rhyolite near Kisii to the south, radiometrically dated at between 546 and 1221 million years old (Cahen & Snelling, 1984). It is likely that Mozambique Belt gneisses were originally similar to these rocks of the Kisii Group of the Bukoban System, but have been deformed by overthrusting of the Mozambique Belt onto the older Nyanzian and Kavirondian basement (Sanders, 1965). This deformation has been interpreted as a continent-to-continent collision similar to that now producing the European Alps (Vail, 1983). This was one of several collisions that assembled the supercontinent of Gondwana (Scotese, Gahagan, & Larson, 1979).

Between the gneisses and the overlying Miocene volcaniclastic sequence is a very strongly developed paleosol. The whole profile was not exposed in any single outcrop, but it is thicker than 5 m along the railway west of Koru. Much of this thickness is yellow saprolite, but above that is a purple-to-orange, very broadly mottled zone with corestones, then a red claystone at least a meter thick. This paleosol has been locally exhumed south of Muhoroni, where it has been mapped as the Kapchure sandy loam and sandy clay soils, and identified as a "groundwater laterite" (Thorp *et al.*, 1960), which in the U. S. classification (Soil Survey Staff, 1975) is a Plinthic Paleudalf. This paleosol is on the "sub-Miocene" or "mid-Tertiary" erosion surface (McFarlane, 1976), which has been mapped widely as a geological datum plane in East Africa (Bishop & Trendall, 1966; Saggerson, 1965).

4.1.1.2 Muhoroni Agglomerate. This pale-brown to yellow volcanic breccia contains abundant angular nonfenitized blocks of gneissic basement up to 1 m across, as well as blocks and lapilli of carbonatite and crystals of mafic minerals (Pickford &

Andrews, 1981). The Muhoroni Agglomerate has been radiometrically dated using the K/Ar method on biotite as 23.8 ± 0.6 and 26.3 ± 0.6 million years (Bishop, Miller, & Fitch, 1969, corrected by method of Dalrymple, 1979). Thus it is latest Oligocene in age (following the time scale of Berggren, Kent, & Van Couvering, 1985). The Muhoroni Agglomerate crops out poorly around Koru, where calculations based on observed dips and width of outcrop yield an estimated thickness of about 200 m. At least 14 m of Muhoroni Agglomerate is well exposed in gullies and banks of the Meswa River just north of the road between Muhoroni and Songhor (Muhoroni 1:50,000 map 117/1, grid reference 408794). Here the basal 7 m of the exposure is a massive, matrix-supported volcanic breccia with little or no bedding. This is overlain by several biotite-rich tuffs (totaling only 10 to 15 cm), and then a sequence of well-bedded lapillistone tuffs containing large (10 cm in diameter) pieces of uncharred fossil wood, root casts, leaves, gastropods, and mammal bones. Much of the bone was found within tuffs filling a shallow paleogully up to 3-m wide and 3-m deep, sloping down to the south southwest in these strata that dip shallowly (5 degrees) northwest (Pickford, 1984, 1986a; Tassy & Pickford, 1983).

The basal breccia along the Meswa River may have resulted from the explosive clearing of a domed volcanic vent. The overlying units may be airfall lapilli tuffs. The runneled and water-worked tuffs were colonized by trees and animals. The Muhoroni Agglomerate is very similar to the tuffs of small (200-400 m across) Pleistocene (?) explosion craters near Basotu and elsewhere in Tanzania (Dawson, 1964; Downie & Wilkinson, 1962).

The paleogully exposed along the Meswa River dips in such a way as to indicate a source to the north, now buried under younger volcanic rocks (Fig. 4.4). Another paleogully in the Muhoroni Agglomerate within the area mapped in detail 1.5 km west of Koru (Muhoroni 1:50,000 sheet, grid reference 505815) has a dip/strike of 46 degrees W/239 degrees: steeper than the enclosing beds at 6 degrees W/172 degrees. This indicates a slope down to the northwest but a source to the southeast, now buried under younger volcanic rocks. A third possible source is a small (600-m by 330-m) carbonatite plug at Buru Hill, 5.5-miles southwest of Muhoroni (408794 on the Muhoroni sheet), which consists mainly of sövite, with a microfoyaite dike, intruding brecciated gneiss (Deans & Roberts, 1984). The plug at Buru Hill cannot be directly related to any nearby effusive rocks and is capped by a thick, deeply weathered soil. This carbonatite plug is at least as old as the Muhoroni Agglomerate, and could be much older. Other carbonatite plugs in eastern Uganda and western Kenya have yielded radiometric ages of about 42 to 4 million years old (Woolley, 1989).

In the uppermost Muhoroni Agglomerate is a deeply weathered red paleosol at least 1-m thick (seen best in gullies 2-km north of Koru, at grid reference 507821 on the Muhoroni sheet). This indicates a hiatus before cover by the overlying Koru Formation. Angular discordances at this contact have been taken as evidence of folding of the Muhoroni Agglomerate prior to cover by the Koru Formation (Pickford, 1984, 1986a). Considering the interpretation presented here of eruption from a series of small cinder cones, it is more likely that these discordances represent original topography of several eroded vents.

4.1.1.3 Koru Formation. This lithologically heterogeneous unit includes red, brown, yellow, gray, and white conglomerates, sandstones, siltstones, and claystones. The coarse-grained beds are cross bedded and laminated. The clayey layers are massive, varicolored, and riddled with root traces. The clayey layers also contain casts of stumps up to 10 cm in diameter, and fossil leaves, snail shells, and mammal bones.

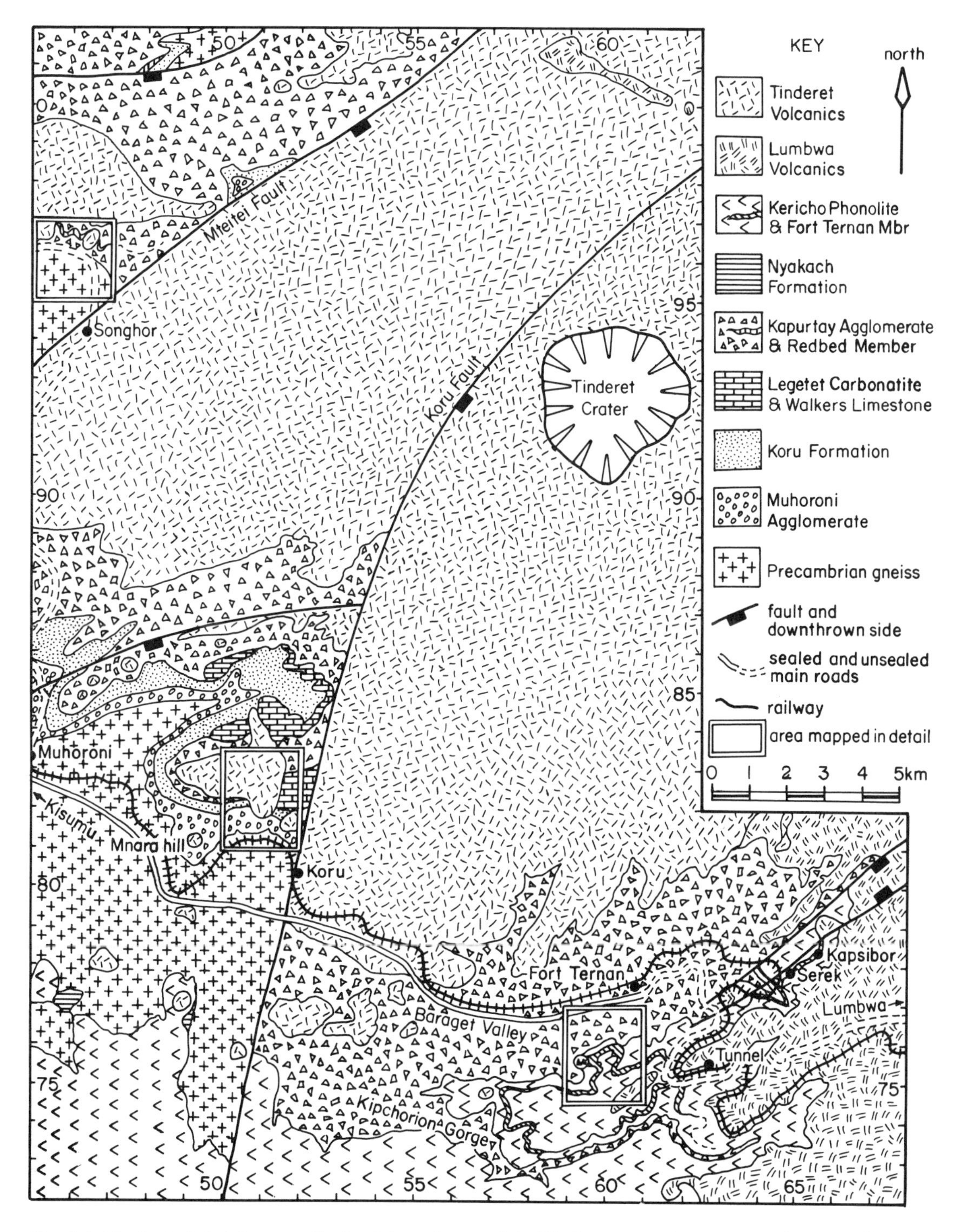

Fig. 4.4. Geological sketch map of bedrock in and around the Tinderet volcanic center, southwestern Kenya, showing also areas studied in more detail (compiled from Pickford, 1984, 1986a; and mapping of author). Grid of 1 km squares from Lumbwa and Muhoroni 1:50,000 sheets.

TABLE 4.1. Stratigraphic nomenclature for Miocene rocks in the area around Koru, Songhor, and Fort Ternan, southwestern Kenya

FORMATION/ Member	Composition	Type section	Radiometric age (Ma)	Original author
TINDERET VOLCANICS	basanite, melanephelinite, and tephrite flows; some tuff and agglomerate	Legetet Hill, quarry, and cliffs, 3 km NW of Koru (g.r.496838)		Shackleton, 1951
Basanite Member	basanite and tephrite flows	Tinderet summit (g.r.917623)	5.6±1.3(c) 5.8±1.3(c) 6.0±1.8(c)	Binge, 1962
Melanephelinite Member	melanephelinite flows	Legetet Hill, quarry, and cliffs, 3 km NW of Koru (g.r.496838)	10.2±0.5(b)	Binge, 1962
LUMBWA VOLCANICS	phonolite and phonolitic nephelinite flows; some basanites, nephelinites, tuffs, and agglomerates	Nyando Valley south of railroad crossing 4.5 km ENE of Lumbwa (g.r.786795)	9.7±0.6(f) 12.7±0.3(f)	Shackleton, 1951
Kipsegi Nephelinite Member	nephelinite flows	hill at Kipsegi farm, 5 km W of Lumbwa (g.r.695786)	9.1±0.5(c) 9.4±0.5(c) 9.7±0.5(c)	Pickford & Andrews, 1981
Polymict Agglomerate Member	nephelinitic agglomerate with enormous blocks (up to 5 m)	cuttings along railway 2 km SW of Lumbwa (g.r.723773)		Pickford & Andrews, 1981; see Pickford, 1984, 1986a
Tunnel Tuff Member	green-to-gray nephelinitic, lapilli tuff, and agglomerate	cuttings along railway near Tunnel siding (g.r.632737)		Pickford & Andrews, 1981
KERICHO PHONOLITE	sparsely porphyritic, biotite-bearing phonolite (Losuguta type)	Mumek Gorge, 20 km ENE of Kericho	12.1±0.3(b) 12.1±0.2(b) 12.4±0.2(b) 12.9±0.4(b) 13.0±0.7(b) 13.6±0.3(b) 13.8±0.4(b)	Binge, 1962
Fort Ternan Member	gray nephelinitic lapilli tuffs, agglomerates, sandstones, and brown claystones	main fossil quarry and vicinity, Fort Ternan National Monument (g.r.603759)	13.9±0.3(e) 14.4±0.2(b) 14.4(a) 15.1±0.1(b)	Van Couvering & Walker for Shipman *et al.*, 1981
Baraget Phonolite Member	porphyritic phonolite with phenocrysts of nepheline, sanidine, and biotite	track below Fort Ternan National Monument (g.r.605761)	15.0±0.1(e)	Van Couvering & Walker for Shipman *et al.*, 1981

FORMATION/ Member	Composition	Type section	Radiometric age (Ma)	Original author
NYAKACH FORMATION	conglomerate, sandstone, and claystone; some white tuff	gullies below scarp near Nyakach, 40 km SW of Muhoroni (g.r.145587)		Saggerson, 1952
KAPURTAY AGGLOMERATE	green to gray nephelinitic agglomerate; some carbonatite tuff, gray sandstone, and red claystone	gullies 4 km NE of Songhor (g.r.503980)		Pickford & Andrews, 1981
Cliff Agglomerate Member	massive, green, nephelinitic agglomerate	scarp 5 km NE of Songhor (g.r.504987)	18(e)	Pickford & Andrews, 1981
Chamtwara Member	interbedded calcareous tuffs and red claystone	gullies near Chamtwara Stream, 4 km N of Koru (g.r.521863)		Pickford & Andrews, 1981
Grey Sandstone Member	gray nephelinitic sandstone	gullies south of Songhor National Monument (g.r.461963)		Pickford & Andrews, 1981
Red Bed Member	red claystone, with thin calcareous tuffs	bluffs in main gullies of Songhor National Monument (g.r.460964)		Pickford & Andrews, 1981
Calcified Tuff Member	gray carbonatite tuffs and thin red marls	rock platforms north of main gullies at Songhor National Monument (g.r.459966)	20.2±0.5(b) 20.4±0.6(b)	Pickford & Andrews, 1981
WALKERS LIMESTONE	yellow bedded calcareous siltstone, some sandstone and conglomerate	lime quarry 3 km N of Koru (g.r.512844)		Pickford & Andrews, 1981
LEGETET CARBONATITE	thick gray carbonatite tuffs, interbedded with red claystones	lime quarry 3 km N of Koru (g.r.512844)		Le Bas & Dixon, 1965
KORU FORMATION	red, brown, and yellow, sandstones and claystones	gully 1 km NW of Koru (g.r.508018)	20.0±0.3(b) 20.1±0.3(b)	Shackleton, 1951
MUHORONI AGGLOMERATE	pale yellow and brown agglomerate with many gneiss blocks	Meswa River banks near bridge 2 km N of Muhoroni (g.r.453852)	23.8±0.6(b) 26.3±0.6(b)	Pickford & Andrews, 1981
BURU HILL CARBONATITE	carbonatite plug	Buru Hill, 5.5 km SW of Muhoroni (g.r.408794)		Pulfrey for Binge, 1962

Note: Radiometric dates from (a) Evernden & Curtis, 1965; (b) Bishop, Miller, & Fitch, 1969; (c) B. H. Baker *et al.*, 1971; (d) Fitch, Hooker, & Miller, 1978; (e) Hooker & Miller for Shipman *et al.*, 1981; (f) W. B. Jones & Lippard, 1979. All were converted to new constants by method of Dalrymple (1979).

TABLE 4.2. Mineral composition (volume percent) of volcanic rocks in southwestern Kenya, determined by point counting (500 points) of petrographic thin sections

rock type	Tinderet melanephelinites						Kipsegi nephelinite	Kericho phonolites		Baraget phonolite	Tinderet basanite
origin	dike	flow	dike	flow top	flow base	xenolith in pipe	flow	flow	flow	flow	pipe
locality	road cut west of Koru	ridge near Brook's house, Koru	road cut west of Koru	knoll north of Fort Ternan National Monument	creek northeast of Fort Ternan National Monument	knoll west of Fort Ternan National Monument	hill south of Fort Ternan National Monument	plateau above Fort Ternan National Monument	creek south-west of Fort Ternan National Monument	road cut at Fort Ternan National Monument	knoll west of Fort Ternan National Monument
map sheet	Muhoroni	Muhoroni	Muhoroni	Lumbwa	Lumbwa	Lumbwa	Lumbwa	Lumbwa	Lumbwa	Lumbwa	Lumbwa
grid ref.	501817	514829	505816	604765	608768	594756	607749	602759	596752	606760	594756
specimen	KO-R112	KO-R113	KO-R111	FT-R164	FT-R204	FT-R201	FT-R203	FT-R166	FT-R202	FT-R165	FT-R201
pyroxene	51	47	54	62	48	51	34	9	4	8	21
nepheline	31	26	27	29	45	37	34	14	16	18	13
feldspar	-	-	-	-	-	-	-	40	39	42	14
opaque oxide	12	10	10	6	3	8	10	1	tr	2	16
glass	-	-	-	-	-	-	20	34	41	26	22
analcite	3	1	-	1	2	-	-	1	-	3	-
olivine	2	3	5	1	1	2	tr	-	-	-	13
biotite	-	-	-	-	-	-	-	1	tr	1	-
red clay	1	7	4	-	-	-	-	-	-	-	-
cancrinite	-	-	-	-	-	-	1	-	-	-	-
natrolite	-	6	-	-	tr	2	-	-	-	-	1
green phyll.	-	-	-	1	-	-	1	1	-	-	-
calcite	-	-	-	-	1	-	-	-	-	-	-

Note: Among minor minerals, red clay replaces mafic minerals, cancrinite replaces nepheline phenocrysts, natrolite fills vesicles, green phyllosilicate replaces olivine. Calcite crystals appear to be xenoliths of the carbonatite variety sövite.

Also seen in an old fossil quarry 2 km northwest of Koru (at grid reference 822507) was a thin (10 cm) bed of light green tuff with concentrically banded accretionary lapilli (type B of Reimer, 1983) up to 4 mm in diameter. Unlike the underlying Muhoroni Agglomerate, the Koru Formation is more brightly colored, is more conspicuously bedded, and has fewer gneiss blocks. The Koru Formation also may be much younger: fresh biotite has yielded K/Ar ages of 20.0 ± 0.3 and 20.1 ± 0.3 million years (Bishop, Miller, & Fitch, 1969, corrected by method of Dalrymple, 1979).

In the area around Koru, this formation crops out very poorly and has not been a target for quarrying. Its general appearance is that of fluvial paleochannels and paleosols, with some thin airfall tuff beds (volcanic arena facies of Pickford, 1986b).

4.1.1.4 Legetet Carbonatite. The lime quarries near Koru have selectively exposed a sequence of about six, thick (each 5-20 m) yellow to dark-gray carbonatite tuffs (Deans & Roberts, 1984). Separating these calcareous tuffs are massive red claystones, which in places fill up to 5.8 m of paleokarst relief on the tuffs (see Fig. 2.24). The Legetet Carbonatite has a very restricted distribution (Fig. 4.3), forming a concentrically bedded oval some 1.5 by 0.75 km in size north of Legetet Hill, with an erosionally dissected sheet a short distance to the north and a small faulted outlier to the south. The base of the formation is not seen within the central oval, but in the outlying outcrops it overlies the Koru Formation more or less conformably.

The carbonatite tuffs are strongly cemented, fragmental rocks, consisting mostly of carbonatite lapilli, scoria, alvikite, and sövite, with lesser amounts of fenite, apatite, and magnetite (Deans & Roberts, 1984). In places the carbonatites are intruded by alnöite (Pulfrey, 1953). Both carbonatites and paleosols developed

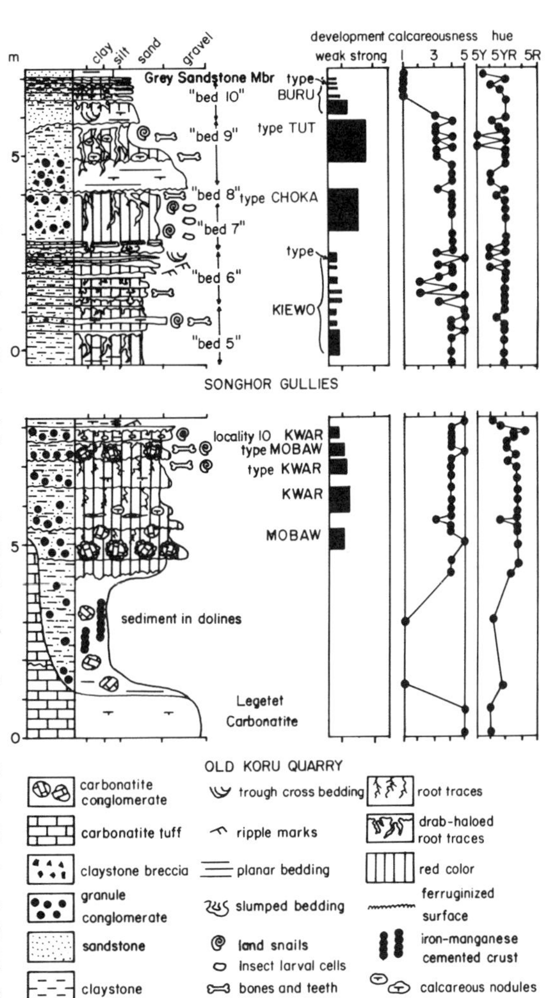

Fig. 4.5. Measured sections of paleosols in the north central wall of an abandoned lime quarry 3 km north of Koru (below) and in the westernmost exposures at Songhor National Monument, Kenya (above). Localities and beds are labelled using scheme of Pickford and Andrews (1981); scale of development and calcareousness from Retallack (1988a); hue from Munsell color chart (1975).

on them were examined in detail in the abandoned quarry south of the Brooks' residential compound (grid reference 516839 on Muhoroni sheet). The carbonatite tuffs examined there vary in thickness from 9 to 15 m over a distance of only 150 m (Pickford, 1984, 1986a). Both upper and lower original surfaces of the carbonatite tuffs are visible, despite karstificat-

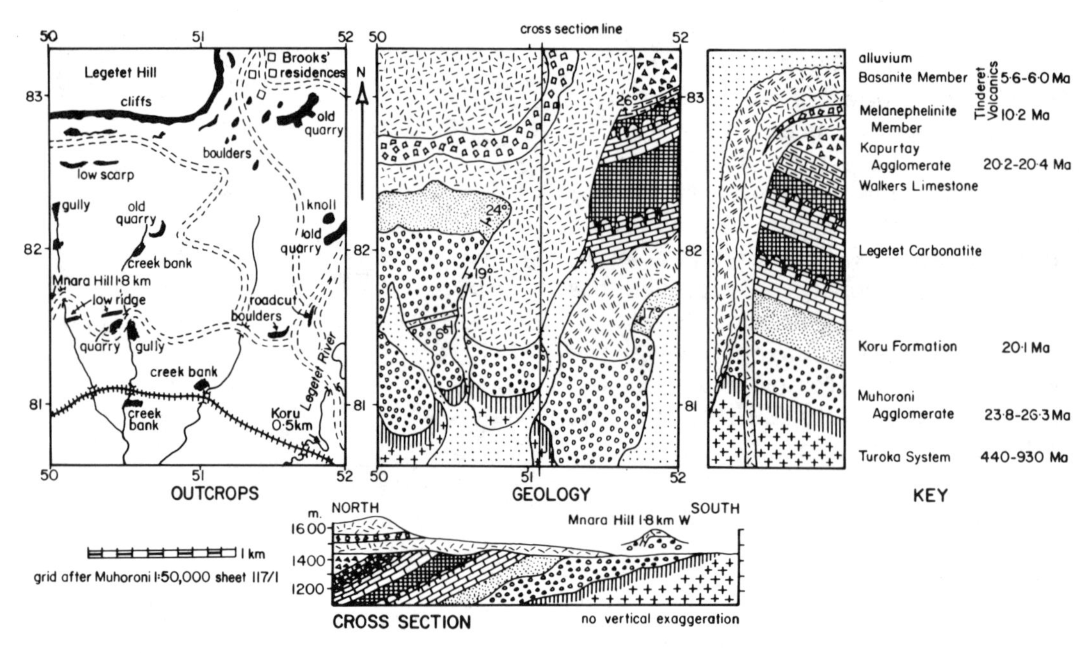

Fig. 4.6. Geological sketch map, outcrop map and cross-section of the area of Miocene fossil localities northwest of Koru, Kenya. Grid from Muhoroni map sheet 117/1, 1:50,000.

ion of the upper surface (Fig. 2.24). The bottoms of the paleokarst grikes contain early Miocene fossils and boulders of carbonatite, indicating that the rock was indurated during early Miocene time. Most of the rock is a tuff of sand to pebble size volcanic fragments, in two thick inversely graded units. The upper unit is weathered yellow, even in the freshest exposures of the quarry, whereas the lower unit is black to gray in color. In thin section the upper unit consists mainly of rounded carbonatite lapilli (mostly type 2b, but some type 1 and 2a as illustrated by Deans & Roberts, 1984) in a sparry calcite cement. The lower unit of tuff here consists of globules of opaque vesicular carbonatite, which are partly fused and contain phenocrysts of rhomboidal and lath-shaped calcite and pyroxene. Some of the calcite crystals may be pseudomorphs after nyerereite, biotite, or melilite (Deans & Roberts, 1984). Vesicles within, and gaps between, the lava globules are filled with sparry, void-filling calcite.

Both these welded and unwelded textures are similar to those documented from the Miocene Kaiserstuhl volcanic vent in West Germany (Deans & Roberts, 1984; Keller, 1981), the Plio-Pleistocene Laetoli Beds in Tanzania (Hay, 1978), and the active natrocarbonatite volcanoes Oldoinyo Lengai and Kerimasi in Tanzania (Dawson *et al.*, 1987; Hay, 1989; Mariano & Roeder, 1983). The surface white tuffs of Oldoinyo Lengai are products of historic Vulcanian-style eruptions (Dawson *et al.*, 1987; Nyamweru, 1988), but these are not so thick as the Black Tuffs and Agglomerates, some 1250 to 2000 years old, or the Yellow Tuffs and Agglomerates some 30,000 to 400,000 years old, on this active volcano. Like these tuffs, the upper unit of the Legetet Carbonatite in the quarry south of Brooks' houses has much of the appearance of a pyroclastic fall deposit. The lower unit at Koru, with its welded lapilli, may be an agglutinate deposit. During recent observations within the crater of Oldoinyo Lengai (Krafft & Keller, 1989; Nyamweru, 1988), black natrocarbonatite lava was observed to boil and form small spatter cones, but the petrographic nature of these materials remains unreported. If a similar origin is likely for the welded carbonatite tuff in the abandoned quarry near Brooks' houses, then this tuff would have been close to the vent. This is suggested anyway by the great variation in thickness of the tuffs. The concentrically outcropping carbonatites north of Legetet Hill may represent the vent itself. Perhaps it was a small cinder cone similar to those already discussed as possible modern analogs for the Muhoroni Agglomerate, or to those dotting the flanks of Oldoinyo Lengai and Kerimasi today (Hay, 1983, 1989).

The massive red claystones between the lithified carbonatite tuffs are interpreted as paleosols developed on the tuffs, but the paleosols also contain features of volcanic interest. Unlike the carbonatite tuffs, the red paleosols and their parent material contain appreciable amounts of nephelinite lapilli, books of vermiculite, and grains of quartz and microcline, as well as carbonatite lapilli (Fig. 2.13B). Some of the carbonatite grains in these red paleosols are coated with fine-grained or radial fibrous calcite (as in Fig. 2.13A), which as already discussed, predate soil formation. Coated carbonatite lapilli have been found among the Plio-Pleistocene ashes at Laetoli (Hay, 1978) and around Oldoinyo Lengai in Tanzania (Hay, 1989), where they are thought to form by rolling of airfall lapilli during transport on the ground by wind and water. They could also have formed during eruption, as seems more likely for the radial-fibrous coated-grains from Quaternary Tanzanian cinder cones (Downie & Wilkinson, 1962). Unlike the thick, locally erupted carbonatite tuffs of the Legetet Carbonatite, these reworked lapilli of the tuffaceous parent materials for the red paleosols may have come from a larger and more complex nearby stratovolcano of mixed nephelinitic and natrocarbonatitic composition, like the modern volcanoes Oldoinyo Lengai and Kerimasi of Tanzania.

4.1.1.5 Walkers Limestone. These sediments fill a local, small (3.5 km in diameter) basin and paleogullies eroded into the underlying Legetet Carbonatite. The Walkers Limestone is a well-bedded to flaggy, calcareous, yellow to brown volcaniclastic siltstone, up to 60-m thick (Pickford & Andrews, 1981). In deep paleogullies at the base of the formation there is a conglomerate of nephelinite and carbonatite boulders.

The Walkers Limestone has been interpreted as a waterlain, perhaps partly lacustrine, deposit of reworked carbonatite-nephelinite ash (Deans & Roberts, 1984). Although these rocks do yield fossil leaves and land snails, no aquatic fossils have yet been found (Pickford, 1984, 1986a). This is not a fatal objection to interpreting the limestone as a deposit within a crater lake. Some of the Quaternary crater lakes near Basotu in Tanzania support animal life including hippos, but others are barren of life (Downie & Wilkinson, 1962). Finely laminated parts of the Walkers Limestone may indeed have been deposited in a crater lake (Pickford & Andrews, 1981). The conglomeratic and flaggy parts of the unit, including the layers overlying the fossil soils studied in detail here, appear more like pedisediments mantling a dry crater floor, as in the breached Quaternary cinder cone Lashaine in Tanzania (Dawson, 1964).

4.1.1.6 Kapurtay Agglomerate. A light-green, nephelinitic, volcanic breccia crops out in an old quarry northeast of Legetet Hill (at grid reference 517838 on the Muhoroni sheet), and this may extend south under a mantle of soil and forest into the area north of Koru mapped in detail (Fig. 4.6). Kapurtay Agglomerate has been mapped widely to the north and west overlying the Walkers Limestone and other formations (Pickford, 1984, 1986a). At Songhor and Fort Ternan there are excellent exposures of this lithologically heterogeneous formation, that form a better basis for its paleoenvironmental interpretation, to be discussed in due course.

4.1.1.7 Melanephelinite Member, Tinderet Volcanics. Much of Legetet Hill and its forested flanks are covered by at least two flows of melanephelinite lava (identified by Campbell-Smith for Kent, 1944). These are black and porphyritic lavas with abundant euhedral, zoned augite, and lesser amounts of olivine and analcite, in a matrix of nepheline. The lavas are nearly flat-lying and overlie with marked angular discordance both the Kapurtay Agglomerate and older rocks. Similar lavas nearby are overlain by basanite lavas that form the eroded peak of Tinderet volcano (Binge, 1962), which is the presumed source of both melanephelinites and basanites. Similar melanephelinitic lavas unconformably overlying the Kapurtay Agglomerates near Songhor Gullies have been radiometrically dated by the K/Ar whole-rock method at 10.2 ± 0.5 million years (Bishop, Miller, & Fitch, 1969, corrected by method of Dalrymple, 1979; rock identification following J. A. Van Couvering for Shipman *et al.*, 1981).

Among the melanephelinite lavas on Legetet Hill is probably also an horizon of green nephelinitic agglomerate (Binge, 1962), but this was not well enough exposed to be mapped confidently (Fig. 4.6).

A dike of melanephelinite up to 3-m wide was mapped from near the road bend 2-km west of Koru (grid reference 505816 on the Muhoroni sheet) for 400 m to the west toward Mnara Hill. A quarry on the northwest flank of Mnara Hill (grid reference 492813) shows a dike of melanephelinite brecciating and deforming the underlying Koru Formation and then passing upward into a flow that forms the top of Mnara Hill. Other melanephelinite flows such as those on Legetet Hill may also have emerged from local fissures or parasitic cones on the flanks of the stratovolcano that predated the basanitic rebuilding of Tinderet peak. Similarly, the active carbonatite-nephelinite volcano Oldoinyo Lengai, in Tanzania, has melanephelinite flows on its western upper slopes, as well as scoriaceous melanephelinitic parasitic cones on its northern footslopes (Dawson, 1962; Hay, 1989).

4.1.1.8 Basanite Member, Tinderet Volcanics. Overlying and in some cases onlapping the greenish-black melanephelinite lavas are blue-black basanite lavas. The basanites are difficult to distinguish from the melanephelinites in the field, because the feldspar that would indicate less alkaline composition is in the groundmass with analcite. Most of the phenocrysts are augite and olivine. Similar basanitic lavas form the great mass of Tinderet, whose summit is capped by a distinctive feldspar-phyric variety (Binge, 1962). Basanites from the summit of Tinderet have yielded K/Ar whole-rock radiometric ages of 5.6 ± 1.3, 5.8 ± 1.3, and 6.0 ± 1.4 million years (B. H. Baker *et al.*, 1971, corrected using method of Dalrymple, 1979). Latest Miocene to Pliocene eruption of these basanites from Tinderet concluded the eruptions of this now deeply eroded volcano.

4.1.2 Geology of the area around Songhor National Monument

The northern slopes of Songhor Hill are clothed in a secondary wooded grassland, like that found in climatically similar nearby Koru. This modern vegetation may have been degraded by human clearance from original forest (Thorp *et al*, 1960; White, 1983), an idea also suggested by the nature of the surface soils (mainly Planosols or Alfisols; Sombroek, Braun, & Van Der Pouw, 1982). Erosional gullies, perhaps also related to human disturbance, are widespread on toeslopes northeast of Songhor Hill (Figs 4.4 and 4.7), where they expose at least 7 m of red clayey sediments in badlands between gray sandy beds. The red beds have yielded enormous numbers

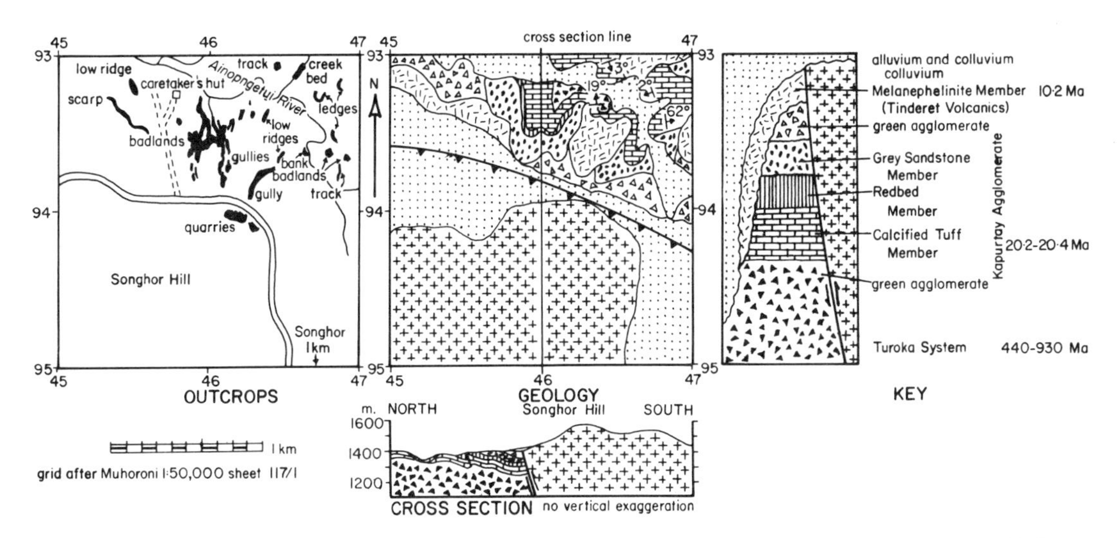

Fig. 4.7. Geological sketch map, outcrop map and cross section of the area around Songhor National Monument, Kenya. Grid from Muhoroni map sheet 117/1, 1:50,000.

of millipede, snail, reptile, bird, and mammal fossils (Pickford & Andrews, 1981). They are protected as a national monument by caretakers for the National Museums of Kenya. An excellent detailed map of the fossiliferous beds and paleosols has been published by Pickford and Andrews (1981). The mapping reported here (Fig. 4.7) was designed to establish the general stratigraphic context of the red beds. Only the upper portion of the red beds was sufficiently well exposed for detailed chemical and petrographic study of the paleosols.

4.1.2.1 Turoka System. Songhor Hill itself consists entirely of light-gray, late Precambrian to Cambrian gneisses of the Turoka System of the Mozambique Belt, like rocks already discussed for the area around Koru. Songhor Hill has been interpreted as an exhumed Miocene inselberg, onlapped by the early Miocene sequence exposed in Songhor gullies to the north (Shackleton, 1951). Such a paleogeographic reconstruction now seems unlikely for a variety of reasons. Clasts of gneiss are rare in the Miocene red beds, especially compared with their abundance in modern colluvium of this steep high hill (Pickford & Andrews, 1981). The hill is lightly wooded and has thin (1-m), brown, sandy loam soils (Thorp *et al.*, 1960), unlike either the case-hardened surface typical of inselbergs (Twidale, 1982) or the thick clayey sub-Miocene paleosols of Kenya and Uganda (McFarlane, 1976).

It is more likely that Songhor Hill is a horst (Pickford, 1984, 1986a), uplifted after eruption of the Tinderet basanites to the north about 10 million years ago (Bishop, Miller, & Fitch, 1969; Dalrymple, 1979; J. A. Van Couvering for Shipman *et al*, 1981). From regional tectonic considerations (W. B. Jones & Lippard, 1979), it is likely that faulting occurred between about 9 and 6 million years ago. The northern boundary fault of the horst is not well exposed but can be placed within 3 m in the deep gully 200 m southeast of the National Monument (grid reference 463962 on the Muhoroni sheet). For at least 500-m east and west of these exposures there is a rounded break in slope about 2 m high, which may represent an active fault scarp.

4.1.2.2 Kapurtay Agglomerate. Most of the Miocene rocks in the area around Songhor are part of this lithologically heterogeneous formation, but only the most distinctive units have been named as members. The most characteristic lithology for the formation is a green, matrix-supported, nephelinitic volcanic breccia. Such rocks crop out poorly at the base of the exposed Miocene sequence near Songhor, on a forested slope near the river northeast of the National Monument (grid reference 463966 on Muhoroni sheet). Similar lithologies of undifferentiated Kapurtay Agglomerate are better exposed near Fort Ternan. They are interpreted there as volcaniclastic and pyroclastic products of a carbonatite-nephelinite stratovolcano like the modern Oldoinyo-Lengai, Tanzania.

4.1.2.3 Calcified Tuff Member, Kapurtay Agglomerate. Overlying the green nephelinitic agglomerates is about 15 m of dip-slope-forming, gray to yellow tuff, firmly cemented with calcite (Pickford & Andrews, 1981). The Calcified Tuff Member is a well-stratified fragmental unit. Most of its grains are alvikite, showing laths of calcite in a trachytoidal arrangement. Also found are fragments of sövite, globules of calcite, and grains of biotite, quartz, feldspar, and opaque oxides. Near the base of the tuffs are two layers about 50 cm apart bearing blocks of fenite, basement gneiss and carbonatite lapilli (Deans & Roberts, 1984). Near the top of these tuffs are red clayey horizons with fossil gastropods and polygonally cracked tuffs with accretionary lapilli (Pickford & Andrews, 1981). About 5 m from the top there is a thin (10 cm) tuff with impressions of leaves and books of biotite that were radiometrically dated using the K/Ar method at 20.2 ± 0.6 and 20.4 ± 0.5 million years (Bishop, Miller, & Fitch, 1969, corrected by method of Dalrymple, 1979).

The red clayey horizons in the Calcified Tuffs are paleosols of the Kiewo Series. Most of this member shows little sign of soil formation, although the polygonal fracturing and calcite veining of thin biotite-rich and accretionary lapilli tuffs are similar to veining attributed to initial weathering of fresh airfall tuffs of carbonatite-melilitite composition in the Plio-Pleistocene Laetoli Beds and historic ejecta of Oldoinyo Lengai, Tanzania (Hay, 1978, 1989). Red paleosols and thin calcite-veined tuffs are a minor part of the Calcified Tuff Member: most of it is thick fragmental tuffs, like those already discussed for the Legetet Carbonatite. Songhor probably was further from the source vent, or vents, than Koru. Like similar tuffs 6 km to the north of Songhor, near Ngeron, Kipsesin, and Kipingai (Deans & Roberts, 1984), these tuffs may record pyroclastic falls from several small cinder cones on the flanks of the large nephelinitic stratovolcano responsible for the bulk of the Kapurtay Agglomerate. Similarly, a variety of small cinder cones dot the flanks of the active carbonatite-nephelinite stratovolcano, Oldoinyo Lengai, Tanzania (Hay, 1989).

4.1.2.4 Red Bed Member, Kapurtay Agglomerate. The main fossiliferous strata of Songhor National Monument are these calcareous red claystones up to 10-m thick (Pickford & Andrews, 1981). The westward extension of these red beds is lost under a cover of younger alluvium and lava. To the east across the river, the red beds pinch out between the underlying Calcified Tuff and overlying Grey Sandstone Members (Fig. 4.7).

The red beds show abundant drab-haloed root traces, diffuse clayey horizons, calcareous nodular horizons, blocky peds, and clay skins. They are interpreted as a succession of paleosols of at least four different kinds (Kiewo, Choka, Tut, and Buru Series). The red beds have also a variety of volcanogenic and sedimentary features and can be divided into three distinct sedimentary facies: (1) interbedded red claystone and gray sandstone; (2) thinly bedded, fine-grained, calcareous layers; and (3) massive thick beds of sandy claystone.

Beds of red claystone interbedded with

weakly calcareous sandstones (beds 1, 6, 9, and 10 of Pickford & Andrews, 1981) are interpreted as Buru and Tut Series paleosols. The weakly calcareous sandstones of their parent materials show ball-and-pillow structures, scour-and-fill, trough cross bedding, ripple marks, and parting lineation. Some of these interbedded sequences dip at low angles to adjacent bedding planes. In thin section, the sandstone grains are mostly red-brown, deeply weathered fragments of nephelinite or melanephelinite. Quartz and microcline are common components also, but there is little carbonatite or pyroxene. These are all features of the deposits of meandering streams draining a deeply weathered volcanic terrain. Nearby water also is suggested by fossil turtles and otters found at Songhor, but not at Koru or Fort Ternan (Pickford, 1986a, 1986b; Schmidt-Kittler, 1987). No large paleochannels were seen within the red beds, but the overlying Grey Sandstone Member probably was such a channel responsible for eroding away the red beds to the east. During accumulation of the red beds this area may have been part of a radial drainage through the toeslopes (apron to arena facies of Pickford, 1986b) and cinder cones of a large carbonatite-nephelinite stratovolcano, responsible for the bulk of the Kapurtay Agglomerate.

Fine-grained highly calcareous layers are scattered through the red beds (Kiewo Series paleosols and beds 2, 4, 6, and 8 of Pickford & Andrews, 1981). Those analyzed (Appendix 6) had unusually high amounts of barium, strontium and lanthanum, found in volcanic ashes of carbonatitic affinity (Bishop, 1968). In some cases these calcareous layers were riddled with fine calcareous veins, here called "pervasively displacive fabric", and interpreted as a result of the early weathering of carbonatite-melilitite ash (Fig. 2.17). Other thin carbonatite-melilitite ashes (capping both Tut and Choka Series paleosols; bed 8 and uppermost bed 9 of Pickford & Andrews, 1981) are less altered and furnish clues for understanding how they were formed. In thin section, these ashes contain almost equal proportions of silt-sized carbonatite and fresh nephelinite. Less common are quartz, microcline feldspar, calcite globules, and calcite fibers, perhaps pseudomorphous after melilite. Grains of fenite, pyroxene and opaque oxides are rare. Many of the grains are coated with micritic or radial fiber calcite that predates soil formation (Fig. 2.13A). These distinctive compositions and textures can be matched closely among carbonatite-melilitite tuffs of the Plio-Pleistocene Laetoli beds and active volcano, Oldoinyo Lengai, in Tanzania (Hay, 1978, 1989). Most of the Red Bed Member at Songhor probably originated as airfall tuffs of this kind, but much of it was altered beyond recognition by soil formation. Even the freshest ashes show mineral pseudomorphs and grain coats taken to indicate some weathering or reworking. The spectrum of alteration found at Songhor is greater than documented by Hay (1978) at Laetoli, and this may be related to a paleoclimate more humid and deeply weathering at Songhor during the early Miocene than at Laetoli during the Plio-Pleistocene.

Also in the Red Bed Member in the Songhor gullies are two thick (about 1-m) beds of sandy claystone forming parent materials to Kiewo and Choka Series paleosols (beds 5 and 7, respectively of Pickford & Andrews, 1981). Pedogenic features of both beds include drab-haloed root traces, red clay skins, and calcareous nodules. In thin section most of the grains are carbonatite clasts of the kind with trachytoidal pseudomorphs of nyerereite. Nephelinite, quartz, and microcline are rare. Like the very similar tuffs of the Legetet Carbonatite and Calcified Tuff Member (Deans & Roberts, 1984), these may represent particularly voluminous ash falls from nearby eruptions of carbonatite cinder cones. Unlike the Legetet Carbonatite and Calcified Tuff, these tuffs were unwelded and uncemented, as indicated by their abundant and deeply penetrating root traces and burrows.

4.1.2.5 Grey Sandstone Member, Kapurtay Agglomerate. The Red Bed Member in the Songhor gullies is overlain conformably by about 10 m of gray-green volcaniclastic sandstone, which is weakly cemented and contains only sparse pebbles, and so crops out poorly. In a few outcrops it shows parallel lamination and low-angle cross bedding. In thin section the sandstone consists for the most part of claystone granules formed by deep weathering of nephelinite and melanephelinite, with common grains also of quartz, microcline, and biotite. Little carbonatite or pyroxene was seen. Because of this composition the member's original name ("Grey Tuff" of Pickford & Andrews,1981) is incorrect.

The Grey Sandstone Member is petrographically identical with thin sandstone layers that form parent material to Buru Series paleosols in the uppermost part of the underlying Red Bed Member. A thin fossiliferous red claystone layer about 5 m above the base of the Grey Sandstone Member also is a Buru Series paleosol. In addition, the Grey Sandstone Member fills a scour that removed the entire Red Bed Member to the east across the river from the Songhor gullies. These and other aspects of the Grey Sandstone Member are understandable if it was the paleochannel of a large meandering stream whose levees have been preserved in the upper part of the Red Bed Member. Most of its sediment was derived from deeply weathered nephelinitic volcanics, like those that form the bulk of the Kapurtay Agglomerate in the Mteitei Valley to the northeast.

Overlying the Grey Sandstone Member is a thick unit of gray-green nephelinitic agglomerates. This is a very characteristic lithology of most of the Kapurtay Agglomerate, and is not regarded as distinctive enough to warrant its naming as a member. As will be discussed for similar, better-exposed rocks near Fort Ternan, these are probably products of a large carbonatite-nephelinite stratovolcano.

4.1.2.6 Melanephelinite Member, Tinderet Volcanics. Filling a paleogully within tilted Kapurtay Agglomerates are black lava flows of melanephelinite, similar to those already described near Koru. A radiometric date of about 10 million years, already discussed, was obtained from the low outcrops in the field south-east of Songhor gullies. The distribution of these lavas within the area around Songhor and Koru indicates that they were derived from vents near or under the basanitic later lavas of Tinderet peak (Binge, 1962; Deans & Roberts, 1984).

4.1.3 Geology of the area around Fort Ternan National Monument

On the hillside south of the Baraget Valley and Fort Ternan railway siding is another prehistoric site administered by the National Museums of Kenya. The monument itself is a large quarry (Figs 4.8 and 4.9) excavated for fossils over the period 1961 to 1967 by L. S. B. Leakey and his associates (Andrews, 1981a; Bishop & Whyte, 1962; Churcher, 1970; S. Cole, 1975; Shipman, 1981). A southern extension of the headwall was excavated by a team under the direction of A. Walker (Shipman, 1977; Shipman *et al.*,1981). This quarry has been a natural focus for geological investigations, because outcrops beyond it are obscured by moist wooded grassland, Afromontane forest (White, 1983), and deep-brown soils (mainly Phaeozems or Argiudolls: Sombroek, Braun, & Van Der Pouw, 1983). In addition, stratigraphic relationships are complex. For these reasons, successive geological maps of the area have shown little agreement (Binge, 1962; Pickford, 1984, 1986a; J. A. Van Couvering for Shipman *et al.*, 1981). The new mapping reported here included only the 5 km^2 immediately around the main fossil quarry. Detailed section measuring was restricted

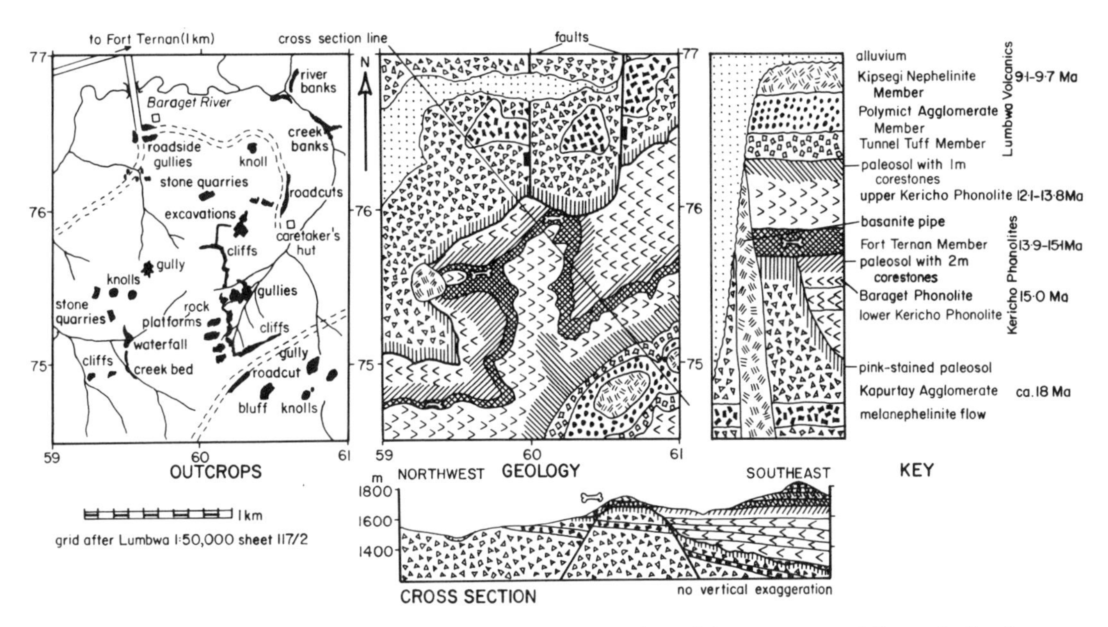

Fig. 4.8. Geological sketch map, outcrop map and cross section of the area around the main fossil quarry of Fort Ternan National Monument, Kenya. Grid from Lumbwa map sheet 117/2, 1:50,000.

to rocks little affected by recent weathering in and around the main fossil quarry (Fig. 4.10).

4.1.3.1 Kapurtay Agglomerate. These green nephelinitic breccias and at least one interbedded melanephelinitic lava form the floor of the Baraget Valley north of Fort Ternan National Monument. The boulders of the breccias consist largely of melanephelinite and nephelinite. Ijolite and basement rocks are less common. A similar clast of agglomerate found as a xenolith in the overlying Baraget Phonolite yielded a stepwise $^{39}Ar/^{40}Ar$ degassing age of about 18 million years (Hooker & Miller for Shipman *et al.*, 1981). This radiometric date is not especially well founded, but it is compatible with the age of the bulk of the Kapurtay Agglomerate overlying the Red Bed Member in Songhor gullies.

A variety of useful observations were made in a narrow trench excavated through the upper 25 m of the agglomerates below the National Monument (by Van Couvering for Shipman *et al.*, 1981, not including their "A beds" here regarded as part of the Fort Ternan Member of the Kericho Phonolites) and in several small abandoned quarries for building stone at this stratigraphic level on the hillsides west of the main quarry. In these exposures, the most voluminous rocks are massive, poorly sorted, green volcanic breccias, up to 11 m thick. Some of these thick beds have the fragmented angular clasts, grain-supported fabric, and normal grading of pyroclastic flows. Others are cemented with calcite and minor zeolite, and have matrix-supported clasts, and reverse grading common in volcanic mudflows (Cas & Wright, 1987). Also found within the upper Kapurtay Agglomerate, although thin and rare, are stratified lapilli tuffs. One distinctive tuff contains crystals of melanite garnet and nepheline in a fine-grained matrix (J.A. Van Couvering for Shipman *et al.*, 1981). In contrast to the Red Bed Member of the Kapurtay Agglomerate and the Fort Ternan Member of the overlying Kericho Phonolites, no clear evidence of ancient soil formation was seen in the Kapurtay Agglomerate near Fort Ternan. Some part of the Kapurtay

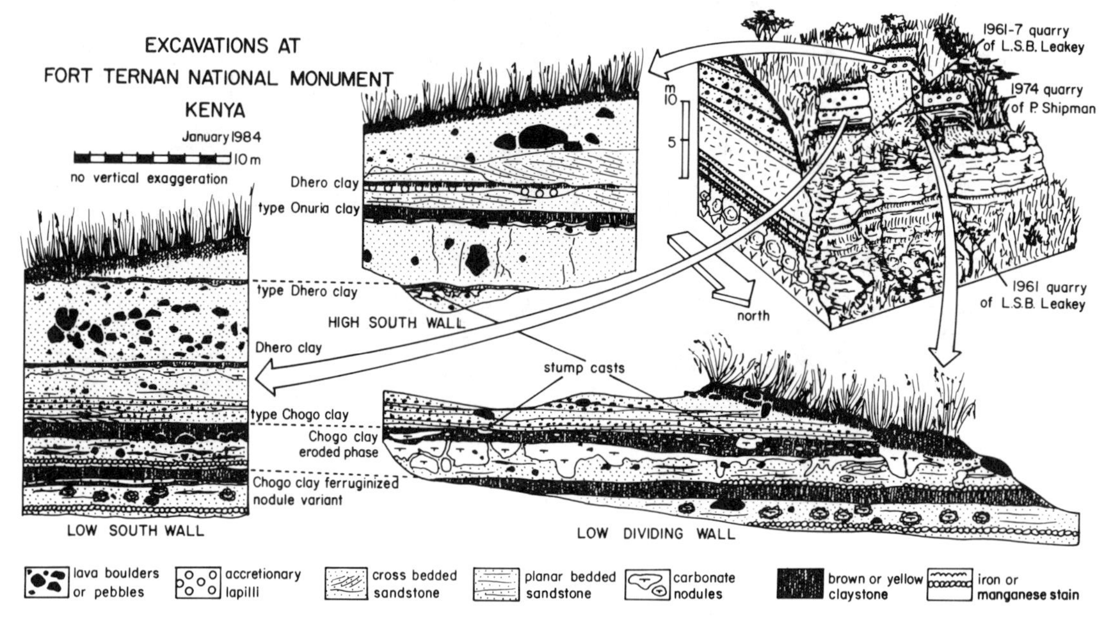

Fig. 4.9. Plan of excavation and field sketch of exposed faces (as of January 1984) in the main fossil quarry at Fort Ternan National Monument, Kenya.

Agglomerate here may correlate with a similarly massive portion of the formation called the Cliff Agglomerate Member, because of its prominent outcrops in the Mteitei Valley northeast of Songhor (Pickford, 1984, 1986a; Pickford & Andrews, 1981). This especially massive and thick horizon of volcaniclastic rocks may represent an unusually sustained burst of activity from a nearby nephelinitic stratovolcano.

A very dark-green locally vesicular lava cropping out in low knolls north of the main fossil quarry (grid reference 603764 on the Lumbwa 1:50,000 sheet 117/2) can be seen in creek bed exposures to the east (grid reference 608766) to be interbedded within the Kapurtay Agglomerate. In thin section it proved to be a medium to fine grained melanephelinite, with abundant zoned phenocrysts and laths of augite and phenocrysts of olivine in a matrix of nepheline (Table 4.2). This may be the same flow that has long been known to be interbedded within the agglomerates in railway cuttings near Fort Ternan 2 km to the northeast (Binge, 1962; Neilson, 1921).

This flow is a useful stratigraphic marker, disrupted by faulting of the Kapurtay Agglomerate into a horst about 600 m wide and displaced some 20-m above the surrounding terrain (Fig. 4.8). The horst is oriented north-south, in a radial arrangement to what remains of the caldera of Tinderet, and also to the original stratovolcano of the Kapurtay Agglomerate. Such narrow horsts are seen today flanking several Kenyan volcanoes (near Mt Suswa as mapped by R. W. Johnson, 1969; and northeast spur of North Ruri as shown by Le Bas, 1977), as well as in nonvolcanic areas (J. B. Wright & Rix, 1967). They have been attributed to volcanic doming, crustal extension in rift valleys, or both.

The faulted surface of the Kapurtay Agglomerate is modified by a thick (2-m) pink paleosol. The upper part of this paleosol was not exposed in available outcrops. Only pink-stained agglomerate corresponding to saprolite beneath the soil solum was seen. In the area around Kapurtay north of Songhor, a thick red surface soil on plateaus near the contact of the Kapurtay Agglomerate and Tinderet Lavas has been mapped as the Sossok clay loam soil and identified as a Brown Forest Soil (by Thorp *et al.*, 1960), which would be a Typic Tropudalf (of Soil Survey Staff, 1975). Elsewhere along this ancient land surface are thin boulder beds, which have been interpreted as a colluvial and alluvial mantle to this landscape (Pickford, 1984, 1986a). Thus both weathering and sheet-wash softened the harsh faulted outlines of this quiescent buried volcanic landscape.

4.1.3.2 Lower Kericho Phonolites and Baraget Phonolite Member. Overlying the paleosol on the Kapurtay Agglomerate below the main fossil quarry is the Baraget Phonolite Member of the Kericho Phonolites. This is a medium-grained black lava with abundant pilotaxitic microlites of feldspar in the matrix and phenocrysts of nepheline, but few sanidine and biotite phenocrysts. This flow has been radiometrically dated near the fossil quarry at 15.0 ± 0.1 million years using $^{39}Ar/^{40}Ar$ stepwise degassing (Hooker & Miller for Shipman *et al.*, 1981).

Despite the abundant phenocrysts of nepheline in the Baraget Phonolite, point counting (Table 4.2) has shown a greater abundance of feldspar, mainly as microlites. Thus this lava is not a nephelinite (contrary to J. A. Van Couvering for Shipman *et al.*, 1981), but has a composition including traces of biotite that indicates affinities with Kericho Phonolites (Lippard, 1973a) and other so-called Losuguta-type phonolites (Binge, 1962; Neilson, 1921; W. C. Smith, 1931). More typical Kericho Phonolites were found in the scarp 1 km southwest of the main fossil quarry (grid reference 596752 on the Lumbwa sheet). Unlike the Baraget Phonolite, this underlying flow is a fine-grained hyalopilitic lava, very sparsely porphyritic with subhedral sanidine and biotite. At least three flows of this kind underlie the Baraget Phonolite in Kipchorion Gorge south of Fort Ternan. Beneath these in the gorge is a pink sanidine crystal tuff, unconformably overlying a paleosol developed in the uppermost Kapurtay Agglomerate (Pickford, 1984, 1986a).

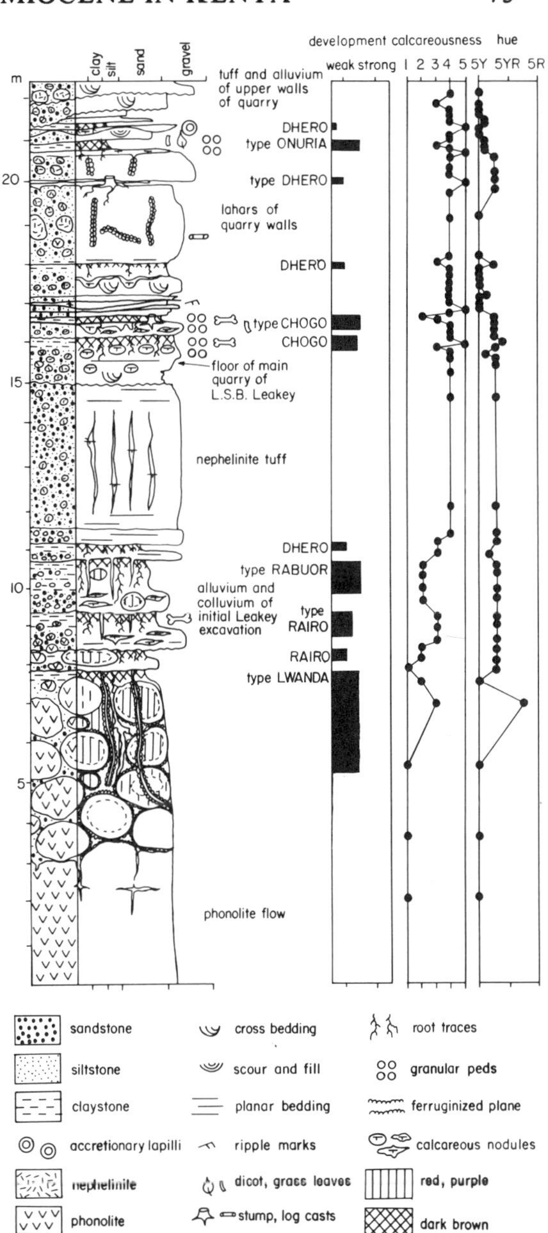

Fig. 4.10. Geological section of paleosols exposed in the main fossil quarry at Fort Ternan National Monument, Kenya. Scales of paleosol development and calcareousness from Retallack (1988a); hue from Munsell color chart (1975).

There is thus a dramatic thinning of the lower Kericho Phonolites from 120 m in Kipchorion Gorge to locally pinched out around the buried ridge (former horst) near the main fossil quarry only 3 km to the north. This thickness change is interpreted to reflect ponding of the lavas against the slopes of the eroding stratovolcano of Kapurtay Agglomerate to the north.

Kericho Phonolites cover an enormous area (at least 2600 km^2) of southeastern Kenya, extending from the Mau Escarpment all the way west to Kisumu and south to the Kisii Hills (Pickford, 1982). Together with other similarly extensive plateau phonolites in Kenya (Lippard, 1973a, 1973b), they erupted from fissures and shield volcanoes associated with initial faulting of the western margin of what was later to become the Gregory Rift (B. H. Baker, 1987).

The glassy and vesicular upper part of the Baraget Phonolite below the main fossil quarry has been obscured by development of a thick (up to 4 m) red paleosol of the Lwanda Series, to be described in a future section. This paleosol has enormous (2 m in diameter) spheroidally weathered corestones, and these form an easily mappable horizon.

4.1.3.3 Fort Ternan Member. Overlying the deeply weathered Baraget Phonolite and the underlying flows of the upper Kericho Phonolites is a sequence of brown claystones and gray tuffs and breccias called the Fort Ternan Member. At the fossil quarry of Fort Ternan National Monument this member is 30-m thick. These fossiliferous brown and gray volcaniclastic sediments can be traced northeast all the way to Serek and Kapsibor, and to the southwest around the ridge of the National Monument into Kipchorion Gorge, a total straight-line outcrop extent of 9.6 km. Over this distance the Fort Ternan Member thickens and thins with the underlying paleotopography on the Baraget Phonolite and ancient ridges of the Kapurtay Agglomerate. The member is thinner (20 m) to the west than in the fossil quarry, and thins also to the northeast at Kapsibor (12 m) and to the south in Kipchorion Gorge (6 m; Pickford, 1984, 1986a). Biotites from tuffs and paleosols in the Fort Ternan Member dated by the K/Ar method yielded ages of 14.4 ± 0.2 and 15.1 ± 0.7 million years (Bishop, Miller, & Fitch, 1969; Evernden & Curtis, 1965; corrected by method of Dalrymple, 1979). This is middle Miocene (Berggren, Kent, & Van Couvering, 1985). This is reasonable considering the nature of the mammalian fauna at Fort Ternan, which is well enough known to be a biostratigraphic standard for African mammalian faunas (Pickford, 1981).

The Fort Ternan Member is best understood in the fossil quarry, and the following account emphasizes sedimentary and volcanic features exposed in the quarry (Fig. 4.9). The basal 3.5 m of the member at the National Monument were exposed during the first excavations there (Andrews, 1981a; Bishop & Whyte, 1962), because this was the part of the slope where the most bone was found loose at the surface. Soon it was realized that these brown, weakly calcareous claystones contained very few fossils, and that the bones and teeth on the surface had been washed down from a higher stratigraphic level now extensively quarried. From the outset, this lower part of the excavation was regarded as paleosol (Bishop & Whyte, 1962). Here it is divided into four superimposed profiles of Rairo, Rabuor, and Dhero Series. The parent material of all these paleosols is a granule-bearing sandstone of deeply weathered nephelinite and phonolite rock fragments, with occasional augite, biotite, quartz, and feldspar. Also found in the paleosols and their parent material are large (up to 23-cm) boulders of subrounded vesicular Baraget Phonolite, which generally are pink and deeply weathered. Taken together with the nature of the paleosols discussed later, this lower part of the Fort Ternan Member is interpreted as a colluvial mantle and soils on the footslopes of a quiescent nephelinite stratovolcano near the break in slope to an extensive lava plain of onlapping phonolites.

Overlying this basal brown clayey interval is a thick (5 m) cliff-forming nephelinite tuff. Most of its sand grains and granules are nephelinite and melanephelinite, but some are crystals of pyroxene and biotite, like the phenocrysts in the rock fragments. The rock fragments are irregularly shaped to rounded and the crystals are subrounded, but few show weathering rinds. Some blocks of nephelinite are up to 15-cm across, and most of these larger blocks are scoriaceous. Both the vesicles and the intergranular spaces are firmly cemented by local fibrous zeolite and abundant sparry calcite. The base and the top of this thick tuffaceous unit are bedded, but the central part is massive and fissured by crudely parallel, vertical calcite veins spaced at a few tens of centimeters.

The distinctive vertical veins may have been cooling structures, but they are not as regular as the columnar jointing seen commonly in basalt flows, nor so large or tubular as degassing pipes of pyroclastic flows (both well illustrated by Cas & Wright, 1987). No textural evidence of welding could be detected, but this tuff was cemented during middle Miocene time, because a redeposited block of it was found within the type Chogo clay paleosol. The large blocks in the tuff appear entrained in coarse layers, rather than ballistically emplaced. The original contacts of this tuff thin from 5 m in the quarry to 2.2 m in badlands 200 m to the southeast and to 1.5 m in gullies 150-m northwest of the main fossil quarry (Pickford, 1984, 1986a; Shipman, 1977). This tuff was not seen at all in badlands exposures 400-m southwest of the quarry (at grid reference 597757 on Lumbwa sheet). Thus it did not mantle topography in an even layer, as do airfall ashes. Instead it has many of the characteristics of a pyroclastic flow deposit, preceded by a ground surge and followed by local fallout and water worked tuffs (as characterized by Cas & Wright, 1987). In its composition, this Miocene tuff from Fort Ternan is most like the Gray Tuffs from parasitic cones around Oldoinyo Lengai, Tanzania (Dawson, 1962), probably some 2,000 to 30,000 years old (Hay, 1989). Their mode of eruption has not been studied. They are similar to flows produced by the collapse of a Plinian-style eruption column, like that seen during the 1940-1941 eruptions of Oldoinyo Lengai (Richard, 1942). Such an origin is plausible also for the Miocene tuff at Fort Ternan, which probably was erupted from a vent nearby to the north or northeast, considering its restricted distribution in a shallow paleovalley and its lack of phonolitic rock fragments.

The 8-m sequence exposed in the deepest part of the main fossil quarry above the thick nephelinitic tuff shows several examples of each of three distinctly different volcano-sedimentary facies: (1) massive, matrix-supported tuff breccias; (2) cross-bedded, grain-supported volcaniclastic sandstones; and (3) thin stratified tuffs. These form parent materials to paleosols of three different kinds, described later as Dhero, Chogo, and Onuria paleosol series.

Especially impressive in the main quarry are the two massive tuff breccias that form most of the high wall. The lower one is 2-m thick and forms parent material to a Dhero paleosol: the upper one is 70-cm thick under an Onuria paleosol. Both consist of blocks of nephelinite and melanephelinite up to 2-m across floating in a matrix of nephelinitic lapilli and tuff, strongly cemented by calcite (excellent photographs have been published by Andrews, 1981a; Andrews & Walker, 1976; Churcher, 1970). In both tuff breccia beds, the largest blocks are concentrated within the massive center, and grain size decreases both upward and downward into more clearly bedded sandstones. Both beds also are traversed by irregular veins of weakly ferruginized calcite, near vertical and spaced at intervals of one or two meters. The ferruginization of these veins is most pronounced immediately below the paleosols capping the tuff breccia beds, an observation suggesting that they were fissures in moderately cemented beds during Miocene time, rather than geologically younger joints. Other indications of early cementation of these beds are enclosed external molds of fossil tree trunks. These are uncrushed by compaction and are filled with clay and sparry calcite. Trunks 8.8, 7.4, and 4.5 cm in diameter were measured from these tuff breccias. All had uncharred bark impressions. These tuff breccia beds also are the most likely source of a fossil chameleon head and shoulders with uncharred skin impressions (Hillenius, 1978), found loose on the quarry floor (Shipman *et al.*, 1981). Finally, these tuff breccias are laterally impersistent. They thin considerably and are at different stratigraphic levels in sections measured only a few hundred meters to the northwest, southwest, and northeast of the main fossil quarry (Pickford, 1984, 1986a; Shipman, 1977).

These various lines of evidence for rapid, violent, low-temperature emplacement and early cementation confirm the suggestion of J. A. Van Couvering (for Shipman *et al.*, 1981) that the tuff breccia beds were lahars. Such volcanic mudflows have been studied extensively on andesitic and dacitic stratovolcanoes (R. V. Fisher & Schminke, 1984), although the Yellow Tuffs and Agglomerates some 30,000 to 400,000 years old (Hay, 1989) on Oldoinyo Lengai, Tanzania, include several similar beds (Dawson, 1962). Lahars can be triggered by eruptions, by earthquakes, by rainstorms, or by a combination of these during an eruptive cycle. Only involvement of hot pyroclastic flows or surges seems unlikely for these Miocene examples, considering indications of low-temperature emplacement. Judging from the orientation of log casts and the laterally restricted distribution of these tuff breccias, their source was a large volcanic edifice to the north or northeast.

Closely associated with the tuff breccias and capped by two Chogo and two Dhero paleosols is a second facies of locally fossiliferous, cross-bedded to plane-bedded, nephelinitic, gravelly sandstones. The coarse grains are mostly 2 to 5 mm across, but some are up to 26 cm. These large blocks were not ballistically emplaced but were entrained between depositional units with low-angle cross-bedding, ripple marks, normal grading, and plane bedding. Individual beds and parts of beds vary greatly in color and weathering of their clasts, ranging from brown and pink deeply weathered nephelinite and phonolite clasts to fresh examples dark green and black in color. Even the most weathered clasts show sharp boundaries to the ubiquitous calcite cement in thin section. Some cement of zeolite fibers, with the x-ray diffraction pattern of phillipsite, was seen lining calcite-filled cavities that postdate earlier calcite cement. Shallow gullies with a narrow U- or V-shaped profile have been well documented from these nephelinitic sandstones where they form the parent material to the highly fossiliferous Chogo paleosols (site plans and photographs of Andrews, 1981a; Andrews & Walker, 1976; Churcher, 1970: Shipman, 1977, 1981; Shipman *et al.* 1981). A steep-sided broad paleogully can also be seen where cross-bedded sandstone grades up into tuff breccia in the high south wall of the quarry (Fig. 4.9).

These nephelinitic sandstones are similar to waterlain tuffs on the active volcano Oldoinyo Lengai, for example the Black Tuffs and Agglomerates, some 1250 to 2000 years old (Dawson, 1962; Hay, 1989). This is not to imply that they were laid down in rivers or lakes. No permanently aquatic sedimentary facies or fossil fauna has been found at Fort Ternan (Pickford, 1985). Nor is pyroclastic base surge a likely explanation for these beds, considering their numerous rounded blocks with weathering rinds and the lack of ballistically emplaced blocks or hyaloclastic chilled or fractured grains. Instead, these Miocene sandstones were deposited by sheet wash on an erosion-furrowed slope, as envisaged for the Black Tuffs and Agglomerates of Oldoinyo Lengai. Some of the Miocene sandstones, especially those with large blocks and paleogullies, and associated with tuff breccias, may have been deposits of hyperconcentrated stream flow near the runout zone of rainstorm lahars, as are well known on andesitic-dacitic stratovolcanoes (Rodolfo, 1989).

These nephelinitic sandstones also contain evidence of their source direction, particularly in the bed containing flattened fossil grasses that may have been rooted in the underlying Onuria clay paleosol. Two sets of paleocurrent measurements have been made on these fossil grasses (by Dugas, 1989). In the field, 35 fossil grasses were found to dip toward a vector mean azimuth of 349 degrees, with a vector magnitude of 90 percent, indicating a very consistent orientation (Potter & Pettijohn, 1963). A large oriented block of grass-bearing matrix was shipped to the laboratory and carefully excavated with dental tools to reveal 50 fossil grasses dipping toward a vector mean azimuth of 351 degrees, with a vector magnitude of 82 percent. These results agree well with the north-south orientation of a small paleogully containing unusually abundant bones in the Chogo clay eroded-phase paleosol (Shipman, 1977, 1981). Other paleogullies, cross-bedding, and ripple marks in the main quarry were not sufficiently pronounced or abundant to be taken as reliable paleocurrent indicators, but they do indicate a source terrain to the north or northeast.

A third facies in the main fossil quarry is calcareous tuffs (especially at 20 and 21.3 m in Fig. 4.10). These are micritic in thin section, with common silt-sized crystals of nepheline, biotite, and augite, and rarer melanite, melilite, and wollastonite. Granule-sized globules of calcite and carbonatite-melilitite accretionary lapilli are common in some layers. These tuffs are unusually rich in barium, strontium, and lanthanum, an indication of carbonatite affinities (Bishop, 1968). This is also apparent from their petrographic similarity with carbonatite-melilitite airfall tuffs of the Plio-Pleistocene Laetoli and Olduvai Beds (Hay, 1976, 1978) and 600-year-old Footprint Tuff of Oldoinyo Lengai, Tanzania (Hay, 1989).

The Tanzanian calcite globules have been interpreted (by Hay, 1978) as filled bubbles, comparable to the vesicular structure of modern desert soils. Calcite globules in these paleosols show no preference for soil surface horizons and may be confined to narrow zones. Nor can the paleosols at Fort Ternan be regarded as desert soils. Climate probably was even wetter during formation of paleosols with rare calcite globules at Koru and Songhor. It is more likely that calcite globules are original volcanic grains or pseudomorphs: perhaps of the carbonatite variety sövite.

Accretionary lapilli in these tuffs were 8 to 14 mm in diameter and concentrated in thin (3-8-cm) bands at the base of normally graded beds of tuff. Some of these have a thin, fine-grained coating around a massive interior of silt-sized grains (type A of Reimer, 1983) and others have a thick rind with concentric darker and lighter bands (type B of Reimer, 1983). Sharply broken accretionary lapilli also were seen in thin section. Their interiors and matrix are petrographically the same as their carbonatite-melilitite matrix. Accretionary lapilli of this kind have not yet been reported from the active carbonatite-nephelinite volcanoes of Tanzania, but are widespread in Miocene volcanic rocks of this composition in southwestern Kenya: in the Chamtwara Member of the Kapurtay Agglomerate near Koru (Deans & Roberts, 1984) and the Rangwa Banded Tuffs on the mainland near Rusinga Island (Le Bas, 1977).

Accretionary lapilli are usually interpreted as records of airfall from a turbulent eruption cloud onto land. Immersion in water generally destroys them (especially the delicate type A of Reimer, 1983). Their preservation in carbonatite-nephelinite tuff is not surprising, considering the tendency of these tuffs to early cementation that is capable of preserving fossil grasses, dung beetles, larval cells of bees, and footprints (Barker & Nixon, 1983; Hay, 1986). Indeed, the sharp-edged broken rounds of accretionary lapilli seen in thin section are evidence of early cementation of these Miocene examples.

These airfall tuffs probably were erupted from Vulcanian-style volcanic activity, as has been observed at Oldoinyo Lengai during historic eruptions (Dawson, Bowden, & Clark, 1968; Nyamweru, 1988). The fresh ash of this volcano contains a much greater proportion of natrocarbonatite minerals than seen in Holocene or Plio-Pleistocene airfall tuffs. Isotopic studies of these recent and ancient ashes have shown that natrocarbonatite minerals such as nyerereite and gregoryite are converted to zeolites such as pirssonite and gaylussite, and finally replaced by calcite that cements the tuffs in less than 600 years (Hay, 1989). Such early cementation of the tuffs in the Fort Ternan Member may also explain the preservation of stump casts in very weakly developed paleosols (Dhero Series) including little-weathered airfall tuffs.

Other parts of the Fort Ternan Member outside the main fossil quarry include the same three kinds of facies in addition to a variety of paleosols. Overlying the sequence exposed in the quarry are several more thick nephelinitic lahars, with interbedded nephelinitic sandstones (Van Couvering for Shipman *et al.*, 1981). Tuff breccias, volcanic sandstones, lapilli tuffs, and brown paleosols also were seen in badlands 1 km southwest of the main fossil quarry (grid reference 597756 on Lumbwa sheet) and 600-m south (grid reference 603753). At Serek (grid reference 649779) and Kapsibor (656785), respectively 5 and 6 km northeast of the fossil quarry, volcano-sedimentary beds are thinner, less coarse grained, and dominated by brown paleosols (Pickford, 1984, 1986a).

Overall, the Fort Ternan Member formed a piedmont (or volcanic apron in the terminology of Pickford, 1986b) between a volcano to the north or northeast and an extensive plateau of phonolite to the south and west. During accumulation of the lower Fort Ternan Member the high ground to the north may have been the eroded remnant of a volcano remaining from the paroxysmal terminal eruptions of the Kapurtay Agglomerate, several million years before (Pickford, 1982, 1984, 1986a). By 14 million years ago, however, volcanic activity had resumed very close at hand to the north or northeast. The thick nephelinitic tuff could have been produced by a small parasitic cone, but the thick lahars more likely represent the building of a new stratovolcano either in or near the eroded caldera of Kapurtay Agglomerate. At about the same geological time there were also massive outpourings of melanephelinite in a rejuvenation of the stratovolcano Kisingiri, near Rusinga Island, 160 km to the southwest of Fort Ternan (Le Bas, 1977). There were also active volcanic centers 50 km northeast of Fort Ternan, now buried by late Miocene Kapkut and Timboroa volcanoes (W. B. Jones & Lippard, 1979). Airfall tuffs in the Fort Ternan Member could have come from any of these volcanoes, as well as from parasitic cones around them.

From this most recent round of geological work at Fort Ternan, two early ideas concerning the paleoenvironment at Fort Ternan can now be discounted: the poison spring hypothesis for the accumulation of the bones (Churcher, 1970) and the interpretation of stones in the paleosols as tools of early human ancestors (L. S. B. Leakey, 1968). In the paleoenvironment already outlined, large stones of nephelinite, phonolite, and even Precambrian basement rocks, transported by sheet wash, lahars, and hyperconcentrated outflow would not be out of place even in clayey paleosols. There is thus no need to propose that they were transported by early human ancestors. In addition, the damage to bone used as evidence in support of this idea is more like the damage caused by trampling and carnivores than that of humans (Binford, 1981; Pickford, 1986c).

The idea of a poison volcanic spring at Fort Ternan deserves more careful consideration. No sulfur or metal mineralization, ferruginized veins, silicification, faults, or other evidence of volcanic springs were seen in the main fossiliferous layers at Fort Ternan National Monument. The fault of the underlying horst of Kapurtay Agglomerate closest to the fossil site ceased activity and was obscured by deep weathering long before accumulation of the Fort Ternan Member. Other faults of small displacement in and around the main fossil quarry (Pickford, 1984, 1986a) do not show changed thickness of beds on either side, as would be expected of faults active during accumulation of these beds. Instead they appear to postdate both the Fort Ternan Member and the upper Kericho Phonolites. Perhaps more revealing is sedimentological evidence, combined with that later to be discussed from the paleosols, that the gullies at Fort Ternan were shallow and dry most of the time. No sideritic, pyritic, or laminated shales of lakes, nor sandy facies of paleochannels, were seen. There are no clearly aquatic fossils either (Pickford, 1985). It could be argued that a crater lake nearby periodically released poison clouds of carbon dioxide, as in the lethal outbursts from Lakes Monoun and Nyos, Cameroon (Kling, Tuttle, & Evans, 1989; Sigurdsson, *et al.*, 1987), which also are a threat from Lake Kivu, Ruanda (Degens *et al.*, 1973). However, nearby crater lakes are unlikely considering the dissimilarity between the Fort Ternan tuffs and those from hydroclastic eruptions.

Also arguable is the mass killing of animals at Fort Ternan by mofettes exhaling only carbon dioxide. Seepage of carbon dioxide through the soil can be fatal to burrowing animals (Allard, Dajlevic, & Delarue, 1989; Tazieff, 1989). In Zaire, such areas are also barren of vegetation (J. Lockwood, personal communication, 1989), probably because high carbon dioxide levels prevent root respiration. The Zairean mofette seen by Lockwood is littered with bones, including many of vultures, scavengers, and predators. At Fort Ternan, with the exception of two articulated skeletons of juvenile mongoose, probably in a burrow, bones are disarticulated attritional accumulations rather than mass death assemblages, and vulture and carnivore remains are rare (Shipman, 1977, 1981; Shipman *et al.*, 1981). Fossil root traces and stumps at Fort Ternan also are evidence of more vegetation than is likely to have tolerated soil regularly saturated with carbon dioxide. Occasional killing of animals by volcanic gases may have occurred at Fort Ternan, but is not a satisfactory explanation for the great abundance of bone there. With regard to abundance, it may not be a coincidence that the most fossiliferous paleosols (Chogo clays) are also the most calcareous, and so best suited to preservation of bone (Pickford, 1986b; Retallack, 1984a). Other less calcareous paleosols developed on phonolite or on lahars that include possible degassing structures (Lwanda, Dhero, and Onuria paleosols), and so more likely candidates for mofettes, have not yet yielded fossil bone.

4.1.3.4 Upper Kericho Phonolites. Overlying the Fort Ternan Member at the National Monument is a single flow of black phonolite, which forms a prominent topographic bench throughout this area. In thin section it is hyalopilitic, with fine microlites of feldspar, and very sparse euhedral phenocrysts of nepheline and subhedral phenocrysts of sanidine and biotite. This flow has been radiometrically dated by $^{39}Ar/^{40}Ar$ total fusion, $^{39}Ar/^{40}Ar$

stepwise degassing and K/Ar whole rock methods (Bishop, Miller, & Fitch, 1969; Fitch, Hooker, & Miller, 1978, using correction of Dalrymple, 1979), with remarkably consistent results: 12.6 ± 0.7, 12.5 ± 0.4, and 12.9 ± 0.15 million years old.

This flow is petrographically identical to other Kericho lavas (Lippard, 1973a) of the Losuguta type (W. C. Smith, 1931), underlying the Fort Ternan Member and widespread elsewhere in southwestern Kenya. Radiometric ages close to 12 million years are common for these flood lavas at localities ranging from Mt Blackett east of Fort Ternan (W. B. Jones & Lippard, 1979), south to Kericho, and west to Kisumu (B. H. Baker *et al.*, 1971).

The upper Kericho phonolites mapped by Pickford (1984, 1986a) are only 6-m thick above the main fossil quarry, but are 15-m thick 3-km south in Kipchorion Gorge, at least 7-m thick some 5-km to the northeast at Serek, and are 12-m thick 6-km to the northeast at Kapsibor. They overlie paleosols of quite varied development (Shipman *et al.*, 1981). Probably they onlapped a nephelinitic stratovolcano to the north of the National Monument in the same way as the lower Kericho Phonolites.

On the uppermost surface of the upper Kericho Phonolite is a paleosol possibly of the Lwanda Series. The upper part of the paleosol is poorly exposed, but its saprolite of corestones about 1 m in diameter are an easily mappable horizon.

4.1.3.5 Tunnel Tuff Member. This green Miocene lapilli tuff with blocks of nephelinite up to 20 cm across, crops out best in the west-facing slope of the knoll 1-km south of the main fossil quarry (at grid reference 604747 on the Lumbwa sheet). Here it is plane bedded and contains casts of uncharred twigs and logs. In other places it is a massive tuff breccia forming conspicuous balds of soft green rock. At the knoll the Tunnel Tuff is only 50-m thick, but it thickens toward the east to 120 m along the railway near Tunnel (at grid reference 632737), beyond which it is covered by an overlying sequence of nephelinitic lavas (Binge, 1962; Pickford, 1984, 1986a). These tuffs are similar in character to the 5-m nephelinitic tuff in the main fossil quarry, and like this the Tunnel Tuff may represent ash flows and waterlain tuffs from a nephelinitic vent east of the present cone of Tinderet. The Tunnel Tuff may have been erupted from a parasitic vent or volcano that produced the Lumbwa Phonolitic Nephelinite lavas southwest of Londiani town, some 20-km east of Fort Ternan (D. J. Jennings, 1971). These lavas have been dated by the whole-rock K/Ar method at 12.7 ± 0.3 million years old (W. B. Jones & Lippard, 1979, corrected by method of Dalrymple, 1979).

4.1.3.6 Polymict Agglomerate. Overlying the Tunnel Tuff in the northeastern bluffs of twin knolls 1-km southeast of the main fossil quarry (grid reference 609749) is a thick (at least 25-m) nephelinitic tuff breccia (Pickford, 1984, 1986a). This breccia has been mapped eastward into the hills around Lumbwa, and is especially well exposed in the railway cuttings from Lumbwa to Tunnel station, where it contains blocks of nephelinite up to 6-m long (Binge, 1962). On the basis of these exposures beyond the area mapped here, the Polymict Agglomerate has been interpreted as a giant lahar (Pickford, 1984, 1986a).

4.1.3.7 Kipsegi Nephelinite. Overlying the Polymict Agglomerate in the hills around Lumbwa are several nephelinite lava flows (Binge, 1962, Pickford, 1984, 1986a). These Kipsegi Nephelinites have not been radiometrically dated. Overlying phonolitic nephelinites of the Lumbwa Phonolites have yielded K/Ar whole rock ages of 9.1 ± 0.5, 9.4 ± 0.5, 9.7 ± 0.5, and 9.6 ± 0.6 million years (B. H. Baker *et al.*, 1971; W. B. Jones & Lippard, 1979). These dated lavas are interbedded with tuffs northwest of Timboroa, but form a thick tuff-free succession southwest of Londiani (W. B. Jones & Lippard, 1979). This may have been near the vents for these lavas, as well as the Kipsegi Nephelinite and the Polymict Agglomerate.

A lava petrographically similar to the Kipsegi Nephelinite caps the flat top of the hill 1-km southeast of the main fossil quarry (grid reference 607749). The main outcrop is a low bench with horizontal bands of fresh pyroxene phenocrysts. This is unlike even the largest clasts of the underlying Polymict Agglomerate, which are tilted and weathered so that only surface pockmarks remain of the phenocrysts.

4.1.3.8 Basanite plug, (?)Tinderet Volcanics. The low knoll at the end of the ridge 1-km southwest of the main fossil quarry (grid reference 593755) is a small (about 250-m diameter) basanite plug. There are excellent exposures on the north side of the ridge where fresh basanite boulders form a talus slope. In thin section this black intrusive rock showed numerous medium-grained laths of feldspar, euhedral phenocrysts of pyroxene, and subhedral to embayed olivine and opaque oxides, in a matrix that includes some nepheline and a low-relief mineral with hints of cubic shape, identified as analcite. Also in this intrusion were numerous xenoliths up to 6-cm across, which proved to be melanephelinite, with abundant euhedral, zoned phenocrysts of augite in a matrix largely of nepheline (Table 4.2).

This plug intrudes Kapurtay Agglomerate in the lower hillside and the whole sequence of Baraget Phonolite, Fort Ternan Member, and upper Kericho Phonolite along the ridge to the east. Thus it is at least younger than 12 million years. Possibly it was emplaced some 5 to 6 million years ago, during eruption of the basanites that form the present summit of Tinderet, as already discussed for the area around Koru. The melanephelinite xenoliths could have been carried up from the flow mapped within the Kapurtay Agglomerates, some 18 million years old in this area (Hooker & Miller for Shipman *et al.*, 1981). There also is a thick series of melanephelinite lavas dated at about 10 million years old, on Legetet Hill and other foothills of Tinderet (Bishop, Miller, & Fitch, 1969; Dalrymple, 1979; Kent, 1944; Shackleton, 1951). Although less likely, the xenoliths could be foundered blocks of these later lavas now eroded from the area around Fort Ternan. Until further radiometric studies are conducted, the plug and xenoliths are tentatively regarded as about 5 and 18 million years old, respectively.

4.2 Description and interpretation of the paleosols

In the following pages are detailed descriptions and interpretations of the different kinds of paleosols so far recognized in the stratigraphic sections measured near Koru, Songhor, and Fort Ternan. These have been named as paleosol series (Retallack, 1990a), taken to be ancient equivalents of mapping units used for surface soils (Soil Survey Staff, 1951, 1962). Many of the local place names are already in use for surface soil series (Thorp *et al.*, 1960), rock formations, and fossil localities (Pickford, 1984, 1986a), so in most cases the paleosol series have been named from simple descriptive terms in Dholuo, a local language of Nilotic affinities. These names may thus seem like nicknames to local people, but in English scientific usage they are meant to represent objective, mappable kinds of paleosols.

A standardized graphic and descriptive format has been used for representing the salient data and interpretation of each paleosol (following Retallack, 1988a). Colors were estimated by comparison with the charts of Munsell Color (1975), taken within a few minutes of exposure of the naturally moist samples. Lithological sections were logged using a scheme that includes a graphic representation of mean grain size, in order to represent field assessment of likely clayey subsurface (Bt) horizons. The grain size scale of soil science has been used (Soil Survey Staff, 1975), rather than the Wentworth scale commonly used in geological studies. Quantitative grain size and mineral compositional data were obtained from 500 points counted in petrographic thin sections using a Swift Automatic Point Counter by G. J. Retallack (specimens SO-R115 to 145, FT-R1 to 40) and G. S. Smith (other specimens; Appendices 3 and 4). This gives an abundance of common components with an error (2σ) of about 2 percent by volume (Friedman, 1958; Murphy, 1983). The point count data are portrayed in columns that may seem narrow and cramped in some cases, so that only components approaching 2 percent or more can be plotted and only differences of 2 percent or more will appear as significant trends. Clay has been placed to the left in both textural and mineral plots to give visual assessment of how well the two counts agree. In the mineral composition plots, easily weathered minerals in the schemes of Goldich (1938) and M. L. Jackson *et al.* (1948) are placed to the left, and progressively more weather-resistant minerals are placed to the right. In comparing these results with similar data for modern soils it is well to beware of changes during burial, as already discussed. There also is a systematic overestimation of clay in thin sections compared with seive analysis for grain size of soils (Murphy & Kemp, 1984), because of edge effects in thin sections and the persistence of clay aggregates in seives.

A selection of molecular weathering ratios (Appendix 7) based on whole-rock chemical analyses (Appendices 5 and 6) also are plotted, and their likely pedogenic significance is indicated (following Retallack, 1990a). Chemical analyses were performed using ICP by A. Irving (University of Washington, Seattle: specimens FT-R1 to FT-R18) and using AA by C. McBirney (University of Oregon, Eugene: other specimens), with error calculated from counting statistics (for ICP) and for 10 replicate analyses of standard rock W2 (for AA). Organic carbon values were determined using the Walkley-Black titration by D. Horneck (Oregon State University, Corvallis), with error estimated from 50 replicate analyses of a standard soil. Bulk density was calculated from weight difference suspended in water of paraffin-coated clods by G. J. Retallack, with error estimated from 10 replicate analyses of specimens FT-R14.

Descriptions of the paleosol profiles follow the conventions and terminology of Roy Brewer (1976; Brewer, Sleeman, & Foster, 1983) and the Soil Survey Staff (1951, 1962, 1975) of the U.S. Soil Conservation Service. These technical descriptions are important to future classification of Kenyan paleosols, but they are tedious to read and so have been set in a smaller type. Only a concise interpretation of likely alteration of each kind of paleosol after burial is given for each paleosol, because a fully referenced general discussion has been offered in Chapter 3. Other interpretations also are offered in the following pages: a reconstruction of the paleosol as it may have been during formation; attempts to identify the paleosols in classifications for surface soils; their former ecosystem as indicated by their fossil content and other features; and their paleoclimate, topographic setting, parent materials, and duration of soil formation. These interpretations have changed in my own mind as the various studies and analyses reported here have been completed. No doubt some of the interpretations will change again as other investigators become intrigued by these fascinating paleosols. To this end, perhaps the most important goal of this work has been to document the named paleosol series with sufficient precision that they can be relocated and investigated by researchers who can extend these results to the thousands of paleosols in Kenya remaining undescribed.

The paleosol series are listed in order of stratigraphic appearance, so that interpretation of their paleoenvironments presents a detailed narrative of the Miocene prehistory of southwestern Kenya.

4.2.1 *Mobaw Series (new name)*

4.2.1.1 Diagnosis. Weakly developed, red (5YR) paleosols on bouldery colluvium of yellow to gray carbonatite tuff.

4.2.1.2 Derivation. Mobaw is Dholuo for patched (Gorman, 1972). The name refers to the striking color contrast between red paleosol matrix and lightly colored carbonatite boulders in it.

4.2.1.3 Description. The top of the type Mobaw silty clay loam paleosol (Figs 2.24, 4.5, and 4.12) is 38 cm below the top of the sequence of red paleosols above the karstified tuff of early Miocene, Legetet Carbonatite in the central part of the abandoned quarry (grid reference 517829 on 1:50,000 Muhoroni sheet 117/1), 3 km north of Koru, Kenya (Figs 4.4, and 4.6). This is Legetet fossil locality 10 of Pickford (1984, 1986a).

+6 cm; C horizon of overlying paleosol (Kwar silty clay); granule-bearing silty claystone, red (2.5YR4/6), with common prominent medium light-gray (10YR7/2) mottles and associated common distinct fine black (5YR2/1) mangans: large clasts white (10YR8/1), yellowish-brown (10YR6/5), brownish-yellow (10YR6/3), and olive (5YR5/4); strongly calcareous; contains fossil snails; intertextic insepic in thin section, with abundant clasts of carbonatite (both alvikite and sövite) and only slightly less abundant nephelinite, common quartz, and microcline, rare and altered pyroxene; abrupt, wavy contact to

0 cm; A horizon; granule-bearing clayey siltstone; dark reddish-brown (5YR3/3); common clasts 1 to 2 cm across of white (5Y8/2) carbonatite have a sharp outer boundary of red (2.5YR5/4) stain grading outwards into the matrix as a diffusion sesquan; other clasts are black (5Y2/1), olive (5Y5/4), brownish-yellow (10YR6/6), white (10YR8/1), and red (10R6/6); medium subangular blocky peds; few subhorizontal reddish-brown (5YR4/3) isostriotubules, 4 mm in diameter; rare mangans; strongly calcareous; calciasepic intertextic in thin section, with almost all clasts carbonatite and these mainly alvikite, some sövite, and calcified nephelinite, rare quartz, microcline, books of vermiculite, and small snail shells; also common 1-mm diameter crystal tubes of calcite after fine root traces; gradual, smooth contact to

-13 cm; Bt horizon; granule-bearing silty claystone; dark reddish brown (5YR3/3); with clasts as in overlying horizon; common bone fragments of white (5YR8/1); snail shell stained red (2.5YR4/8), with a red (2.5YR4/6) diffusion halo and filled by light-gray (5Y7/2) claystone with black (5Y2/1) mangans; strongly calcareous; insepic intertextic in thin section; clasts mainly of carbonatite, as in overlying horizon, but some with rounded outlines like "Pelee's tears"; common fine calcite crystal tubes; gradual, wavy contact to

-33 cm; C horizon; pebble-bearing clayey siltstone; red (2.5YR4/6); with abundant pebbles of light gray (2.5Y7/2) carbonatite, speckled with black (5Y2/1) opaque oxides; these clasts are irregularly shaped and grain-supported and have a thin (1-2 mm) red (2.5YR4/8) sesquan; very strongly calcareous; insepic intertextic in thin section, which shows clasts of carbonatite tuff in a matrix of rounded carbonatite lapilli; abrupt, irregular contact to

-47 cm; A horizon of underlying paleosol (type Kwar clay); silty claystone; yellowish red (5YR4/6) with scattered sand-sized clasts of white (10YR8/2), black (10YR2/1), and olive gray (5Y5/2); coarse granular peds, outlined by argillans and few black (10YR2/1) mangans; strongly calcareous; insepic intertextic in thin section; with abundant alvikite, common nephelinite, fenite, quartz, microcline, rare books of vermiculite; few calcite crystal tubes.

4.2.1.4 Further examples. Another profile, the Mobaw bouldery clay, was found lower in the same measured section (Fig. 4.11), within the paleokarst grikes (Fig. 2.24). This profile formed on carbonatite colluvium overlying a thick black and brown clayey zone that formed within slowly drained parts of the paleokarst depressions. A similar paleosol, also within a paleokarst on carbonatite tuff, was seen in the ridge west of the abandoned quarry 2 km north of Koru (grid reference 519822 on Muhoroni sheet). This is the "maize crib" fossil locality of H. L. Gordon, E. J. Wayland and L. S. B. Leakey, and Legetet locality 14 of Pickford (1984, 1986a).

4.2.1.5 Alteration after burial. These clayey profiles could have been compacted to as much as 75 percent of their former thickness, considering their depth of burial. However, the carbonatite clasts and

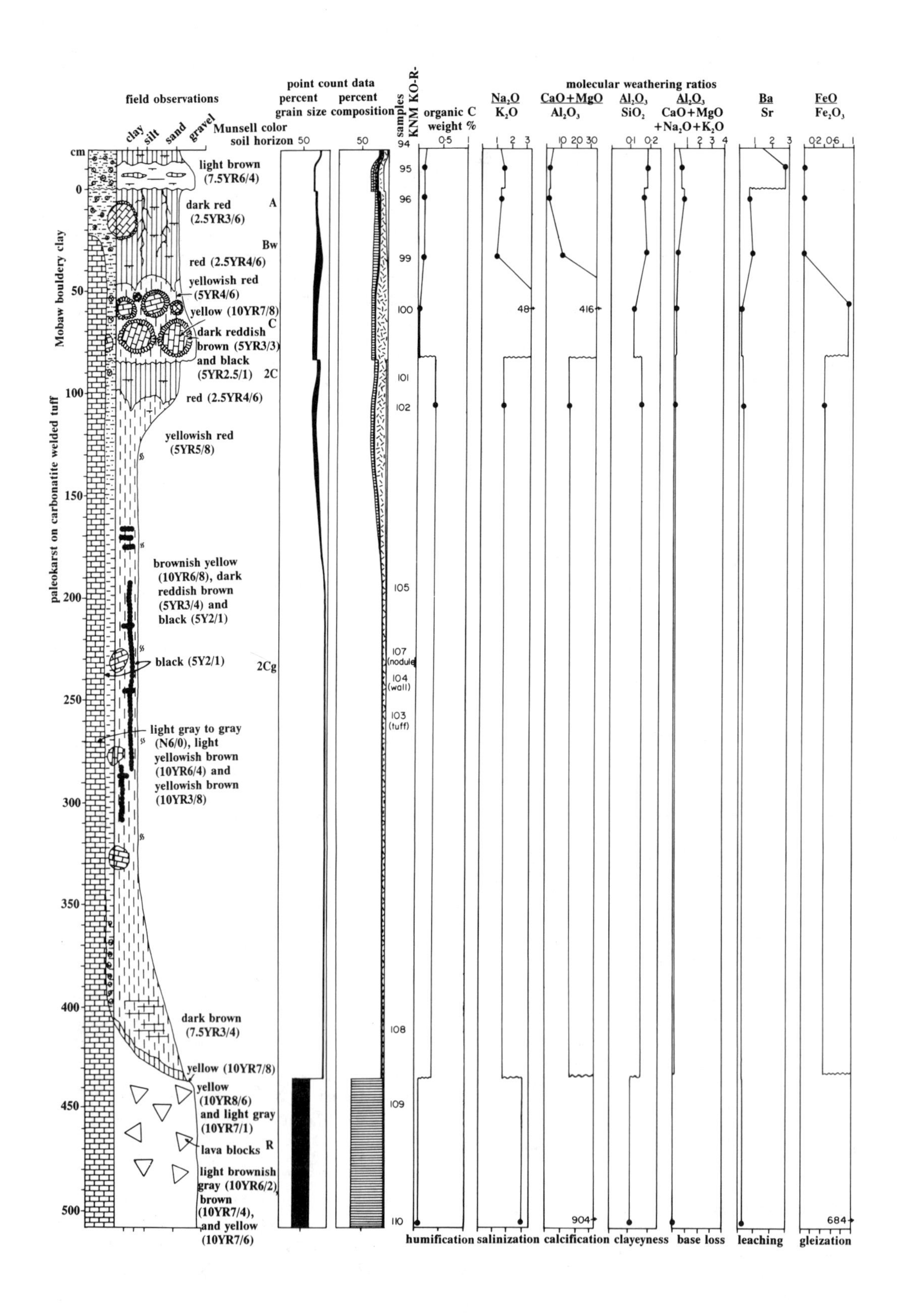
field observations
clay
silt
sand
gravel
Munsell color
soil horizon
point count data
percent grain size
percent composition
samples KNM KO-R-
organic C weight %
molecular weathering ratios
Na₂O / K₂O
CaO+MgO / Al₂O₃
Al₂O₃ / SiO₂
Al₂O₃ / CaO+MgO+Na₂O+K₂O
Ba / Sr
FeO / Fe₂O₃
cm
Mobaw bouldery clay
paleokarst on carbonatite welded tuff
light brown (7.5YR6/4)
dark red (2.5YR3/6)
A
Bw
red (2.5YR4/6)
yellowish red (5YR4/6)
yellow (10YR7/8)
C
dark reddish brown (5YR3/3) and black (5YR2.5/1)
2C
red (2.5YR4/6)
yellowish red (5YR5/8)
brownish yellow (10YR6/8), dark reddish brown (5YR3/4) and black (5Y2/1)
black (5Y2/1)
2Cg
light gray to gray (N6/0), light yellowish brown (10YR6/4) and yellowish brown (10YR3/8)
dark brown (7.5YR3/4)
yellow (10YR7/8)
yellow (10YR8/6) and light gray (10YR7/1)
lava blocks
R
light brownish gray (10YR6/2), brown (10YR7/4), and yellow (10YR7/6)
94
95
96
99
100
101
102
105
107 (nodule)
104 (wall)
103 (tuff)
108
109
110
48
416
904
684
humification
salinization
calcification
clayeyness
base loss
leaching
gleization

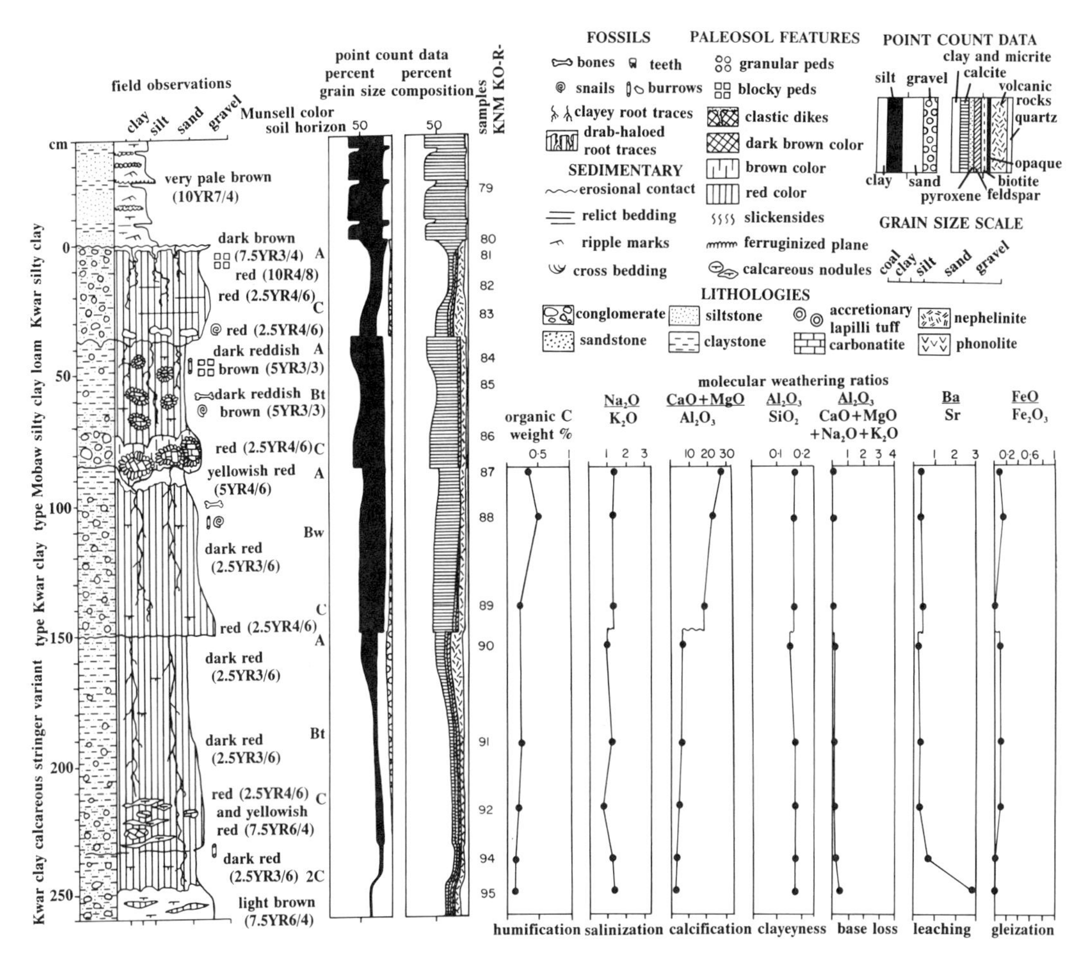

Fig. 4.12. Measured section, Munsell colors, soil horizons, grain size, mineral composition, organic matter, and selected molecular weathering ratios of the Kwar clay calcareous-stringer variant, type Kwar clay, type Mobaw silty-clay loam, and Kwar silty-clay paleosols, in an abandoned lime quarry 3 km north of Koru, Kenya. This section is 5.4 to 8.3 m in Fig. 4.5.

Fig. 4.11 (to left). Measured section, Munsell colors, soil horizons, grain size, mineral composition, organic matter and selected molecular weathering ratios of the Mobaw bouldery-clay paleosol and underlying paleokarst on carbonatite welded tuff, in an abandoned lime quarry 3 km north of Koru, Kenya. This section is 0.2 to 5.4 m in Fig. 4.5. Lithological key as for Fig. 4.12.

calcite crystal tubes are little deformed, and there is petrographic evidence of early cementation by calcite. Mobaw paleosols also have lost most of their organic matter and probably have suffered burial dehydration of ferric hydroxides, so that their original color may have been browner than their present red.

4.2.1.6 Reconstructed soil. The original Mobaw soils probably had a reddish-brown, weakly humified, fine blocky A horizon over a slightly more clayey Bt horizon that did not qualify as argillic. Scattered in the A horizon and common in subsurface horizons was a rubble of deeply weathered and ferruginized boulders of carbonatite tuff.

These soils probably were well drained (low ferrous iron/ferric iron ratios), with alkaline soil reaction and high cation exchange capacity (high alkaline earths to alumina), although the brown to yellow clay beneath the Mobaw bouldery clay has iron-manganese nodules already discussed as evidence of former slow drainage. Even here, however, ferric mottles and skelmasepic plasmic microfabric are evidence of periodic drying out. The soda/potash ratios of Mobaw profiles are high and change little within profiles, but are lower than for the carbonatite parent materials. Thus they probably do not represent salinization during soil formation, but rather limited desalinization of the contained carbonatite and nephelinite clasts.

4.2.1.7 Classification. Boulders of carbonatite in these paleosols are more thoroughly weathered than would be expected in Orthents of the U.S. taxonomy (Soil Survey Staff, 1975), or Lithosols of the Food and Agriculture Organization (1974) World Map and Australian soil classification (Stace *et al.*, 1968). This degree of weathering is more like that of a U.S. Inceptisol. Classification as a volcanic ash soil or Andept in the U.S. taxonomy was considered but rejected, because Andepts are found on basaltic to rhyolitic ash that weathers to distinctive low density and noncrystalline materials, unlike the Mobaw paleosols rich in silicates and carbonates. Many surface soils on peralkaline ash in Kenya today also have bulk densities too high for Andepts (Weilemaker & Wakatsuki, 1984). Among other Inceptisols, Tropept is more likely than Ochrept, considering the lack of evidence for strongly seasonal climate, as seen in Pakistani Miocene paleosols. Thus Mobaw paleosols are best regarded as Lithic Ustropepts in the U.S. taxonomy, and as Calcic Cambisols in the FAO World Map of soils.

In the Australian classification (Stace *et al.*, 1968) there is no separate provision for volcanic soils, and Mobaw paleosols are most like Red Calcareous Soils. These form by dissolution of limestone bedrock and by accumulation of eolian dust, and so tend to be less calcareous than Mobaw paleosols, which have a geologically unusual parent material. In the South African binomial soil classification (MacVicar *et al.*, 1977), Mobaw profiles are most like soils on limestone bedrock, such as Mispah muden. In the Northcote (1974) key, they are Um5.11.

4.2.1.8 Paleoclimate. Mobaw Series paleosols compare best with calcareous African soils of dry regions (Food and Agriculture Organization, 1977a). This is because of their unusual carbonatite parent material and is at variance with indications of a wet paleoclimate from the degree of weathering of mafic minerals, alteration of lithoclasts, and the deeply incised associated paleokarst (Fig. 2.24). The deep grikes, locally aggregated into complex depressions, indicate a paleoclimate that was subhumid or wetter (Choquette & James, 1987). Iron-manganese nodules in claystone within the grikes are indications of slow drainage and a moist subsoil.

The high local relief (5.8 m) and close spacing of the grikes in the paleokarst could be taken as an indication of a tropical temperature regime (Sweeting, 1973). However, the tabular surface of the carbonatite tuff has persisted between the grikes. This is a feature of both temperate and young karst landscapes (J. N. Jennings, 1985), so the paleoclimatic interpretation of this paleokarst remains uncertain.

4.2.1.9 Ancient vegetation. These weakly developed profiles on carbonatite colluvium contain both large root traces and numerous fine calcareous crystal tubes. These, together with evidence for a well drained paleotopographic position, are indications of upland colonizing forest with a good ground cover. The cinder cones on which they formed were not so high that Afromontane vegetation would be expected.

Surface soils with which these paleosols compare in Australia (Stace *et al.*, 1968),

United States (Soil Survey Staff, 1975) and Africa (Food and Agriculture Organization, 1977a) support vegetation ranging from dry forest to grassy woodland, wooded grassland and desert shrubland. The drier kinds of vegetation grow in soils that are calcareous even on noncalcareous parent materials, so that subhumid to humid forest is more likely for Mobaw paleosols that have been decalcified from an unusually calcareous parent material.

In Uganda today there are two different kinds of colonizing forest vegetation (Lind & Morrison, 1974): (1) colonizing forest of umbrella trees (*Maesopis emini*) on deep rich soils and (2) more open colonizing forest on shallow, rocky, usually lateritic soils. This second kind includes *Albizzia* spp., *Croton* spp., *Dombeya mukole*, *Olea welwitschii*, *Phyllanthus* spp., *Sapium ellipticum* and *Spathodea campanulata*. Very few of these genera have an early Miocene fossil record in Africa (only *Olea* of Chaney, 1933; *Albizzia*-type pollen of Muller, 1981), and there is no direct evidence of the taxonomic composition of the ancient vegetation on Mobaw Series paleosols. Nevertheless, these distinctive carbonatite soils may have been as difficult for plants as nutrient-poor lateritic soils. Nutrient cations were not lacking in carbonatites, but may have been difficult to obtain from volcanic ashes whose initial weathering tended to produce salts and calcareous cement (Hay, 1978, 1986). Widespread killing of crops and native grasses was observed after a light dusting of ash from the eruptions of Oldoinyo Lengai during 1940-1941 (Richard, 1942). Phosphate was abundant in these carbonatite soils, but would have been difficult to obtain in the presence of so much free carbonate (as indicated by a variety of studies on plant nutrition; Stevenson, 1986). Thus the second type of open colonizing forest on laterites now found in Uganda is a reasonable modern analog for the likely ancient vegetation of Mobaw paleosols. A more convincing modern analog would be provided by comparable African soils on carbonatite parent materials in a subhumid climate, but I have been unable to discover published reports of such a case.

4.2.1.10 Former animal life. Mobaw paleosols are less abundantly fossiliferous than associated Kwar paleosols in the abandoned quarry 3 km north of Koru (locality 10 of Pickford, 1984, 1986a), but yield a similar fauna. A more reliable guide to their former animal life is another locality closer to Koru ("maize crib," or Legetet locality 14 of Pickford, 1984, 1986a), where only a Mobaw paleosol is developed on paleokarst of carbonatite tuff. Here the Mobaw paleosol contains a variety of land snails and millipedes (Appendix 8). The vertebrate fauna is dominated numerically by remains of deerlike ungulate, *Walangania africanus*, and an early ancestor of chevrotains, *Dorcatherium songhorense*. Early Miocene chevrotain ancestors were much more diverse than living chevrotains, so it should not be assumed that these also were necessarily creatures of wet forests. Also found in Mobaw paleosols were remains of chameleons, monitor lizards, apes (*Micropithecus clarki*, *Kalepithecus songhorensis*, *Proconsul africanus*, *P. major*), archaic carnivore (*Teratodon spekei*), mongoose (*Kichechia zamanae*), and hedgehog (*Gymnurechinus leakeyi*).

Also found in Mobaw paleosols were a variety of invertebrate trace fossils: subhorizontal galleries referable to the ichnogenus *Planolites* (Häntzschel, 1975), tear-shaped internal casts of *Celliforma* (Retallack, 1984b), and ovoid internal casts of a type unnamed, but similar to trace fossils illustrated by Freytet and Plaziat (1982). The galleries are like those of termites (Bown, 1982; Lee & Wood, 1971; Sands, 1987) and the *Celliforma* chambers have internal polish like the larval cells of bees (Retallack, 1984b). The ovoid internal casts found widely in Miocene paleosols of southwestern Kenya have been interpreted as cocoons (Pickford, 1984, 1986a). They are very similar to cocoons of beetles such as chrysomelids, but a variety of African moths and large flies also make similar ellipsoidal cocoons in the ground (Skaife, 1953).

4.2.1.11 Paleotopographic setting. Local paleorelief on the paleokarst below the Mobaw bouldery clay was measured at 5.8 m (Fig. 2.24). The whole Legetet Carbonatite is only 60-m thick (Pickford & Andrews, 1981). The cinder cone that it formed probably was at least several hundred meters above its surroundings. At the time the Mobaw bouldery-clay paleosol formed, there was at least 1 m of local paleokarst protruding locally from the paleosol. By the time of the Mobaw silty-clay loam, local paleokarst relief was covered by colluvial and volcanic rocks, but the weathered carbonatite blocks within it are an indication of paleokarst exposed further upslope.

Every part of Mobaw paleosols was above the water table, because they are strongly ferruginized. Even the weakly permeable brown clays within the grikes of the paleokarst have strongly developed illuviation argillans and skelmasepic plasmic microfabric, as indications of periodic drainage and drying out. Thus it is likely that Mobaw paleosols accumulated on the flanks of carbonatite tuff cones well above the influence of permanent streams or lakes.

4.2.1.12 Parent material. Boulders within and paleokarst below Mobaw paleosols are of partly agglutinated carbonatite tuffs, which were hard cemented rocks at the time these paleosols formed. Some of the sand and silt sized grains of carbonatite in these paleosols also may have been derived from these tuffs, but there also is a component of nephelinite, quartz, microcline, and vermiculite not seen in thin sections of the carbonatite tuffs. A substantial nephelinitic component near the surface of these paleosols also is indicated by higher silica and alumina and lower lime and barium near the top compared with the base of the Mobaw bouldery-clay (Fig. 2.19; Appendices 5 and 6). There are similar chemical differences between local nephelinite lavas and carbonatite tuffs (Binge, 1962; Deans & Roberts, 1984; Richard, 1942). Thus there were additions of nephelinitic airfall ashes, probably from a separate nearby volcano of that composition. The microcline and quartz grains are more coarse-grained than could have been lifted by wind, and more angular and broken than fluvially transported grains, as seen in the Grey Sandstone Member of the Kapurtay Agglomerate. Probably they represent fragments of fenitized Precambrian wall rocks of the carbonatite-nephelinite vents, broadcast during explosive eruptions of nephelinite or carbonatite. Mobaw paleosols formed on a mixture of these alkaline volcanic rocks.

4.2.1.13 Time for formation. Although showing discernible clay formation and reddening, Mobaw paleosols lack well-differentiated horizons, such as argillic or calcic horizons (in the sense of Soil Survey Staff, 1975). They are thus only weakly developed (Retallack, 1990a), and so represent a relatively short time of formation. These paleosols are at least as weathered as the 30,000-400,000-year-old Yellow Tuffs and Agglomerates, and certainly are more weathered than the 2000-30,000-year-old Gray Tuffs and 1250-2000-year-old Black Tuffs and Agglomerates in the desertic climate around the active carbonatite-nephelinite volcano Oldoinyo Lengai, Tanzania (Hay, 1989). Mobaw paleosols are less developed than Alfisols on phonolitic colluvium less than 40,000 years old in the humid and cool lower Teleki Valley of Mt Kenya (profile above paleosol TV23 of Mahaney, 1989; Mahaney & Boyer, 1989). None of these modern soils are similar in all of their soil forming factors to Mobaw paleosols, but they do constrain its time of formation to a few tens of thousands of years.

4.2.2 Kwar Series

4.2.2.1 Diagnosis. Weakly developed, clayey, red (2.5YR) paleosols, on carbonatite-nephelinite tuffs.

4.2.2.2 Derivation. Kwar is Dholuo for red (Gorman, 1972).

4.2.2.3 Description. The top of the type Kwar clay (Figs 2.24, 4.5, and 4.12) is 85-cm below the top of the sequence of red paleosols exposed above the karstified tuff of early Miocene Legetet Carbonatite in the central part of the abandoned quarry (grid reference 517829 on Muhoroni sheet), 3-km north of Koru (Figs 4.4 and 4.6).

+14 cm; C horizon of overlying paleosol (type Mobaw silty-clay loam); pebble-bearing clayey siltstone; red (2.5YR4/6); with abundant, grain-supported pebbles of light-gray (2.5Y7/2) carbonatite speckled with black (5Y2/1) opaque oxides, and with irregularly shaped margins, bearing a thin (1-2 mm) red (2.5YR4/8) sesquan; very strongly calcareous; insepic intertextic in thin section; most grains are subangular to rounded clasts of carbonatite tuff; abrupt, irregular contact to

0 cm; A horizon; silty claystone; yellowish red (5YR4/6), with scattered, sand-sized clasts of white (10YR8/2), olive gray (5Y5/2) and black (10YR2/1); common calcareous and clayey root traces; coarse granular peds, outlined by argillans and few black (10YR2/1) mangans; strongly calcareous; insepic intertextic in thin section, with abundant grains of alvikite carbonatite, common grains of nephelinite, fenite, quartz, and microcline, and rare books of vermiculite; few calcite crystal tubes after root traces; this surface layer is distinctly more yellowish than the rest of the profile and has an irregular clear contact to

-8 cm; Bw horizon; granule-bearing clayey siltstone; dark red (2.5YR3/6); with clasts white (10YR8/2), reddish brown (5YR5/4) and olive (5Y5/4); common fine root traces, and occasional stout ones, filled with dark red (2.5Y3/6) clay and red (2.5YR4/6) micrite; common fossil bones, teeth, and snails; few fine distinct black (10YR2/1) mottles of iron-manganese; sparse books of vermiculite ; some burrows (ortho-isotubules) 4 to 11 mm in diameter; indistinct medium angular blocky peds, defined by thin clay skins (sesquiargillans); strongly calcareous; calciasepic intertextic in thin section; most clasts are alvikite with trachytoidal-textured laths of calcite after nyerereite; common nephelinite, quartz, and microcline; fine calcite crystal tubes and fine ferric nodules; irregular gradual contact to

-52 cm; C horizon; granule-bearing clayey siltstone; red (2.5YR4/6); with clasts up to 4-mm across of reddish brown (5YR5/4), brown (10YR4/3), white (10YR8/1), and black (10YR2/1); granular texture with indistinct bedding; few indistinct ortho-isotubules as in horizon above; strongly calcareous; insepic intertextic; abundant carbonatite clasts in the form of "Pelee's tears"; common quartz, microcline, biotite, and books of vermiculite; clear, irregular (around pedotubules penetrating down) contact to

-65 cm; A horizon of underlying paleosol (Kwar clay calcareous-stringer variant); sandy and silty claystone; dark red (2.5YR3/6); with grains of dusky red (10YR3/3), yellow (10YR7/6), white (10YR8/1), black (10YR2/1), and dark greenish gray (5BG4/1); weakly defined coarse granular peds and abundant fine clayey root traces; strongly calcareous; insepic to calciasepic intertextic in thin section; grains almost equally alvikite and nephelinite; common vermiculite, quartz, and microcline; scattered irregular opaque mottles after fine root traces.

4.2.2.4 Further examples. The type Kwar clay has the best preserved A horizon and is also the most fossiliferous layer in the sequence of similar weakly developed paleosols above the paleokarst on the carbonatite tuffs in the abandoned quarry 3 km north of Koru (Figs. 2.24 and 4.5). Underlying it is another similar profile, the Kwar clay calcareous stringer variant. This is an intergrade toward Kiewo Series paleosols in its scattered displacive veins of bladed calcite, but it lacks the calcareous pervasively displacive concentrations of veins found in Kiewo Series. The Kwar silty clay is another comparable profile at the top of the sequence of paleosols exposed in the quarry. Both the Kwar silty clay and Kwar clay calcareous stringer variant have a distinct clayey subsurface horizon (Fig. 4.12), but the paucity of argillans and other soil structures in these profiles are indications that it probably did not qualify as argillic. These Kwar paleosols can be regarded as intergrades toward Choka paleosols in which argillic horizons are present.

4.2.2.5 Alteration after burial. All the alterations outlined for associated Mobaw Series paleosols also affected Kwar profiles, for which losses of organic matter and dehydration of ferric hydroxides may have been more profound. In addition, the yellowish color of the surface horizon of the type Kwar clay may be a result of burial gleization.

4.2.2.6 Reconstructed soil. Kwar paleosols can be envisaged as reddish-brown soils on clayey carbonatite-nephelinite ash, with dark and moderately organic A horizons over a more reddish Bw horizon, and with limited subsurface reorganization of carbonate. They were very calcareous even in areas away from the early cement of sparry calcite, so probably had an alkaline pH and high cation exchange capacity. Molecular weathering ratios (Fig. 4.12) reveal negligible leaching, base depletion, or gleization. Soda/potash ratios are uniformly high, and some increase slightly toward the surface. This may reflect slight salinization from addition of fresh volcanic ash.

4.2.2.7 Classification. Kwar paleosols do not have large boulders of carbonatite but are similar in many ways to Mobaw Series. Both their subsurface clayey (Bw) and calcareous (Bk in Kwar clay calcareous stringer variant only) horizons are weakly expressed. For much the same reasons outlined for Mobaw paleosols, Kwar paleosols are best identified as Ustropepts (of Soil Survey Staff, 1975), Calcic Cambisols (of Food and Agriculture Organization, 1974), and Calcareous Red Soils (of Stace *et al.*, 1968). In the Northcote (1974) key, Kwar paleosols would have been Uf5.31. Among South African soils (MacVicar *et al.*, 1977), they are most like Oakleaf makulek.

4.2.2.8 Paleoclimate. Similar surface soils in Africa today (Food and Agriculture Organization, 1977a) are distributed in climates ranging from humid temperate to hot tropical deserts, with an emphasis on the drier end of the spectrum (less than 500-mm mean annual rainfall). However, most of these soils accumulated their carbonate in a dusty dry climate, whereas Kwar paleosols inherited carbonate with their unusual carbonatite-nephelinite parent materials. Evidence of a wet climate for Kwar paleosols includes about half the carbonate content and soda/potash ratios in some of these profiles (Kwar clay calcareous-stringer variant and Kwar silty clay) compared with slightly less-developed profiles (type Kwar clay). To this may be added the stout root traces, clay skins, weathered grains, and remobilization of carbonate seen in thin sections of these paleosols. On the other hand, their paleoclimate probably was not humid either. Quaternary carbonatite tuff has been deeply decalcified under secondary grassland near Fort Portal, Uganda, where mean annual rainfall is 1400 to 1500 mm (Barker, 1989; Harrop, 1960; White, 1983). Similarly decalcified are red soils formed under calcareous coastal sands some 125,000 to 240,000 years old under humid forests of Kenya and Somalia receiving 1200 mm per annum (map unit Qc37-1a of Food and Agriculture Organization, 1977a; Braithwaite, 1984; Muchena & Sombroek, 1982; Müller, 1982). In addition, the dominance of smectite over other clay minerals in the type Kwar clay is an indication of mean annual rainfall of less than 1200 mm, by comparison with other East African volcanic soils (Mizota & Chapelle, 1988; Mizota, Kawasaki, & Wakatsuki, 1988). Thus mean annual precipitation was probably within the subhumid range.

Climate also is likely to have been warm, although definitive evidence for this was not found. Suggestive hints include the abundance of burrows, including some attributable to termites, and of tree root traces, indicating a productive ecosystem. The surface of the vermiculite books and other volcanic grains with unoxidized interiors are strongly ferruginized. Chemically also (Fig. 4.12), the thorough oxidation of such weakly developed soils is impressive, as is the case for tropical soils (Birkeland, 1984).

4.2.2.9 Ancient vegetation. These brick red paleosols contain large, deeply penetrating root traces and incipient development of a clayey subsurface (Bt or Bw) horizon, as is typical for forested soils. The root traces are not drab haloed, and thus are not conspicuous among the abundant burrows. Kwar paleosols are less developed than otherwise comparable paleosols at Songhor (Choka Series), and some of them are more strongly developed than associated Mobaw paleosols. They probably supported upland, but not montane, forest intermediate in character between colonizing and old-growth forest.

Comparable surface soils in the United States (Soil Survey Staff, 1975), Australia (Stace *et al.*, 1968) and Africa (Food and Agriculture Organization, 1977a) support dry woodland, wooded grassland or desert shrubland. Unlike most of these soils, however, formation of Kwar paleosols involved destruction of carbonate rather than its accumulation. The uppermost horizons of Kwar paleosols show blocky structure, rather than the granular structure of grassland soils (Duchafour, 1985), or paleosols (Onuria and Chogo Series). These comparisons with soils of open vegetation overlook the unusual carbonatite-nephelinite parent material of Kwar paleosols. A better comparison is with red soils on Quaternary calcareous sands of coastal Kenya and Somalia (map unit Qc37-1a of Food and Agriculture Organization, 1977a; Muchena & Sombroek, 1981) under tropical semideciduous forest (vegetation unit 16b of White, 1983). Typical large tree species in this area include *Aningeria pseudoracemosa*, *Antiaris toxicaria*, *Burttdavya nyasica*, *Chlorophora excelsa**, *Khaya nyasica*, *Laroa swynnertonii*, *Maranthus (Parinari) goetzeniana*, *Newtonia buchananii**, *Parkia filicoides**, *Ricinodendron heudelotii**, *Sterculia appendiculata* and *Terminalia zambeziaca*. Paleobiogeographically this is not a far-fetched comparison, because there are numerous shared species (25.8% of forest trees, including those indicated above by asterisk) between these coastal forests of Kenya and those of the Guineo-Congolian floristic realm to the west. A mid-successional forest similar to that suggested for Mobaw Series paleosols also is possible for Kwar profiles. These general comparisons are useful constraints, but vegetation of Kwar paleosols was probably an open forest as unique botanically as its volcanic parent material is petrographically.

The Kwar silty clay paleosol has yielded a single plant fossil: a hackberry (*Celtis*) endocarp (Kenyan National Museum specimen KO-F22651). This specimen is larger (7.4-mm long by 6.8 mm in diameter) than early Miocene (about 17 million years old) endocarps of *C. rusingensis* previously described from Rusinga Island, Kenya (Chesters, 1957). The Koru fossil also shows four distinct radial ribs. Such ribs are found in *C. durandii*, which has fruits about half the size of the fossil, and in *C. mildbraedii*, which has the most similar fruits among living plants. *Celtis mildbraedii* is a seasonally deciduous tree with a buttressed trunk. It grows to 45-m tall in lowland (elevation 300-1600 m) forest of East Africa (Polhill, 1966). This and other species of hackberry dominate mixed forests intermediate in successional status between colonizing forest and old-growth forest of ironwood (*Cynometra alexanderi*) in East Africa (Lind & Morrison, 1974). These are within the drier peripheral Guineo-Congolian floristic region (of White, 1983), and in Uganda include the following common trees: *Alstonia congoensis*, *Celtis mildbraedii*, *Chrysophyllum* spp., *Cynometra alexanderi*, *Khaya anthotheca*, and *Trichilia prieuriana*. Some 40 to 50 species of trees are common in these mixed, mid-successional forests, and comparable diversity is likely also for the vegetation of Kwar paleosols. Even if hackberry endocarps were common in Kwar paleosols it should not be assumed that these were prominent in the original vegetation, because their siliceous and calcareous endocarps are preferentially preserved over remains of other plants

(Retallack, 1984a).

From evidence of fossil plants found elsewhere in Africa, at least a few of the genera of open colonizing and mixed mid-successional forest in Uganda and of semideciduous forests of coastal Kenya are known to have existed during early Miocene time. Leaflets of *Cynometra* have been reported from Oligocene shales in Egypt (Bown *et al.*, 1982) and fossil wood similar to that of *Newtonia buchananii* from early Miocene (probably about 19 million years old, based on faunal correlations by J.A. Van Couvering, 1972) beds at the base of Bogoro Scarp, Lake Mobutu (formerly Lake Albert; Lakhanpal & Prakash, 1970). Locally abundant fruits and seeds of *Celtis*, *Terminalia*, and *Sterculia* some 18 million years old have been found on Mfangano Island, Kenya (Chesters, 1957; Collinson, 1983). The sequence on Mfangano Island includes red beds that may prove to be similar to Kwar paleosols. Considering the potential for preservation of organic remains in carbonatite-melilitite ash (Hay, 1986), it may be possible with detailed studies there to establish more convincingly the botanical nature of ancient vegetation on red paleosols like the Kwar Series.

4.2.2.10 Former animal life. Kwar paleosols, especially the type Kwar clay and Kwar clay calcareous-stringer variant, are abundantly fossiliferous with snail shells, millipede exuviae, and bones of lizards, snakes, birds, and mammals (Appendix 8: Legetet locality 10 of Pickford, 1984, 1986a). Especially common snails are the very high-spired subulinids (Subulininae sp. indet.), large achatinids (*Burtoa nilotica* and *Achatina leakeyi*), ovoid streptaxids (*Gonaxis protocavalii*), and small pupiform streptaxids (*Primigulella miocenica*). Reptile and bird remains have not yet been studied in detail, but may include chameleons, monitor lizards, pythons, cobras, pigeons, passerines, turacos, hornbills, and guinea fowl. The most abundant vertebrate remains are deerlike ungulates, *Walangania africanus*, and chevrotains, *Dorcatherium songhorense*. Other large animals include gomphothere elephants (cf. *Archaeobelodon*) and chalicotheres (*Chalicotherium rusingense*). Common insectivores include giant elephant shrew (*Miorhynchocyonclarki*), hedgehog (*Amphechinus rusingensis*), tenrec (*Protenrec tricuspis*), and golden mole (*Prochrysochloris miocaenicus*). Fossil bats found include insectivorous tomb bats and large fruit bats. Archaic carnivores are represented by *Kichechia zamanae*, and apes include *Micropithecus clarki*, *Kalepithecus songhorensis*, *Limnopithecus legetet*, *Proconsul africanus* and *P. major*.

Kwar paleosols also contain a variety of invertebrate trace fossils. Especially distinctive are irregularly shaped, calcareous, ovoid masses, 6-8-cm across by 2-3-cm thick, attached to a network of calcareous tubules, 3 to 4 mm in diameter, that ramify in all directions through the paleosol. These traces are generally similar to the ichnogenus *Termitichnus*, but are flatter than described examples thought to be fungus gardens of termites (Bown, 1982). The Koru trace fossils are more like food storage chambers of harvester termites (such as *Hodotermes*) than the large fungus gardens of mound-building termites (such as *Macrotermes*; Sands, 1987; Skaife, 1953).

Most of the subhorizontal galleries, 3 to 4 mm in diameter, referable to *Planolites* (of Häntzschel, 1975), probably also are the work of termites, because some connect to *Termitichnus* structures. Other near vertical, unbranched burrows, up to 10 mm in diameter and also referable to *Planolites*, could equally be the work of insects, spiders, or millipedes, which are known to create similar traces (Ratcliffe & Fagerstrom, 1980).

Tear-shaped internal casts with polished surfaces also are common in Kwar paleosols (Kenyan National Museum field numbers LG450, LG451, collection 10/660). These are similar to the ichnogenus *Celliforma*, thought to be the work of wasps and bees (Retallack, 1984b). Those seen in the field were isolated rather than arranged in complex patterns seen, for example, in nests of living sweat bees (Sakagami & Michener, 1962) and similar Miocene fossil examples from Rusinga Island, Kenya (Thackray, 1989). The Koru examples, with their polish, may be larval cells of solitary bees.

Also found were ovoid internal casts 12-17-mm long and 4 to 6 mm in diameter (comparable to Kenyan National Museum specimen LG453) as well as spherical casts 5 to 6 mm in diameter. The spherical casts may be large eggs of achatinid snails or of geckos, also recorded from these paleosols (Hirsch & Harris, 1989). Some of the more common ellipsoidal casts show small ridges and tears that indicate a flexible, silklike coat, rather than the gelatinous cover of snail eggs or the brittle and thin coat of gecko eggs. Most of these ellipsoidal casts probably are the internal fill of insect cocoons, known also in other paleosols of Kenya (Pickford, 1984, 1986a) and France (Freytet & Plaziat, 1982). Living chrysomelids and other beetles make very similar cocoons, as do a variety of African moths and flies (Skaife, 1953).

4.2.2.11 Paleotopographic setting. Kwar paleosols probably formed on the flanks of the same carbonatite cinder cones envisaged for associated Mobaw paleosols. Kwar paleosols, however, formed in areas of less prominent paleokarst, probably around the footslopes of the cinder cones where older welded carbonatite tuffs were mantled with nephelinitic airfall ash from a nearby stratovolcano. No geochemical or petrographic traces of gleization were detected in Kwar Series paleosols (Fig. 4.12), which were probably all well drained.

4.2.2.12 Parent material. This also was similar for both Kwar and Mobaw paleosols, although Kwar profiles had fewer boulders of carbonatite tuffs. Individual profiles vary in the mix of nephelinitic airfall ash to the carbonatite. The type Kwar clay is dominated by carbonatite, whereas the Kwar clay calcareous-stringer variant is chemically more like local Kenyan nephelinites (Fig. 2.19 and Appendices 4 and 5, compared with analyses reported by Binge, 1962; Deans & Roberts, 1984; Richard, 1942).

4.2.2.13 Time for formation. The type Kwar clay paleosol is developed to about the same extent as associated Mobaw paleosols. The Kwar silty-clay and Kwar clay calcareous-stringer variant are a little more developed, but still weakly developed (in the sense of Retallack, 1988a), and well short of the development seen in Choka Series. For reasons outlined in the discussion of these other paleosols, Kwar paleosols probably represent a few tens of thousands of years of soil formation.

4.2.3 Kiewo Series (new name)

4.2.3.1 Diagnosis. Thin, clayey, calcareous, red (5YR) paleosols, with conspicuous relict bedding and shallow calcareous hardpans.

4.2.3.2 Derivation. Kiewo is a Dholuo word for striped (Gorman, 1972). The name is a reference to the calcareous bands of these paleosols.

4.2.3.3 Description. The type Kiewo clay paleosol (Fig. 4.13) is below the type Choka clay (below 5.5 m in Fig. 4.5) within the early Miocene Red Bed Member of the Kapurtay Agglomerate, in the westernmost exposure of Songhor gullies (area 1 of Pickford & Andrews, 1981; or grid reference 459965 on Muhoroni sheet), 2.5 km north of Songhor village, Kenya (Figs 4.4 and 4.5).

+11 cm; Ck horizon of overlying paleosol (type Choka clay); clayey siltstone; light brownish-gray (10YR6/2); clear relict bedding, and some ball-and-pillow structure (soft sediment deformation); strongly calcareous; crystic porphyroskelic in thin section, with common nephelinite rock fragments, quartz, microcline, and pyroxene in clayey matrix riddled with sparry pervasively displacive calcite; abrupt wavy contact to

0 cm; A horizon; sandy claystone grading down to clayey medium to coarse grained sandstone; reddish-brown (5YR4/4), with sandy burrows and root traces of light gray (10YR7/2) and grayish-brown (10YR5/2); moderately calcareous; skelmosepic porphyroskelic in thin section, with abundant quartz, microcline, pyroxene, and ferruginized grains of nephelinite; common calcite crystal tubes, after roots; some circumgranular displacive sparry calcite; clear, wavy contact to

-12 cm; Ck horizon; medium to coarse grained sandstone; light gray (10YR7/2); local stain (neomangans) of dark grayish-brown (10YR4/2) and root traces of grayish brown (10YR5/2); few interbeds of reddish-brown (5YR4/4) claystone, this alternation of clayey and sandy beds shows low-angle cross bedding and ripple-drift cross lamination in places; root traces and burrows abundant in upper horizons are rare and deflected in the calcareous sandy beds; strongly calcareous overall; crystic porphyroskelic in thin section, with abundant deeply ferruginized fragments of nephelinite, rare quartz, and pyroxene, riddled with veins of pervasively displacive sparry calcite; abrupt, wavy boundary to

-36 cm; A horizon of underlying paleosol (Kiewo clay single-stringer variant); clayey coarse-grained sandstone; reddish-brown (5YR4/4); common root traces and burrows of very pale brown (10YR7/3) sandstone, with thin (2-3 mm) drab haloes of grayish-brown (10YR5/2) claystone; traces of relict bedding are extensively disrupted by burrows and root traces; moderately calcareous; skelmasepic agglomeroplasmic in thin section, with common nephelinite rock fragments, sövite, quartz, microcline, and feldspar in a ferruginized clayey matrix.

4.2.3.4 Further examples. Many additional Kiewo paleosols underlie the type profile in the measured section in Songhor gullies (Fig. 4.5), but only two were studied in sufficient detail to receive distinctive names: in order down from the type Kiewo clay, the Kiewo clay single-stringer variant and the Kiewo clay thin-surface phase (Fig. 4.13). The calcareous stringers that distinguish these paleosols from Buru Series have prominent relict bedding and pervasively displacive fabric, unlike nodular and massive carbonate in Tut and Choka paleosols.

4.2.3.5 Alteration after burial. Kiewo paleosols probably have lost much organic matter, have been reddened from burial dehydration of ferric hydroxides, and were locally discolored by burial gleization around root traces. Compaction of clayey parts of paleosols at Songhor may have been to as much as 72 percent of former thickness. Compaction and calcite cementation during burial may not have been as significant for these as for other paleosols at Songhor. The calcareous stringers with pervasively displacive fabric underwent dramatic expansion unlikely under thick overburden (Fig. 2.17), and they are avoided by many root traces and burrows. These stringers can be regarded as an originally cemented tuffaceous layer of parent material not yet disrupted by soil formation. In addition, the outcrops at Songhor have been disrupted by faults of only a few centimeters of displacement into blocks a few tens of meters in size (mapped by Pickford & Andrews, 1981). Some these minor fault planes have been injected by silty and sandy sediment. This brittle deformation may have been associated with uplift of nearby Songhor Hill some 9 to 6 million years ago (W. B. Jones & Lippard, 1979; Pickford, 1982).

4.2.3.6 Reconstructed soil. Kiewo paleosols can be reconstructed as thin reddish-brown clayey soils with shallow calcareous hardpans. Their A horizons are identifiable from abundant root traces and burrows. Their calcareous hardpans show relict bedding and pervasively displacive fabric related to their carbonatite parent material. Thus they are best designated Ck rather than Bk horizons. Molecular weathering ratios support the idea that these were weakly developed soils: barium/strontium ratios show marked leaching of these sensitive elements, but ratios involving alumina and bases show little depletion of alkalis and alkaline earths and only modest clay formation. From their abundant carbonate they were probably alkaline and base rich. Organic carbon in Kiewo paleosols, while low for soils, is nevertheless high for red Kenyan paleosols (Fig. 4.13). This and a moderately high ferrous iron/ferric iron ratio and abundant drab-haloed root traces and burrows may be

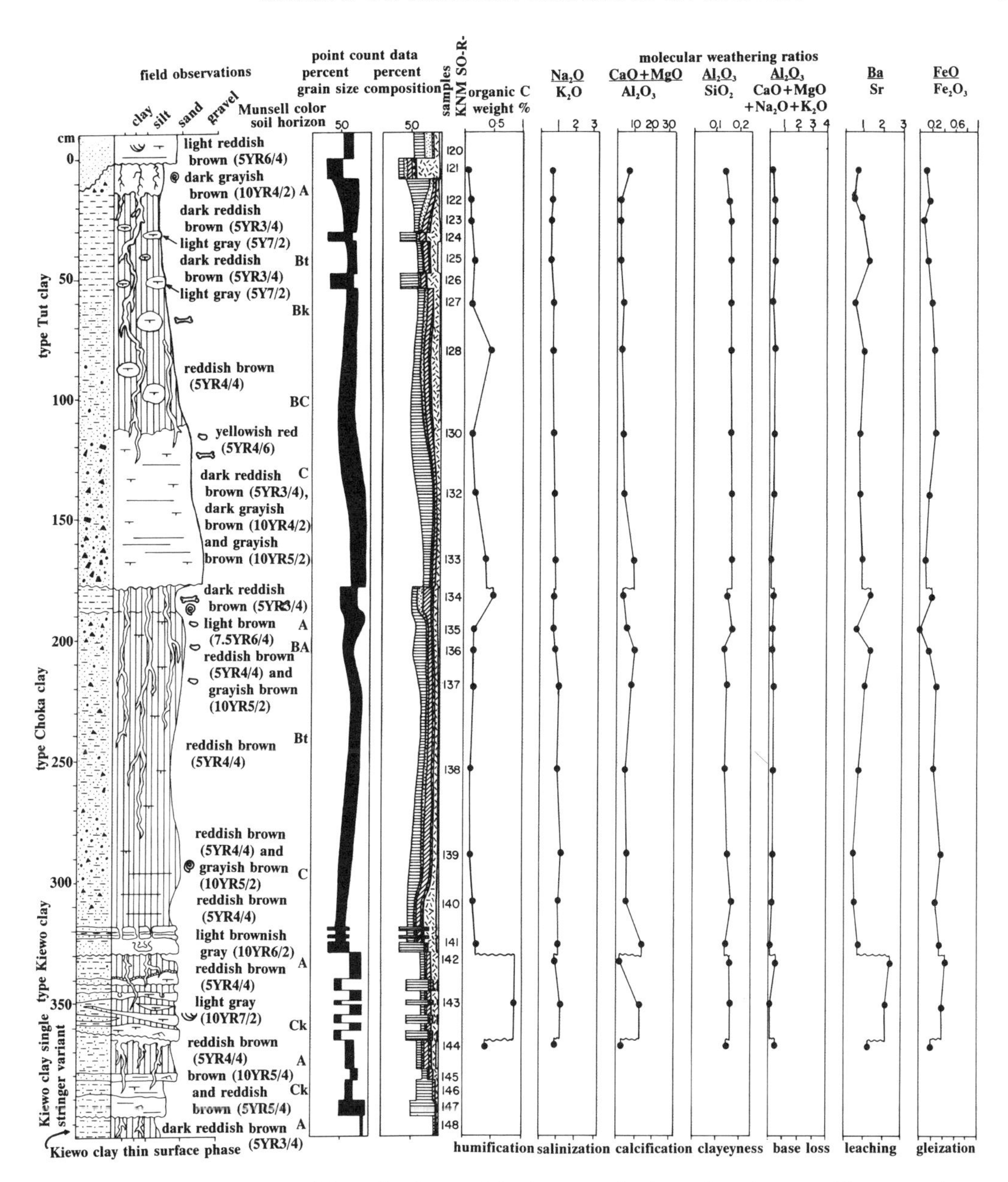

Fig. 4.13. Measured section, Munsell colors, soil horizons, grain size, mineral composition, organic matter, and selected molecular weathering ratios of the Kiewo clay thin-surface phase, Kiewo clay single-stringer variant, type Kiewo clay, type Choka clay and type Tut clay paleosols, in the westernmost exposures at Songhor National Monument, Kenya. This section is 2.2 to 6.5 m in Fig. 4.5. Lithological key as for Fig. 4.12

indications of some waterlogging. However, their red and oxidized state is an indication that they were dry for most of the year. No trace of salinization can be seen from the soda/potash ratios, which decline up-profile and so show instead leaching of salts.

4.2.3.7 Classification. Kiewo paleosols with their unusual pervasively displacive calcareous stringers do not fit comfortably in soil classifications. They have few, rather than 60 percent volcanic fragments as required for Andepts in the U.S. taxonomy (Soil Survey Staff, 1975), but then the pervasively displacive fabric could be regarded as evidence for high porosity and low bulk density of the soil. This idea could only be sustained if the bladed sparry calcite were regarded as a cement formed early after burial, and this does not seem to be the case, as previously discussed (see also Fig. 2.16). Carbonatites and nephelinites probably did not weather to low-density amorphous materials, and their soils are as unlikely to qualify as Andepts or Andosols as East African surface soils on peralkaline volcanic rocks (Weilemaker & Wakatsuki, 1984). Kiewo paleosols are better identified as Fluvents (of Soil Survey Staff, 1975), Calcaric Fluvisols (Food and Agriculture Organization, 1974), Alluvial Soils (of Stace *et al.*, 1968), Uf1.33 (of Northcote, 1974), and Dundee dundee (of MacVicar *et al.*, 1977). These identifications stress their relict bedding and lowland alluvial setting, rather than volcanic contributions to their parent material.

4.2.3.8 Paleoclimate. Such weakly developed paleosols were not formed for long enough to be a good record of paleoclimate. Their prominent mottling, high organic carbon content and moderately high ferrous iron/ferric iron ratio, may indicate some seasonal waterlogging. The pervasively displacive fabric also could be taken as evidence for more humid weathering than the formation of less pervasively displacive fabrics in calcite-cemented tuffs with distinctive cracking patterns in the dry modern climate around Oldoinyo Lengai, and the dry Plio-Pleistocene paleoclimate envisaged for the Laetoli Beds, in Tanzania (Hay, 1978, 1989). I could not find a report on pervasively displacive fabric in surface soils, but reexamination of soils on carbonatite-nephelinite volcanics in humid southwestern Uganda would be a possible test of this hypothesis.

4.2.3.9 Ancient vegetation. These red paleosols with their prominent relict bedding, subtle geochemical signs of periodic waterlogging, and scattered large drab-haloed root traces probably supported early successional lowland woody vegetation. This may have lined streams, where flooding and alluvial deposition were frequent.

Comparison of these paleosols with surface soils is not especially revealing, because they do not have close modern analogs and because such calcareous alluvium is unusual outside dry climates. Calcaric Fluvisols in Africa now support Somali-Masai *Acacia-Commiphora* wooded grassland and desert scrub of thorntrees and succulents (Food and Agriculture Organization, 1977a). In the United States such calcareous Fluvents are widespread in grassland and desert regions. As for Kwar and Mobaw Series paleosols, these comparisons do not take into account the unusual carbonatite-nephelinite parent material of Kiewo paleosols.

A more likely vegetation of Kiewo paleosols is very early stages in the ecological succession toward the dry forests envisaged for associated Tut and Choka Series paleosols. Considering their prominent relict bedding and other signs of youth, Kiewo soils probably supported vegetation cover even less developed than that for Kwar and Mobaw Series paleosols. A typical ecological succession on disturbed ground, such as river banks, in forested regions of East Africa begins with an elephant grass (*Pennisetum purpureum*) thicket. This is then invaded by shrubs such as *Acanthus pubescens*, *Vernonia* spp., *Clausena* spp., and *Acalypha* spp. Later in the succession colonizing trees appear; usually *Albizzia zygia* or *Maesopis emini*, together with *Celtis africana*, *Dombeya mukole*, *Margaritaria discoides*, *Prunus africanus*, *Polyscias fulva*, *Sapium ellipticum*, *Teclea nobilis*, *Blighia unijugata*, and *Markhamia platycalyx* (Lind & Morrison, 1974). As discussed for Mobaw Series paleosols, the carbonatite-nephelinite parent material and hardpans of Kiewo paleosols would have been difficult for plants. Its shrubs and trees may have been more scattered than for early successional vegetation on nephelinitic sandstones more deeply weathered of salts and hardpans, such as Buru Series paleosols.

Of the various living plants mentioned, *Celtis* certainly was present in East Africa during early Miocene time, as noted for Kwar paleosols. Leaves of *Markhamia* and *Teclea* have been found as old as 12.2 million years near Ngorora, Kenya (B. F. Jacobs & Kabuye, 1987). Fossil wood and seed pods similar to those of *Albizzia* have been reported from Zaire and Ethiopia (by Lakhanpal, 1966; Lemoigne & Beauchamp, 1972) in deposits of uncertain age, but probably late Miocene (3-11 and about 8 million years, respectively, to judge from radiometric dating by Bellon & Pouclet, 1980; W. H. Morton *et al.*, 1979). Although the fossil record of plants likely to have grown in Kiewo paleosols is poor, the nature of Miocene vegetation early in plant succession toward forest may yet be revealed in detail because of the unusual preservation of organic remains in some carbonatite-nephelinite ashes (Hay, 1986). At Bukwa in Uganda, in volcaniclastic sediments some 23 million years old (J. A. Van Couvering, 1972), there are preserved natural casts of dicot flowers and rhizomes and leaves of grasses or sedges (A. C. Hamilton, 1968). Paleosols at Bukwa remain unstudied, but the enclosing tuffaceous sediments are superficially similar to Dhero and Kiewo Series paleosols.

4.2.3.10 Former animal life. Fossil snails and bones in Kiewo paleosols (beds 5 and 6 of Pickford & Andrews, 1981), like most fossils from Songhor, appear to have been attritional accumulations of remains of creatures dying, disarticulating, and weathering on the soils (Andrews & Van Couvering, 1975). Some of the rodent fossils, especially of cane rats (*Diamantomys luederitzi*) and mole rats (*Bathyergoides neotertiarius*), were found articulated (Lavocat, 1973), and may have been buried in burrows.

Snail fossils (with numbers of specimens in parentheses from Pickford & Andrews, 1981) include the large high-spired *Achatina leakeyi* (16), the low-spired *Tayloria* sp. (5), elongate high spired subulinids (4), and the large ovoid streptaxid *Gonaxis* sp. (1). Fragments of turtles, snakes, and birds have been reported. The most common mammals were cane rats (*Diamantomys luederitzi*, 45) and small apes (*Limnopithecus legetet*, 8). Among large mammals, the most common were extinct deerlike ungulates (*Walangania africanus*, 3) and chevrotain (*Dorcatherium songhorense*, 3). Kiewo paleosols yielded at least 14 species of bats, insectivores, and rodents in addition to *Diamantomys*, including cane rat (*Paraphiomys pigotti*, 3), mole rat (*Bathyergoides neotertiarius*, 3), flying squirrel (*Paranomalurus bishopi*, 3), hedgehog (*Gymnurechinus camptolophus*, 3) and spring hare (*Megapedetes pentadactylus*, 2). Primates in these paleosols in addition to the arboreal ape *Limnopithecus*, include bush babies (*Komba minor*, 3; *K. robustus*, 1; and *Progalago dorae*, 1), arboreal ape (*Dendropithecus macinnesi*, 1), small generalized apes (*Rangwapithecus gordoni*, 4; *Nyanzapithecus vancouveringorum*, 1; and *Proconsul africanus*, 3) and large ape (*P. major*, 4).

The only trace fossils seen in Kiewo paleosols were locally abundant, near-vertical, simple tubes 6 to 11 mm in diameter. The most obvious examples were filled with material more sandy and drab colored than the red clayey matrix. These can be referred to the ichnogenus *Skolithus*, like similar trace fossils from paleosols elsewhere (Retallack, 1976). Burrows of this general form are made by a wide variety of insects and spiders (Ratcliffe & Fagerstrom, 1980).

4.2.3.11 Paleotopographic setting. Kiewo

paleosols include relict bedding with scour-and-fill, low-angle heterolithic cross bedding and ball-and-pillow structures typical of the deposits of river levees (for example, Potter & Pettijohn, 1973). Such a lowland setting, intermittently flooded, also is indicated by their high ferrous iron/ferric iron ratios and organic carbon content (Fig. 4.13), compared with other Kenyan Miocene paleosols. Judging from the scale of these sedimentary structures and the composition of the paleosols, these were broad, loosely sinuous creeks that drained carbonatite tuff cones. They may have been tributaries of the large meandering stream envisaged for the Grey Sandstone Member of the Kapurtay Agglomerate and associated Buru Series paleosols.

4.2.3.12 Parent material. The Ck horizon of the type Kiewo clay has a chemical composition most like carbonatite, with a minor contribution of nephelinite or aluminosilicate weathering products (Fig. 2.20; compared with analyses of Binge, 1962; Deans & Roberts, 1984; Richard, 1942). Quartz and feldspar seen in thin section may have been from fenitized wall rocks of the vents, but most of the carbonatite has been altered to the pervasively displacive calcite veins that characterize this horizon. The A horizon of the Kiewo clay, in contrast, is more nephelinitic in composition and has more deeply weathered clasts of nephelinite, as is common also in the Grey Sandstone Member of the Kapurtay Agglomerate and in associated Buru Series paleosols. This material may have been introduced by streams draining nearby deeply weathered nephelinitic volcanics. A varying mix of carbonatite ash and nephelinite alluvium is likely for other Kiewo paleosols also.

4.2.3.13 Time for formation. Abundant relict bedding, little mixed by burrows and root traces, is an indication that Kiewo paleosols formed over a short period of time. Cementation of the general kind seen in Kiewo paleosols is known from the Footprint Tuff only 600 years old on Oldoinyo Lengai, Tanzania (Hay, 1989). Kiewo paleosols probably do not represent as much time as Kwar and Mobaw Series paleosols formed in similar parent materials and probably also similar climate and vegetation. Only a few tens to perhaps thousands of years would have been needed to form Kiewo paleosols.

4.2.4 Choka Series (new name)

4.2.4.1 Diagnosis. Thick (1-m or more), moderately developed, reddish-brown (5YR to 7.5YR), strongly calcareous and copiously burrowed paleosols, with subsurface clayey (Bt) horizon, on carbonatite tuffs.

4.2.4.2 Derivation. Choka is Dholuo for whitewash (Gorman, 1972), in reference to the pale color of this calcareous paleosol among darker reddish brown Kiewo and Tut profiles.

4.2.4.3 Description. The type Choka clay paleosol (Figs 4.13 and 4.14) is near the middle of the early Miocene, Red Bed Member of the Kapurtay Agglomerate (below 190 cm in Fig 4.5; beds 7 and 8 of Pickford & Andrews, 1981) in the westernmost bluff of Songhor gullies (area 1 of Pickford & Andrews, 1981; grid reference 459965 on Muhoroni sheet) 2.5 km north of Songhor village, Kenya (Figs 4.4 and 4.7).

+29 cm; C horizon of overlying paleosol (type Tut clay); coarse-grained sandstone; grayish-brown (10YR4/2), with clasts and clayey layers of dark reddish-brown (5YR3/4); prominent relict bedding; strongly calcareous; calciasepic porphyroskelic in thin section, with common deeply weathered nephelinite rock fragments, floating within a matrix of micrite; few grains of fenite, biotite, quartz, and microcline; abrupt, wavy contact to

+12 cm; volcanic tuff, a late addition to A horizon; silty claystone; dark reddish-brown (5YR3/4); with clasts of light brown (7.5YR6/4), bluish-gray (5B4/1), and black (5Y2/1); common bones are pale brown (10YR8/4 to 10YR7/3) and yellow (10YR7/6), and tooth enamel of common rodents is yellow (2.5Y7/6) and brownish-yellow (10YR7/6); common root traces and burrows (metagranotubules, 6 to 8 mm in diameter) are light gray (2.5Y7/2), with haloes of dark grayish-brown (10YR4/2), extending down into paleosol below; moderately calcareous; mosepic agglomeroplasmic; clasts mostly alvikite; common quartz, microcline, and pyroxene; few calcite globules, sövite clasts, biotite crystals, fenite fragments, and unzoned accretionary lapilli; clear, irregular contact to

0 cm; A horizon; granule-bearing claystone, a prominent ledge-forming layer; light brown (7.5YR6/4); common distinct medium mottles of very pale brown (10YR7/3) and reddish-brown (5YR4/4), after abundant drab-haloed root traces, and burrows 1 cm in diameter and pupal cells 2-cm long (Fig. 2.8); common clasts up to 4 mm in diameter of dark reddish-brown (5YR3/4), strong brown (7.5YR4/6), and dark yellowish-brown (10YR4/4); few vertical white (5Y4/4) calcite crystal sheets (calcans), with local black (5Y2/1) spots of iron-manganese; strongly calcareous; mosepic agglomeroplasmic in thin section; matrix of mosepic clay, with deeply weathered clasts of alvikite and sövite, and common quartz and microclinc; burrows and root holes filled with grains of rounded alvikite similar to "Pelee's tears" in a displacive cement of sparry calcite; gradual, irregular contact to

-8 cm; A horizon; granule-bearing silty claystone; mottled equally reddish-brown (5YR4/4) and grayish-brown (10YR5/2), around abundant drab-haloed root traces, burrows, and pupal cells as in overlying horizon; also seen was a large (8.4-cm across by 13.2-cm long) bulbous chamber, filled with pale-olive (5Y6/3) to pale-yellow (5Y7/3) calcareous siltstone, and with an entrance burrow 12 cm in diameter, probably a small mammal den; few clasts up to 6-cm long by 3-cm deep of light reddish-brown (5YR6/4) calcareous claystone; strongly calcareous; mosepic porphyroskelic in thin section of clayey matrix and some burrows (ortho-isotubules) with well preserved clay skins, these areas also showing highly weathered grains of nephelinite, quartz, and microcline; other areas calciasepic porphyroskelic, with micritic matrix and common fine calcite crystal tubes; gradual smooth contact to

-27 cm; BA horizon; granule-bearing silty claystone; mottled equally reddish-brown (5YR4/4) and grayish-brown (10YR5/2); with some large, tuberose, and irregular clasts of light gray (10YR7/2) to grayish brown (10YR5/2) carbonatite tuff, with weathering rinds (diffusion sesquans) of dark grayish-brown (10YR4/2); abundant drab-haloed root traces and burrows and pupal cells as in overlying horizon; strongly calcareous; mosepic to calciasepic porphryoskelic in thin section; carbonatite tuff fragments are calcite-cemented aggregates of alvikite of the "Pelee's tears" form; few calcite crystal tubes and irregular tubes of iron-manganese after root traces; few veins of displacive sparry calcite; gradual, smooth contact to

-63 cm; Bt horizon; silty claystone; reddish-brown (5YR4/4); few sand-sized grains of black (5Y2/1); ortho-isotubules extending down from above are light brownish gray (10YR6/2), with an oxidized rim (diffusion sesquan) of grayish brown (10YR5/2); strongly calcareous; mosepic porphyroskelic in thin section, with common deeply weathered nephelinite, quartz, and microcline; also common are clasts of alvikite carbonatite, in many cases with a selvadge of sparry calcite, often chipped off and lying nearby in the matrix; gradual, smooth contact to

Fig. 4.14. Excavation into type Tut clay paleosol (between stratified sandy benches) and underlying type Choka clay paleosol in the westernmost bluff at Songhor National Monument, Kenya. Hammer at base of Tut clay for scale.

-98 cm; Bk horizon; granule-bearing clayey sandstone; mottled equally reddish-brown (5YR4/4) and grayish-brown (10YR5/2); with clasts up to 5-mm across of pale olive (5Y6/4), white (5Y8/2), and black (5Y2/1); massive in appearance, with indistinct patches of relict bedding; strongly calcareous; calciasepic porphyroskelic in thin section, with extensive displacive sparry calcite and replacive micrite; many clasts are alvikite with calcite selvadges as in overlying horizon; quartz and microcline also are common; outside calcareous areas are mosepic porphyroskelic red claystone within burrows (meta-isotubules); clear, smooth contact to

-125 cm; C horizon; silty claystone; reddish-brown (5YR4/4); with granules of dark reddish-brown (5YR3/4), yellow (10YR7/8), pale olive (5Y6/3), and black (5Y2/1); few grayish-brown (10YR5/2) ortho-isotubules after root traces and burrows, as in overlying horizon; distinct relict bedding and very poor sorting; strongly calcareous; calciasepic granular in thin section; clasts well rounded and with weathering rinds (diffusion sesquans), mainly nephelinite, common pyroxene, and alvikite; abrupt, irregular (along pedotubules from above) contact to

-142 cm; Ck horizon; siltstone; light brownish gray (10YR6/2); clear relict bedding, and ball-and-pillow structure (soft sediment deformation); strongly calcareous; calciasepic porphyroskelic in thin section; common nephelinite, quartz, microcline, and pyroxene in a ferruginized clayey matrix; also riddled with pervasively displacive sparry calcite; abrupt, wavy contact to

-153 cm; A horizon of underlying paleosol (type Kiewo clay); sandy claystone grading down into clayey, medium to coarse grained sandstone; reddish-brown (5YR4/4); with sand-filled and drab-haloed root traces and burrows of light gray (10YR7/2) and grayish-brown (10YR5/2); moderately calcareous; skelmosepic porphyroskelic in thin section; with abundant deeply weathered grains of nephelinite, quartz, microcline, and pyroxene; few calcite crystal tubes after fine root traces; some local sparry calcite filling displacive circumgranular cracks.

4.2.4.4 Further examples. No other paleosols like the type Choka clay were seen, although some Kwar and Kiewo profiles are thick and have incipient subsurface clayey horizons approaching the form of the type Choka clay.

4.2.4.5 Alteration after burial. The type Choka clay probably has lost much organic matter, been reddened by burial dehydration of ferric hydroxides, and been locally discolored by burial gleization around organic matter of root traces and burrows. Clayey parts of paleosols at Songhor could have been compacted to only 72 percent of their former thickness. This is an unlikely maximum in view of the evidence from calcite crystal tubes filling root holes, for early cementation. The late Miocene faulting and jointing, outlined for Kiewo paleosols, also affected the type Choka clay.

4.2.4.6 Reconstructed soil. The original Choka soil was probably a friable silty clay with a tendency to become hard setting because of additions of carbonatite airfall ash. A surface silty reddish-brown (A) horizon was riddled with burrows and root traces and graded down into a similarly bioturbated clayey subsurface (Bt) horizon. Below that was an incipient horizon (Bk) of pedogenic carbonate nodules, showing both displacive and replacive textures in thin section. Below that again there were nephelinitic sandstones and calcareous stringers with pervasively displacive fabric, which may represent an overprinted paleosol of the Kiewo Series, above the type Kiewo clay. Calcification, clay formation, and humification of the type Choka clay were modest but significant. The pH of the type Choka clay was probably alkaline throughout, and its cation exchange capacity and base saturation high are high. Although soda/potash ratios were high (close to 1, Fig. 4.13), they decrease toward the surface, so that salts did not accumulate, but were leached. Similarly, ferrous iron/ferric iron ratios could be taken as evidence of periodic gleization at depth, but this ratio declines up profile, where the soil was probably well drained most of the time. The water table in this lowland soil may have fluctuated to within some 80 cm of the surface.

4.2.4.7 Classification. The subsurface clayey horizon of the type Choka clay qualifies as argillic. This together with its calcareous composition, distinguishes the type Choka clay as a Haplustalf in the U.S. soil taxonomy (Soil Survey Staff, 1975). Equivalent categories are Calcic Luvisols (of Food and Agriculture Organization, 1974), Red-Brown Earths (of Stace *et al.*, 1968), Gc2.12 (of Northcote, 1974), and Valsrivier marienthal (of MacVicar *et al.*, 1977). Such soils are well known on Precambrian basement rocks in dry regions of Kenya (units A11, Pt1, Y2 of Sombroek, Braun, & Van Der Pouw, 1982), Zimbabwe (unit Lk4-2ab of Food and Agriculture Organization, 1977a) and South Africa (Botha, 1985; Gertenbach, 1983; MacVicar *et al*, 1977; Venter, 1986). Although taxonomically similar in their profile form, these soils are different in many ways from the carbonatitic and nephelinitic Choka clay paleosol, for which I am unaware of a good modern analog.

4.2.4.8 Paleoclimate. Surface soils of profile form comparable to the type Choka clay are found mainly in dry climates. In these soils carbonate has accumulated, compared with their weakly calcareous parent material. In the Choka clay, however, soil formation has decalcified the soil compared with its carbonatite-nephelinite parent material and has in addition created a subsurface horizon of clay accumulation, as in humid to subhumid climates. A subhumid range of mean annual precipitation also is in evidence from the depth of its incipient calcareous horizon: some 98 cm below the surface now and perhaps as much as 136 cm before compaction during burial. In soils of the Great Plains of North America (Jenny, 1941), the calcareous horizon is this deep in regions receiving a mean annual precipitation of 660 to 1200 mm. Surprisingly similar estimates for the former rainfall at Songhor are gained by comparing the depth of the calcareous horizon in this paleosol with that in soils of the Mojave Desert of California (Arkley, 1963), the Serengeti Plains of Tanzania (de Wit, 1978), and the Indo-Gangetic Plains of Punjab, India (Sehgal, Sys, & Bhumbla, 1968).

No firm evidence for paleotemperature or seasonality was found. The abundance and depth of burrows and root traces are compatible with, but not compelling evidence for, a productive tropical ecosystem. Ferrous iron/ferric iron ratios, organic content, manganese spotting, and local iron-rich haloes round drab-haloed root traces could be evidence of periodic waterlogging that may have been seasonal. Clastic dikes in these paleosols could be taken as evidence of seasonal cracking, but this explanation is unconvincing. The Choka clay paleosol is well exposed throughout the Songhor gullies, and so are the dikes, which are near vertical, are straight, and penetrate the whole sequence like associated faults of minor displacement (Pickford & Andrews, 1981). Not seen was any deformation of soil horizons typical of swelling clay soils of seasonal climates (mukkara structure of Paton, 1974). Lack of evidence for seasonality in the early Miocene Mobaw, Kwar, Kiewo, and Buru Series paleosols can be excused on the grounds that these are weakly developed profiles. However, various indications of seasonality found in middle and late Miocene paleosols of Pakistan and Kenya are lacking in the early Miocene Choka and Tut Series, perhaps because monsoonal circulation was not yet pronounced in East Africa.

4.2.4.9 Ancient vegetation. This thick, red paleosol with its clayey subsurface (Bt) horizon and abundant large, drab-haloed root traces probably supported forest vegetation. Compared with parent material, carbonate has been leached to depths of at least a meter, where there was some accumulation of carbonate. This would be unlikely under rain forest, where in addition root systems tend to be tabular (Sanford, 1987), unlike those of the Choka clay. Nor does this paleosol have such a shallow calcareous horizon as soils under dry woodland, nor the granular peds of grassland soils.

Soils of comparable profile form in Australia (Stace *et al.*, 1968) and the United States (Soil Survey Staff, 1975) support sclerophyll or deciduous forests or grassy woodlands. In Kenya, such soils support *Acacia-Commiphora* wooded grassland and bushland (units A11, L28, Ps28, Vt1, Y2 of Sombroek, Braun, & Van Der Pouw, 1982). Broadly comparable soils in South Africa, Botswana, and Zimbabwe support dry Zambezian deciduous woodland dominated by *Colophospermum mopane* (map unit Lk4-2ab of Food and Agriculture Organization, 1977a; Botha, 1985; Gertenbach, 1983; Venter, 1986), although many mopane soils are more sandy and salinized than the type Choka clay (B. S. Ellis, 1950). These comparisons are unfair, because most of these soils are formed on very ancient landscapes of noncalcareous Precambrian rocks, unlike the parent material of the Choka clay. Those on Neogene marls under mopane woodland and thorntree wooded grassland also are relict soils, combining the effects of ancient and modern weathering. This also is true of the soils on a carbonatite intrusion near Phalarborwa, South Africa (Frick, 1986). Compared with the mopane woodland around it, woodland on the intrusion is much richer in species and includes *Acacia nigrescens*, *Dichrostachys cinerea*, *Ormocarpum trichocarpum*, *Zizyphus zeyheriana*, *Lannea stuhlmanii*, *Grewia hexamita*, and *G. bicolor* (M. M. Cole, 1986).

If these are somewhat drier vegetation types than likely for mildly decalcified Choka paleosols, how much lusher then is African vegetation capable of strongly decalcifying such calcareous parent materials? Thick noncalcareous soils (Acrisols, map units Af2, Af12-1/2a, Af15, Af31, Ao67-2bc of Food and Agriculture Organization, 1977a) are found in many parts of Africa on Cretaceous sandstones and marls. They support vegetation ranging from wooded grassland to tropical semideciduous rain forest. However, these soils also are very ancient and combine effects of past climates. This is less problematic for Kenyan and Somali coastal coral limestones and calcareous sandstones some 125,000 to 240,000 years old (Braithwaite, 1984), which have been converted to noncalcareous soils (Cambic Arenosols, Ferric Acrisols, and Pellic Vertisols in unit Qc37-1a of Food and Agriculture Organization, 1977a; Muchena & Sombroek, 1981) under tropical semideciduous rain forest. Also strongly decalcified are soils under humid secondary grassland on late Quaternary carbonatite tuffs near Fort Portal, Uganda (Barker, 1989; Harrop, 1960; Nixon & Hornung, 1973). From these comparisons, the Choka clay is likely to have supported forest as unique in botanical composition as its carbonatite-nephelinite parent material, and neither as open as wooded grassland nor as dense as tropical rain forest.

Fossil plants of early Miocene age are indications that many East African forest types are at least that ancient (Axelrod & Raven, 1978). Guineo-Congolian rain forest is represented by such genera as *Antrocaryon* and *Entandrophragma* among fruits and seeds some 18 million years old from Mfangano Island, Kenya (Chesters, 1957; Collinson, 1983). Fossil trunks of *Dipterocarpus* and *Cyathea* (Bancroft, 1932a, 1933) from volcanics some 22 million years old on Mt Elgon, Uganda (Bishop, Miller, & Fitch, 1969), could have been part of an early version of wet Afromontane forest.

Fossil wood and leaves (Beauchamp, Lemoigne, & Petrescu, 1972; Lemoigne & Beauchamp, 1972; Lemoigne, Beauchamp, & Samuel, 1974) from Ethiopian volcaniclastic rocks dated at some 23 to 28 million years old (Justin-Visentin *et al.*, 1974; I. McDougall, Morton, & Williams, 1975) are taxonomically similar to living plants of Sudanian grassy woodlands (as defined by White, 1983). A fossil flora some 25 million years old (J. A. Van Couvering, 1972) near the base of the sequence on the Uganda side of Mt Elgon incudes *Acrostichum*, *Olea*, *Maranthus (Parinari)*, *Pittosporum*, and *Terminalia*, as well as such characteristic genera of Zambezian dry deciduous forests and woodlands as *Bauhinia*, *Berlinia* and *Dalbergia* (Chaney, 1933). Another fossil flora from the foot of the Bogoro Scarp near Lake Mobutu (formerly Lake Albert) has been dated by associated mammal fossils at about 19 million years old (J. A. Van Couvering, 1972). It contains fossil plants allied to *Dichrostachys*, *Newtonia*, *Isoberlinia*, *Brachystegia* and *Baphia* (Lakhanpal & Prakash, 1970), an assemblage especially suggestive of Zambezian miombo woodland (of White, 1983). Both of these early Miocene floras of Zambezian floristic affinities were found close to lateritic paleosols, of a kind now widely exhumed and still supporting floristically similar vegetation in dry regions of Africa (M. M. Cole, 1986). Finally, there are the fossil fruits and seeds from carbonatite-nephelinite alluvial deposits of Rusinga Island, Kenya, now known to be about 17 million years old (Drake *et al.*, 1988). Common genera likely to have been trees and shrubs are *Lannea*, *Terminalia*, *Zizyphus*, *Grewia*, *Cordia* and Annonaceae. Fossils likely to have been shrubs include *Cnestis*, *Berchemia* and Sapindaceae. There also was a great variety of plants likely to have been vines: *Cissus*, *Cavratia*, *Lagenaria*, *Stephania*, *Cissampelos*, *Triclisia*, Menispermaceae, and Icacinaceae (Chesters, 1957; Collinson, 1983). This assemblage compares well in its floristic composition with drier types of peripheral Guineo-Congolian rain forest and Zambezian riparian woodlands (of White, 1983).

Of all the Miocene fossil assemblages known from Africa, these kinds of dry semievergreen forest and woodland are likely to have been closest to the vegetation of the Choka clay. Some floristic affinity with Zambezian woodlands of lateritic soils is to be expected, considering that dry woodlands on laterites already were in existence as a source of propagules during the early Miocene, and the problems for plant nutrition presented by both laterites and carbonatite-nephelinite tuffs, as outlined for Mobaw paleosols.

4.2.4.10 Former animal life. Compared with Kiewo and Tut Series paleosols at Songhor, the Choka clay is not so abundantly fossiliferous (beds 7 and 8 of Pickford & Andrews, 1981). Only fossils that were very common in other paleosols were found in the Choka clay, including a variety of snails (*Ligatella*, Subulininae, *Tayloria*, *Gonaxis*, *Primigulella*), as well as cane rats (*Diamantomys luederitzii*) and deerlike ungulates (*Walangania africanus*). Considering that this paleosol is at least as calcareous as associated Kiewo and Tut Series and is better developed than Kiewo Series paleosols, it should have been a good preservational environment for bones and shells, at least in theory (Retallack, 1984a). Perhaps animals avoided this soil because it was hard setting or highly saline or alkaline during the incorporation of the carbonatite airfall ash that caps the profile.

The carbonatite-nephelinite ashy nature of this paleosol also may be responsible for the preservation of abundant, diverse and little-compacted trace fossils. The Choka clay paleosol can be readily identified by abundant tear-shaped to ellipsoidal internal casts of structures interpreted as pupal cells of insects (Fig. 2.8). Some (23) of them were measured at 19.2-23.6-mm long (mean 20.7, standard deviation 1.1 mm) by 12.1-15.2-mm wide (mean 13.2 mm, standard deviation 0.8 mm). Most of these (16) had a broken end, where the insect presumably emerged. This former orifice was rather ragged, but ranged from 7.7 to 10.9 mm in diameter (mean 9.5 mm, standard deviation 0.7 mm). The remaining casts measured were smooth ellipsoids. Within the paleosol most of these structures lay more or less horizontally, but a variety of other orientations was also seen. Some also were seen to be connected into pinnate systems of galleries, including near vertical shafts.

These ellipsoidal trace fossils are similar to the ichnogenus *Pallichnus* (Retallack, 1984b), thought to be the work of geotrupine or onthophagine dung beetles with relatively simple nesting behavior for dung beetles (pattern I of Halffter & Edmonds, 1982). In living beetles of this kind, eggs are laid in elongate chambers that are packed with dung carried down a vertical shaft from a large dung pat immediately overhead. The larvae hatching from the eggs consume their dung and then pupate in ovoid to spherical cells formed from larval feces and clay. Most of the pupae hatch into young beetles that burrow out to the surface through the loosely packed fill of the entrance burrow, but some pupal cells remain intact because the pupae have been killed by parasites or fungi. Common living African dung beetles showing this general pattern of nesting are *Liatongus*, *Onitis*, *Chironitis*, and *Onthophagus* (Halffter & Edmonds, 1982). Fossil dung beetles from the early Miocene (about 18 million years old according to Pickford, 1984, 1986a), Kiahera Formation of Rusinga and Mfangano Islands, Kenya, have been identified (by Paulian, 1976) as extinct species in the extant genera *Anachalcos*, *Copris*, and *Metacatharsius* (this last regarded by some taxonomists as a subgenus of *Catharsius*). The nesting behavior of these genera today is elaborate and involves considerable parental care (patterns V, II, and III, respectively, of Halffter & Edmonds, 1982). These more elaborate nesting patterns are more common in open grassy vegetation, whereas the onthophagine pattern seen in the pupal cells of the Choka clay is thought to be more primitive for dung beetles and typical of woodland and forest species.

Other trace fossils of the Choka clay are near-vertical burrows 6 to 11 mm in diameter, referable to *Skolithus*. Some of these are associated with *Pallichnus* cells, but others could have been created by a variety of insects and spiders, as indicated for similar trace fossils in associated Kiewo paleosols.

Also found were a few large (59.2-mm long by 20.2-mm wide) elongate-conical internal casts with smooth but cracked and partly crushed outer surface. These may have been internal casts of the pupal cells of a large elongate insect. Hawk and emperor moths (families Sphingidae and Saturniidae) make large earthen cells of comparable size, and other African butterflies and moths make silk-lined cocoons in the ground (Skaife, 1953). A fossil caterpillar of comparable size has been found in carbonatite tuff some 18 million years old on Mfangano Island, Kenya (L. S. B. Leakey, 1952).

4.2.4.11 Paleotopographic setting. The type Choka clay was part of the same alluvial landscape outlined for Kiewo paleosols, but was probably a little better drained on an unusually thick carbonatite tuff. Compared with Kiewo paleosols the Choka clay is well oxidized above a depth of 90 cm (Fig. 4.13) and is much thicker and more deeply penetrated by root traces, burrows, and clay skins. Judging from its moderate development of a clayey subsurface (Bt) and incipient calcareous (Bk) horizon, it was removed from frequent disturbance by floods that affected Kiewo paleosols. Probably it formed on a terrace of tuff, elevated from the main stream channel.

4.2.4.12 Parent material. The most abundant grains in the Choka clay are alvikite similar to those of the Legetet Carbonatite and the Calcified Tuff Member of the Kapurtay Agglomerate (Deans & Roberts, 1984). However, the clayey tuffaceous layer overlying and partly incorporated into the profile is more like a nephelinite tuff, similar to Plio-Pleistocene examples from Tanzania (Hay, 1978, 1989) and to the thick tuff below the main fossiliferous level at Fort Ternan National Monument (Figs 4.9 and 4.10). This nephelinitic material has penetrated the burrows of the paleosol so that both carbonatite and nephelinite clasts are found throughout the profile. Carbonatite is more abundant near the base and nephelinite near the top, judging from chemical composition (Fig. 2.20), particularly silica, alumina, and lime, in comparison with the composition of local carbonatites and nephelinites (Binge, 1962; Deans & Roberts, 1984; Richard, 1942).

Parent material of the Choka clay was thus mainly a thick airfall carbonatite tuff. Some thin nephelinite tuffs and sandstones may have been added and incorporated during soil formation. To this was later added a thin carbonatite tuff, and then nephelinitic sandstones.

4.2.4.13 Time for formation. The Choka clay is moderately developed, with subsurface clayey (Bt) and incipient calcareous horizons that indicate a longer period of soil formation than envisaged for Mobaw, Kwar, or Kiewo Series paleosols.

Calcareous horizons develop over only 2000 to 9000 years in carbonatite-nephelinite ashy soils of Tanzania (Hay & Reeder, 1978). However, the Choka clay probably formed in a more humid climate than these Tanzanian soils, and its Bk horizon is less expressed than the Bt horizon, which is probably a better guide to its time for formation.

Radiocarbon-dated soils in phonolitic colluvium, till, and loess on Mt Kenya have developed perceptible Bt horizons in as few as 1940 years, and these qualify as argillic after only 40,000 years (Mahaney, 1989; Mahaney & Boyer, 1989). These Mt Kenyan soils formed in a much wetter and cooler climate than envisaged for the Choka clay, so perhaps a better comparison is with clayey paleosols developed in carbonatite-nephelinite ashes as young as 40,000 years on phonolite bedrock in the climatically drier northern Serengeti woodlands of Tanzania (profile Lamai 2 of Jager, 1982).

Similar general estimates can be gained also by comparison with carefully studied North American soil chronosequences. Both calcic and argillic horizons developed to the extent seen in the Choka clay paleosol are found in soils some 25,000 to 75,000 years old in quartzofeldspathic colluvium within the desert of Las Cruces, New Mexico (Gile, Hawley, & Grossman, 1980). Argillic horizons develop to this extent in 40,000 to 130,000 years under wooded grassland in quartzofeldspathic alluvium in the San Joaquin Valley of California (Harden, 1982). Although none of these soils is exactly like the Choka clay, these comparisons constrain its time for formation to 50,000 years or so.

4.2.5 Tut Series (new name)

4.2.5.1 Diagnosis. Thick (50 cm or so), moderately calcareous, red (5YR) profiles with distinct subsurface clayey (Bt) and calcareous (Bk) horizons.

4.2.5.2 Derivation. Tut is Dholuo for deep (Gorman, 1982), in reference to the thickness of these paleosols.

4.2.5.3 Description. The type Tut clay paleosol (Figs 4.13 and 4.14) is near the top of the early Miocene, Red Bed Member of the Kapurtay Agglomerate (below 6.2 m in Fig. 4.5; bed 9 of Pickford & Andrews, 1981) in the westernmost bluff of Songhor gullies (area 1 of Pickford & Andrews, 1981; grid reference 459965 on Muhoroni sheet), 2.5 km north of Songhor village, Kenya (Figs 4.4 and 4.7).

+40 cm; paleochannel and C horizon of overlying profile (unnamed Buru Series); coarse-grained sandstone; light reddish-brown (5YR6/4); planar bedding and cross bedding; few 7-20-mm diameter burrows (meta-isotubules) of pale brown (5YR6/4) claystone, with inner 1-mm halo of pale yellow (2.5Y7/4) and outer 3-mm halo of light brownish-gray (10YR6/2), including small black (10YR2/1) nodules of iron-manganese; few, 5-10-mm wide, pale yellow (10YR7/3) clastic dikes filled with silt; moderately calcareous; calciasepic granular in thin section, with grains mainly reddish-brown, highly oxidized nephelinite; common quartz and microcline; few carbonatite clasts; local cement and crystal tubes of sparry calcite; abrupt, wavy contact to

+9 cm; volcanic ash overlying and partly incorporated into paleosol; clayey siltstone; dark grayish-brown (10YR4/2); with scattered grains of brown (10YR4/2) and black (5Y2.5/1); massive granular structure; common pale yellow (5Y8/3), 10-11-mm diameter metagranotubules, after root traces and burrows; few large snail shells; strongly calcareous; calciasepic intertextic in thin section; grains equally of fresh alvikite carbonatite and nephelinite; common scoriaceous lava, biotite, quartz, and microcline; few calcite globules, pyroxene, vermiculite, and calcite selvadges around alvikite grains; rare wollastonite and natrolite detected in XRD traces; clear, irregular contact to

0 cm; A horizon; clayey sandstone; dark brown (7.5YR3/2); abundant 6-8-mm diameter, light yellowish-brown (10YR6/4) metagranotubules, with a drab halos 2-3-mm wide of dark grayish-brown (10YR4/2), after root traces, as well as 9-11-mm diameter, dark grayish-brown (10YR4/2) metagranotubules, after burrows; moderately calcareous; insepic agglomeroplasmic in thin section; abundant carbonatite and nephelinite clasts, common

quartz, and microcline; few calcite crystal tubes after roots; gradual, irregular contact to

-3 cm; A horizon; sandy claystone; dark reddish-brown (5YR3/4); common 6-9-mm diameter, pale yellow (2.5Y8/4) metagranotubules, with 3-mm haloes of dark grayish-brown (10YR4/2), after root traces; texture dominated by grains, but there are indistinct subangular blocky peds within larger (very coarse) angular-blocky peds outlined by local black (5YR2.5/1) mangans; moderately calcareous; mosepic agglomeroplasmic in thin section, with grains mostly deeply weathered nephelinite and alvikite; common quartz and microcline; few pyroxene and biotite; few calcite crystal tubes after roots; abrupt, broken contact to

-24 cm; Bt horizon; sandy claystone; dark reddish-brown (5YR3/4); very coarse angular-blocky peds, outlined by scattered black (5YR2.5/1) mangans; moderately calcareous; mosepic porphyroskelic in thin section; clasts mainly nephelinite and quartz, with few pyroxene and sövite; few calcite crystal tubes after roots; this horizon also has a few prominent, coarse, tuberose to tubular, calcareous nodules that are light gray (5Y7/2), with a 2-3-mm halo of dark grayish-brown (10YR4/2); nodules are strongly calcareous and calciasepic porphyroskelic in thin section, with clasts of nephelinite, pyroxene, feldspar, and quartz in a matrix of replacive micrite, locally penetrated by calcite-filled displacive veins; this horizon also penetrated by clastic dikes of light gray (5Y7/2) calcareous siltstone that extend through the whole sequence here; abrupt, wavy contact to

-42 cm; Bk horizon; siltstone; light gray (5Y7/2); this horizon has many more nodules similar to those found isolated in the overlying horizon, and they appear concentrated along a textural change from the clayey upper part to the silty lower part of the profile; strongly calcareous; calciasepic porphyroskelic in thin section, similar to nodules in horizon above, but with fine, opaque mottles of iron-manganese after root traces; abrupt, wavy contact to

-48 cm; BC horizon; granule-bearing sandy claystone; reddish-brown (5YR4/4); common 5-mm diameter, light gray (5Y7/2) metagranotubules of silty claystone, with 2-3-mm haloes of dark grayish-brown (10YR4/2), after root traces; clastic dike seen in upper horizons extends also to this level; moderately calcareous; mosepic agglomeroplasmic in thin section, with common alvikite, nephelinite, quartz, and microcline; many alvikite clasts have a thin selvadge of calcite (as in Fig. 2.13A); this horizon also has a few distinct, strongly calcareous, light reddish-brown (5YR6/4) nodules, with calciasepic to insepic agglomeroplasmic microfabric; grains in the nodules are mainly nephelinite and alvikite, with common pyroxene, microcline, and quartz; gradual, irregular contact to

-102 cm; BC horizon; granule-bearing sandy claystone; yellowish-red (5YR4/6); common distinct calcareous nodules of light reddish-brown (5YR6/4) and light red (2.5YR6/6), with 3-4-mm haloes of light gray (5YR7/2); common claystone clasts of reddish-brown (5YR5/4); few thin, strong-brown (7.5YR5/8) clay skins defining indistinct coarse blocky angular peds; some clastic dikes up to 4-mm wide, light gray (5Y7/2) with central zone (sesquan) of strong brown (7.5YR5/6), as in overlying horizons; few 3-4-mm diameter, light yellowish-brown (5YR6/4) metagranotubules, with marginal reddish-brown (5YR5/4) clay skins, after burrows; few *Pallichnus* cells; matrix and nodules both strongly calcareous; calciasepic to masepic porphryoskelic; mostly with grains of highly weathered nephelinite; common alvikite, pyroxene, and quartz; common crystal tubes of calcite and irregular tubular features of opaque iron-manganese, both after root traces; gradual, irregular contact to

-122 cm; C horizon; granule-bearing clayey sandstone; dark reddish-brown (5YR3/4); with many distinct carbonatite clasts, of grayish-brown (10YR5/2) with a dark grayish-brown (10YR4/2) weathering rind (diffusion sesquan); indistinct relict bedding; strongly calcareous; calciasepic porphyroskelic in thin section, with clasts mainly deeply weathered nephelinite; few grains of sövite, alvikite, vermiculite, quartz, microcline, and opaque oxides; few crystal tubes of sparry calcite; gradual, smooth contact to

-148 cm; C horizon; granule-bearing clayey sandstone; grayish-brown (10YR4/2); with angular grains of dark reddish-brown (5YR3/4) claystone; prominent relict planar bedding; strongly calcareous; calciasepic porphyroskelic in thin section, with grains mainly deeply weathered nephelinite; few sövite, quartz, and microcline; many grains coated with micritic rinds; one granule consisted of little-weathered fenite, with intergrown coarse-grained pyroxene, biotite, quartz, and microcline; abrupt, smooth contact to

-165 cm; volcanic tuff addition to A horizon of underlying paleosol (type Choka clay); silty claystone; dark reddish-brown (5YR3/4); with clasts of light brown (7.5YR6/4), bluish-gray (5B4/1), and black (5Y2/1); common, 6-8-mm diameter metagranotubules of light gray (2.5Y7/2), with haloes of dark grayish-brown (10YR4/2), after burrows and root traces; moderately calcareous; mosepic agglomeroplasmic in thin section; clasts mainly alvikite; common quartz, microcline, pyroxene; few calcite globules, sövite, biotite, fenite, and unzoned accretionary lapilli.

4.2.5.4 Further examples. Only the type Tut clay has so far been recognized as an example of this series. The Kwar silty-clay paleosol near Koru shows some differentiation of a clayey subsurface horizon, but neither this nor its subsurface calcareous horizon are developed to the extent seen in the type Tut clay.

4.2.5.5 Alteration after burial. Loss of organic matter, local burial gleization, reddening due to compactional dehydration of ferric hydroxides, and brittle deformation into fault blocks, as outlined for the Choka clay paleosol, probably also affected this paleosol. The Tut clay, however, is less well cemented with early burial cement of calcite and more clayey overall than the Choka clay. Compaction of clayey parts of the Tut profile may have reduced thickness to the 72 percent estimated for Songhor, bringing the Bk horizon to 42 cm below the overlying ashy bed from an original depth of about 58 cm.

4.2.5.6 Reconstructed soil. The Tut clay was buried by a thin (9 cm) airfall tuff of carbonatite-nephelinite, which was incorporated into the soil to a limited extent before burial by thick nephelinitic sandstones. Before burial by either of these materials it probably had a silty, brown A horizon over a clayey reddish-brown Bt horizon with nodules that formed an irregular hardpan at a depth of about 58 cm, above bedded, sandy, volcaniclastic alluvium. Its pH was probably alkaline. Also high were Eh, cation exchange capacity, and base saturation, judging from ratios of alkaline earths/alumina and ferrous iron/ferric iron (Fig. 4.13). Declining amounts of organic matter and ratios of ferrous iron/ferric iron and soda/potash toward the surface are evidence against appreciable gleization or salinization.

4.2.5.7 Classification. The Tut clay formed on parent materials richer in weathered nephelinitic sandstone than the type Choka clay and is a little better developed, but in taxonomically significant features the two profiles are similar. For reasons already explained for the Choka clay and possible analogs among surface soils, the Tut clay also can be identified as a Haplustalf (of Soil Survey Staff, 1975), Calcic Luvisol (of Food and Agriculture Organization, 1974), Red Brown Earth (of Stace *et al.*, 1968), Gc2.12 (of Northcote, 1974) and Valsrivier marienthal (of MacVicar *et al.*, 1977).

4.2.5.8 Paleoclimate. Given the highly alkaline and calcareous parent material of this paleosol, its amelioration into weakly calcareous clay of the Bt horizon is an indication of a warm or rainy climate or long time of formation (Birkeland, 1984). It is unlikely to have been humid, however, because surface soils on late Quaternary carbonatite-nephelinitic volcanics in humid regions of Uganda, Tanzania and Ruanda (Barker, 1989; Harrop, 1960; Mizota & Chapelle, 1988; Mizota, Kawasaki, & Wakatsuki, 1988) are more deeply weathered, rich in kaolinite and halloysite rather than smectite, and lack a horizon of carbonate accumulation like that seen in the Tut clay paleosol. The depth of carbonate nodules in this paleosol, compaction corrected to 58 cm, when compared with that in soils of the North American Great Plains (Jenny, 1941), is evidence of mean annual precipitation of 400 to 700 mm (semiarid to subhumid). This is the total range of precipitation received by soils with a Bk horizon at similar depth. Similar results can be gained by comparing this paleosol with surface soils of the Mojave Desert of California (Arkley, 1963), the Indo-Gangetic Plains of India (Sehgal, Sys & Bhumbla, 1968), and the Serengeti Plains of Tanzania (de Wit, 1978). The Tut clay, however, may have received increments of airfall tuff at a greater rate than these surface soils. The mildly altered carbonatite-nephelinite tuff overlying the profile is a good example. These additions of ash also may be an explanation for occasional calcareous nodules quite high in the profile. For these reasons, as well as the rainfall indicated by the associated Choka clay, the subhumid end of the estimated range seems most reasonable. Again as for the Choka clay, climate may have been warm and mildly seasonal, but there is little firm evidence for these aspects of paleoclimate in this paleosol.

4.2.5.9 Ancient vegetation. The Tut clay paleosol has many of the features expected of a forested soil: a thick red profile with abundant large root traces (now drab haloed) and a broad zone of clay enrichment. It probably supported a dry tropical forest generally similar to that envisaged for the Choka clay. Unlike this other paleosol, the Tut clay has clearly defined calcareous nodules in a less-calcareous clayey matrix derived in part from deeply weathered nephelinitic volcanics, and it is unlikely to have suffered problems of salt accumulation, high alkalinity, or hard setting. For these reasons, forests on the Tut clay probably were taller and lusher, though they probably were not rain forest, for the same reasons offered for the Choka clay.

Among living forests, the Tut clay may have supported vegetation like that in dry peripheral areas of the Guineo-Congolian floristic region (of White, 1983), such as the ironwood (*Cynometra alexanderi*) forests of Budongo in Uganda (Lind & Morrison, 1974). Vegetation of the Tut clay may also have included a component of Zambezian dry woodland plants, likely to have been more prominent on the Choka clay. Afromontane elements are unlikely on either of these alluvial lowland paleosols. Among fossil floras, comparison may be best with assemblages including *Entandrophragma* and *Antrocaryon* some 18 million years old on Mfangano Island, Lake Victoria (Collinson, 1983) and the assemblage including *Bersama*, *Cola*, Euphorbiaceae, and Malvales some 22 million years old at Bukwa, Uganda (A. C. Hamilton, 1968). Living relatives of these plants are widespread in rain forests, but also range out into dry forests peripheral to the Guineo-Congolian rainforest. Additional study of these fossil floras may prove to be a useful guide to the floristically unique dry forests that probably colonized lowlands around early Miocene carbonatite-nephelinite eruptive centers.

4.2.5.10 Former animal life. Fossils preserved as an attritional assemblage (Andrews & Van Couvering, 1975) with the possible exception of a few articulated rodent remains (Lavocat, 1973) are abundant and diverse in the Tut clay (bed 9 of Pickford & Andrews, 1981). These include snails (mainly high-spired Subulininae and large *Achatina leakeyi*), turtles, snakes, and birds. The mammal fauna is numerically dominated by remains of cane rats (*Diamantomys luederitzii*, 38 specimens), mole rats (*Bathyergoides neotertiarius*, 8), and flying squirrels (*Paranomalurus bishopi*, 6). There are in addition at least 11 other species of small insectivores and rodents. Large mammals include creodont carnivore (*Hyaenodon pilgrimi*, 1), short faced pig (*Nguruwe kijivium*, 2), chevrotain (*Dorcatherium songhorense*, 2), and deerlike ungulate (*Walangania africanus*, 2). There also was a variety of primates, including bush babies (*Komba minor*, 3; *K. robustus*, 3), small arboreal apes (*Dendropithecus macinnesi*, 1; *Limnopithecus legetet*, 4), generalized ape (*Rangwapithecus gordoni*, 1), and large ape (*Proconsul major*, 2).

The most common trace fossils seen in this paleosol are near-vertical tubules 6 to 11 mm in diameter, filled with material more sandy than their matrix (metagranotubules of Brewer, 1976). These can be referred to the ichnogenus *Skolithus* and, as in the Kiewo Series, may have been excavated by insects or spiders. Also found was a smooth-walled, tear-shaped internal cast referable to *Celliforma*. Like similar remains in Mobaw and Kwar Series paleosols, this may have been a larval cell of a solitary bee.

4.2.5.11 Paleotopographic setting. The Tut clay paleosol was probably part of a lowland riverine landscape. It is overlain by Buru paleosols and the Grey Sandstone Member of the Kapurtay Agglomerate, interpreted as levee and paleochannel facies (respectively) of a meandering stream. The Tut clay is underlain by Choka and Kiewo paleosols interpreted to form the floodplain of loosely sinuous creeks draining local cinder cones. Within this landscape, the Tut clay was removed from areas of frequent flooding and the cumulative surface horizons introduced by floods. It also is chemically more oxidized than associated paleosols, and so was probably better drained. Probably it formed on a terrace on the toeslopes of a nearby volcanic edifice, only a little elevated above the level of streams flanked by Kiewo and Buru Series paleosols.

4.2.5.12 Parent material. The most common grains forming the planar-bedded C horizon of the Tut clay paleosol are deeply weathered nephelinite. At one level, there are common, irregularly shaped pebbles of carbonatite tuff, like that of the Calcified Tuff Member of the Kapurtay Agglomerate. This was probably a water-worked tuffaceous sandstone eroded from deeply weathered nearby carbonatite cinder cones and nephelinitic stratovolcanoes. Other clasts, especially little-weathered accretionary lapilli and calcite globules, are evidence of additions of volcanic airfall tuff during soil formation. These volcanic materials were not supplied at such a rate that they dominate the fabric of upper parts of the paleosol, as they tend to do in the associated Choka clay. In overall chemical composition (Fig. 2.20) the Tut clay is more nephelinitic than the Choka clay, with substantial carbonatite components only near the base and in the thin covering ash bed.

4.2.5.13 Time for formation. The Tut clay is a little better developed than the Choka clay, but is comparable in development to surface soils of known age discussed for the Choka clay. Like it, the Tut clay may represent about 50,000 years of soil formation.

4.2.6. Buru Series (new name)

4.2.6.1 Diagnosis. Thin, clayey, weakly calcareous, red (5YR) soils with root traces and conspicuous relict beds of sandstone, siltstone, and claystone.

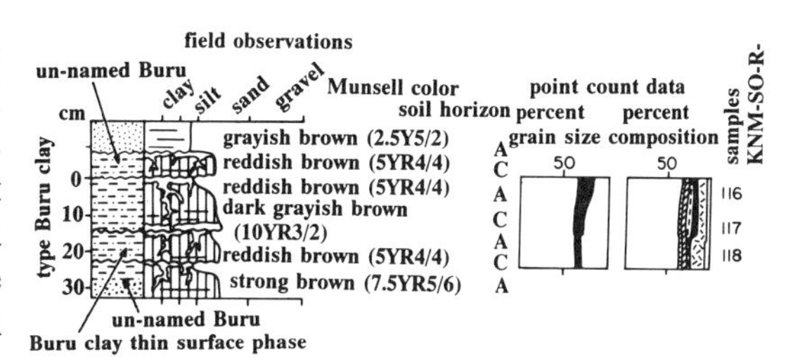

Fig. 4.15. Measured section, Munsell colors, soil horizons, grain size, and mineral composition of the type Buru clay, Buru clay thin-surface phase and other Buru Series paleosols, in the westernmost exposures at Songhor National Monument, Kenya. This section is 7.2 to 7.6 m in Fig. 4.5. Lithological key as for Fig. 4.12.

4.2.6.2 Derivation. Buru is Dholuo for dirt (Gorman, 1972), which in this part of the world is mostly red and volcanogenic (Thorp *et al.*, 1960).

4.2.6.3 Description. The type Buru clay paleosol (Fig. 4.15) is 10 cm below the base of the early Miocene, Grey Sandstone Member of the Kapurtay Agglomerate (below 5.3 m in Fig. 4.5; or 7.5 m in the measured section of Pickford & Andrews, 1981), in the westernmost bluff of Songhor gullies (area 1 of Pickford & Andrews, 1981; grid reference 459965 on Muhoroni sheet), 2.5 km north of Songhor village, Kenya (Figs 4.4 and 4.7).

+7 cm; A horizon of overlying paleosol (unnamed Buru Series); clayey fine sandstone; reddish-brown (5YR4/4); with common light gray (5Y7/2) metagranotubules, surrounded by grayish-brown (10YR5/2) haloes, after burrows and root traces; noncalcareous; skelmosepic porphyroskelic in thin section, mainly with grains of deeply weathered nephelinite; common quartz and microcline; rare pyroxene and biotite; abrupt, wavy contact to

0 cm; A horizon; clayey fine-grained sandstone, but with a distinctly clayey parting from the overlying bed; reddish-brown (5YR4/4); common light gray (5Y7/2), metagranotubules, with grayish-brown (10YR5/2) haloes, after burrows and root traces; crude platy peds in upper part of horizon pass downward into distinct relict bedding; noncalcareous; skelmosepic porphyroskelic in thin section, with most grains deeply weathered nephelinite; common quartz and microcline; rare biotite and pyroxene; clear, smooth contact to

-11 cm; C horizon; medium-grained sandstone; dark grayish-brown (10YR3/2); this unit is disrupted into a tesselated pavement of blocks about 10-15-cm across by red clay skins and some drab-haloed root traces from above; within these blocks there is clear relict bedding; noncalcareous; skelmosepic intertextic in thin section, with abundant clasts of weathered nephelinite; common quartz, microcline, and pyroxene; abrupt, smooth contact to

-13 cm; A horizon of underlying paleosol (Buru clay thin-surface phase); clayey medium-grained sandstone; reddish-brown (5YR4/4); common light gray (5Y7/2) metagranotubules with grayish-brown (10YR5/2) haloes, after root traces and burrows; distinct relict beds of grayish-brown (10YR3/2) siltstone, reddish-brown (5YR4/4) claystone, and dark grayish-brown (10YR3/2) sandstone; some ripple-drift cross-lamination; noncalcareous; skelmosepic intertextic in thin section, with mainly weathered nephelinite grains.

4.2.6.4 Further examples. In the sequence of five Buru profiles between the type Tut clay paleosol and the Grey Sandstone Member of the Kapurtay Agglomerate, the type profile is the second down from the top of the Red Bed Member, and the Buru clay thin-surface phase paleosol is third from the top. These paleosols lack the pervasively displacive calcareous stringers of Kiewo Series paleosols but are otherwise similar. Buru Series paleosols also have been seen in the Hiwegi Formation on Rusinga Island, Kenya, within the badlands north of Hiwegi Hill, where they are a little more red and calcareous than examples from Songhor (Thackray, 1989).

4.2.6.5 Alteration after burial. All the alterations after burial outlined for the type Tut clay paleosol apply also to Buru paleosols: loss of organic matter, local burial gleization, reddening by dehydration of ferric hydroxides, compaction to 72 percent of thickness in clayey parts, to 85 percent in sandy parts, and small scale faulting.

4.2.6.6 Reconstructed soil. These soils can be envisaged as very weakly developed. They are little more than nephelinitic alluvium penetrated by a few roots and burrows. They are weakly calcareous to noncalcareous, but still retain some carbonate and mafic minerals, so they probably were near neutral in pH. Judging from surface soils of this pH (Birkeland, 1984), they would have been base saturated but with only modest cation exchange capacity. Their red color and abundant opaque oxides are evidence of a soil that was well drained for most of the year.

4.2.6.7 Classification. These were soils of minimal profile development, with little more than penetration of parent alluvium by root traces and burrows. They are best identified as Fluvents (of Soil Survey Staff, 1975), Calcaric Fluvisols (of Food and Agriculture Organization, 1974), Alluvial Soils (of Stace *et al.*, 1968), Uf1.23 (of Northcote, 1974) and Dundee dundee (of MacVicar *et al.*, 1977).

4.2.6.8 Paleoclimate. Buru paleosols are surprisingly clayey, red, and noncalcareous for soils developed on such alkaline parent materials. These could be taken as evidence of a warm, humid paleoclimate, but most of these features probably came with the parent alluvium derived from other soils. Paleosols as weakly developed as Buru Series cannot be regarded as adequate gauges of paleoclimate.

4.2.6.9 Ancient vegetation. The mixture of stout and fine deeply penetrating root traces along with relict bedding are indications that Buru paleosols supported woody vegetation early in the ecological succession toward forest represented by the associated Tut clay. No large stumps were seen, nor root traces larger than 2 cm in diameter, so large trees are unlikely. Root traces are abundant, and much of the thin (2-3-cm) surface horizon of these paleosols is drab colored, perhaps due to burial gleization of a moderately humified surface horizon. Buru paleosols lack evidence of salinization, high alkalinity, or hardpans likely for Kiewo paleosols. Compared with these paleosols, Buru paleosols probably supported more lush early successional vegetation.

Among the various kinds of early successional vegetation found in East Africa today, as outlined in discussion of Kiewo and Mobaw Series paleosols, Buru paleosols are likely to have supported vegetation comparable to the dense thickets dominated now by umbrella trees (*Maesopis emini*).

4.2.6.10 Former animal life. Only turtle scrap was found in these paleosols in Songhor gullies (bed 10 of Pickford & Andrews, 1981). Such a paucity of shell and bone is to be expected because these paleosols are weakly calcareous to non-calcareous and very weakly developed: it takes time and alkaline chemical conditions for such fossils to accumulate in soils (Pickford, 1986b; Retallack, 1984a). Many of the same kinds of animals found in the associated Tut clay paleosol may have passed over Buru profiles on their way to streams for a drink.

Buru paleosols do contain some vertical sand-filled burrows, 6 to 11 mm in diameter, referable to the ichnogenus *Skolithus*. As in associated Tut, Choka and Kiewo Series paleosols, these could have been excavated by insects or spiders.

4.2.6.11 Paleotopographic setting. Buru paleosols cap individual sandstone interbeds of a large heterolithic low-angle cross set in the uppermost Red Bed Member of the Kapurtay Agglomerate. This facies has been interpreted here as an ancient levee of a meandering stream that drained deeply weathered nephelinitic volcanics. The abundant drab mottles attributed to burial gleization and the shallow oxidized horizon above little-weathered green sandstone in Buru paleosols could reflect a high water table in such a stream-side location. Frequent flooding beside a stream also would account for the very weak development of these paleosols.

4.2.6.12 Parent material. Buru Series paleosols developed on volcaniclastic sandstones, consisting largely of prior-weathered nephelinite grains, and lesser amounts of pyroxene, microcline, and quartz. These materials were derived by river erosion of soils on nearby nephelinitic volcanics.

4.2.6.13 Time for formation. These shallow profiles with their clear relict bedding formed in a very short period of time, like Kiewo paleosols. They probably represent only a few to a few hundred years.

4.2.7 Lwanda Series (new name)

4.2.7.1 Diagnosis. Thin, brown (10YR), clayey surface over deeply weathered, white to red (10R), mottled corestones of phonolite.

4.2.7.2 Derivation. Lwanda is Dholuo for boulder (Gorman, 1972), in reference to the conspicuous phonolite corestones in this paleosol.

4.2.7.3 Description. The type Lwanda clay (Fig. 4.16) is in the gully, at the base of the middle Miocene, Fort Ternan Member and down into the Baraget Phonolite Member of the Kericho Phonolites (7.8 m in Fig. 4.10), 3-m below the prominent nephelinitic tuff (Fig. 4.9) at Fort Ternan National Monument (grid reference 603759 on Lumbwa sheet), Kenya (Figs 4.4 and 4.8).

+39 cm; C horizon of overlying paleosol (Rairo bouldery clay); boulder-bearing claystone; brown to dark brown (10YR4/3); with scattered granules of strong brown (7.5YR5/8), pale yellow (5Y8/3), gray (5Y5/1), and black (5Y2.5/1); few large boulders of deeply weathered, vesicular phonolite, light brownish-gray (2.5Y6/2) overall, with a mix of white (2.5Y8/2) vesicle fills and pseudomorphs of phenocrysts, and a grayish-brown (2.5Y5/2) matrix; indistinct coarse prismatic peds and fine angular peds defined by slickensided clay skins; weakly calcareous; bimasepic porphyroskelic in thin section; some recognizable volcanic rock fragments, but most are very deeply weathered; few irregular zones of iron stain (ses-

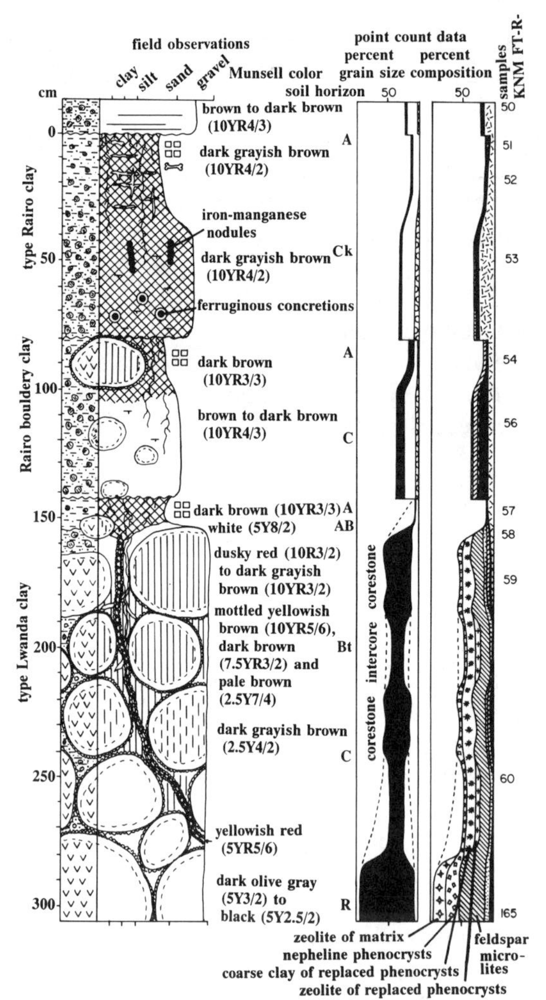

Fig. 4.16. Measured section, Munsell colors, soil horizons, grain size, and mineral composition of the type Lwanda clay, Rairo bouldery-clay and type Rairo clay paleosols below the main fossil quarry at Fort Ternan National Monument, Kenya. This section is 6.2 to 9.4 m in Fig. 4.10. Lithological key as for Fig. 4.12.

quans) after fine root traces; gradual, smooth contact to

0 cm; A horizon; boulder-bearing claystone; dark brown (10YR3/3); with granules of dark reddish-brown (5YR3/4), brownish-yellow (10YR6/6), yellow (2.5Y7/6), pale olive (5Y6/3), and greenish-gray (5G6/1); indistinct very coarse-prismatic and coarse-granular peds; few fine black (10YR2/1) mottles of iron-manganese, after root traces; noncalcareous; trimasepic porphyroskelic in thin section; a very clay-rich fabric with scattered rounded clay pellets (fecal?) and lithorelicts of deeply weathered phonolite; some quartz and opaque grains; clear, irregular(penetrating underlying corestones) contact to

-15 cm; AB horizon; upper portion of deeply weathered corestones of phonolite; white (5Y8/2); with white (5Y8/1), deeply weathered phenocrysts and vesicles; grading down to a level where the matrix includes patches of olive gray (5Y5/2) and phenocrysts and vesicles of white (5YR8/1) and pinkish-gray (5YR7/2); this horizon is penetrated by a distinct 1-3-cm wide, near-vertical, clastic dike of dark yellowish-brown (10YR4/6) claystone breccia; weakly calcareous; crystic porphyroskelic in thin section; matrix of the phonolite corestones shows numerous relict laths of feldspar, and there are some relict phenocrysts of nepheline, sanidine, pyroxene, and opaque oxides; most phenocrysts are replaced by a distinctive, highly birefringent clay, but some phenocryst cavities and many vesicles are filled with phillipsite and natrolite; gradual, irregular contact to

-23 cm; Bt horizon; phonolite corestone occupying most of trench; dusky red (10R3/2); with white (10R8/1) and pale red (10R6/2) deeply weathered phenocrysts arranged in trachytoidal texture; toward the interior of the corestone colors are dark reddish-brown (5YR2.5/2), with white (5YR8/1) and pinkish-white (5YR8/2) phenocrysts; grading further into the center of the corestone into very dark grayish-brown (10YR4/4) matrix, with deeply weathered phenocrysts of white (5YR8/1) and pinkish-white (5YR8/2), and scattered streaks of dark yellowish-brown (10YR4/4) stain (sesquans); small areas of deeply weathered claystone outside corestone are red (2.5YR4/6); this horizon also is penetrated by clastic dikes of yellowish-red (5YR5/6) claystone breccia; moderately calcareous overall; crystic porphyroskelic in thin section; the ferruginized and clayey phonolite matrix has light colored streaks after feldspar laths, and a few persistent phenocrysts of nepheline, sanidine, pyroxene, and opaque oxides, but most of these crystals are altered to a highly birefringent clay; gradual, irregular contact to

-52 cm; Bt horizon; phonolitic corestones and an area of intercore claystone occupy the trench; intercore matrix mottled yellowish-brown (10YR5/6), dark brown (7.5YR-3/2), and pale yellow (2.5Y7/4); adjacent corestones are dark olive-gray (5Y3/2) with areas of black (5Y2/1) and light olive-brown (2.5Y5/4), this is flanked by a weathering rind some 30-cm thick of weak red (2.5Y4/2) and another rind outside that, some 3-cm thick of weak red (2.5YR-4/2); also in corestones but not matrix are narrow veins of calcite that are light gray (5Y6/1) with outer layers of white (5Y8/1) and black (5Y2.5/1); moderately calcareous overall; gradual irregular contact to

-83 cm; C horizon; phonolite corestones; dark grayish-brown (2.5Y4/2), with weathered phenocrysts of light yellowish-brown (2.5Y6/4) and white (2.5Y8/2); in places gray (5Y7/2) with phenocrysts of white (5Y8/1); few calcite veins, light gray to gray (5Y6/1), with an outer crystal sheet of white (5Y8/1) and local mangan of black (5Y2.5/1); weathered rims to corestones at this level include a dark yellowish brown (10YR4/4) quasisesquan within a weak red (2.5YR4/2) neosesquan; this horizon also penetrated by clastic dikes of yellowish red (5YR5/6) claystone breccia; moderately calcareous overall; porphyroskelic crystic in thin section, with abundant relict, silt-sized feldspar laths and also some areas of zeolitic matrix unclotted by reddish clay; some relict phenocrysts of nepheline, sanidine, pyroxene, biotite, and opaque oxides, but many phenocrysts are altered to highly birefringent coarse clay or low-birefringent fibrous zeolite; gradual, irregular contact to

-160 cm; R; phonolite; dark olive-gray (5Y3/2) to black (5Y2.5/2), with phenocrysts of light gray (10YR7/2); few joint surfaces are stained reddish gray (5YR5/2), with phenocrysts of pink (7.5YR7/4) and black (5YR2.5/1); noncalcareous overall; even here some weathering was detected; fresh rock was not found in the trench but in the road bed of the vehicular track 200 m to the northeast; fresh parent material is a trachytoidal-textured phonolite with a matrix of glass and feldspar microlites, and abundant phenocrysts of nepheline, fewer phenocrysts of sanidine, pyroxene, and opaque oxides (Table 4.2).

4.2.7.4 Further examples. The type Lwanda paleosol is not well exposed outside the gully where the description above was made. A layer of 2-m-diameter corestones like those in the Lwanda clay can be mapped widely on top of the Baraget Phonolite (Fig. 4.8), and may represent similar paleosols, but the top of the profile was not seen elsewhere to verify this. A similar mappable layer of 1-m-diameter corestones on top of the upper Kericho Phonolite may represent other Lwanda profiles. These were not completely excavated either.

4.2.7.5 Alteration after burial. Losses of organic matter during burial, as in other Kenyan paleosols studied, are also likely for the Lwanda clay. Subsurface horizons of this paleosol may also have been reddened by burial dehydration of ferric hydroxides, but the surface horizon is neither red nor reddened.

Cementation by calcite is seen locally at depth. The calcite veins penetrate weathered phonolite, and thus postdate some weathering. The last filling of the veins containing calcite is opaque iron manganese similar to that seen associated with compacted root traces higher in the profile. This observation may be evidence for calcite cementation during soil formation or very early after burial. Many of the phenocrysts of the parent phonolite are pseudomorphed by fibrous zeolite, probably phillipsite, and a highly birefringent clay, probably mixed-layer montmorillonite-vermiculite (Fig. 2.33), but these are cut in places by calcite veins and by clastic dikes of claystone breccia. Much of this replacement was thus completed by the time of soil formation. It could be argued that the contact between an impermeable clayey layer and the rubbly phonolite was a favored location for intrastratal alteration during burial. However, these effects were minor, considering the relationship of pseudomorphed phenocrysts and vesicles to veins and clastic dikes, as already outlined. Only pedogenically generated microfabrics and altered grains were seen in the surface horizon.

Compaction of the phonolite corestones, even those deeply weathered, was probably slight. Their vesicles and pseudomorphed phenocrysts are little distorted compared with those in fresh phonolite. The clayey upper portion of the profile however may well have suffered reduction to 74 percent of former thickness estimated from the likely depth of burial of the Fort Ternan Member. This clayey layer contains few and scattered rigid grains and is noncalcareous, so it is not surprising that the clastic dike filled with claystone breccia seen low in the profile is distorted beyond recognition in the A horizon.

4.2.7.6 Reconstructed soil. The Lwanda clay paleosol can be envisaged as a bouldery top to a deeply weathered phonolite flow, with a surface (A) horizon of varying thickness between the corestones, of noncalcareous, dark-brown, granular-structured clay. The subsurface (Bt) horizon was deeply weathered, brown to red (10R to perhaps as brown as 5YR), but not so clayey or well structured as in an argillic horizon (of Soil Survey Staff, 1975). The thick saprolite (C horizon) of the soil probably was colorfully variegated around large (2-m) corestones. The profile was prone to cracking, which created clastic dikes of claystone breccia, penetrating at least to the level of the saprolite. Its pH was probably near neutral at the surface and mildly alkaline below that, to judge from the distribution of carbonate and easily weathered minerals. These observations also are compatible with high base saturation and moderate to high cation exchange capacity. The highly oxidized nature of the Bt horizon and the great depth of the saprolite in this paleosol are evidence of generally good drainage, with water table well below 1.6 m from the surface.

4.2.7.7 Classification. In the field this paleosol appears very deeply weathered compared with its parent phonolite, as would be expected in Vertisols, Ultisols, or Oxisols. However, the subsurface clayey (Bt) horizon does not qualify as argillic, is moderately calcareous, and retains much relict vesicular and crystalline structure of the parent phonolite. Its clastic dikes are narrow and not associated with deformation of soil horizons found in Vertisols (as described by Paton, 1974). The Lwanda clay is best classified as an Inceptisol, too deeply weathered for an Ochrept, and lacking the obvious pedogenic carbonate that might be expected in an Ustropept. A better identification for this paleosol is Vertic Eutropept (of Soil Survey Staff, 1975), which is equivalent to Vertic Cambisol (of Food and Agriculture Organization, 1974). Among Australian soils (Stace *et al.*, 1986), the Lwanda clay is most like Chocolate Soils or Ug5.32 in the Northcote (1974) key. In the South African scheme (MacVicar *et al.*, 1977) it is similar to Glenrosa ponda.

These kinds of surface soils are found on mafic lavas in Zimbabwe (map unit Bv13-3a of Food and Agriculture Organization, 1977a) and South Africa (Botha, 1985; Gertenbach, 1983; Venter, 1986). A broadly similar soil is developed on Miocene phonolite in the northeastern part of Serengeti National Park, Tanzania (profile Lamai 1, a Petroferric Haplustoll of Jager, 1982), but that profile includes plinthite and quartz gravel from a geologically longer and more complex history than envisaged for the type Lwanda clay. Also different are surface soils on Kericho Phonolite today (Humic Nitosols and Humic Cambisols of map units Nh2-2c and Bh4-3c of Food and Agriculture Organization, 1977a), which are more ferruginized, more humified, and less calcareous than the Lwanda clay.

4.2.7.8 Paleoclimate. The depth of weathering and degree of decalcification of the Lwanda clay are about what might be expected in a subhumid climate. Surface soils on Kericho Phonolites around Fort Ternan (Food and Agriculture Organization, 1977a) enjoying 1500 mm a year in rainfall (Binge, 1982) are now completely decalcified. Also noncalcareous are soils forming on phonolite flows and colluvium in the humid climate of Mt Kenya (Mahaney, 1989; Mahaney & Boyer, 1989). In contrast, several soils in and near phonolite in the northern part of Serengeti National Park, where mean annual rainfall is 1000 mm (Lamai profiles of Jager, 1982), are slightly more calcareous than the Lwanda clay. These Tanzanian soils are on erosional landscapes at least as old as Miocene, and include products of weathering under ancient as well as the modern climate. The type Lwanda clay also is smectitic (Fig. 2.33), and so unlikely to have received more than 1200 mm of mean annual rainfall, by comparison with other East African volcanic soils (Mizota & Chapelle, 1988; Mizota, Kawasaki, & Wakatsuki, 1988). These comparisons serve broadly to constrain paleoclimate of the Lwanda clay to subhumid.

The depth and thoroughness of weathering of the phonolite could be taken as an indication of a tropical climate (Birkeland, 1984). This is especially impressive here because of the relict crystal textures remaining within the weathered corestones from the parent phonolite; an indication that the paleosol formed over a modest interval of time.

Clastic dikes filled with claystone breccia, including clasts like the surface horizon, may have formed as deep cracks during a dry season. These features are not so pronounced as in Vertisols (Paton, 1974), but if weathering had continued to reduce the phonolite more completely to clay, these paleosols may well have become Vertisols. A pronounced dry season may well have been established in East Africa by middle Miocene time.

4.2.7.9 Ancient vegetation. No fossil plants and only a few large root traces were seen in the Lwanda clay paleosol. The root traces, the thorough weathering around corestones, and the subsurface clayey (Bt) horizon are features of forested soils. On the other hand, the Lwanda clay has a brown, clayey, granular-structured surface horizon, deep clastic dikes, and some subsurface carbonate, as is more typical of grassland soils. The grassland-related features would have been more readily destroyed under forest than the forest features under grassland, if Miocene relict paleosols at the surface today are a guide (Jager, 1982; MacFarlane, 1976). Thus a forest cover of the Lwanda clay probably would have been early in its history, if at all. Just before burial it probably supported grassy woodland or wooded grassland.

Generally comparable surface soils in the United States (Soil Survey staff, 1975), Australia (Stace *et al.*, 1968) and Africa (Food and Agriculture Organization, 1977a; Gertenbach, 1983; Venter, 1986) support vegetation ranging from grassy woodland to wooded grassland, open grassland, and desert shrubland. Forested soils on phonolite, such as those at the surface today on the Kericho Plateau south of Fort Ternan (map units Nh2-2c, Bh14-3c of Food and Agriculture Organization, 1977a) are more decalcified than the Lwanda clay. Confining attention to comparably calcareous soils on mafic rocks in Africa, their vegetation is more consistent: dry grassy woodland to wooded grassland of the Sudanian and Zambezian floristic realm (of White, 1983). This modern vegetation includes such trees as *Acacia*, *Combretum*, *Terminalia*, *Adansonia*, and *Sclerocarya*.

As already outlined for Choka Series paleosols, dry woodland vegetation of Sudanian and Zambezian floristic affinities existed in Africa as long ago as early Miocene, judging from fossil plant assemblages of that age (Beauchamp, Lemoigne, & Petrescu, 1973; Chaney, 1933; Chesters, 1957; Collinson, 1983; Lakhanpal & Prakash, 1970; Lemoigne, Beauchamp, & Samuel, 1974). Among the living plants just mentioned, pollen of *Acacia* and grasses (Gramineae) are known from Eocene marine rocks of Cameroon (Salard-Cheboldaeff, 1979), and in Tertiary rocks elsewhere in Africa (Bown *et al.*, 1982; Kedves, 1971). *Acacia* also may be represented by twigs with curved stipular spines found in the Chogo paleosols at Fort Ternan (Kenyan National Museum specimen, FT1962'167). Grass pollen (Bonnefille, 1984) and sheathed culms also have been found in the Chogo clay, and well-preserved grasses in the Onuria clay paleosol at Fort Ternan (Dugas, 1989). The Lwanda paleosol is not much removed stratigraphically or in geological age from Chogo and Onuria paleosols, and it is likely that vegetation had already changed to grassy woodland and wooded grassland by the middle Miocene formation of the Lwanda clay, from the dry forests indicated by early Miocene Tut and Choka series paleosols.

4.2.7.10 Former animal life. No animal fossils or their traces were seen in the Lwanda clay paleosol in the single gully examined. Nor would snail shells or mammal bone be expected, considering their noncalcareous surface horizon (Retallack, 1984a).

4.2.7.11 Paleotopographic setting. The strong oxidation and deep corestone-style weathering of this paleosol are indications of water table well below 1.6 m from the surface. There are clasts of quartz and deeply weathered nephelinite in the surface horizon as well as some large boulders, probably derived from a hillslope nearby. This also has been revealed by small-scale mapping (Fig. 4.8; see also Pickford, 1984, 1986a), which shows the Baraget Phonolite banked against a weathered ridge of nephelinitic agglomerate that was part of a quiescent stratovolcano to the north. The Lwanda clay formed on the toeslopes of this volcano near where it flattened out into the extensive lava plains to the south.

4.2.7.12 Parent material. Most of this profile has relict crystalline structures of its parent material of vesicular Baraget Phonolite, but the surface clayey horizon has nephelinite and quartz probably derived from the nearby nephelinite stratovolcano and xenoliths of Precambrian basement. This and clays between corestones and within clastic dikes may have been derived by sheetwash down the slope that also introduced some large boulders into the surface clayey horizon.

4.2.7.13 Time for formation. An upper limit of about a million years for the type Lwanda clay is provided by radiometric dating of its phonolitic parent material at about 15 million years old and of tuffs above four additional paleosols that are about 14 million years old (Shipman *et al.*, 1981). Near Kilombe, east of Mt Londiani in the dry Gregory Rift of Kenya, spheroidally weathered trachyphonolite radiometrically dated, using K/Ar on whole rock, at 1.7 ± 0.05 million years is overlain by tuffs that are magnetically reversed and so more than 700,000 years old (Bishop, 1978). These limits seem overgenerous in view of the preservation of relict volcanic structures in the Lwanda clay. In wet and cool parts of Mt Kenya, phonolitic lavas and colluvium have been weathered to soils with B horizons showing relict igneous and colluvial structures within 1940 years, but accumulation of clay in Bt horizons to levels that qualify as argillic has taken 40,000 years or more (Mahaney, 1989; Mahaney & Boyer, 1989). From these comparisons, a time for formation for Lwanda paleosols of a few thousands to a few tens of thousands of years is most likely.

4.2.8 *Rairo Series (new name)*

4.2.8.1 Diagnosis. Thin (less than 30 cm), brown (10YR) paleosols, with relict bedding and deep cracks on clayey volcaniclastic sediments.

4.2.8.2 Derivation. Rairo means gray in Dholuo (Gorman, 1972).

4.2.8.3 Description. The type Rairo clay paleosol (Fig. 4.16) is in the middle Miocene, lower Fort Ternan Member of the Kericho Phonolites, within the gully 1-m below the prominent nephelinite tuff (9.3 m in Fig. 4.10) at Fort Ternan National Monument (grid reference 603759 on Lumbwa sheet), Kenya (Figs 4.4, 4.8).

+70 cm; C horizon of overlying paleosol (type Rabuor clay); clayey, granule conglomerate; brown to dark brown (10YR4/3); with granules of pale olive (5Y6/4), greenish-gray (5G5/1 and 5BG5/1), yellow (2.5Y7/6), strong-brown (7.5YR5/8), and black (5Y5/1); prominent relict bedding is picked out by subhorizontal and also some vertical, 2-3-mm-wide stringers of white (10YR8/2) to light gray (10YR7/2) sparry calcite; few scattered, 8-mm-diameter nodules of micrite, also white (10YR8/2) to light gray (10YR7/2); few bone chips, white (N8/0), locally stained very pale brown (10YR8/3); weakly calcareous; skelsepic intertextic in thin section; mainly clasts of deeply weathered nephelinite; few pyroxene and quartz; abrupt, smooth contact to

0 cm; A horizon; granule-bearing claystone; dark grayish-brown (10YR4/2); with granules up to 4 mm of yellow (5Y7/6 and 2.5Y7/6), strong-brown (7.5YR4/6), dark greenish-gray (5G4/1), and black (5Y2/1); few angular clasts of carbonatite tuff, light brownish-gray (2.5Y6/2), with a 2-3-mm halo of very dark grayish-brown (10YR3/2); few root traces up to 4-mm across, some of which are replaced by black (10YR2/1) iron-manganese; few friable fragments of bone as in overlying layer; structure is angular blocky to platy, and defined by 3-4-mm thick subhorizontal and vertical, brownish-gray (10YR6/2) veins

filled with sparry calcite; noncalcareous; porphyroskelic skelmosepic in thin section; common deeply weathered and micritized rock fragments of nephelinite and phonolite; few calcite crystal tubes after roots; gradual, smooth contact to

-13 cm; A horizon; granule-bearing claystone; dark grayish-brown (10YR4/2); with granules up to 4 mm of yellow (10YR7/6), olive (5Y4/4), reddish-gray (5YR5/2), bluish-gray (5B6/1), and black (5Y2/1); common, distinct, light brownish-gray (10YR6/2) calcans define angular-blocky to platy peds; few slickensides; moderately calcareous; intertextic skelinsepic; common rock fragments of phonolite and nephelinite; few micritized grains of biotite and pyroxene; diffuse, smooth contact to

-26 cm; C horizon; clayey granule conglomerate; dark grayish-brown (10YR4/2); with clasts up to 2 cm in diameter of strong-brown (7.5YR5/8), pale yellow (2.5Y7/4 and 5Y7/4), dark greenish-gray (5GY4/1), and black (10YR2/1); distinct near-vertical very dark grayish-brown (10YR3/2) to black (10YR2/1) planes of iron-manganese (mangans), along with bedding planes, outline irregular, subangular-blocky peds in the upper part of this horizon; few diffuse iron-manganese concretions, up to 3 cm in diameter, also are very dark grayish-brown (10YR3/2) to black (10YR2/1); moderately calcareous; intertextic skelinsepic in thin section; many rock fragments of nephelinite and phonolite, ranging from fresh to extremely weathered; few biotite and pyroxene; abrupt, wavy contact to

-80 cm; A horizon of underlying paleosol (Rairo bouldery-clay); claystone; dark brown (10YR3/3); with clasts up to 1 m across of deeply weathered, yellow (10YR8/8 and 5Y7/6), dark bluish-gray (5B4/1), and black (5Y2/1) nephelinite and phonolite, the largest boulders being only phonolite; indistinct, coarse-prismatic and medium-granular peds defined by clay skins (argillans) and local slickensides; few distinct iron-manganese mottles of very dark gray (10YR3/4); few irregular drab mottles of grayish-brown (2.5Y5/2) ; few fine and coarse root traces, in some cases black or drab, but mostly areas of dark brown (10YR3/3) disturbed claystone; very weakly calcareous; porphyroskelic trimasepic in thin section, with phonolite and nephelinite rock fragments varying from fresh to extremely weathered.

4.2.8.4 Further examples. Another similar profile, the Rairo bouldery-clay paleosol, underlies the type Rairo clay and overlies the type Lwanda clay paleosol in the gully below the main fossil quarry at Fort Ternan (Figs 4.9, 4.10, and 4.16).

4.2.8.5 Alteration after burial. As for the surface horizon of the Lwanda clay paleosol, Rairo paleosols may have lost much organic matter but were little affected by burial dehydration of ferric hydroxides or cementation during early burial. The clayey Rairo paleosols also could have been compacted to 74 percent of their former thickness, considering their clayey, noncalcareous, and slickensided nature. Thus the surface horizon of the type Rairo clay may have been as thick as 35 cm.

4.2.8.6 Reconstructed soil. The original Rairo soil may have had a dark brown, well humified A horizon, with indistinct structure beyond relict bedding, over more clearly bedded sandy nephelinitic parent material. The iron-manganese spots, mottles, and concretions are indications of slow drainage, perhaps because of footslope seepage encouraged by the less permeable clayey surface horizons of underlying paleosols (Rairo bouldery-clay and Lwanda clay). The degree of chemical weathering, carbonate content, and bioturbation of these paleosols are compatible with an original neutral to mildly alkaline pH, high base saturation, and moderate to high cation exchange capacity.

4.2.8.7 Classification. These weakly developed paleosols have relict bedding obscured to a greater extent than in Entisols. These paleosols are darker than Ochrepts, and may have been darker originally. Given their tropical lowland location in the middle Miocene, compatible with evidence of local waterlogging, and their deeply weathered boulders of phonolite, Rairo paleosols are best identified as Vertic Ustropepts (of Soil Survey Staff, 1975). In the Food and Agriculture Organization (1974) classification these are Cambisols, probably Vertic Cambisols judging from evidence of surface cracking. Brown Clays (of Stace *et al.*, 1968) and Ug1 (of Northcote, 1974) are the most similar surface soils found in Australia, but these categories include a variety of profiles, few of which are very similar to Rairo clays. In the South African scheme (of MacVicar *et al.*, 1977), Rairo paleosols are most like Inhoek coniston.

4.2.8.8 Paleoclimate. Weakly developed, partly gleyed paleosols like the Rairo clays cannot be regarded as compelling evidence for paleoclimate. Vertic Cambisols on mafic parent materials in Africa (map units Bv2, Bv8-3a, Bv13-3a of Food and Agriculture Organization, 1977a) are found widely in semiarid to subhumid, monsoonal, tropical to subtropical regions of Mali, Nigeria, and Zimbabwe. A dry but not desertic climate also is compatible with the moderately calcareous composition and dark swelling clays of Rairo paleosols. Slickensides, surface cracks, reduction spots, and iron-manganese mottles and concretions are evidence of shrink-swell behavior and waterlogging, which may both have occurred during a wet season.

4.2.8.9 Ancient vegetation. These paleosols contain scattered stout root traces among more common fine ones. Their relict bedding is not completely obscured by roots and burrows, so they may have supported vegetation early in plant succession, colonizing sites disturbed by colluvium or alluvium. If further time were available, soils and vegetation of Rairo paleosols may have become similar to those envisaged for the associated Rabuor clay.

Comparable modern soils (Soil Survey Staff, 1975; Food and Agriculture Organization, 1977a; Stace *et al.*, 1968) form under vegetation ranging from grassy woodland to wooded grassland and desert shrubland. Desert vegetation is unlikely for reasons outlined already for paleoclimate of these paleosols. Nor is an old-growth woodland likely, considering the paucity of stout root traces and the persistence of relict bedding. Analogous soils to Rairo paleosols are found in the southern Serengeti Plains of Tanzania, where there are weakly developed, dark, clayey soils on carbonatite-nephelinite tuff from Oldoinyo Lengai volcano (Anderson & Talbot, 1965; de Wit, 1978; Food and Agriculture Organization, 1977a). These support wooded grassland, with scattered trees of *Acacia tortilis*, within mid-length to short grasses of *Pennisetum mezianum*, *P. stramineum*, *Cynodon dactylon*, and *Andropogon greenwayi* (White, 1983). As already outlined for the type Lwanda clay paleosol, the fossil record of grasses and *Acacia* extends at least back to the Eocene in Africa, and there is evidence of both from Chogo and Onuria Series paleosols at Fort Ternan.

4.2.8.10 Former animal life. Some friable fragments of unidentifiable bone were found in the surface of the type Rairo clay paleosol and in the overlying type Rabuor clay. The paucity of bone in these paleosols is to be expected considering their weakly calcareous composition (Retallack, 1984a). Some fossils found previously at Fort Ternan could have come from this stratigraphic level, but if so, were more likely from the type Rabuor clay and so discussed there. No trace fossils were seen in Rairo clays. Although direct evidence is lacking, Rairo clays are not greatly different from associated Rabuor, Chogo, and Onuria paleosols, and so may have shared generally similar animal life.

4.2.8.11 Paleotopographic setting. Rairo clays show many features of well-drained soils: deeply penetrating root traces, clay skins, and a brown oxidized color. There also are features of periodic waterlogging: reduction spots, surface cracks, and iron-manganese concretions, as in some rice soils today (Rahmatullah *et al.*, 1990). Considering their highly clayey composition, slow drainage would have been expected.

These paleosols probably formed near the break in slope between a nephelinitic stratovolcano and an extensive phonolitic lava plain, as indicated by geological mapping (Fig. 4.8). Footslope seepage may have been perched on the clayey surface of the underlying Lwanda paleosol. Concentration of groundwater at this level could be part of the reason why Rairo paleosols are so clayey compared with other paleosols at Fort Ternan. Anomalously large, deeply weathered blocks of volcanic rock in these clayey paleosols may have been derived by colluvial sheetwash down the sides of the quiescent volcano.

4.2.8.12 Parent material. Clasts within and below Rairo paleosols indicate that they formed on volcaniclastic pebbley sandstones of mixed nephelinitic and phonolitic composition. Most of the large boulders in these paleosols are of Baraget Phonolite, also deeply weathered, and so probably derived from soils upslope. Rare, little-weathered crystals of biotite and pyroxene in Rairo paleosols may have been introduced by thin dustings of airfall tuff.

4.2.8.13 Time for formation. Rairo paleosols are similar to the type Rabuor clay paleosol, but are thinner and only weakly developed. Better developed, weakly calcareous Inceptisols have formed on low coastal terraces of coral about 25,000 years old along the wet and hot Kenyan coast (Braithwaite, 1984; Sombroek, Braun, & Van Der Pouw, 1983). More like Rairo paleosols are Inceptisols on phonolitic colluvium radiocarbon dated at 1940 to 3610 years old in the cool, wet upper Teleki Valley of Mt Kenya (Mahaney, 1989; Mahaney & Boyer, 1989; Mahaney & Spence, 1989). Rairo paleosols similarly represent only a few thousand years.

4.2.9 Rabuor Series (new name)

4.2.9.1 Diagnosis. Thick, brown (10YR) clayey paleosols, with abundant slickensides, on clayey volcaniclastic colluvium.

4.2.9.2 Derivation. Rabuor means brown in Dholuo (Gorman, 1972).

4.2.9.3 Description. The type Rabuor clay (Fig. 4.17) is in the middle Miocene, lower Fort Ternan Member, 35-cm below the prominent nephelinitic tuff (at 10.7 m in Fig. 4.10) in the gully (grid reference 603759 on Lumbwa sheet) below the main fossil quarry at Fort Ternan National Monument, Kenya (Figs 4.4 and 4.8).

+20 cm; C horizon of overlying paleosol (Dhero clay loam); granule-bearing sandstone; light olive-brown (2.5Y5/4); with granules of strong-brown (7.5YR4/6), white (10YR8/2), olive-yellow (2.5Y6/6), pale olive (5Y6/4), greenish-gray (5G5/1), dark bluish-gray (5B4/1), and black (5Y2/1); clear relict bedding; few prominent burrows (meta-isotubules) up to 2 cm in diameter of white (10YR8/2) micrite, with a diffusion mangan of dark grayish brown (10YR4/2); few calcite crystal sheets, penetrating to this level from overlying massive nephelinitic tuff; moderately calcareous; granular skelsepic; abundant grains of weathered nephelinite; little pyroxene, calcite and opaque oxides; tongues of this material 1-1.5-cm wide extend for up to 6 cm into the underlying paleosol, making an irregular, clear contact to

0 cm; A horizon; claystone; dark grayish-brown (10YR4/2); with scattered grains of yellow (2.5Y7/6), pale olive (5Y6/4), and dark gray (5Y4/1); few, large (up to 13-cm) clasts of weathered phonolite, reddish brown (5YR4/3), with phenocryst pseudomorphs of white (5YR8/1) and a weathering rind of pinkish white (5YR8/2); common root traces, 3-1-mm in diameter, filled with strong-brown (7.5YR4/6) claystone; few coarse, distinct, light brownish-gray (2.5Y6/2) mottles with thin (2-mm) clay skins of dark brownish-gray (10YR3/2); indistinct coarse granular peds, outlined by strong-brown (7.5-YR4/6) clay skins; weakly calcareous; skelmosepic porphyroskelic in thin section; few weathered fragments of nephelinite; rare pyroxene, quartz, and opaque oxides; wavy, gradual contact to

-17 cm; A horizon; claystone, brown to dark brown (10YR4/3); with scattered granules of strong-brown (7.5YR5/8), very pale brown (10YR7/4), greenish-gray (5GY6/1), dark bluish-gray (5B4/1), and black (5Y2/1); common root traces, 4-5-mm in diameter, filled with strong-brown (7.5YR4/6 and 7.5YR5/8) claystone; few strong-brown (7.5YR4/6) zones of iron stain (sesquans) and incipient concretions up to 7 cm in diameter; few light olive-gray (5Y6/2) nodules with white (5Y8/2) core, forming local drab haloes to root traces; indistinct angular-blocky peds outlined by root traces and impersistent black (5YR2.5/1) areas of iron-manganese (mangans); weakly calcareous; skelmasepic porphyroskelic in thin section; clasts mainly weathered nephelinite, with some phonolite; common root traces filled with opaque iron-manganese; few grains of quartz and pyroxene; gradual wavy contact to

-41 cm; Bt horizon; claystone; dark brown (10YR3/3); with scattered granules of yellow (10YR7/6 and 5Y7/6), pale yellow (5Y7/4), and black (10YR2/1); also a few weathered boulders, up to 23-cm across, of phonolite, dark brown (7.5YR3/2), with phenocryst pseudomorphs of white

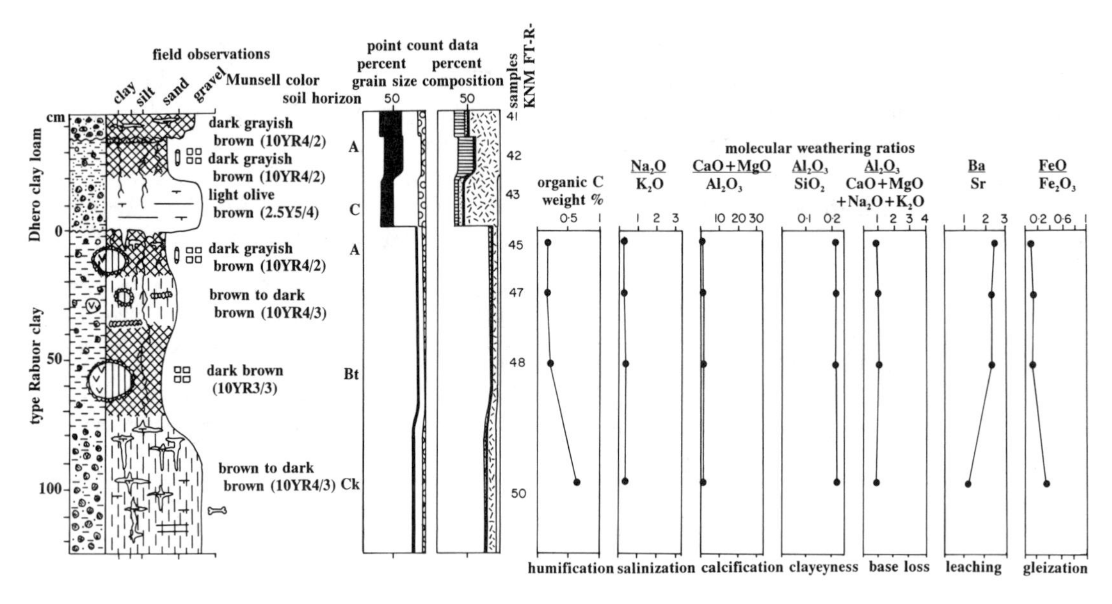

Fig. 4.17. Measured section, Munsell colors, soil horizons, grain size, mineral composition, organic matter, and selected molecular weathering ratios of the Dhero clay-loam and type Rabuor clay paleosols below the main fossil quarry at Fort Ternan National Monument, Kenya. This section is 9.5 to 11.2 m in Fig. 4.10. Lithological key as for Fig. 4.12.

(7.5YR8/1) and pinkish-white (7.5YR8/2), as well as a weathering rind that is dark grayish-brown (10YR4/2) with phenocryst pseudomorphs of white (10YR8/1 and 10YR-8/2), and an outermost, thin, ferruginized zone (sesquan) of yellow (2.5Y8/6); common root traces, 1-4-mm in diameter, filled with strong brown (7.5YR5/8) clay; crude prismatic peds, with slickensided clay skins; weakly calcareous; skelmasepic porphyroskelic in thin section; grains largely nephelinite, with some phonolite; few quartz, pyroxene, or opaque oxides, the latter showing wispy ferrans from weathering in place; smooth, gradual contact to

-75 cm; C horizon; clayey granule conglomerate; brown to dark brown (10YR4/3); with common clasts of strong-brown (7.5YR5/8), yellow (2.5Y7/6), pale olive (5Y-6/4), greenish-gray (5G5/1 and 5BG5/1), and black (5Y-5/1); massive, with clear relict bedding; common narrow (2-3-mm), calcite crystal sheets, that fill impersistent subhorizontal and vertical cracks; few light gray (10YR7/2) to white (10YR8/2) calcareous nodules up to 8 mm in diameter; rare splinters of friable fossil bone, white (N8/0) and stained very pale brown (10YR8/3); weakly calcareous overall; skelinsepic intertextic in thin section; common grains of deeply weathered nephelinite; few pyroxene, quartz, and opaque oxides; wavy, abrupt contact to

-125 cm; A horizon of underlying paleosol (type Rairo clay); granule-bearing claystone; dark grayish-brown (10YR4/2); with granules up to 4 mm of yellow (5Y7/6 and 2.5Y7/6), strong brown (7.5YR4/6), dark greenish-gray (5G4/1), and black (5Y2/1).

4.2.9.4 Further examples. No other paleosols of this kind were identified with confidence. Superficially similar, thick, brown claystones are widespread in the Fort Ternan Member, in the gully 500-m west of the main fossil quarry, and in the hillside outcrops at Serek and Kapsibor to the northeast of the quarry (Fig. 4.4).

4.2.9.5 Alteration after burial. As for other paleosols at Fort Ternan, the Rabuor clay probably has lost much organic matter but was not reddened appreciably by dehydration of ferric hydroxides. This is a very clayey profile, so that compaction during burial may have been significant: perhaps to only 74 percent of its original thickness, from an original depth of 101 cm to the base of horizon Bt. There may also have been some calcite cementation into cracks low in the profile. This probably formed soon after burial in cracks that would have been closed under a great weight of overburden.

4.2.9.6 Reconstructed soil. The Rabuor clay can be envisaged as a thick dark-brown soil. It had deep cracks, which may have opened during the dry season, but this structural instability was ameliorated by scattered large blocks of phonolite and by a dense root mat that created granular peds at the surface. The degree of leaching of bases was slight, as indicated by ratios of alumina/bases and barium/strontium (Fig. 4.17), so pH was probably neutral to alkaline. Base saturation probably was high and cation exchange capacity moderate. Soda/potash ratios decline up profile and are evidence against salinization. The paleosol was oxidized, with low values of, and an up-profile decrease in, ferrous iron/ferric iron ratios. Thus it was not permanently waterlogged in horizon Bt or above. Nevertheless, the high content of organic matter, incipient iron-manganese concretions, and iron-manganese filling root traces are indications of slowly draining or seasonally waterlogged soil.

4.2.9.7 Classification. Hand specimens taken from this paleosol were examined by I. W. Cornwall (for Bishop & Whyte, 1962) and compared with Braunlehm soils (of Kubiena, 1970). This is equivalent to Orthox (of Soil Survey Staff, 1975) and Xanthic Ferralsol (of Food and Agriculture Organization, 1974); both soil types much more depleted in bases than the type Rabuor clay paleosol (Fig. 4.17). This thick clayey paleosol has clastic dikes to a depth of 6 cm and a crude system of slickensided joints, but these structures and the degree of development of the paleosol fall short of the lentil peds and deformed horizons found in Vertisols. The Rabuor clay is best classified as a Vertic Argiustoll (of Soil Survey Staff, 1975), similar to profiles around the western and northwestern margins of the Serengeti Plains of Tanzania (Girtasho II profile of de Wit, 1978; and Mbalageti 1 profile of Jager, 1982). Also similar are soils in the transition zone between "vertisols" and "eutrophic brown soils" in the footslopes of Ngorongoro Crater, Tanzania (Anderson & Herlocker, 1973) and the Saka Series on carbonatite tuff of the Fort Portal volcanic field, southwestern Uganda (Harrop, 1960). The Ugandan soils are deeper, darker, and more decalcified. The type Rabuor clay is most similar in profile form and chemical depth functions to the Serengeti soils, but is more mafic in overall composition because its parent material was not diluted with quartzo-feldspathic debris from kopjes of Pre-cambrian gneiss as are soils on the Serengeti Plains.

In the Food and Agriculture Organization (1974) classification, soils with granular structured dark surface horizons are, in order of depth and degree of development, Kastanozems, Chernozems, and Phaeozems. The Rabuor clay is most like Luvic Kastanozems, which are uncommon soils on the Food and Agriculture Organization (1977a) soil map of Africa: for example, in Morroccan intermontane basins on Jurassic sandy marls (map unit K1135-3c). Soils with dark, organic, granular-structured surface horizons (mollic epipedons) are under-represented on the FAO map, because its compilers thought that soil organic matter is much too rapidly destroyed in tropical climates for the formation of soils as rich in organic matter as mollic soils of temperate regions. Subsequent work on soils of the Serengeti Plains (de Wit, 1978; Jager, 1982) has shown that these soils really are mollic, and qualify as Kastanozems, Chernozems and Phaeozems, rather than only Cambisols, as they are represented on the FAO map (1977a, unit Bk25-2a).

In the Australian classification (Stace *et al.*, 1968), the Rabuor clay is more calcareous than otherwise similar Chocolate Soils, Brown Earths, and Non-calcic Brown Soils, not so vertic or dark as Black Earths, nor so calcareous and finely structured as Chernozems. This leaves Prairie Soils and Ug3.3 in the Northcote (1974) key as the most similar surface soils.

In the South African classification (MacVicar *et al.*, 1977), the Rabuor clay is most like Bonheim glengazi. Less calcareous but otherwise similar profiles (Bonheim bonheim) are widespread on basalts in northern Kruger National Park, South Africa (Botha, 1985; Gertenbach, 1983; Venter, 1986).

4.2.9.8 Paleoclimate. The depth of the calcareous horizon of the Rabuor clay, here reconstructed to be about 101 cm, is similar to that found in soils of the Serengeti Plains receiving more than 800-mm mean annual rainfall (de Wit, 1978; Jager, 1982). However, this interpretation is undermined by the weak development of calcareous nodules in and below horizon Bt, where most of the carbonate may be cement formed during early burial. The thickness of decalcified Rabuor clay may have accumulated as colluvial slopewash at the foot of a slope, again as in many Serengeti soil catenas (for example, the Barsek catena of de Wit, 1978). The lighter colored, near surface horizon between 17 and 41 cm could have been a new layer of parent material, later incorporated into the profile. By this interpretation the carbonate is now only 34 cm from the surface, and this would have been about 46 cm from the surface originally, as in Serengeti soils receiving mean annual rainfall of about 550 mm. This can be taken as a minimum estimate, and the other estimate given can be regarded as an unlikely maximum.

The total amount of rainfall received by the Rabuor clay is difficult to estimate, but there are several indications that rainfall was seasonal in distribution through the year: surface cracks filled with material fallen in from above, common slickensides, and sparse, small, iron-manganese concretions. These features all are better expressed in Vertisols of seasonal climates (Duchafour, 1982).

There is little secure evidence for paleotemperature. The degree of weathering of volcanic clasts showing relict crystal structure and abundant root traces are compatible with a productive ecosystem and warm temperatures.

4.2.9.9 Ancient vegetation. These kinds of unstable clayey soils with thick, dark, granular surface horizons, abundant fine root traces, and common stout ones, commonly support grassy woodland or wooded grassland.

Comparable surface soils of the Australian (Stace *et al.*, 1968), U.S. Soil Conservation Service (Soil Survey Staff, 1975) and Food and Agriculture Organization (1974) classifications support vegetation ranging from grassy woodland to wooded grassland and desert shrubland, as for Lwanda paleosols. Lwandalike profiles on phonolite may have been upslope, contributing the weathered boulders found in the Rabuor clay. As discussed for Lwanda paleosols, desert shrubland is unlikely for the Rabuor clay. The parent material of the Rabuor clay is best matched in the Serengeti region of Tanzania, where deeply weathered carbonatite-nephelinite tuffs locally form a mantle over nephelinitic and basaltic lavas (de Wit, 1978; Food and Agriculture Organization, 1977a; Jager, 1982). These deep, cracking soils support a tall grassland of *Andropogon greenwayi*, *Digitaria macroblephora*, *Cynodon dactylon*, and *Eustachys paspaloides*, with scattered trees of *Acacia tortilis* (Anderson & Talbot, 1965; White, 1983). As already discussed for the Lwanda clay, both *Acacia* and a variety of grasses were in existence in East Africa by middle Miocene time.

4.2.9.10 Former animal life. Some unidentifiable, friable splinters of long bones were found in the Rabuor clay. A similar brown clay in the gullies 500-m west of the main fossil quarry yielded rhinoceros remains (Pickford, 1984, 1986a), probably *Paradiceros mukirii* (Hooijer, 1968). Some of the earliest fossils collected from the surface at Fort Ternan (L. S. B. Leakey, 1962) could have come from the Rabuor clay, including the type specimens of *Kenyapithecus wickeri*. Early excavations into the Rabuor and Rairo clay paleosols were soon abandoned, when it was realized that much of this loose material came from the highly fossiliferous Chogo paleosols higher in the slope. Another fossil jaw of *Kenyapithecus* was found loose on the surface of the Rabuor and Rairo Series, but was presumed to have been washed down from the excavation higher up the slope (Andrews, 1971). This is the safest assumption, because *Kenyapithecus* and the other fossils species all have been reported in place from Chogo paleosols (Shipman, 1977, 1981; Shipman *et al. 1981*), but few fossils have been found in the Rabuor clay. No distinctive trace fossils of animals were found in Rabuor paleosols.

4.2.9.11 Paleotopographic setting. A low and declining upward ferrous iron/ferric iron ratio (Fig. 4.17), brown hue, and weakly developed calcareous nodules all are evidence against a permanent water table within or near the profile. Nevertheless, this paleosol would have had slow drainage, as indicated by its clayey texture and iron-manganese concretions and mottles, which are common in rice soils today (Rahmatullah, Dixon, & Golden, 1990).

Seasonal puddling may have been exacerbated by topographic position mapped to be at the break in slope between a nephelinitic volcanic edifice and an extensive plateau of phonolite (Fig. 4.8; Pickford, 1984, 1986a). Such a location would allow introduction of the anomalous large, deeply weathered boulders of Baraget Phonolite seen in the Rabuor clay paleosol, as colluvial additions. Similarly thick clayey and bouldery soils form on footslopes inside the caldera at Ngorongoro, northern Tanzania (Anderson & Herlocker, 1973).

4.2.9.12 Parent material. Recognizable rock fragments in the Rabuor clay include large boulders of Baraget Phonolite and many clasts of nephelinite like those in the Kapurtay Agglomerate. In chemical composition (Fig. 2.21), the Rabuor clay is close to local nephelinites (Binge, 1962; Richard, 1942), with slight enrichment in silica, alumina, and alkalis, and depletion in iron and magnesia that could be due to mixing of small amounts of local phonolite (Lippard, 1973a). No chemical trace could be seen of carbonatites like those known locally (Deans & Roberts, 1984). All of the volcanic fragments are deeply weathered, and some of the nephelinite clasts have been altered to red and brown claystone with transparent patches after phenocrysts, even in the C horizon. These volcanic fragments probably had already been weathered in other soils on the volcanic slopes. The parent material of the Rabuor clay was mostly volcaniclastic colluvium derived from soils on nephelinitic and phonolitic volcanics.

4.2.9.13 Time for formation. The Rabuor clay is best considered only weakly developed, because its weathered clasts and perhaps much of its clay were derived from preexisting soils. The lighter and darker layers may also be relict beds, as discussed for interpretation of its likely paleoclimate. The Rabuor clay is not developed quite to the extent of soils identified as Natrustolls, Pellusterts, and Argiustolls on 40,000-year-old Mbalageti Ash on the Serengeti Plains of Tanzania (Jager, 1982). On the other hand, the Rabuor clay is better developed than the various Orthents some 600 to 2050 years old around the active Tanzanian volcano Oldoinyo Lengai (Hay, 1989). The Rabuor clay also is significantly better developed than the associated Rairo paleosols and comparable surface soils. An age of a few tens of thousands of years is most likely for the Rabuor clay.

4.2.10 Dhero Series (Retallack et al., *1990)*

4.2.10.1 Diagnosis. Thin, gray to brown (10YR) paleosols with root traces and well-bedded carbonatite-nephelinite sandstone and siltstone.

4.2.10.2 Derivation. Dhero means thin in Dholuo (Gorman, 1972), in reference to the slight depth of alteration of these paleosols.

4.2.10.3 Description. The type Dhero clay paleosol (Fig. 4.18) is in the middle Miocene, middle Fort Ternan Member, between two thick tuff breccia beds (Fig. 4.9) forming the walls 5-m above the floor of the main fossil quarry (20 m in Fig. 4.10; grid reference 608759 on Lumbwa sheet) at Fort Ternan National Monument, Kenya (Figs 4.4 and 4.8).

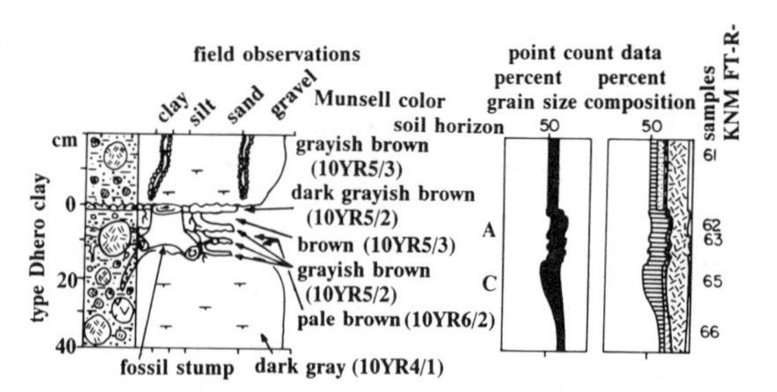

Fig. 4.18. Measured section, Munsell colors, soil horizons, grain size, and mineral composition of the type Dhero clay paleosol in the main fossil quarry at Fort Ternan National Monument, Kenya. This section is 19.7 to 20.2 m in Fig. 4.10. Lithological key as for Fig. 4.12.

+65 cm; lahar deposit, below C horizon of overlying paleosol (type Onuria clay); boulder breccia; grayish-brown (10YR5/2); with clasts of strong-brown (7.5YR4/6), brownish-yellow (10YR5/8), pale yellow (5Y8/3), dark bluish-gray (5B4/1), and black (7.5YR2/1); boulders were observed up to 37-cm across and were concentrated in a zone about 30-cm above the base of this unit; boulders are supported by a matrix of sand and granule size; common near-vertical veins filled with sparry calcite, of pinkish-white (7.5YR8/2), flanked by iron stain (diffusion ferrans) of strong-brown (7.5YR5/8) and red (2.5YR4/6); strongly calcareous overall; undulic intertextic in thin section; rock fragments are mainly weathered nephelinite, with some phonolite and sövite; matrix mainly clay, but some calcite crystal tubes, after roots; abrupt, wavy contact to

0 cm; A horizon; lapilli tuff; dark grayish-brown (10YR4/2); with clasts of strong-brown (7.5YR5/6), dark grayish-brown (10YR4/2), dark bluish-gray (5B4/1), and black (7.5YR2/1); this single bed of tuff pinches and swells with the underlying wavy contact, as if it were a train of ripple marks; strongly calcareous; crystic agglomeroplasmic, with nephelinite rock fragments, some pyroxene and sövite; some grains are partly replaced by micrite; few calcareous crystal tubes, after root traces; abrupt, wavy contact to

-2 cm; A horizon; clayey siltstone; brown (10YR5/3); with granules of dark bluish-gray (5B4/1) and olive (5Y5/4); this horizon includes stump casts up to 31-cm across, with multiple boles of the order of 10 cm in diameter, rooted in lower layers of the paleosol; the clayey bark impressions of the stumps are dark grayish-brown (10YR4/2), and the stumps are surrounded by and filled with white (10YR8/2) and light yellowish-brown (2.5Y6/4) sparry calcite; conspicuous relict bedding is distorted around the stump casts; very strongly calcareous; granular crystic in thin section; grains are mainly nephelinite; common concentrically zoned accretionary lapilli; few grains of pyroxene and alvikite; common calcite crystal tubes after fine root traces, in places showing displacive fabric; clear, broken contact to

-5 cm; A horizon; granule sandstone; grayish-brown (10YR5/2); this is the coarse base of an upward-fining unit that includes the layer above; both layers are disturbed around stump casts; very strongly calcareous; insepic agglomeroplasmic in thin section; with abundant nephelinite rock fragments, most of them weathered; common fine calcite crystal tubes after root traces; abrupt, broken contact to

-7 cm; A horizon; clayey siltstone; pale brown (10YR6/3); this is the upper part of an upward-fining bed disrupted by stump casts, as in interval 2 to 5 cm; very strongly calcareous; clear, broken contact to

-8 cm; A horizon; granule sandstone; grayish brown (10YR5/2); this is the lower part of a upward-fining bed, including the layer above; an additional upward-fining bed is preserved locally around the base of the stump casts and in deep, irregular extensions of these fine grained tuffaceous layers into underlying layers, following burrows or root traces 3-12-cm across and 10-cm into the underlying layer; very strongly calcareous; broken, abrupt contact to

-9 cm; C horizon; boulder breccia; dark gray (10YR4/1); with clasts of pale yellow (5Y8/3), yellow (2.5Y7/8), pale olive (5Y6/4), dark bluish-gray (5B4/1), and black (5Y2/1); a nephelinite boulder 9 cm in diameter was dark greenish gray (5B4/1) with pale olive (5Y6/4) phenocrysts in its little-weathered center, but had a 1-cm-wide weathering rind of olive (5Y5/3) with pale yellow (5Y7/4) phenocrysts, as well as an outer 2-mm-thick oxidized zone (neosesquan) of strong-brown (7.5YR5/6), and an outermost clay skin (sesquiargillan) of light yellowish-brown (10YR6/4); this is the massive top of a deposit interpreted as a lahar, with its boulders concentrated in the middle of the bed some 60 cm below; strongly calcareous; agglomeroplasmic inundulic in thin section; many rock fragments of nephelinite; few pyroxene and sövite; common calcite crystal tubes, after fine root traces.

4.2.10.4 Further examples. Three additional Dhero profiles were recognized in the main fossil quarry at Fort Ternan

National Monument (at 11.2, 18 and 21.3 m in Fig. 4.10), but only two were studied in sufficient detail to be given names. The lowest of them, the Dhero clay-loam, is developed on resorted nephelinitic tuff overlying the type Rabuor clay paleosol (Fig. 4.17). The uppermost paleosol in the quarry, the Dhero clay lapillistone-variant (Fig. 4.22) formed on a very fresh carbonatite-nephelinite ash with abundant, concentrically zoned, accretionary lapilli up to 1 cm in diameter.

4.2.10.5 Alteration after burial. The type Dhero clay and Dhero clay lapillistone-variant are strongly cemented by sparry calcite of a type that could be imagined as an early burial cement. Nevertheless, they also have many of the petrographic features of carbonatite-nephelinite ashes that are cemented within a few hundred years of eruption from Oldoinyo Lengai, Tanzania. Uncompacted tabular stump casts and root traces also are evidence for cementation before burial. There is little oxidation of iron-bearing minerals that might have been involved in dehydration during burial. The main alteration after burial may have been loss of organic matter, including the loss of cellular structure from the stump casts. The paleosol itself shows scattered roots and little soil structure, so that even losses of organic matter may not have been great.

The Dhero clay-loam in contrast, does not show evidence of fresh volcanic ash, but instead developed on colluvium of weathered nephelinitic sandstone. It may have suffered marked compaction and loss of organic matter, as postulated for better developed and more clayey associated Rairo paleosols.

4.2.10.6 Reconstructed soil. Dhero paleosols are little-altered volcanic ash or colluvium, with well-preserved bedding only sparsely penetrated by roots. They may have been hard-setting, as indicated by the preservation and deflection of stumps and their roots, and as also documented for early weathering of comparable historic carbonatite-nephelinite ashes from the Tanzanian volcano Oldoinyo Lengai (Hay, 1989). The brown hues and deep root traces of Dhero paleosols are evidence that they were generally well drained. They were certainly alkaline, and probably also had high cation exchange capacity and base saturation, considering their abundant carbonate. They may have had a pH at times approaching 9.5, as well as salt efflorescences, like modern ashes of this kind (Hay, 1978).

4.2.10.7 Classification. If it were not for their traces of roots and stumps, these would not be regarded as paleosols, but as tuff beds. Many of the root traces run along the surface of underlying boulder breccias or between the tuff layers, as if these were cemented by calcite during soil formation. This phenomenon is seen also in early weathering of petrographically comparable modern ashes (Hay, 1978, 1989). Although quite tuffaceous, these soils have less than 60 percent glassy pyroclastic material and are too thin for Andepts or Andosols. For these reasons, Dhero paleosols are best classified as Orthents (of Soil Survey Staff, 1975), Calcaric Regosols (of Food and Agriculture Organization, 1974) and Lithosols (of Stace *et al.*, 1968). Broadly similar soils (Eutric Regosols) are found with Lithosols and Andosols in the area around Oldoinyo Lengai, Tanzania (map unit I-Re-T-c of Food and Agriculture Organization, 1977a). In the Northcote (1974) key, these paleosols are Uf1.31, and in the South African system (MacVicar *et al.*, 1977) they are Mispah kalkbank forms.

4.2.10.8 Paleoclimate. Such weakly developed paleosols are not adequate guides to paleoclimate. Nevertheless, Dhero paleosols with their early calcite cementation are texturally very similar to modern and Plio-Pleistocene slightly weathered carbonatite-nephelinite ashes in dry regions of Tanzania (Hay, 1978, 1989). They do not have pervasively displacive fabric of Kiewo Series paleosols, here proposed to have formed from similar parent materials under a humid to subhumid climate (Fig. 2.16). Whether this difference has paleoclimatic significance will be better established when, and if, surface soils with pervasively displacive fabric are discovered.

4.2.10.9 Ancient vegetation. These very weakly developed paleosols may have supported vegetation early in the ecological colonization of disturbed ground. They contain common fine fossil root traces, but these have not obscured bedding. Two fossil stumps were found rooted in the type Dhero clay: one 31 cm in diameter on the eastern wall of the southern alcove, and another 23 cm in diameter, in the southern wall of the alcove (Fig. 4.9). Both consist of three or four separate boles of lesser diameter, with a common root base. These stumps, and isolated logs of comparable diameter in the lahar, all preserved impressions of rough vertically-fluted bark. Multiple bole trees with similar bark are found in a variety of living plants of grassy woodland and wooded grassland: *Cassia* (Caesalpinoidea), *Combretum*, *Terminalia* (Combretaceae) and *Commiphora* (Burseraceae). The fossil bark impressions are rougher than those usually found in *Acacia*, and very different from East African succulents, such as the candelabra cactus (*Euphorbia candelabra*) and the smooth-barked buttressed trees common in East African forests (Lind & Morrison, 1974; personal observations). From these observations the most likely vegetation of Dhero paleosols is early successional wooded grassland or grassy woodland in areas disturbed by volcanic tuffs and mudflows.

Comparable surface soils of Africa (Calcaric Regosols of Food and Agriculture Organization, 1977a), of the United States (Ustorthents and Xerorthents of Soil Survey Staff, 1975), and of Australia (Lithosols of Stace *et al.*, 1968) support vegetation ranging from deciduous and sclerophyll woodland to wooded grassland and desert shrubland. The drier kinds of vegetation are indicated principally by calcareous composition of these paleosols, which is not pedogenic but rather related to their carbonatite parent material. The range of likely vegetation also can be narrowed by assuming that Dhero paleosols, given time, would have developed into soils like associated Chogo and Onuria Series paleosols, which probably supported a mosaic of grassy woodland and wooded grassland. Possible analogs for all these kinds of soils and vegetation can be found on the Serengeti Plains of Tanzania. In the eastern part of the plains, sparse grassland and wooded grassland colonize fresh carbonatite-nephelinite ash from the active volcano Oldoinyo Lengai. The grasses, mainly *Chloris gayana*, *Dactyloctenium* sp., *Digitaria macroblephora*, *Sporobolus iocladus* and *S. kentrophyllus*, create about 15 to 20 percent basal cover among scattered bushes of *Acacia mellifera* (White, 1983).

As already discussed for Lwanda paleosols, grasses have a fossil record in Africa extending back to Eocene, and fossil grass leaves, culms, and pollen also have been found in Chogo and Onuria Series paleosols at Fort Ternan. Some of the characteristic tree genera of wooded grassland, including *Acacia*, *Terminalia*, and *Cassia*, also have a fossil record in Africa extending back to early Miocene time (Chaney, 1933; Chesters, 1957; Collinson, 1983; Lemoigne, Beauchamp, & Samuel, 1974).

4.2.10.10 Former animal life. No trace fossils, snail shells, or bones were seen in Dhero paleosols, but it is possible that some of the many fossils collected in the main quarry came from these paleosols. With a few exceptions noted below, I doubt this, because all the fossils I have seen from Fort Ternan have surface stains and dark-brown matrix characteristic of Chogo paleosols, rather than the grayish cast of Dhero paleosols. Nevertheless, a naturally mummified anterior portion of a fossil chameleon (*Chameleo intermedius*), found loose on the quarry floor (Hillenius, 1978), may have come from the prominent lahars or their interbedded Dhero paleosols (Shipman *et al.*, 1981). Calcite internal casts of a darkling beetle (Tenebrionidae) and a whip scorpion (Amblypygi) from Fort Ternan (Shipman, 1977) also are preserved in a manner unusual for Chogo paleosols, and may have come from associated tuffs, lahars, or Dhero paleosols. Some of these insect fragments are thought (by Pickford, 1985) to have been misidentified as remains of freshwater crabs (by Shipman *et al.*, 1981). Little-weathered, early Miocene, carbonatite-nephelinite ashes on Rusinga and Mfangano Islands are known to have preserved soft tissues of lizards and a variety of weakly sclerotized insects and spiders (Estes, 1962; Hay, 1986; L. S. B. Leakey, 1952; Paulian, 1976; Wilson & Taylor 1964). Dhero paleosols probably were suitable for the preservation of bones and snail shells, but it takes time for such remains to accumulate to levels that are regarded as fossiliferous (Retallack, 1984a). Fresh ash on these soils may also have deterred animal life, as it has during historic eruptions of Oldoinyo Lengai, Tanzania (Richard, 1942). Although not much trace of it remains, the animal life of Dhero paleosols probably included many of the same species found in the associated Chogo paleosols.

4.2.10.11 Paleotopographic setting. Each of the three Dhero paleosols in the main fossil quarry formed on different kinds of young, well-drained, geomorphic surfaces. The Dhero clay-loam formed on nephelinitic sandstone interpreted as colluvial sheetwash on the footslopes of a nephelinitic volcano, near the break in slope to an extensive phonolitic plateau.

The type Dhero clay and another unnamed Dhero paleosol (at 18 m in Fig. 4.10) formed on volcanic mudflows, again near the footslopes of a volcano, where the toes of the flows may have formed low hills, as seen in Quaternary examples of lahars (Cas & Wright, 1987). The type Dhero clay also received fresh increments of airfall ash of carbonatite-nephelinite composition.

Fresh ashes were thicker in the case of the Dhero clay lapillistone-variant, in which ash mantles waterlain nephelinitic sandstone. This paleosol and overlying ash beds were eroded away locally by a paleochannel of cross-bedded nephelinitic sandstone. The Dhero clay lapillistone-variant may have formed within a tract of low-sinuosity stream channels. These were probably only seasonally flowing, because this paleosol shows no more evidence of gleization than other Dhero paleosols, and these lapilli would not have been preserved in water (Reimer, 1983).

4.2.10.12 Parent material. Dhero paleosols formed on a variety of fresh to little-weathered carbonatite-nephelinite volcaniclastic rocks: nephelinitic lahars for the type profile and an unnamed additional profile (at 18 m in Fig. 4.10), fresh carbonatite-nephelinite ash for the Dhero clay lapillistone-variant, and already weathered, water-worked nephelinitic sandstone for the Dhero clay-loam paleosol. All the Dhero paleosols are little altered from their parent material.

4.2.10.13 Time for formation. Carbonatite-nephelinite ashes around Oldoinyo Lengai in Tanzania show fenestral structure of cracks filled with calcite much more pronounced than in Dhero paleosols after only 600 years (Hay, 1989). Volcanic ashes in this area some 1300 and 2050 years old have calcite-cemented layers better developed than in Dhero paleosols. These young ashy soils formed in a modern climate very similar, though a little drier, than that envisaged for Fort Ternan during the middle Miocene, when it was probably not so humid as envisaged for early Miocene Kiewo Series paleosols. Thus Dhero paleosols probably represent only a few tens to hundreds of years.

4.2.11 Chogo Series (Retallack et al., *1990)*

4.2.11.1 Diagnosis. Moderately thick, clayey, granular, grayish brown (10YR) surface (A) horizon over a calcareous nodular horizon (Bk) in water-worked nephelinitic sandstone.

4.2.11.2 Derivation. Chogo is Dholuo for bone (Gorman, 1972), which has been collected from these paleosols in abundance (Shipman *et al.*, 1981).

4.2.11.3 Description. The type Chogo clay (Fig. 4.19) in the middle Miocene, middle Fort Ternan Member, is the uppermost of two similar brown paleosols exposed in the center of the southern and deepest wall of the main fossil quarry (Fig. 4.9; grid reference 603759 on Lumbwa sheet) at Fort Ternan National Monument, Kenya (Figs 4.4 and 4.8). The top of the profile is 1.6-m above the floor of the quarry, and at 16.6 m in the measured section here (Fig. 4.10).

+10 cm; overlying sediment, granule-bearing nephelinitic sandstone; light olive (5Y6/2); with granules of dark yellowish-brown (10YR5/4), greenish-gray (5GY6/1), and black (5Y2.5/1); this bed is a low angle cross-set and has the largest granules concentrated near the bottom; the lower surface also is iron stained (neosesquan), with 3 cm of brown (7.5YR5/2) stain, grading down into 7 mm of yellowish brown (10YR5/6) at the base; strongly calcareous; granular crystic in thin section; common nephelinite clasts with thin weathering rind (diffusion sesquan), some phonolite clasts, pyroxene, sövite, and alvikite; cement is sparry calcite (Fig. 3.2); abrupt, smooth contact to

0 cm; A horizon; granule-bearing sandy claystone; very dark grayish-brown (10YR3/2); with grains and crystals of yellowish-brown (10YR5/6), light yellowish-brown (2.5Y6/4), light olive gray (5Y6/2), and black (5Y2.5/1); common fine root traces; indistinct coarse-prismatic peds and coarse-granular peds, defined by very dark grayish-brown (10YR3/2) clay skins (argillans); weakly calcareous; porphyroskelic skelinsepic in thin section; grains mostly nephelinite, usually deeply weathered, and occasionally with rims of micrite; few deeply weathered pyroxene, and biotite crystals, and ferruginized opaque oxides; gradual, smooth contact to

-13 cm; A horizon; granule-bearing sandy claystone; dark grayish-brown (10YR4/2); with grains and crystals of reddish-brown (2.5YR4/4), strong-brown (7.5YR5/8), light grayish-brown (10YR5/4), yellowish-brown (10YR5/6), olive (5Y5/3), greenish-gray (5G5/1), and black (5Y2/1); root traces and indistinct granular and prismatic peds as in horizon above; this horizon also contains large boulders of volcanic rocks; one boulder 26 cm long by 14 cm high was Baraget Phonolite, with olive (5Y4/4) matrix and pale yellow (2.5Y7/4) phenocrysts fringed by wispy iron stain (sesquans) of yellowish-red (5YR5/6), as well as an outermost weathering rind of white (5Y8/2); another boulder 23 cm across was light gray (5Y7/2) nephelinitic sandstone, rich in black (10YR2/1) pyroxene, as well as rock fragments of strong-brown (7.5YR5/6), yellowish-brown (10YR7/2), and olive-gray (5Y7/2) nephelinite showing varying degrees of weathering; moderately calcareous overall; porphyroskelic skelmosepic in thin section; with weathered nephelinite rock fragments; few pyroxene and opaque oxides; rare biotite; wavy (around boulders), clear contact to

-24 cm; Bk horizon; granule-bearing sandstone; grayish-brown (10YR5/2); with grains of yellowish-brown (10YR5/4, 10YR5/8), brownish-yellow (10YR6/8), pale olive (5Y6/3), greenish-gray (5G6/1), dark greenish-gray (5GY4/1), grayish-green (5G5/2), and light gray (N6); prominent in this horizon are distinct, white (10YR8/2) calcareous stringers and nodules picking out relict bedding and crude, very thick platy peds; a large clast 19-cm long by 8-cm high of Baraget Phonolite is black (5YR2.5/1) and very dark grayish-brown (10YR3/2), with weathered phenocrysts of yellowish-brown (10YR5/4); another clast 15 cm long by 6.4 cm high of nephelinite was grayish-brown (5G5/2) with veins and phenocrysts of pinkish-gray (5YR6/2), yellowish-brown (10YR5/6), and yellow (10YR8/6); strongly calcareous overall; granular crystic and calciasepic in thin section; with abundant sparry and micritic calcite forming circumgranular crack fills and displacive cavity fills, and replacing volcanic grains and pyroxene crystals; clasts mainly nephelinite and melanephelinite; few biotite and alvikite; rare phillipsite detected in XRD traces; gradual, wavy contact to

-42 cm; Bk horizon; granule-bearing sandstone; dark grayish-brown (10YR4/2); with grains and crystals of strong-brown (7.5YR4/6), olive (5Y5/6), dark olive-gray (5Y5/2), and black (5Y2.5/1); common distinct calcite crystal sheets (calcans), subparallel with relict bedding define very thick platy peds; strongly calcareous overall; granular crystic and calciasepic porphyroskelic in different areas of thin sections; mainly clasts of nephelinite, with some phonolite, pyroxene, biotite, and opaque oxides; gradual, wavy contact to

-57 cm; C horizon; granule-bearing sandstone; very dark gray (10YR3/1); with grains of this color, as well as yellowish-brown (10YR5/4), pale olive (5Y6/3), greenish-

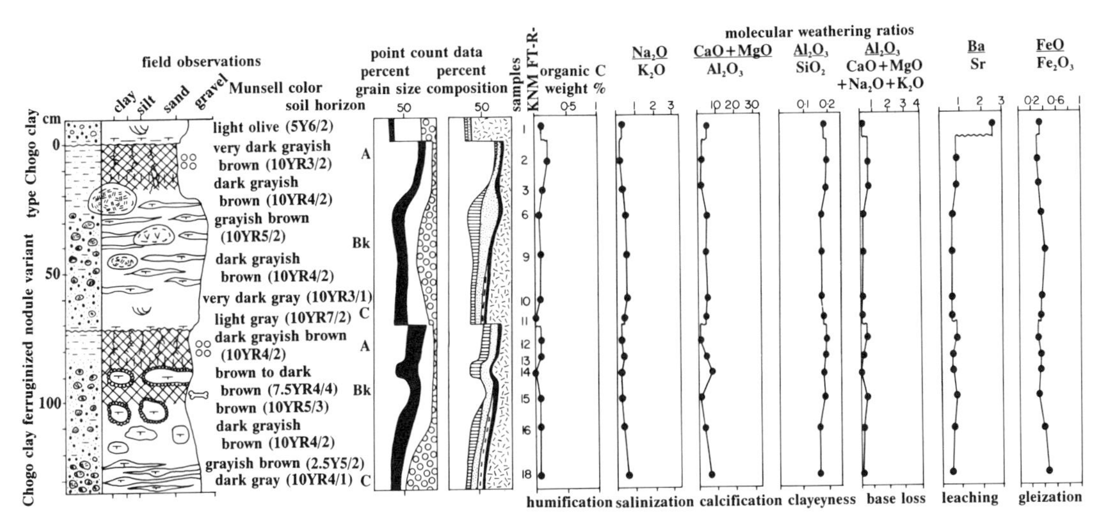

Fig. 4.19. Measured section, Munsell colors, soil horizons, mineral composition, organic matter, and selected molecular weathering ratios for the type Chogo clay and Chogo clay ferruginized-nodule variant paleosols from the high south wall of the main fossil quarry at Fort Ternan National Monument, Kenya. This section is at 15.4 to 16.8 m in Fig. 4.10. Lithological key as for Fig. 4.12.

gray (5G5/1), and black (10YR2/1); relict bedding, with indistinct indications of low-angle cross-bedding; strongly calcareous overall; granular crystic in thin section; abundant fragments of nephelinite; common pyroxene, biotite, alvikite, and opaque oxides; cement of sparry calcite; rare books of vermiculite and pseudomorphs of pyroxene; gradual, smooth contact to

-68 cm; C horizon; granule-bearing sandstone; light gray (10YR7/2); with clasts of brownish-yellow (10YR6/6), pale green (5G6/2), and black (10YR2/1); relict bedding planes are picked out by iron stain (diffusion sesquans) of brownish-yellow (10YR6/6); strongly calcareous overall; granular crystic in thin section; abundant nephelinite rock fragments; common pyroxene, biotite, and opaque oxides; rare alvikite; cement of micrite and sparry calcite; abrupt, wavy contact to

-71 cm; A horizon of underlying paleosol (Chogo clay ferruginized nodule variant); granule-bearing silty claystone; dark grayish-brown (10YR4/2); with grains of yellowish-red (5YR4/6), brownish-yellow (10YR6/6), greenish-gray (5GY6/1), and black (5Y2.5/1); abundant fine root traces; indistinct coarse granular peds, defined by dark grayish-brown (10YR4/2) argillans; few, fine, white (10YR8/2) and very pale brown (10YR8/3) calcite crystal tubes after roots; strongly calcareous overall; porphyroskelic skelmosepic in thin section; clasts mainly deeply weathered nephelinite; common pyroxene and opaque oxides, and few grains of biotite, all very deeply weathered.

4.2.11.4 Further examples. Three distinct kinds of Chogo paleosols can be recognized in the two superimposed ancient land surfaces that are the main fossiliferous horizons at the base of the quarry at Fort Ternan National Monument (Fig. 4.9). The type Chogo clay passes laterally over a distance of about 7 m into the Chogo clay eroded-phase paleosol (Fig. 4.20), which is the upper profile exposed in a low dividing wall within the excavation (Fig. 4.21). This is a similar profile overall, but its surface (A) horizon is disrupted by a small paleogully filled with nephelinitic sandstone and claystone, and by nodular masses that may represent small tree trunks degraded by termites. Its subsurface (Bk) horizon has large bulbous burrows filled with calcareous claystone breccia. Underlying both the type Chogo clay and the Chogo clay eroded-phase is the Chogo clay ferruginized-nodule variant. This profile also has a dark grayish-brown (10YR4/2) clayey surface (A) horizon over a grayish brown (2.5Y-5/2) calcareous subsurface (Bk) horizon, but the contact between the two horizons is irregular. Especially distinctive features of its Bk horizon are ellipsoidal to tuberose, calcareous nodules up to 6-cm across, which are brown (10YR5/3) to pale brown (10YR6/3) in

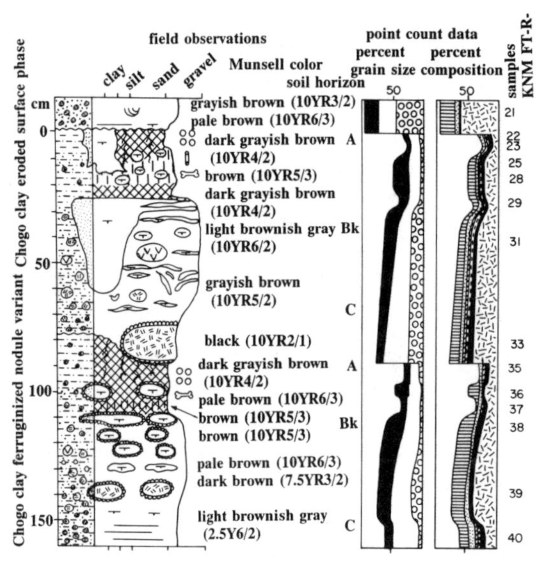

Fig. 4.20. Measured section, Munsell colors, soil horizons, grain size, and mineral composition of the Chogo clay eroded-phase and Chogo clay ferruginized-nodule variant paleosols in the low dividing wall, west of the high walls, of the main fossil quarry at Fort Ternan National Monument, Kenya. This section is a 15.1 to 16.8 m in Fig. 4.10. Lithological key as for Fig. 4.12.

color and have thin (1-2-mm) surface stain (diffusion sesquans) of yellowish-red (5YR4/6) and strong-brown (7.5YR4/6).

4.2.11.5 Alteration after burial. These profiles are about as brown as they could be considering their oxidized chemical condition (Fig. 4.19). Thus reddening due to burial dehydration of ferric hydroxides is unlikely. They may, however, have lost a lot of organic matter during burial, because this is present now in much lower amounts than one would expect in soils with granular structured surface horizons. Compaction of their clayey surface horizons may have been to as much as 74 percent of their former thickness, so that the Bk horizon of the type Chogo clay was formerly more like 33-cm down into the profile, rather than its present depth of 24 cm. Some calcite cement may have been added shortly after burial, but this was seen more commonly in subsurface (Bk) horizons, where it is distinct from pedogenic micrite. Little cement was seen in the clayey surface horizons. Indistinct broad prismatic peds in surface horizons may be due to drying in the modern outcrop, because no indication of salinization, often associated with such structure in soils (Northcote & Skene, 1972), can be found in the soda/potash ratios (Fig. 4.19).

4.2.11.6 Reconstructed soil. Chogo paleosols can be envisaged as moderately developed soils with dark brown, granular-structured, clayey A horizons over gray to white calcareous nodular Bk horizons on nephelinitic sandstones. Lack of leaching revealed by molecular ratios of alumina/bases and barium/strontium, as well as by abundant carbonate, indicate high pH, cation exchange capacity, and base saturation. Prismatic peds are not developed to the extent found in salt-affected soils, nor is there evidence of sodium saturation in the upward declining soda/potash ratios (Fig. 4.19). Good drainage and high Eh is indicated by the low ferrous iron/ferric iron ratios, deeply penetrating root traces, and warm hue of the profiles.

4.2.11.7 Classification. Hand specimens of Chogo paleosols were examined by I. W. Cornwall (for Bishop & Whyte, 1962) and compared by him with Braunerde soils (of Kubiena, 1970). In the U.S. taxonomy (Soil Survey Staff, 1975) these would be equivalent to Ochrepts or Tropepts, and in the Food and Agriculture Organization (1974) classification to Cambisols. These are soils less developed than Chogo paleosols with their prominent calcareous nodules. Chogo paleosols are more like Mollisols, although close to being excluded on the thickness criterion. Even disregarding compaction effects, the surface dark horizon of some of these paleosols is only 10-cm thick, but the top 18 cm mixed with this would still meet the color and chemical requirements of a mollic epipedon. The crude prismatic peds do not have domed tops or clear margins found in natric soils, and there

Fig. 4.21. The Chogo clay eroded-phase (above) and Chogo clay ferruginized-nodule variant (below) paleosols in the low dividing wall, looking west, within the main fossil quarry at Fort Ternan National Monument, Kenya. The structures here interpreted as stumps and mongoose dens are sketched and labelled in Fig. 4.9. Hammer above upper paleosol gives scale.

is no chemical evidence of salinization in soda/potash ratios. Nor do these paleosols contain 60 percent glassy volcanic material or evidence of low bulk density required of andic soils. Excluding these as possibilities, Chogo paleosols with their shallow calcareous (Bk) horizons are most like Haplustolls (of Soil Survey Staff, 1975).

Surface soils with similar profile form and chemical depth functions, also identified as Haplustolls, are widespread on carbonatite-nephelinite ashes in the central Serengeti Plains of Tanzania (for example, NaNo-A profile 28A and NaLag profile 46 of de Wit, 1978). Tanzanian soils comparable to the Chogo clay are a little more sandy and more felsic in overall chemical composition because flanked by quartzofeldspathic gneisses, rather than by the phonolite and nephelinite up slope from Chogo paleosols. These central and eastern Serengeti soils are nevertheless more mafic and carbonatitic, and thus chemically closer to Chogo paleosols, than soils of the western Serengeti (de Wit, 1978; Jager, 1982).

In the Food and Agriculture Organization (1974) classification, Chogo paleosols are less rich in organic matter and are lighter colored than Chernozem. This probably is not merely a result of changes due to burial, because the granular structure of Chogo paleosols is not developed nearly to the extent seen in Chernozems. A more likely choice is Calcic Kastanozem, recorded for example, on the Food and Agriculture Organization (1977a) soil map from Quaternary lacustrine and alluvial deposits in Morrocco and Tunisia. On this same map, the Serengeti Plains are mapped as Calcic Cambisols (unit Bk25-2a), associated with Eutric Cambisols, with inclusions of Pellic Vertisols and Eutric Planosols. Considering the subsequently collected chemical and other data on these soils (de Wit, 1978), this Serengeti unit on the FAO map would now be better labeled Haplic Kastanozems, with associated Calcic and Luvic Kastanozems, and inclusions of Mollic Solonetz and Pellic Vertisols.

Among Australian soils (Stace *et al.*, 1968), Chogo paleosols are most like Chernozems, even to the extent of their coarser structure and lower organic content than Chernozems as usually understood in the northern hemisphere (Food and Agriculture Organization, 1974). Prairie Soils, Black Earths, Non-calcic Brown Soils, Chocolate Soils and Brown Earths are all thicker and more leached of carbonate than Chogo paleosols. In the Northcote (1974) key for Australian soils, Chogo paleosols are Uf5.11.

In the South African scheme (MacVicar *et al.*, 1977), Chogo paleosols are most like Inhoek drydale. Similar less calcareous profiles are widespread on basalts of northern Kruger National Park, South Africa (Botha, 1985; Gertenbach, 1983; Venter, 1986).

4.2.11.8 Paleoclimate. In the Great Plains of North America, soils with calcareous horizons as shallow as 33 cm are found in regions receiving some 320-590-mm mean annual rainfall (Jenny, 1941). Similar estimates of a semiarid climate can be gained from comparable climosequence studies in the Mojave Desert of the United States (Arkley, 1963), the Indo-Gangetic Plains of India (Sehgal, Sys, & Bhumbla, 1968) and the Serengeti Plains of Tanzania (de Wit, 1978). Less than 1200 mm of mean annual precipitation is indicated by the smectitic, rather than kaolinitic and halloysitic, composition of the type Chogo clay (Fig. 2.33), in comparison with surface soils on volcanic ash in Ruanda and Tanzania (Mizota & Chapelle, 1988; Mizota, Kawasaki, & Wakatsuki, 1988).

There is no clear evidence of paleotemperature in these paleosols, whose root traces and humification are compatible with a productive tropical or temperate ecosystem.

Seasonal climate is indicated for Chogo paleosols by several features. There is a dense network of fine root traces near the surface, but also a few stout and deeply penetrating root traces, some of which extend very deeply into the underlying thick nephelinitic tuff (between 11.6 and 15.0 m in Fig. 4.10). Similar distribution of root traces is common in seasonally dry savanna ecosystems (van Donselaar-ten Bokkel Huinink, 1966). The ferruginized nodules of the lowermost Chogo profile may have been produced by dry-season oxidation of iron liberated during wet-season weathering. None of these effects of seasonality are as pronounced as they could be, and it is unlikely that seasonal contrast was as severe as it is today in the Serengeti Plains, or as it was in the late Miocene and still is in the Indo-Gangetic Plains of India.

4.2.11.9 Ancient vegetation. Chogo paleosols have many features of grassland soils: simple profile form (A-Bk), a dark, clayey, granular-structured surface horizon, and a clear smooth contact to the subsurface calcareous horizon. Comparable surface soils of the Serengeti Plains of Tanzania now support wooded grassland (de Wit, 1978). In addition, poorly preserved stems, leaves (Dugas, 1989), and pollen (Bonnefille, 1984) of grasses are common in Chogo paleosols, and there is also indication of grasses from the striated microwear seen on fossil antelope teeth from these paleosols (Shipman *et al.*, 1981).

There are also some indications of trees in these paleosols: incipient subsurface (Bt) horizons of orange claystone, gradual or disrupted transition between surface clayey (A) horizon and subsurface calcareous (Bk) horizon, occasional large (more than 1 cm in diameter) root traces, and poorly preserved stump casts at the surface of the profiles. Similar soil features are found in the grassy woodlands north and west of the Serengeti Plains of Tanzania (Jager, 1982). Paleontological indications from Chogo paleosols of trees and woodland include fossil fruits and seeds (Shipman, 1977), pollen (Bonnefille, 1984), snails (Pickford, 1985), and adaptive features of the mammalian fauna (Andrews & Evans, 1979; Andrews, Lord, & Evans, 1979; Andrews & Walker, 1976; Evans, Van Couvering, & Andrews, 1981; Gentry, 1970; Kappelman, 1991; Pickford, 1987; Shipman, 1986a).

Interpretations of the vegetation of these fossiliferous paleosols have varied considerably, with arguments among the publications cited for forest, woodland, bushland, wooded grassland, and open grassland. Probably several of these were present in a mosaic of vegetation, like that found widely in African game parks today. Such a mosaic was not apparent from the various kinds of fossils found, which could have been transported and mixed, and were in any case treated as a single assemblage by some researchers. A mosaic can be interpreted from the paleosols, which are by definition in the place they formed. Each of the three recognized variants of Chogo paleosols probably supported a different element of the mosaic.

The Chogo clay eroded-phase paleosol, the upper profile at the western end of the quarry, has been extensively excavated for fossils (L. S. B. Leakey, 1962; Shipman, 1977, 1981; Shipman *et al.*, 1981). This profile includes a filled gullylike feature oriented south-southeast (illustrated by Andrews, 1981a; Andrews & Walker, 1976; Churcher, 1970). At the surface of the paleosol are two cylindrical and downward-expanded nodular features, interpreted as casts of stumps degraded by termite activity. These are 16 cm and 7.5 cm in diameter and spaced 2.6-m apart. They may have been part of a grassy woodland fringing a system of dry erosional gullies. Such an interpretation also is consistent with fossil plant remains from this paleosol: leaf sheath of Gramineae, seed of *Annonaspermum* sp., locule of *Leakeyia vesiculosa*, fruit of *Cnestis* sp., spiny exocarp possibly of Cucurbitaceae, endocarp of *Icacinocarya* sp., trilobate calyx of *Euphorbiotheca pulchra*, large flat pod of Papilionaceae, curved stipular spine like those of *Acacia* spp., exocarp of *Berchemia pseudodiscolor*, exocarp of *Zizyphus rusingensis*, and endocarp of *Celtis rusingensis* (M. Collinson, personal communication, 1989; Shipman, 1977; personal observations). Related living plants include trees, shrubs, and vines of rain forest to wooded grassland. This combination of genera, however, is most similar to that of Zambezian woodland or wetter parts of the Somali-Masai *Acacia-Commiphora* wooded grasslands, particularly their riparian gallery woodlands (Lind & Morrison, 1974; White, 1983).

The type Chogo clay paleosol is the upper fossiliferous paleosol beneath the high wall of the quarry about 7-m east of the laterally equivalent Chogo clay eroded-phase paleosol. Unlike this ancient land surface to the west, the type Chogo clay has no likely stump casts or large root traces and has a smooth, clear transition between the surface (A) and subsurface calcareous (Bk) horizon. In these features it is similar to soils of wooded and open grassland on the Serengeti Plains of Tanzania (de Wit, 1978). The type Chogo clay is the source of Bonnefille's (1984) poorly preserved (only 284 grains recovered) pollen sample, which was dominated by pollen of grasses (54%) and sedges (27%), with small amounts of *Podocarpus* (podo, 3%) and Urticaceae (stinging nettle family, 3%). Also found were a few grains each of *Botryococcus* (algae), monolete pteridophyte spores, *Juniperus* (juniper), Amaranthaceae or Chenopodiaceae (families of herbs and shrubs such as pokeweed and cockscomb), *Celtis* (hackberry), Combretaceae (bush willow family), *Acalypha* (euphorbiaceous tree or shrub), cf. *Croton* (euphorbiaceous tree or shrub), cf. *Lannea* (Anacardiaceae), *Olea* sp. cf. *O. africana* (olive), *Plantago* sp. cf. *P. coronopus* (plantain), *Anthospermum* (rubiaceous tree or shrub), Compositae or Tubiflorae (daisies and eyebrights), *Potamogeton* sp. cf. *P. pectinatum* (pondweed), and *Typha* (cattail). This assemblage also is one of an open grassy habitat, with nearby dry montane forest (for *Olea*, *Juniperus*, and *Podocarpus*) and marsh (for *Potamogeton*, *Typha*, and *Botryococcus*). The montane forest and marsh may not have been very close at hand, because these indicator plants are known to be copious producers of wind-dispersed pollen, yet are rare in this assemblage. These taxa are unlike *Acacia* and grasses, which are underrepresented in most pollen diagrams (A. C. Hamilton, 1982). The open or wooded grassland envisaged for the type Chogo clay was only 7 m from grassy woodland of the Chogo clay eroded-phase paleosol. Reconstruction of vegetation of these paleosols as an ecotone between woodland and grassland may explain the mixture of woodland and more open country elements in the mammalian fauna from Fort Ternan (Shipman, 1977, 1981; Shipman *et al.*, 1981).

The Chogo clay ferruginized-nodule variant is laterally more consistent in form than either the type Chogo clay or Chogo clay eroded phase above it. The ferruginized-nodule variant has an irregular to disrupted boundary between surface clayey (A) horizon and subsurface calcareous (Bk) horizon. A clayey band below the upper band of nodules (Figs 4.19 and 4.20) may be the surface (A) horizon of an earlier paleosol, buried by about 20 cm of nephelinitic sandstone and then overprinted by additional soil formation. But both these possibly welded paleosols are disrupted by occasional large (more than 1 cm in diameter) root traces. This paleosol is thus similar to the Chogo clay eroded-phase paleosol and to similar Serengeti soils, in showing features of soils formed under grassy woodland. Some poorly preserved woody twigs and grass culms were found in this paleosol. A fossil leaf superficially similar to *Sterculia* was seen in tuffs of the C horizon of this paleosol (Andrews & Walker, 1976). It is possible that this paleosol yielded some of the fruits and seeds found by L. S. B. Leakey and his assistants at Fort Ternan. However, the excavations reported by Shipman (1977, 1981; Shipman *et al.*, 1981) did not extend to this level, and the plant fossils described by her were mostly from the overlying Chogo eroded-phase paleosol. Similar vegetation of both this and the Chogo clay ferruginized-nodule variant is likely considering their overall similarity of profile form and other features.

As a group, Chogo paleosols compare well with soils under "intermediate grassland on calcimorphic soils with soft pans" in the central Serengeti Plains of Tanzania (Anderson & Talbot, 1965; de Wit, 1978; Jager, 1982; White, 1983). The lightly wooded grassland there is dominated by the panicoid grass *Andropogon greenwayi*, and the trees include *Acacia tortilis* and *Balanites aegyptica*.

Several previous hypotheses concerning the middle Miocene vegetation at Fort Ternan can now be reevaluated using the evidence of paleosols. It was unlikely to have been an Afromontane grassland, because soils in such sites are thinner and non-calcareous (Mahaney, 1989; Mahaney & Boyer, 1989). Beautifully preserved fossil grasses from above the associated Onuria clay paleosol do not include the pooid grasses that now dominate Afromontane grasslands. The pollen grains of montane trees, on which Bonnefille (1984) based this opinion, probably blew into this open vegetation from nearby mountains. Nor is there any evidence of permanent waterlogging, such as siderite or peat, in any of the paleosols at Fort Ternan. The Fort Ternan paleosols also differ from black, clayey, vertic soils of seasonally waterlogged depressions supporting dambo grasslands (White, 1983), where there commonly are bambusoid and arundinoid grasses, which also are not present above the associated Onuria clay (Dugas, 1989). No aquatic elements have been found among the fossil snails or mammals either (Pickford, 1985). Thus the rare marsh spores and pollen found by Bonnefille (1984) probably also blew in from elsewhere. None of the paleosols at Fort Ternan show any resemblance to the thick, noncalcareous, often humic, Andepts under forest of the present Mau Escarpment (Mbuvi & Njeri, 1977), an analog suggested (by Andrews & Walker, 1976) on the basis of misidentification of paleosol samples as part of a former Braunerde soil (I.W. Cornwall for Bishop & Whyte, 1962). Also different are the thick red soils on Precambrian basement under dry woodland and wooded shrubland, locally called "nyika", as in Tsavo West National Park, Kenya. This vegetation was suggested as a modern analog for Fort Ternan by Pickford (1987) on the basis of the fossil snail assemblage. These surface soils are in part relict soils from Miocene time, which have continued to form under the very different climatic conditions now prevailing. The interpretation offered here for Fort Ternan as a mosaic of wooded grassland and grassy woodland is most like that proposed by Shipman (1977, 1981; Shipman *et al.*, 1981).

4.2.11.10 Former animal life. The fossil mammals of Chogo clay paleosols are among the best-documented examples anywhere of an assemblage that accumulated in a soil by normal processes of death and decay. Several coordinated studies (by Shipman, 1977, 1981; Shipman *et al.*, 1981) of this large assemblage (4658 identified bones out of a total of 11,000) have shown that few of the bones were articulated. There is little sorting of the bones by skeletal part, by hydrodynamic group, or by size, apart from a slight underrepresentation of small bones that can generally be related to dissolution by rainwater and soil solutions (Retallack, 1988b). Nor do the bones show preferred orientation, either of their long axes or of the surface facing up. The proportion of predator bones was low, as were bones damaged by carnivores. In addition, the age distribution of large collections of common antelope fossils has many young and old individuals, rather than a dominance of adults found in mass-death assemblages fossilized by a catastrophic event.

The most common fossils are antelope of two kinds: *Kipsigicerus labidotus* (444 specimens in "Leakey assemblage" of Shipman, 1977), an early boselaphine with short, backward-swept, vanelike, helically twisted horn cores, and *Oioceros tanyceras* (428 specimens), an early caprine with horn cores curving outward in divergent arcs. Other large animals include additional antelope (*Gazella* sp., 21, and *Capratragoides potwaricus*, 76), four-tusker elephant ancestors (*Choerolophodon ngorora*, ca. 60), two-horn browsing rhinoceros (*Paradiceros mukirii*, 97), large chevrotain ancestor (*Dorcatherium chappiusi*, 28), giraffe ancestors (*Palaeotragus primaevus*, 167, and *Samotherium africanum*, 11), deerlike ungulates (*Walangania africanus*, 5, and *Climacoceras gentryi*, 4), pig (*Lopholistriodon kidogosana*, 7), and aardvark (*Myorycteropus chemeldoi*, 2). Carnivorous animals include an enormous creodont (*Megistotherium osteothalastes*, 5), wolflike creodont (*Dissopsalis pyroclasticus*, 1), bear dog (Amphicyonidae indet., 5), hyaena (*Percrocuta* sp., 5), and mongoose (*Kanuites* sp., 20). The mongoose is represented by two juvenile articulated skeletons and one adult skull. Skeletal articulation is very unusual for Fort Ternan fossils (Shipman, 1977, 1981). The mongooses may have been buried alive in a burrow like those observed in the Chogo clay eroded-phase paleosol. Ape fossils from Fort Ternan (as reassessed by Pickford, 1984, 1985, 1986a) include a loris (Lorisinae indet, 1), small apes (cf. *Oreopithecus* sp., 1; *Micropithecus* sp., 1), and medium-sized apes (*Proconsul* sp. cf. *P. africanus*, 2; *Kenyapithecus wickeri*, 24). There also are at least 12 species of small mammals: bats, insectivores, and rodents (Appendix 9). Common and paleoenvironmentally significant rodents and insectivores include gerbil (*Leakeymys ternani*, 8), cricetid mouse (*Afrocricetodon* sp., 11), flying squirrel (*Paranomalurus* sp., 6), theridomyid cane rat (*Paraphiomys pigotti*, 6) and elephant shrew (*Miorhynchocyon rusingensis*, 3). Ostrich egg shell and vulture bones also have been found. Specimens identified as crocodile teeth have been reinterpreted (by Pickford, 1985) as poorly preserved canine teeth of primates or carnivores.

Some idea of preferences of mammalian species for different parts of the vegetation that existed at Fort Ternan can be gained by comparing the fossil assemblage excavated from the Chogo clay eroded-phase paleosol (Shipman, 1977, 1981; Shipman *et al.*, 1981) with the collections from all Chogo paleosols (the "Leakey assemblage" of Shipman). As interpreted here, these collections reflect the faunas of woodland flanking a system of dry gullies, versus wooded grassland and grassy woodland away from the gullies. The paleosol interpreted to have supported gallery woodland has more reptiles, rodents, carnivores, primates, rhinoceroses, and elephants than the paleosols thought to have been under wooded grassland and grassy woodland away from the gullies, where there were more antelopes and giraffes. Moderately common species (more than 0.9% in each assemblage) that show preference for the gully are *Choerolophodon ngorora* (3.3% in the Shipman assemblage vs. 1.8% in the Leakey assemblage), *Paradiceros mukirii* (4.5% vs. 2.9%), and *Capratragoides potwaricus* (4.5% vs 2.3%). Common species that show some preference for wooded grassland or grassy woodland away from gullies are *Palaeotragus primaevus* (4.9% in Leakey assemblage vs. 2.3% in Shipman assemblage), *Kipsigicerus labidotus* (13% vs. 8.3%) and *Oioceros tanyceras* (13% vs. 9%). In this analysis 0.9 percent is 35 specimens in the Leakey assemblage but only 3 specimens in the Shipman assemblage. The two assemblages are statistically similar at the ordinal taxonomic level (Shipman, 1977, 1981; Shipman *et al.*, 1981). Nevertheless, these slight habitat preferences of species do make some sense when one considers the ecology of related living mammals.

Differences in the antelope assemblages of the three kinds of Chogo paleosols (after Gentry, 1970; Shipman, 1977; Thomas, 1984) also may have some paleoecological significance. The Chogo clay eroded-phase paleosol, here reconstructed to have supported gullyside woodland, yielded 41 percent *Oioceros*, 38 percent *Kipsigicerus*, and 21 percent *Capratragoides*, in a collection of 87 specimens. The Chogo clay ferruginized-nodule variant, thought to have supported grassy woodland, yielded 44 percent *Kipsigicerus*, 39 percent *Oioceros*, 9 percent *Capratragoides*, and 7 percent *Gazella*, among 515 specimens. The type Chogo clay, thought to have supported wooded grassland, had 48 percent *Kipsigicerus*, 43 percent *Oioceros*, 5 percent *Gazella*, and 4 percent *Capratragoides*, among 191 specimens. Judging from tooth microwear and other features, *Oioceros tanyceras* at Fort Ternan has been considered a grazer and *Kipsigicerus labidotus* a browser (Shipman *et al.*, 1981). This is not apparent from their distribution in the paleosols, because they are equally common in each of the kinds of Chogo paleosol. However, both *Capratragoides potwaricus* and *Gazella* sp. show weak preference for the less open, wooded paleosols, as would be expected from the ecology of related living antelope (Thomas, 1984).

The precise habitat of the ape *Kenyapithecus wickeri*, within the likely vegetation mosaic at Fort Ternan is difficult to determine. It is a rare species (0.6% of both Leakey and Shipman assemblages). Several of the best specimens were found loose on the surface, as discussed already for Rabuor Series paleosols. Those specimens, mainly isolated teeth, found in place within paleosols were in the Chogo clay eroded-phase paleosol (Shipman, 1977, 1981; Shipman *et al.*, 1981). For this reason I incline toward regarding it as an ape of the gullyside woodland part of the mosaic. Arguments to the contrary, based on the state of preservation of their remains, were advanced by Shipman (1977; Shipman *et al.*, 1981), and have been countered by Pickford (1985) in a reassessment of the identification of these ape fossils. The idea of *K. wickeri* as a woodland ape is compatible with the generalized arboreal and quadrupedal structure of associated ape limb bones (Senut, 1988).

Fossil snails from Fort Ternan were almost all derived from Chogo paleosols, as can be seen from their brown discoloration, unlike gray Dhero or olive Onuria paleosols. The most common snails are high-spired stenogyrids, similar to living *Subulina*. The fossil snail assemblage is taxonomically most like that of woodland today (Pickford, 1985), although it is much less diverse than the land snail faunas found in early Miocene paleosols near Koru and Songhor in Kenya.

Fossil millipedes also have been found in Chogo paleosols (Shipman, 1977, 1981). A fossil beetle and whip scorpion are preserved as internal casts (Shipman, 1977), unusual for Chogo paleosols, and probably came from associated tuffs, lahars, or Dhero paleosols. Prior records of freshwater crabs from Fort Ternan were based on specimens more likely to have been fragments of insects (Pickford, 1985).

Trace fossils in Chogo paleosols include numerous near-vertical tubes, referable to the trace fossil genus *Skolithus*, and probably the work of insects or spiders (Ratcliffe & Fagerstrom, 1980). Elongate to ovoid, internal casts (Kenyan National Museum FT-F952, 1050, 1076, 1078, 1105, and 1147) are similar to cocoons, probably of beetles, found also in Mobaw, Kwar, and Tut Series paleosols, and elsewhere (Freytet & Plaziat, 1982). Other tear-shaped, smooth walled, internal casts (specimen FT-F1309) are like the ichnogenus *Celliforma*, thought to be the larval cells of wasps or bees (Retallack, 1984b). *Celliforma* also was seen in Mobaw, Kwar, and Tut Series paleosols. The calcareous nodular structures at the surface of the Chogo clay eroded-phase paleosol have a cylindrical, downward flaring shape like stump casts, but lack bark impressions like the stump casts found in the type Dhero clay paleosol. Similar calcareous stumplike nodules have been found in Eocene paleosols in Wyoming, U.S.A (Kraus, 1988). These may have been degraded by the activity of wood-boring termites, whose nests are known to encourage the precipitation of carbonate (Lee & Wood, 1971).

4.2.11.11 Paleotopographic setting. By the time Chogo paleosols formed, the break in slope between the nephelinitic stratocone and phonolitic flood lavas envisaged during formation of the Lwanda clay paleosol would have been softened considerably by colluvium and volcanic tuffs. A small gullylike feature runs south southeast within the Chogo clay eroded-phase paleosol (Andrews & Walker, 1976; Churcher, 1970; Shipman, 1977, 1981; Shipman *et al.*, 1981). This is a small-scale feature, exploited by burrows, probably of mongooses, as already discussed. It is filled with tuffaceous soil material rather than paleochannel sandstone. None of the Chogo paleosols show any chemical or physical evidence of waterlogging. The gully was thus freely drained and dry for most of the time. Similar erosional gullies are abundant on modern volcanoes, including Oldoinyo Lengai, Tanzania (Dawson, 1962; Hay, 1989).

4.2.11.12 Parent material. Chogo paleosols are developed on nephelinitic sandstone, which include some large boulders of weathered phonolite and nephelinite. This material has low-angle cross-bedding and scour-and-fill structures as found in colluvial deposits of sheet wash. The nephelinite and phonolite clasts were probably derived from soils further up slope. There may also have been a contribution of fresh carbonatite airfall ash, as indicated by rare little-weathered grains of biotite, pyroxene, alvikite, and sövite in the paleosols. Weathering of carbonatite-nephelinite ash would have supplied abundant calcite cement for lower horizons of these paleosols. These conclusions, based on observations of thin sections, are supported by the major-element chemical composition of Chogo paleosols (Fig. 2.21; Appendices 5 and 6). They are chemically close to local nephelinitic lavas (Binge, 1962; Richard, 1942), with very slight admixture of phonolite and carbonatite (compared with analyses of local examples of Deans & Roberts, 1984; Lippard, 1973a). The amounts of trace elements and titania are consistent throughout the type Chogo clay and may indicate a single parent material for that paleosol. There are, however, variations in the Chogo clay ferruginized nodule variant, which can be explained by a minor disconformity between weathered and fresh nephelinitic parent material at a depth of 25 cm from the surface. An earlier A horizon at this level may have been buried by nephelinitic sandstone and then obscured by soil formation from this higher level.

4.2.11.13 Time for formation. Chogo paleosols have a moderately developed subsurface calcareous (Bk) horizon (stage II of Gile, Peterson, & Grossman, 1966) of a general form that developed in only 2000 to 9000 years in soils on carbonatite-nephelinite ash of the Naisiusiu Beds around Olduvai Gorge, Tanzania (Hay & Reeder, 1978). In other parts of the Serengeti Plains, similar calcareous nodular horizons are developed in soils no older than 16,000 to 22,000 years on carbonatite-nephelinite ash of the Ndutu beds (de Wit, 1978; Hay, 1976). These estimates of only a few thousand years for the development of calcic horizons in this unusual parent material are much shorter than for the formation of similar horizons in noncalcareous parent materials in the dry southwestern United States (Gile, Hawley, & Grossman, 1980) or in northern Pakistan (Ahmad, Ryan, & Paeth, 1977). The formation of calcareous subsurface horizons in the Serengeti soils is hastened by redistribution of abundant carbonate in the volcanic ash (Hay, 1989), in addition to the influx of calcareous dust usual in dry climates.

4.2.12 Onuria Series (Retallack et al., *1990)*

4.2.12.1 Diagnosis. Moderately thick, clayey, granular, olive-brown (2.5Y) surface (A) horizon over a thin calcareous hardpan (Bk) on nephelinitic tuff breccia.

4.2.12.2 Derivation. Onuria is Dholuo for light yellow (Gorman, 1972), in reference to the distinctive weathered color of this paleosol among gray sediments and other brown paleosols.

4.2.12.3 Description. The type Onuria clay (Fig. 4.22) in the middle Miocene, middle Fort Ternan Member is in the high southern wall, 5.9-m above the base of the main fossil quarry (or 20.9 m in Fig. 4.10; at grid reference 603759 on Lumbwa sheet), at Fort Ternan National Monument, Kenya (Figs 4.4 and 4.8). The exposure examined (in 1984) was last excavated in 1974 (Shipman, 1977) and is separated from the surface soil by 2 m of nephelinitic sandstone and tuff.

+22 cm; sheetwash deposit below C horizon of overlying paleosol (Dhero clay lapillistone-variant); granule-bearing sandstone; light olive-gray (5Y6/2) in sandy layers and olive-gray (5Y5/2) in granule layers; with grains of dark grayish-brown (10YR4/2), yellow (5Y7/6), pale olive (5Y6/4), gray (5Y5/1), and black (5Y2/1); low-angle cross-sets, with 4-6-mm thick, normally-graded laminae; few casts of fossil grasses filled with pale yellow (2.5Y8/4) claystone and calcite, and with pale olive (5Y6/3) clayey exterior; strongly calcareous; granular crystic in thin section; grains mostly nephelinite, with some phonolite; few biotite, sövite, and opaque oxides; clear, smooth contact to

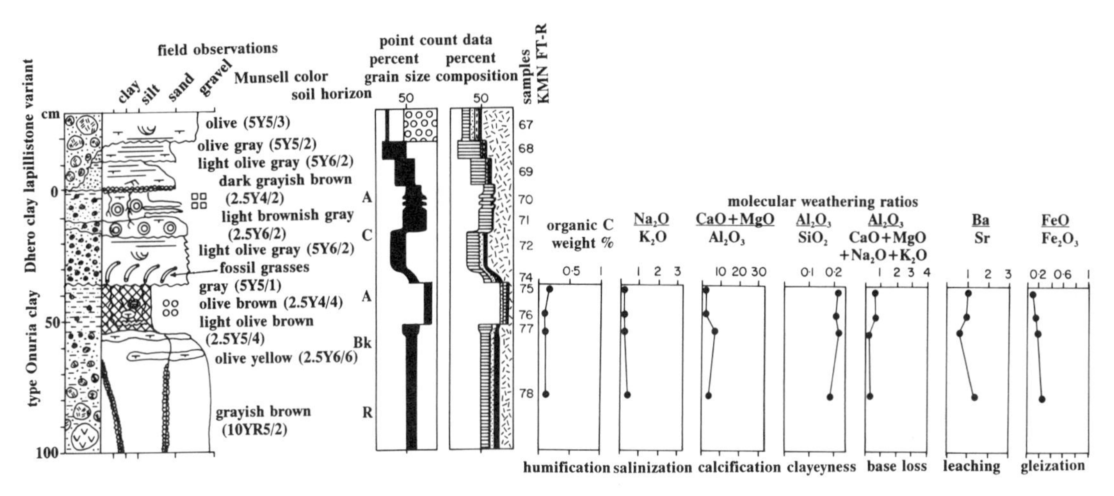

Fig. 4.22. Measured section, Munsell colors, soil horizons, grain size, mineral composition, organic matter, and selected molecular weathering ratios of the type Onuria clay and Dhero clay lapillistone-variant paleosols in the high south wall of the main fossil quarry at Fort Ternan National Monument, Kenya. This section is at 20.5 to 21.8 m in Fig. 4.10. Lithological key as for Fig. 4.12.

+5 cm; sheetwash deposit of "fossil grass bed"; granule-bearing sandstone; gray (5Y5/1); with clasts of strong-brown (7.5YR4/6), yellow (5Y7/6), white (2.5Y8/2), and black (5Y2/1); abundant fossil grasses filled with pale yellow (2.5Y8/4) calcite and with light olive-brown (2.5Y-5/4) and yellowish-brown (10YR5/4) clayey external mold; fossil grasses dominate the fabric, which is massive and unbedded between the culms and leaves; the grasses dip and are oriented toward the north as if pushed over in place of growth by flow from the north; strongly calcareous; granular silasepic in thin section; clasts mainly nephelinite, with some phonolite; few claystone clasts ripped up from underlying paleosol; few pyroxene and sövite grains; local cement of sparry calcite; abrupt, wavy contact to

0 cm; A horizon; claystone; olive-brown (2.5Y4/4); with clasts of red (10R5/6), yellow (2.5Y7/6 and 5Y8/6), and black (5Y2/1); abundant fine (1-2-mm) root traces, filled with olive-yellow (2.5Y6/6) clay; coarse-granular peds; few small (1-2-cm) strong-brown (7.5YR5/6) concretions and mottles; moderately calcareous; porphyroskelic skelmosepic; scattered grains of quartz, opaque oxides, nephelinite, and pyroxene; some of the latter surprisingly fresh; rare wollastonite and analcite detected in XRD traces; gradual, smooth contact to

-5 cm; A horizon; claystone; light olive-brown (2.5Y5/4); with grains of strong-brown (7.5YR5/8 and 7.5YR5/6), yellow (2.5Y7/6), pale green (5G6/2), and black (5Y2/1); common fine root traces, and few ferric concretions, both colored strong-brown (7.5YR5/8); coarse-granular peds; few micritic nodules of olive-yellow (2.5Y6/6); moderately calcareous, even away from nodules; porphyroskelic skelmosepic in thin section of clayey matrix; porphyroskelic calciasepic in calcareous nodules; scattered grains of quartz, opaque oxides, nephelinite, and pyroxene; clear, irregular (around nodules and rock fragments) contact to

-16 cm; Bk horizon; calcareous siltstone; olive yellow (2.5Y6/6); with grains up to 2-cm across of reddish-yellow (7.5YR6/6), dark yellowish-brown (10YR4/4 and 10YR-3/4), olive (5Y5/4), bluish-gray (5B5/1), and black (5Y2/1); this horizon is a thin, laterally impersistent petrocalcic layer that deflects root traces and is nodular and variable in thickness; very strongly calcareous; porphyroskelic calciasepic in thin section; micritic carbonate replaces the edges of nephelinite grains and pyroxene crystals and in places forms displacive circumgranular rims; some sparry calcite fills narrow (1-mm) tubes after root traces; clear, irregular contact to

-19 cm; C horizon; matrix-supported breccia; grayish-brown (10YR5/2); with clasts of brownish-yellow (10YR-6/6), yellowish-brown (10YR5/6), dark bluish-gray (5B4/1), and black (5Y2/1); few fine (1-mm) calcareous white (5Y8/1) root traces; common narrow (2-3-mm) near-vertical calcite crystal sheets of white (5Y8/1) with local manganese spots (mangans) of black (5Y2/1) and iron stain (diffusion sesquans) of reddish-brown (5YR4/4); these calcareous and ferruginized joints penetrate deeply into this thick lahar; strongly calcareous; agglomeroplasmic inundulic in thin section; common grains of nephelinite, phonolite, and pyroxene; few ijolite, sövite, and alvikite; few calcite crystal tubes after fine root traces.

4.2.12.4 Further examples. No other Onuria paleosols were seen.

4.2.12.5 Alteration after burial. The Onuria paleosol is a little more drab colored, more organic, and less calcareous than Chogo clays. It also may have suffered 74 percent compaction, so that the original depth to the calcareous nodules was 22 cm. There are well-preserved fossil grasses in the overlying nephelinitic sandstone and an upward-increasing organic carbon content within the Onuria clay, but much organic matter was lost during burial. The olive-brown color of the surface horizon could have been produced by mild burial gleization of organic matter there, but this discoloration is not nearly so extreme as seen in Tut and Choka Series paleosols in Kenya, or in Khakistari or Pandu Series paleosols in Pakistan. The root traces are more oxidized than their matrix and, along with ferric concretions, are evidence that the slight gleization suffered may well have been during soil formation. Some cementation by carbonate during burial is possible, especially within root traces and joints penetrating the underlying lahar. Nevertheless, the close association between carbonate nodules and the claystone-lahar contact, as well as the paucity of root traces penetrating both nodules and lahar, is evidence that the original soil already had subsurface layers firmly cemented.

4.2.12.6 Reconstructed soil. The Onuria clay paleosol can be imagined as a brown, granular-structured, clayey surface (A) horizon over a nodular (Bk) horizon near the contact with a massive, indurated, bouldery lahar. The molecular ratios of alumina/bases and barium/strontium indicate little leaching of cations (Fig. 4.22), which would have left the soil with an alkaline pH and high cation exchange capacity and base saturation. There is no hint of salinization from soda/potash ratios. Some hints of gleization are given by the olive color of the claystone and the scattered ferric concretions. However, the ferrous iron/ferric iron ratios are low and decline upward within the profile, and there is a well-developed calcic horizon, as in well-drained soils. This clayey soil, with its shallow calcareous hardpan in well-cemented bouldery rock, may have been prone to puddling after rain, but for most of the time it was probably dry and oxidized.

4.2.12.7 Classification. The olive-brown clayey surface horizon of this profile is dark, base rich, and finely structured enough to qualify as a mollic epipedon, which is diagnostic for Mollisols (of Soil Survey Staff, 1975). It is currently a little thin to qualify as a mollic epipedon, but would have been thick enough before burial compaction. In any case, such thin epipedons still qualify as mollic when there is a shallow lithic contact, as is likely for the Onuria clay. Unlike Rendolls, for which this provision was made, the Onuria clay is calcareous even near the surface. It is best classified as a Calciustoll, considering its shallow calcic horizon, abundant clay, and lack of evidence for permafrost or pronounced waterlogging. In the Food and Agriculture Organization (1974) classification, the Onuria clay is most like a Calcic Kastanozem. Among Australian soils (Stace *et al.*, 1968) it is similar to a Chernozem, for reasons outlined for Chogo paleosols. In the Northcote (1974) key it would be Uf5.12 and in the South African scheme (MacVicar *et al.*, 1977) it is a Milkwood graythorne.

The Onuria clay is most like soils on the flank of Lemuta Hill in the eastern Serengeti Plains of Tanzania (particularly Barsek profile 44 of de Wit, 1978). Similar soils in South Africa are found on footslopes of nephelinitic lavas in northern Kruger National Park, South Africa (Botha, 1985; Gertenbach, 1983; Venter, 1986).

4.2.12.8 Paleoclimate. Petrocalcic stringers at the contact between claystone and lahar in the Onuria clay may have been controlled by this lithological break rather than by paleoclimate, as in some surface soils of the Serengeti Plains (de Wit, 1978). Nevertheless, this paleosol also has calcareous nodules above this level, some 16-cm below the surface now and probably 22 cm in the uncompacted soil. Such shallow calcic horizons are found in dry parts of the Serengeti Plains, receiving 450 mm or less of rain per year. In the Mojave Desert of California (Arkley, 1963) and the Great Plains of North America (Jenny, 1941), such shallow calcic horizons are found in soils receiving 280-510-mm mean annual rainfall.

Seasonality of rainfall is in evidence from ferric concretions in the clayey subsurface of the Onuria clay. Seasonality may also explain the paradox of good structure and calcareous nodules indicative of good drainage, but olive tinge and ferric concretions indicative of slow drainage. These indications of seasonality are certainly more pronounced than in early Miocene paleosols in Kenya, but are not nearly so marked as in late Miocene paleosols of Pakistan.

The dense root traces, thorough weathering, granular structure, and humification of the surface horizon of the Onuria clay are evidence of a productive ecosystem. This is compatible with but by no means compelling evidence for a warm paleotemperature.

4.2.12.9 Ancient vegetation. The Onuria clay has the most finely-structured surface (A) horizon of all the paleosols found in the main fossil quarry at Fort Ternan. This, together with its abundant fine root traces and laterally even lower contact with the calcic (Bk) horizon, is very similar to soils formed under grassland. There are a few large root traces and local disruption of the calcareous hardpan as evidence for scattered trees.

Wooded grassland vegetation also is evident from fossil plants preserved in overlying nephelinitic sandstone (Dugas, 1989; Retallack, Dugas, & Bestland, 1990). The fossil grasses are oriented obliquely across bedding, as if pushed over in place of growth by flow from the north, as already discussed in establishing the geological setting of the paleosols at Fort Ternan. The fossil grass culms and leaves have been observed up to 8 mm in diameter and 23-cm long. They are evenly-distributed throughout the basal 17 cm of the bed in its 7 m of lateral exposure in the quarry. From these observations and the nature of the underlying paleosol, the fossil grass bed has preserved a sod-forming, tall to mid-length grassland.

Some of the fossil grasses and a few fragments of dicot leaves in the nephelinitic sandstone have well-preserved areas showing cell outlines of their cuticles and placement of opal phytoliths. The best preparations of these microscopic features were gained from fragments of grass mounted on scanning electron microscope stubs immediately after extraction from the matrix. Specimens exposed to air for as short a time as a few days proved barren of cellular detail, because drying and decay destroyed organic matter of the cuticle and allowed it to exfoliate from a calcite internal mold of the epidermal cells. Five distinct kinds of fossil grasses and one kind of dicot were recognized from 12 useful preparations during microscopic study of the fossil cuticles (Dugas, 1989). By comparison with similar studies of the cuticles of living East African grasses (P. G. Palmer & Tucker, 1981, 1983; P. G. Palmer & Gerbeth-Jones, 1986, 1988; P. G. Palmer, Gerbeth-Jones, & Hutchinson, 1985) and current understanding of taxonomically significant features of grass epidermis (R. P. Ellis, 1987), the five kinds of grass cuticle were identified as three different species of the grass subfamily Panicoideae and as two species of Chloridoideae. These subfamilies dominate wooded grassland in Africa today, with chloridoids more abundant in dry and desert grasslands and panicoids more abundant in grassy woodlands (Clayton, 1970; Clayton, Phillips, & Renvoize, 1974; Clayton & Renvoize, 1982). There was no sign of the pooid grasses that now dominate Afromontane grasslands, nor of bambusoid or arundinoid grasses common in grasslands of seasonally wet depressions (dambos), grassy woodlands, and forests. The fossil grass cuticles also show abundant phytoliths, as usual in living grasses under marked grazing pressure (McNaughton & Tarrants, 1983). Their exposed stomates can be taken as evidence of a grassland that was not desertic, but well watered, at least during the growing season (Metcalfe, 1960).

Surface soils similar to the Onuria clay in the United States supported open grassland before human disturbance (Soil Survey Staff, 1975). Soils with which the Onuria clay is identified in the Australian (Stace *et al.*, 1968) and Food and Agriculture Organization (1974) classification are covered by open grassland, wooded grassland, grassy woodland and forest. Closer modern analogs for the Onuria clay can be found in the eastern Serengeti Plains of Tanzania, where short grassland of sedges (*Kyllinga nervosa*) and grasses (*Sporobolus marginatus*) grows in carbonatite-nephelinite ash altered to clay over a hard calcareous pan of cemented tuff (Anderson & Talbot, 1965; de Wit, 1978; White, 1983). The climate of this area is a little drier and the vegetation somewhat sparser than likely for the Onuria clay and its overlying fossil grass bed.

4.2.12.10 Former animal life. No fossil bone was found in the Onuria clay. Nor were there any distinctive trace fossils, although the claystone appears well bioturbated. The Onuria clay is less calcareous and better developed than the associated fossiliferous Chogo clay paleosols, and these attributes have been found elsewhere to correlate with lower yields of fossil bone (Retallack, 1984a, 1988b). Nevertheless, the Onuria clay is not a hopeless prospect for bones. Some may yet be found with a more dedicated search. Both Chogo and Onuria paleosols supported generally similar ecosystems and may have shared some animal species.

4.2.12.11 Paleotopographic setting. The type Onuria clay paleosol formed on a thick bouldery lahar where the footslope of a nephelinitic stratovolcano gave onto a broad volcanic plateau of phonolites. In addition to local relief of large boulders on lahars, these mudflows commonly form a local hummocky topography near their terminus (Cas & Wright, 1987). Such a landscape position is compatible with the similarity of this paleosol with hillside soils of the Serengeti Plains (profile 44 of de Wit, 1978) and with evidence for good drainage from calcareous nodules and from low ferrous iron/ferric iron ratios. Nevertheless, ferric concretions and olive tinge noted for this paleosol are features associated with slow drainage. This very clayey soil with a shallow hardpan on a firmly cemented, hummocky lahar may have been slow to drain in the wet season, but probably was generally dry.

4.2.12.12 Parent material. Beneath the clayey surface horizon of the Onuria paleosol is a thick nephelinitic tuff breccia, which probably was well cemented with calcite as the soil formed. This lahar deposit has a minor component of carbonatite, as can be seen by comparing its chemical composition (Fig. 2.21; Appendices 5 and 6) with that of local carbonatites and nephelinites (Binge, 1962; Deans & Roberts, 1984). This calcareous volcanic component may have been responsible for its firm cementation after emplacement. Such early cementation has been observed in historic carbonatite-melilitite volcanic ashes in Tanzania (Hay, 1978, 189) and is evident for the Onuria clay from deflection of root traces by this horizon.

This lahar bedrock probably was not the only parent material to the Onuria clay, considering the sharp contact between it and the surface horizon and the abrupt compositional change in chemical elements resistant to weathering (Fig. 2.21). The claystone is chemically like nephelinite, without the carbonatite component seen in the lahar. It may originally have been a nephelinitic sandstone or airfall tuff. Some dustings of fresh nephelinite ash when most of this parent material was already weathered to clay are apparent from occasional little-weathered grains of pyroxene in the claystone (Fig. 2.30A). Addition of small amounts of carbonatite-melilitite tuff may be indicated by traces of wollastonite and analcite detected by x-ray diffractometry (Fig. 2.33). Comparable rejuvenation of soils by volcanic airfall ash has also been observed in surface soils of the Serengeti Plains (Jager, 1982).

4.2.12.13 Time for formation. The laterally extensive petrocalcic horizon (stages II to III of Gile *et al.*, 1966) and thorough weathering of the clayey surface horizon of this paleosol are indications of moderate to strong development and a longer time of formation than for Chogo paleosols. Such thin shallow hardpans form in surface soils in carbonatite-nephelinite ash within the Serengeti Plains in only 2000 to 9000 years (Hay & Reeder, 1978). The higher part of this range is a reasonable estimate for the type Onuria clay.

4.3 Summary and Prospects for Kenyan Miocene Paleosols

Although each paleosol has something to offer in terms of paleoenvironmental information (Tables 4.3 and 4.4), much also can be gained by considering the assemblage of paleosols at a single site as representing different parts of the same ancient landscape. This landscape can then be compared with modern soilscapes considered from the perspective of landscape ecology (Forman & Godron, 1986). The potential interrelationship between paleosols can be explored by plotting them in two dimensions with hue as the dependent axis and degree of development as the independent axis (Fig. 4.23). As already discussed, broad classes of hue reflect degree of drainage and humification. Development, on the other hand, reflects time for formation of the paleosol and stage of ecological succession of its vegetation. Thus weakly developed paleosols can be considered examples of likely precursors to better developed paleosols on the same general parent material and in the same general paleotopographic setting. Although weakly developed paleosols may dominate stratigraphic sections in depositional settings here studied, better developed paleosols probably dominated most of the original landscape.

From this perspective, the suite of paleosols in deposits some 20 million years old around carbonatite cinder cones and the nephelinitic stratovolcano near Songhor and Koru can be imagined as a landscape of Haplustalfs, with Fluvents along streams and Ustropepts on the flanks of cinder cones. There is no modern landscape quite like this because of the rarity of active carbonatite-nephelinite volcanic centers. The most similar unit of the Food and Agriculture Organization (1977a) soil map of Africa is in the area (Lk4-2ab) of Zimbabwe, South Africa, and Botswana along the Limpopo River and its alluvium, derived from Precambrian schists, limestones, and sandstones, with large mafic intrusions. Vegetation of this area is dry deciduous woodland dominated by *Colophospermum mopane* (M. M. Cole, 1986; White, 1983), and climate is subhumid, seasonally dry, and tropical. At Messina, South Africa, for example, mean annual rainfall is 340 mm, with 78 mm in January and 3 mm in July: mean annual temperature is 23°C, but 30°C in January and 18°C in July (Müller, 1982). This can be regarded as drier than likely for the ancient soilscape of Koru and Songhor, because the parent materials of these soils are mainly noncalcareous, and the soils have accumulated salts and carbonate in this dry climate (B. S. Ellis, 1950). The soilscape around the active nephelinite-carbonatite volcanic center of Oldoinyo Lengai (Hay, 1978, 1989) and the nearby Serengeti Plains of Tanzania (de Wit, 1978;

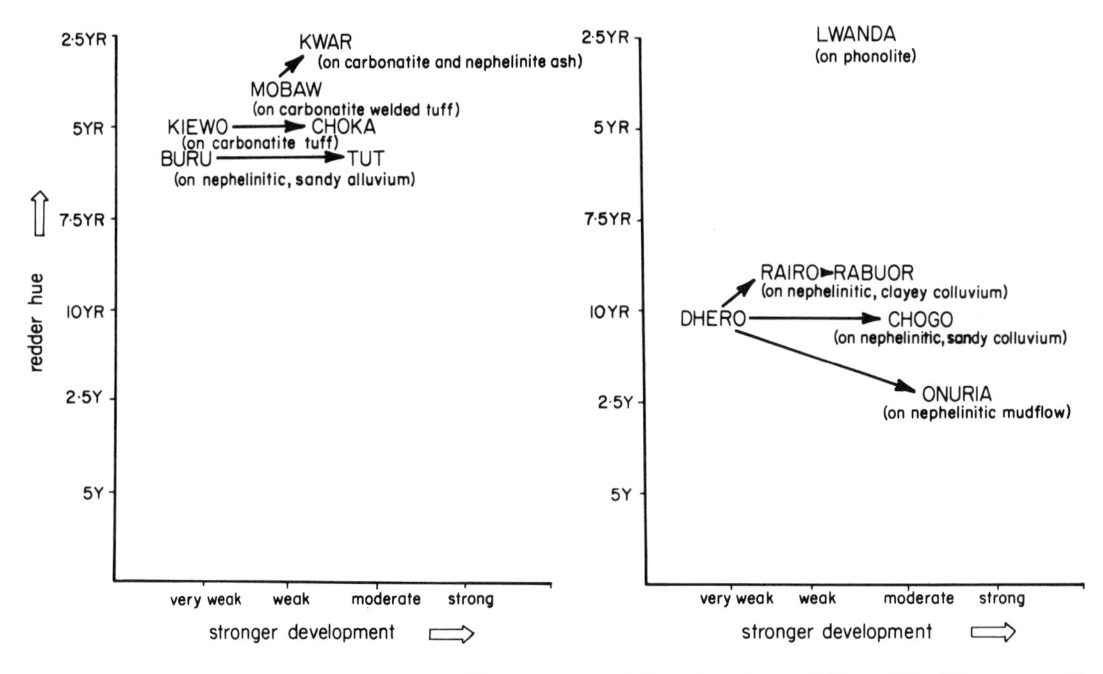

Fig. 4.23. Interpretive relationships between Kenyan paleosols from Songhor and Koru (20 million years old on left) and from Fort Ternan (14 million years on right), in terms of likely former degree of drainage and humification (red versus drab hue) and duration of soil formation (weak versus strong development), and likely direction of ecological succession of plants and of soil development (arrows).

Jager, 1982) is even drier and more calcareous. The paleosols at Songhor and Koru, in contrast, have been decalcified compared with their carbonatite-nephelinite parent materials. However, the paleosols were not nearly so decalcified as red soils on calcareous coastal sands some 125,000 to 240,000 years old in coastal Kenya and Somalia (unit Qc37-1a of Food and Agriculture Organization, 1977a; Braithwaite, 1984; Muchena & Sombroek, 1981), under tropical semideciduous forest in a humid, seasonally dry climate. Nor were they so decalcified as soils around active carbonatitic, nephelinitic, basaltic and trachytic volcanoes of the Western Rift Valley in Uganda, Ruanda, Burundi, and Zaire, where there are weakly calcareous Eutrandepts, Placandepts, Vitrandepts, and Hapludolls (Harrop, 1960; Nixon & Hornung, 1973; Mizota & Chapelle, 1988). These soils support a mosaic of tropical semideciduous, transitional Guineo-Congolian and Afromontane forest, with large areas of secondary grassland and cultivation (White, 1983) in a humid, seasonally dry, montane to tropical climate. For example, at Kisozi in Burundi, mean annual rainfall is 1447 mm, with 167 mm in January and 6 mm in July. Mean annual temperature is 16°C, with 16°C in January and 15°C in July (Müller, 1982). This region is at higher elevation and is much more humid than envisaged for Koru and Songhor. Although there is no precise modern analog for paleosols at Koru and Songhor, their soilscape is constrained by these various comparisons. They are best imagined as moderately to weakly developed, deep, red soils on carbonatite cinder cones and a nephelinitic stratovolcano in a subhumid warm climate with a short dry season, under dry forest that was floristically rich and shared many elements with the Zambezian floristic realm.

TABLE 4.3. Identification in modern soil classifications of Miocene paleosol series of southwestern Kenya

Paleosol Series	Luo Meaning	Diagnosis	U.S. Taxonomy	FAO Map	Australian Handbook	Northcote Key	South African System
Buru	dirt	thin, clayey, bedded, weakly calcareous and red (5YR)	Fluvent	Calcaric Fluvisol	Alluvial Soil	Uf1.23	Dundee dundee
Chogo	bone	thin, clayey, and grayish-brown (10YR), over calcareous (Bk) horizon	Haplustoll	Calcic Kastanozem	Chernozem	Uf5.11	Inhoek drydale
Choka	white-wash	thick, calcareous, and reddish-brown (5YR), with subsurface clayey (Bt) horizon and abundant insect larval cells	Haplustalf	Calcic Luvisol	Red Brown Earth	Gc2.12	Valsrivier marienthal
Dhero	thin	thin, brown (10YR), bedded shale and siltstone	Orthent	Calcaric Regosol	Lithosol	Uf1.31	Mispah kalkbank
Kiewo	striped	thin, red (5YR), and strongly calcareous, with bedded limestone stringers	Fluvent	Calcaric Fluvisol	Alluvial Soil	Uf1.33	Dundee dundee
Kwar	red	thin, clayey, and red (2.5YR), with boulders of carbonatite	Ustropept	Calcic Cambisol	Calcareous Red Earth	Uf5.31	Oakleaf makulek
Lwanda	boulder	thick, clayey, and red (10R), with large phonolite corestones	Vertic Eutropept	Vertic Cambisol	Chocolate Soil	Ug5.32	Glenrosa ponda
Mobaw	patched	thin, clayey, and red (2.5YR), with boulders of carbonatite	Lithic Ustropept	Calcic Cambisol	Red Calcareous Soil	Um5.11	Mispah muden
Onuria	yellow	thin, clayey, and olive brown (2.5Y), over calcareous hardpan	Calciustoll	Calcic Kastanozem	Chernozem	Uf5.12	Milkwood graythorne
Rabuor	brown	thick, clayey, and dark brown (10YR), with deep cracks	Vertic Argiustoll	Luvic Kastanozem	Prairie Soil	Ug3.3	Bonheim glengazi
Rairo	gray	thin, clayey, and dark brown (10YR), with relict bedding and deep cracks	Vertic Ustropept	Vertic Cambisol	Brown Clay	Ug1	Inhoek coniston
Tut	deep	thick and red (5YR), with subsurface clayey (Bt) and calcareous (Bk) horizons	Haplustalf	Calcic Luvisol	Red Brown Earth	Gc2.12	Valsrivier marienthal

TABLE 4.4. Paleoenvironmental interpretation of Miocene paleosols of southwestern Kenya

Paleosol Series	Locality	Age (Ma)	Paleoclimate	Ancient Vegetation
Buru	Songhor	20	too weakly developed to reflect paleoclimate	colonizing forest
Chogo	Fort Ternan	14	320-590-mm annual rainfall, seasonally dry	grassland mosaic of wooded grassland for type Chogo clay and grassy woodland for Chogo clay eroded-phase and ferruginized-nodule variant: including Gramineae, *Annonaspermum* sp., *Leakeyia vesiculosa, Cnestis rusingensis,* Cucurbitaceae, *Icacinocarya* sp., *Euphorbiotheca pulchra*, Papilionaceae, *Berchemia pseudodiscolor, Zizyphus rusingensis*, *Celtis rusingensis*
Choka	Songhor	20	660-1200-mm annual rainfall, seasonally dry, tropical	dry forest
Dhero	Fort Ternan	14	too weakly developed to reflect paleoclimate	colonizing grassy woodland or wooded grassland
Kiewo	Songhor	20	too weakly developed to reflect paleoclimate	open colonizing forest
Kwar	Koru	20	subhumid, seasonally dry, tropical	colonizing dry forest, midway in ecological succession: including *Celtis* sp. cf. *C. mildbraedii*
Lwanda	Fort Ternan	14	subhumid, seasonally dry, tropical	grassy woodland or wooded grassland
Mobaw	Koru	20	humid to subhumid, tropical	colonizing forest
Onuria	Fort Ternan	14	280-450-mm annual rainfall, seasonally dry	wooded grassland, grasses tall to intermediate, including panicoids (3 species) and chloridoids (2 species); less common dicot leaf fragments
Rabuor	Fort Ternan	14	550-800-mm annual rainfall, seasonally dry	grassy woodland or wooded grassland
Rairo	Fort Ternan	14	too weakly developed to reflect paleoclimate	grassy woodland or wooded grassland
Tut	Songhor	20	400-700-mm annual rainfall, seasonally dry, tropical	dry forest

Former Animal Life	Paleotopographic Setting	Parent Material	Time for Formation
tortoise, burrow of insect or spider	levee of meandering stream	nephelinitic sandy and clayey alluvium	300 years or less
diverse mammal fauna dominated by antelope (*Oioceros tanyceras, Kipsigicerus labidotus*); including ape (*Kenyapithecus wickeri*): also with bird, snake, lizard, millipede, beetle, whip scorpion, and snail.	around shallow erosional gully on footslope of nephelinitic strato-volcano above extensive phonolitic lava plain	nephelinitic sandstone washed down from slopes, with occasional boulders of phonolite and light dustings of carbonatite airfall ash	2000 to 9000 years
common cane rat (*Diamantomys luederitzii*) and archaic deer (*Walangania africanus*); also bird, turtle, snake, insect cocoons, and snail	alluvial terrace	carbonatite-melilitite and nephelinite airfall and resorted tuffs	40,000 to 100,000 years
no fossils found in place: may include chameleon (*Chameleo intermedius*), whip scorpion, and beetle	lahar mounds of volcano footslopes	carbonatite-melilitite, nephelinitic tuff, lahar	600 years or less
diverse mammal fauna dominated by cane rat (*Diamantomys luederitzii*) and ape (*Limnopithecus legetet*): also turtle, snake, bird, snail, and burrows	banks of loosely sinuous creeks draining small cinder cones	carbonatite-melilitite and nephelinite airfall tuff	2000 years or less
diverse fauna dominated by early deer (*Walangania africanus*) and chevrotain (*Dorcatherium songhorense*): also bird, reptile, millipede, insect, and snail	paleokarst on carbonatite tuff footslopes of small cinder cone	carbonatite-melilitite and nephelinite airfall tuff on welded tuff	20,000 to 40,000 years
no animal fossils found	lava plain near toeslopes of stratovolcano	phonolite lava with surface addition of nephelinite colluvium	2000 to 20,000 years
diverse fauna dominated by archaic deer (*Walangania africanus*) and chevrotain (*Dorcatherium songhorense*): also lizard, millipede, insect, and snail	paleokarst pits on shoulder or backslope of small cinder cone	bouldery colluvium of carbonatite welded tuff, some airfall tuff	20,000 to 40,000 years
no animal fossils found	lahar mounds at footslope of volcano above lava plain	nephelinitic lahar, with carbonatite-melilitite and nephelinite tuff	2000 to 9000 years
only bone scrap: may have yielded rhino (*Paradiceros mukirii*) and ape (*Kenyapithecus wickeri*)	volcano footslopes above extensive lava plain	nephelinitic colluvium, some phonolite clasts	2000 to 40,000 yrs
unidentified bone scrap	volcano footslopes above extensive lava plain	nephelinitic colluvium, some phonolite clasts	2000 years or less
diverse mammal fauna dominated by cane rat (*Diamantomys luederitzii*), mole rat (*Bathyergoides neotertiarius*), and flying squirrel (*Paranomalurus bishopi*): also bird, reptile, traces of insects, snail	alluvial terrace	mixed carbonatite-nephelinite tuff and nephelinite alluvium	40,000 to 100,000 years

The suite of paleosols some 14 million years old at Fort Ternan, in contrast, included widespread Argiustolls, Haplustolls, and Calciustolls on the colluvial footslopes of a nephelinitic stratovolcano, with Eutropepts on phonolite, Ustropepts on young surfaces of colluvium, and Orthents on cemented lahar and volcanic ash. These are similar in many details to soils of the Serengeti Plains of Tanzania: an extensive rolling terrain of Precambrian gneisses showered with carbonatite-nephelinite ash from the active volcano Oldoinyo Lengai to the east (Hay, 1989).

Climate there is tropical, with a marked dry season, but varies from arid around Oldoinyo Lengai within the Gregory Rift to subhumid in the western Serengeti Plains (Andrews, 1989; de Wit, 1978). Near Seronera to the west of the open plains, mean annual rainfall is 785 mm, with 82 mm in January and 24 mm in July (Anderson & Talbot, 1965). Mean annual temperature there is 22°C, with 23°C in January and 20°C in July (de Wit, 1978). Vegetation also varies from east to west, from desert shrubland to *Acacia-Balanites* wooded grassland and grassy woodland along watercourses and hills to the west (Jager, 1982; White, 1983). There is not evidence of salinization in any of the Fort Ternan paleosols, and their degree and depth of weathering are most like those of soils in the western part of the Serengeti Plains, under tall to mid-length wooded grassland in a mosaic with streamside and hilltop grassy woodland. This part of the plains is mantled with carbonatite-nephelinite ash, most of which is at least 40,000 years old. The weathered ash forms a cover to Precambrian gneissic basement, in places with a thick red relict Miocene paleosol, and in other places emergent as bare-rock inselbergs, or "kopjes" as they are called there. These kopjes are a source for quartz and feldspar in the Tanzanian soils, which have a more felsic overall chemical composition than the paleosols at Fort Ternan on only nephelinitic and phonolitic parent materials. Despite these overall chemical differences, chemical depth functions within the Tanzanian soils (unpublished graphic analysis by E. A. Bestland of data from de Wit, 1978, and from Jager, 1982) are very similar to those seen in the Fort Ternan paleosols.

Paleosols at Fort Ternan are not so thin, calcareous, or zeolitic as soils in the desert around Oldoinyo Lengai (Hay, 1978, 1989). Nor are they so thick, clayey, organic, red, and decalcified as soils forming under Afromontane forest and cool humid climate at high elevations on extinct stratovolcanoes to the east and south of the plains, such as the trachytic Ngorongoro crater rim (Anderson & Herlocker, 1973; Mizota, Kawasaki, & Wakatsuki, 1988). Nevertheless, similar forested soils may have formed on volcanic slopes above the preserved sequence at Fort Ternan. Such soils could have been a source for deeply weathered boulders of nephelinite and phonolite found in the paleosols at Fort Ternan.

Paleosols also are becoming better known from other Kenyan Miocene fossil localities (Pickford, 1986a, 1986b). A reconnaissance survey of paleosols yielding fossil apes such as *Proconsul* and *Dendropithecus* some 17 million years old (Drake *et al.*, 1988) on Rusinga Island has been undertaken by our research group (Bestland, 1990b; Thackray, 1989). Many of these paleosols were Inceptisols and Entisols with abundant, large, drab-haloed root traces, like those reported here from Songhor. There also are sequences of olive and brown calcareous paleosols similar in many respects to those from Fort Ternan. This suite of paleosols may have supported a mosaic of moist Zambezian grassy woodland and wooded grassland and perhaps also some peripheral dry Guineo-Congolian forest.

On Maboko Island, a sequence of claystones some 15 to 16 million years old (Andrews *et al.*, 1981; Harrison, 1989) have yielded fossils of apes (*Kenyapithecus*, *Limnopithecus*, *Micropithecus*, *Nyanzapithecus*, *Mabokopithecus*) and some of the earliest monkeys (*Victoriapithecus*). These remains come from a variety of paleosols, including dark-brown profiles similar to Rabuor and Rairo clays at Fort Ternan, as well as a distinctive gray, clayey paleosol, mottled orange and white, as is found in soils of seasonally dry swamp woodlands. At Ombo, also about 15 to 16 million years old, gray-green siltstones with calcareous nodules have yielded a variety of fossil mammals (Pickford, 1984, 1986a). These also could be paleosols of seasonally dry swamp or marsh.

In the Samburu Hills, 4-km west of Nachola, in tuffaceous sediments originally dated at 10 million years (unit C of Emuruilem Member of Makinouchi *et al.*, 1984), but more likely 15 to 16 million years (Hill & Ward, 1988), there is a pale-yellow paleosol containing unusually abundant remains of *Kenyapithecus* (Ishida *et al.*, 1984; Pickford *et al.*, 1984). This paleosol is capped by a cherty bed containing permineralized wood referred to *Acacia*. It probably supported grassy woodland of these trees, an interpretation compatible with its similarity to the Dhero clay loam paleosol described here.

A molar tooth, which has been compared with that of a chimpanzee, was found in a black layer of iron-manganese, 4-cm thick, beneath 15 cm of green silt and fine-grained sand with root traces, in a part of the Ngorora Formation some 11 million years old, 0.5-km south of Bartabwa in the Tugen Hills (Hill & Ward, 1988). This dark layer could well be the placic horizon of a seasonally dry swamp or marsh paleosol.

In the same region, a mandible of *Australopithecus afarensis*, about 4 million years old, has been reported from a "dark ferruginous horizon" in the Chemeron Formation 2-km north of Rondonin (Hill & Ward, 1988). This specimen may also have come from a paleosol, perhaps similar in general profile form to the Kwar or Tut Series described here. There are thus numerous Miocene paleosols in Kenya, some of them yielding fossils that span the evolutionary gap between fossil apes and early australopithecine ancestors of humans.

The various paleosols described here and identified as worthy of further attention are evidence of a more complex mosaic vegetation and an earlier drier climate for Kenya than envisaged from prior studies of fossil mammals (Evans, Van Couvering, & Andrews, 1981; Kappelman, 1991), snails (Pickford, 1983, 1985), and plants (Collinson, 1983). There are several possible reasons for this discrepancy. Body fossils can be transported and mixed from different parts of the environmental mosaic into the environments in which they are preserved. Around volcanic slopes coursed by lahars and floods, potential fossils could have been transported great distances (Pickford, 1986b). In most of these studies, the mammal, snail, and plant assemblages were pooled from several sites and stratigraphic levels, again obscuring the likely original environmental mosaic, that has been revealed by excavations made in the detailed manner of archeology (Andrews & Van Couvering, 1975; Shipman, 1977, 1981; Shipman *et al.*, 1981).

There are also preservational and evolutionary reasons why mammal, snail, and plant assemblages might indicate more humid conditions than prevailed. In general, plant fossils are preserved in waterlogged soils and sediments (Retallack, 1984a). The casting of organic remains in carbonatite-melilitite ash is a rare and local exception (Hay, 1986) that allows preservation of vegetation of dry soils away from streams and lakes. In most cases, such dry vegetation as grassland and desert shrubland is not represented in the fossil record. In contrast, bones and snails are best preserved in calcareous soils and sediments of dry regions, and so show a bias toward open-country faunas (Retallack, 1984a). Again, this is confounded to a limited extent by the unusually calcareous volcanic parent materials of many Miocene paleosols in Kenya, that preserve some record of dry forest faunas (Hay, 1986; Pickford, 1986b).

In addition, the evolution of open grassy ecosystems during Miocene time was driven by adaptations of an original forest biota to increasingly open vegetation, but the whole suite of adaptations found in modern faunas of open habitats evolved over geological time scales (Retallack, 1982). The land-snail fauna of Africa today is still most diverse in forested regions, with a subset of genera extending into wooded grasslands and a smaller subset into deserts (Van Bruggen, 1978; Heller, 1984). Modern mammals of African wooded grassland now extend widely into desert grasslands and grassy woodlands. Before the evolutionary diversification of antelopes and immigration of horses and other open-country mammals from Eurasia during late Miocene time (J. A. H. Van Couvering, 1980), a mammalian fauna adaptively more like that found now in dry forests may have extended also into grassy woodlands and wooded grasslands.

The interpretation offered here of Kenyan early and middle Miocene vegetation as a mosaic of dry forest, grassy woodland, and wooded grassland is by no means new. It was a common view during early studies of this region's fossil mammals (Whitworth, 1953), plants (Chesters, 1957), lizards (Estes, 1962), and land snails (Verdcourt, 1963). More recently such a view has been supported by studies of limb proportions of giraffe ancestors (Churcher, 1970), taxonomic affinities of fossil dung beetles (Paulian, 1976), lateral variation in mammalian faunas within a single paleosol (Andrews & Van Couvering, 1975), floristic comparisons of fossil plants (Axelrod & Raven, 1978), microwear on fossil antelope teeth (Shipman *et al.*, 1981), assemblages of fossil pollen (Bonnefille, 1984), and subfamilial affinities of fossil grasses (Dugas, 1989). A general picture of a mosaic of dry forest and early succesional woodland during early Miocene time, opening up to wooded grassland, with streamside and upland grassy woodland during the middle Miocene, is now well substantiated by a variety of lines of evidence. That of paleosols adds not only confirmation, but local precision in reconstructing the local mosaic of vegetation and other aspects of the paleoenvironment of our remote East African apelike ancestors.

5. LATE MIOCENE PALEOSOLS IN NORTHERN PAKISTAN

Early pioneers of the study of Siwalik Group sediments of sub-Himalayan Pakistan and India, Hugh Falconer and Proby Cautley, recognized that these were deposits of ancient river floodplains and swamps roamed by an impressive variety of extinct mammals (Falconer, 1868). The idea that the Siwalik Group was a record of alluvial deposition from the ancestral Ganges and Indus Rivers, became strengthened over the years (Cotter, 1933; Pascoe, 1920a; Pilgrim, 1919), but paleosol features, with rare exception (Pilgrim, 1913), were widely overlooked. These pioneers of Indian geology clearly understood that these ancient alluvial sequences included soil paleoenvironments. However, soil science itself was still in its infancy and of little use for their efforts to understand Siwalik mammalian faunas, sedimentary environments, and structural deformation. Since that time much has been learned of soil formation on the modern Indo-Gangetic Plains of India and Pakistan (Ahmad, Ryan, & Paeth,, 1977; Kooistra, 1982; Murthy *et al.*, 1982; Rahmatullah, Dixon, & Golden, 1990; Sehgal & Stoops, 1972; Sehgal, Sys, & Bhumbla, 1968). This expertise now is being applied increasingly effectively to interpreting soil paleoenvironments of the Siwalik Group (Behrensmeyer & Tauxe, 1982; G.D. Johnson, 1977; G.D. Johnson *et al.*, 1981; Retallack, 1985; Visser & Johnson, 1978).

Not only are soil features well preserved in the Siwalik Group, as already discussed in Chapter 2, but paleosols are abundant. In one measured section reported here from northwest of Kaulial village, there are 80 separate paleosols in only 58 m of red clayey colluvium between two thick paleochannel sandstones. A conservative estimate would be at least one paleosol for every meter of the entire Siwalik Group, estimated to be some 6-km thick (by Wadia, 1975).

Siwalik sediments command attention for several reasons. They are exceptionally thick. They range in age from Pleistocene well back into the Miocene. They also are well exposed in badlands created in part by human desertification of the landscape throughout the foothills of the Himalayan ranges from Pakistan into northern India, Nepal, and Bangladesh. This sedimentary sequence is thus an outstanding natural laboratory for the study of a variety of geological processes. Many of these studies could be enhanced by better understanding of paleosols.

Siwalik sediments can be seen as a long record of Himalayan uplift and deformation (Burbank, 1983, 1988; Burbank & Raynolds, 1984; Burbank, Raynolds, & Johnson, 1986; Burbank *et al.*, 1988; N.M. Johnson *et al.*, 1985; Parkash, Sharma, & Roy, 1980; Raynolds *et al.*, 1980). In this respect, paleosols could be used profitably to analyze changing rates of fluvial sedimentation on time scales finer than can be resolved using magnetostratigraphic or radiometric methods, as has been demonstrated in other fluvial sequences (Retallack, 1986a).

Siwalik sediments also have been used to show that the best paleomagnetic signal in red beds is carried by detrital magnetite (Tauxe, Kent, & Opdyke, 1980). The signal appears to be scattered by soil formation, because of weathering of magnetite and other iron-bearing minerals to ferric hydroxides that are realigned with soil structure and then dehydrated to hematite during burial (Tauxe & Badgley, 1984, 1988). As demonstrated here, the degree and kinds of soil formation were very varied during Siwalik sedimentation. The effects of particular kinds of ancient soil formation on the quality of paleomagnetic data from this and other ancient red bed sequences remains to be studied in detail.

Like many fluvial sequences, the Siwalik Group also shows distinctly episodic sedimentation on a variety of time scales, ranging from thousands to hundreds of thousands and millions of years (Badgley & Tauxe, 1990; Behrensmeyer & Tauxe, 1982; N.M. Johnson *et al.*, 1988; Tandon, Kumar, & Singh, 1985). Climatic and tectonic changes are only 2 of at least 13 factors known to cause episodic deposition or erosion in and around modern streams (Schumm, 1977). Paleosols may provide independent evidence of paleoclimate, former topographic setting, and other paleoenvironmental factors useful in unraveling the causes of episodic sedimentation, as has been demonstrated for other alluvial sequences (Retallack, 1986a).

Then there are the famous Siwalik vertebrate faunas: an extraordinary fossil record of Neogene mammalian evolution (Barry *et al.*, 1985; Vasishat, 1985). These fossils include remains of apes, such as *Sivapithecus*, that are still thought by some (Oxnard, 1987; Schwartz, 1984, 1987) to be human ancestors, but are more widely recognized as a side branch to the human evolutionary tree and ancestral to the living orang-utan of Southeast Asia (Andrews & Cronin, 1982; Pilbeam, 1982). There are clues to the former habitats of these creatures from adaptive features of associated fossil mammals (Andrews, 1983; Tatersall, 1969a, 1969b; Vasishat, 1985), from fossil plants (Awasthi, 1982; Y. K. Mathur, 1984; K. N. Prasad, 1982; H. P. Singh, 1982; Vishnu-Mittre, 1984), and from their sedimentary context (Badgley & Behrensmeyer, 1980; Behrensmeyer, 1987, 1988; Behrensmeyer & Tauxe, 1982). To these lines of evidence can now be added that of paleosols (G. D. Johnson, 1977; G. D. Johnson *et al.*, 1981; Retallack, 1985). Paleosols by definition are in the place where they formed, and they can be particularly useful in reconstructing ancient mosaics of vegetation. Mammalian faunas of distinct kinds of paleosols can be a guide to the paleoenvironmental preferences of extinct mammals, as has been demonstrated in other fossiliferous variegated and red beds (Bown & Beard, 1990; Bown & Kraus, 1981; Retallack, 1983, 1988b).

Also of interest is the degree to which the fossil bones and teeth can be considered an accurate reflection of past faunas, considering their potential for transportation, often fragmentary nature, and local occurrence (Badgley, 1982, 1986). Each kind of paleosol is a distinct preservational environment, the quality of which can be judged by those chemical and structural features most relevant for preservation of fossils. In general, bones are found in calcareous paleosols, and plant remains in peaty paleosols (Retallack, 1984a). Assemblages of bones in paleosols, especially bones that have the kinds of preservation, orientation, and population structure to be expected from the accumulation of animals dying from normal causes (Shipman, 1981), also can be considered relatively free of transportation and so are representative of former populations (Behrensmeyer, Western, & Dechant-Boaz, 1979). Much can be learned about the way in which former animals and plants entered the fossil record from the study of paleosols containing them.

Despite the late start of paleopedological studies of the Siwalik Group, there are many and varied paleosols and good reasons to study them.

5.1. Geological Setting

The magnificent wreath of mountain ranges and their alluvial aprons around the northern margin of the Indo-Pakistan subcontinent (Fig. 5.1), were produced by its northward drift from Africa and Madagascar and its collision with the southern margin of Asia. The following paragraphs are an outline of this dramatic geological history, as an introduction to the detailed geological setting of the Miocene paleosols studied in the region near Khaur, northern

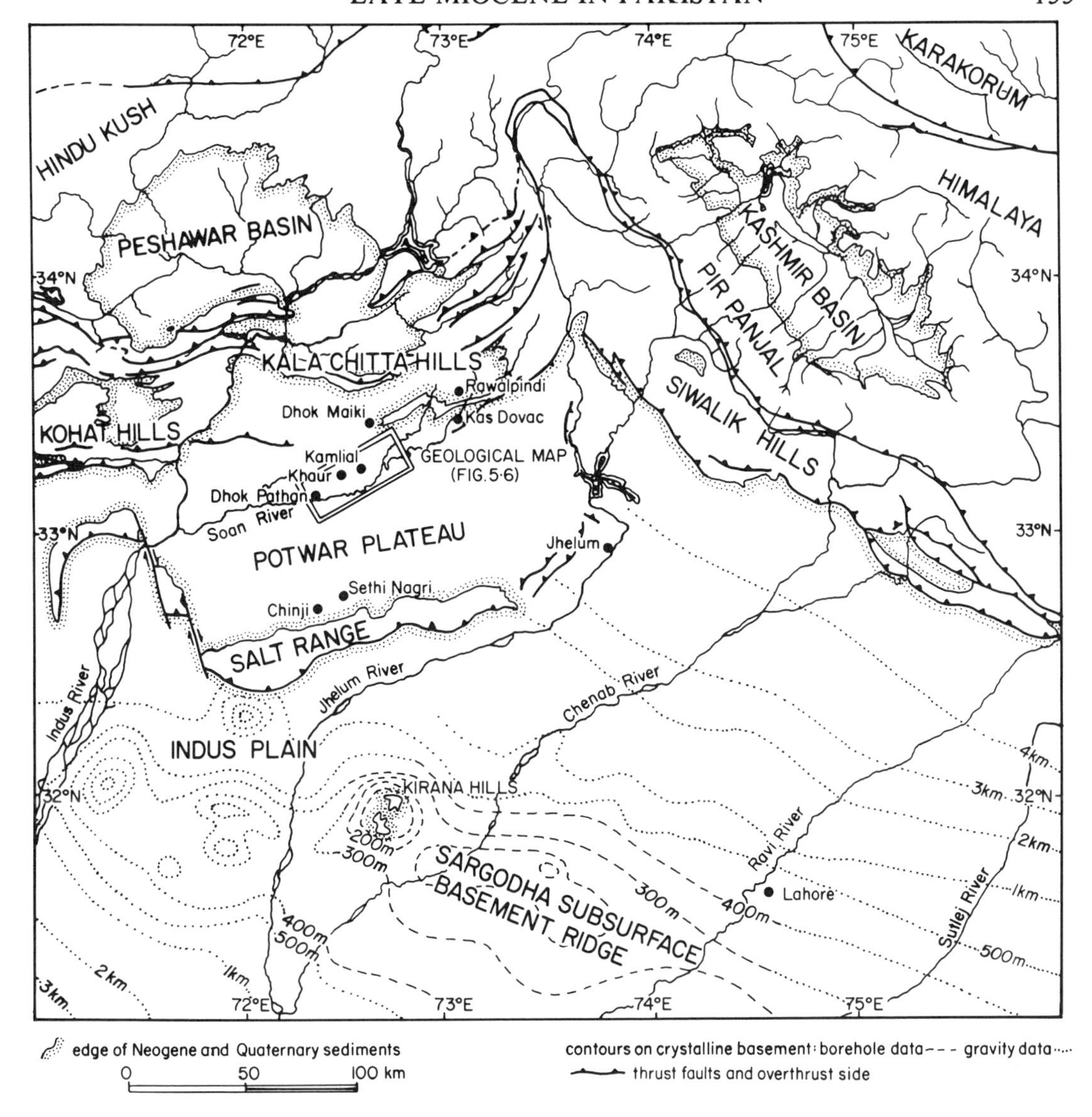

Fig. 5.1. Mentioned localities, major faults, sedimentary basins and contoured Precambrian basement to Indo-Gangetic alluvium in northern Pakistan and India (data from Burbank, 1983; Farah *et al.*, 1977).

Pakistan.

Peninsular India, and low-lying parts of Pakistan and Bangladesh, are a continental shield of Precambrian crystalline and metamorphic rocks. The Aravalli Craton of northwestern India and Pakistan, for example, has a core of crystalline rocks (Banded Gneiss Group and Bundhelkhand Complex) some 3500 to 2000 million years old, flanked by and structurally interleaved with continental margin sediments (Aravalli and Delhi Groups) some 2500 to 1500 million years old. This ancient continental platform is overlain by little disturbed marine rocks (Vindhyan Supergroup) some 1400 to 900 million years old, and a sequence of rhyolites and basalts (Malani Volcanics) some 940 to 740 million years old (Navqui & Rogers, 1987). The northernmost exposures of the Aravalli Craton not caught up in the Himalayan ranges are in the Kirana Hills, near Sargodha, Pakistan. These hills are a deformed sequence of felsic and basaltic volcanics, radiometrically dated at 870 ± 40 million years (Davies & Crawford, 1971).

Paleozoic marine rocks of an old continental margin now crop out in the Salt Range of Pakistan, where they have been thrust off their Precambrian crystalline basement. The classical non-marine Gondwanan sequences of Carboniferous to Triassic age in peninsular India were deposited in a system of rift valleys following sutures between old continental cratons. The distinctive fossil flora and fauna of these rocks have long been recognized as evidence that India was at that time part of the high-latitude supercontinent of Gondwana (Tripathi & Singh, 1987). The Indo-Pakistan continental block formed a northern coast to the supercontinent, flanking the ancient equatorial ocean of Tethys, between what would later drift off to become Madagascar and Western Australia (Scotese, Gahagan, & Larson, 1989). Indo-Pakistan was excised from this neighboring continental crust by a system of rift valleys during Jurassic and early Cretaceous time. Between about 80 million years ago (late Cretaceous) and 53 million years ago (early Eocene), India separated and moved north at an unusually fast rate (average of 15 cm/year) as the Indian Ocean opened to the south (Powell, 1979). Great volumes of basalt were extruded to make this new ocean floor, but there was also at this time an enormous outpouring of continental flood basalts to form the Deccan Traps that dominate the landscape of the modern Western Ghats of India. During later Eocene and Oligocene time, India's northward motion was slowed to less than 3 to 4 mm per year as it collided with continental blocks of the future Asian landmass. Some of the old Indo-Pakistan continent was underthrust, but there also was a good deal of crustal deformation and shortening (Dewey *et al.*, 1988). The deforming sedimentary margins, intervening island arcs, and microcontinents of the Tethyan Ocean created a complex and changing paleogeography. It can be imagined as similar in general to the present Mediterranean region, now at a similarly early stage in the convergence of Africa and Europe.

By early Miocene time (20-22 million years ago) no trace of the old Tethyan seaway remained. Thrusting had developed along the line of the present central crystalline axis of the Himalaya, creating an additional mountain barrier rising through the antecedent streams draining the already hilly Tibetan Plateau to the north (Fig. 5.2). These mountains may have been of alpine proportions, but were not yet so high as the modern Himalayan ranges, as can be seen from their alluvial outwash (Murree and Kamlial Formations), which is more mineralogically mature than geologically younger sub-Himalayan outwash (Chaudhri, 1975), and accumulated at a slower rate (N. M. Johnson *et al.*, 1985). Southeasterly paleocurrents (N. M. Johnson *et al.*, 1985) and a low-grade metamorphic suite of heavy minerals in these sediments (Chaudhri, 1984) are indicators that they were deposited in a forerunner of the Ganges River, at that time reaching well west of its present drainage into mountains of metamorphic rocks in northwestern Pakistan. This extended Ganges drainage may have been bounded to the south by the Sargodha Ridge, a northwest-trending ridge of metamorphic basement rocks well above that part of the Precambrian shield underthrusting the Himalayan ranges (Fig. 5.1). The Sargodha Ridge crops out only in the Kirana Hills, but can be mapped beneath cover of younger alluvium from data of boreholes and gravity surveys (Farah *et al.*, 1977). The Indus River during early Miocene time was a separate drainage among the rising fold mountains of far western Pakistan. It drained into the Sind Sea, a shallow extension of the Indian Ocean, that covered most of present southern Pakistan, north almost as far as the area of the Bugti Hills (Raza & Meyer, 1984).

By the time of the middle to late Miocene boundary (about 11 million years

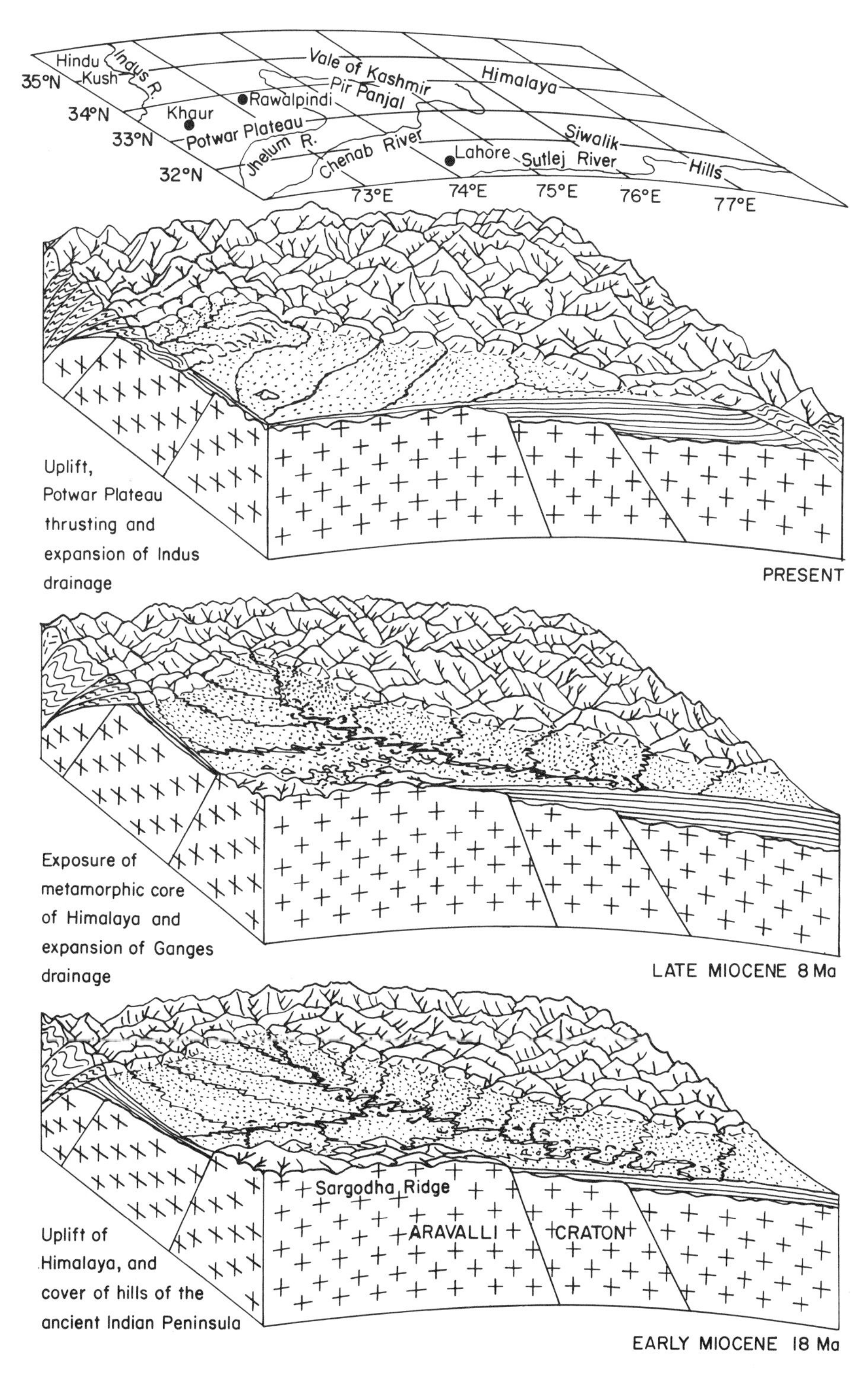

Fig. 5.2. Interpretative geological history of alluvial deposition with uplift of the Himalaya, Karakorum and Hindu Kush ranges (data from Burbank, 1983; Dewey *et al.*, 1988; Farah *et al.*, 1977; G.D. Johnson *et al.*, 1986; Yeats & Lawrence, 1984).

ago), there was a marked uplift of the Himalayan ranges. This is indicated by an increased rate of accumulation of fluvial outwash, apparent from the greater proportion of paleochannel sandstones over paleosols in both the ancestral Ganges (transition from Chinji to Nagri Formation of N. M. Johnson *et al.*, 1985) and Indus drainages (from Vitrow to Litra Formations of Hemphill & Kidwai, 1973). At about this time also, kyanite and blue hornblende appears for the first time in these sediments (Chaudhri, 1984; N. M. Johnson *et al.*, 1985). This middle to late Miocene uplift exposed the ultramafic and metamorphic core of the Himalayan ranges. It was also a time of deformation of the Himalayan foothills, resulting in the erosion and redeposition of older alluvial rocks (Gill, 1952). Middle to late Miocene uplift also is apparent from influx of sediment in the extensive submarine fans offshore from the mouths of the Indus River (Weser, 1974) and the Ganges-Brahmaputra River (Moore *et al.*, 1974).

Another great pulse of uplift during the Pliocene (5 million years ago) to present, dispersed gravelly sandstones (Soan Formation) over mildly deformed sequences richer in paleosols (Dhok Pathan Formation). This second great coarsening in grain size of the outwash was associated with the appearance of the high-grade metamorphic mineral sillimanite in the alluvial outwash (Chaudhri, 1984). New systems of thrusts, faults, and folds deformed Miocene alluvium south of the rising mountain fronts. For example, movement of the Salt Range Thrust some 5 million years ago, and again 2 million years ago, created the Potwar Plateau of Pakistan, as a broad syncline of Miocene alluvial rocks (Burbank, 1988; Burbank & Raynolds, 1984; Burbank, Raynolds, & Johnson, 1986; Burbank *et al.*, 1988). Associated alluvial outwash buried most of the Sargodha basement ridge.

These events allowed capture of former Gangetic tributaries by the Indus River, which was now a major axial drainage for increasingly high fold mountains of western Pakistan (Gill, 1952; N.M. Johnson *et al.*, 1985; Raynolds, 1982: contrary to prior reconstructions of Miocene sub-Himalayan drainage by Pascoe, 1920b; Pilgrim, 1919). Capture by the Indus River of the Jhelum, Chenab, Ravi, and Sutlej Rivers may have been aided by outwash from uplift of the Pir Panjal, beginning some 4.5-5 million years ago (Beck & Burbank, 1990; Raynolds, 1982). A strong swing to the west in paleocurrents of the now-dry Ghaggar River, near Chandigarh, India, some 1.7 million years ago (Tandon, Kumar, & Singh, 1985) may signal the last of these former tributaries of the Ganges to be captured by the Indus. The course of the Indus River itself was shifted eastward by deformation west of the Salt Range some 0.5 million years ago (J. W. McDougall, 1989). Rivers in the Indo-Gangetic Plains have continued to shift course in historic times (Mani, 1974a; Oldham, 1893; Shroder, 1989). For example, the recapture of the Yamuna River by the Ganges River diverted the flow formerly carried into the Indus by the now-dry bed of the Ghaggar River.

5.1.1 Geology of the area around Khaur

Although the Siwalik Group is named for the Siwalik Hills of northern India, the type localities of many of the formations of the Group are scattered over the Potwar Plateau of Pakistan, especially around the villages of Chinji and Khaur (Fig. 5.1). Part of the reason for this is the better exposures in the Potwar Plateau than in the Siwalik Hills. There are extensive badland outcrops of claystone and strike ridges of sandstone in the deep canyons draining across bedding into the Soan River, which runs down the axis of a broad syncline of the Miocene rocks of the Potwar Plateau (Figs 5.3 and 5.4). Vegetation is a sparse, overgrazed thorn scrub (vegetation unit 6B/C2 of Champion & Seth, 1968). Surface soils are thin, dry and calcareous (mainly

Fig. 5.3. Badlands exposures 1.2 km west of Kaulial village, Potwar Plateau, Pakistan. Part of the lower measured section of Dhok Pathan Formation (15 to 36 m of Fig. 5.9) was in a trench within the shadowed face to the left of the low sandstone-capped spur.

Orthents or Calcaric Regosols of Food and Agriculture Organization, 1977b, map unit Rc40-2b). The sparseness of vegetation in the Khaur area is due in large part to human abuse, because rainfall of the area is adequate for grassy woodland and wooded grassland. Mean annual rainfall at Rawalpindi is 960 mm, with 64 mm in January and 233 mm in July (Rao, 1981). Another reason for geological scrutiny of the Khaur area was the discovery and production of oil from the core of a small anticline (Gill, 1952; Pascoe, 1920a).

The Potwar Plateau also commanded attention because of its abundant mammalian fossils, which were used to subdivide the Siwalik Group into "faunal zones" (Pilgrim, 1910, 1913, 1934) named after villages such as Chinji, Nagri, and Dhok Pathan, that were near especially productive fossil localities. During geological mapping of the Potwar Plateau by Cotter (1933), these faunal zones, essentially biostratigraphic units, became used as stages, or chronostratigraphic units. Following additional geological research in India and Pakistan (Lewis, 1937), these stages became formalized with definitions based on rock types as formations of the Siwalik Group. This is how they are widely used today (Fatmi, 1974; Shah, 1977), despite the use of these same names as very different kinds of stratigraphic units in the past. They could still be used as chronostratigraphic units, provided they are given a distinguishing suffix, for example Dhok Pathanian or Nagrian. Such units defined on faunal content and rock sections are in principle and practice no different than land mammal stages proposed for North America (Woodburne, 1987) or elsewhere (Berggren, Kent, & Van Couvering, 1985). A fresh start on biostratigraphy of these

Fig. 5.4. Badlands exposures along Kaulial Kas, 1.5 km southeast of Kaulial village, Potwar Plateau, Pakistan. Human figure for scale is sitting in trench of measured section (Fig. 5.10), above prominent light-colored sandstone band in lower part of cliff.

beds has been made with the proposal of a series of biozones, including for example the *Selenoportax lydekkeri* interval zone (Barry, Behrensmeyer, & Monaghan, 1980; Barry, Lindsay, & Jacobs, 1982). The mammalian assemblages of these interval zones are similar to those of the old faunal zones of Pilgrim, and the boundaries of the new biozones are now accurately placed within long measured sections calibrated for geological age by magnetostratigraphic methods (Barndt *et al.*, 1978; N. M. Johnson *et al.*, 1985; Tauxe & Opdyke, 1982), and in a few places by means of radiometrically-dated volcanic ashes (G. D. Johnson *et al.*, 1982). The ages of most of the magnetozones were estimated by correlation with those already dated from deep sea basalts (by La Breque, Kent, & Cande, 1977), and these estimates now need to be revised slightly in view of refined understanding of the geological time scale (Berggren, Kent, & Van Couvering, 1985; Flynn *et al.*, 1990; Tauxe & Badgley, 1988). There are thus at least four separate stratigraphic schemes (Fig. 5.5) useful for understanding these Miocene alluvial sediments: lithostratigraphy, magnetostratigraphy, biostratigraphy, and chronostratigraphy.

The Khaur area is on the northern flank of the broad, open Soan Syncline, and south of the upthrust Kala Chitta Hills and other foothills flanking the Hindu Kush and Himalaya. There are some local anticlines around Khaur and Dhulian sympathetic with thrust faults of the foothills, but other areas expose long homoclinal sequences of alluvial sediments (Fig. 5.6). An important section for magnetostratigraphic (Tauxe & Opdyke, 1982) and biostratigraphic (Barry, Lindsay, & Jacobs, 1982) studies is exposed in Kaulial Kas (Fig. 5.7). Kas is a Punjabi

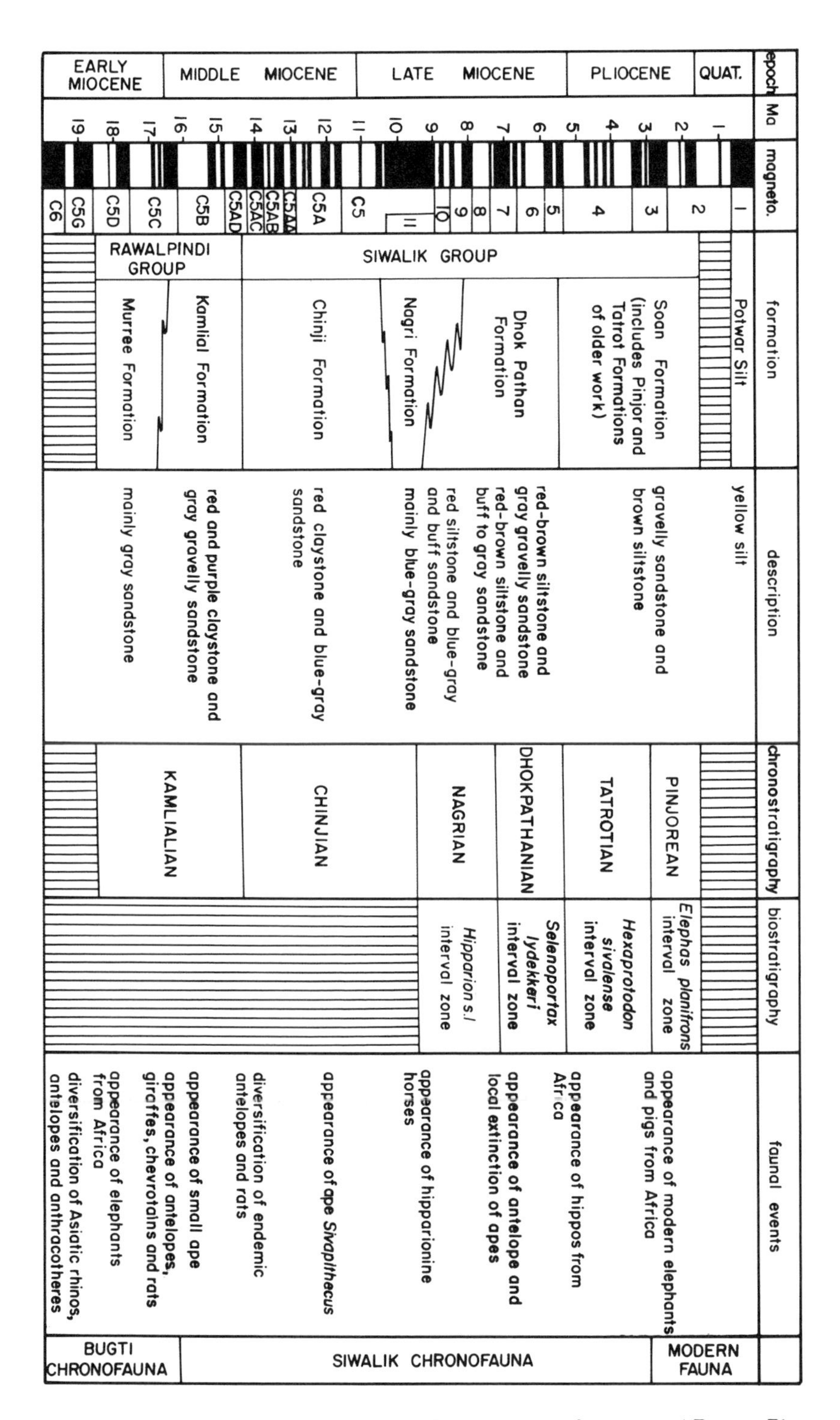

Fig. 5.5. A stratigraphic framework for Miocene rocks of the central Potwar Plateau, Pakistan (data from Barry *et al.*, 1982, 1985; Barry & Cheema, 1984; Berggren, Kent, & Van Couvering, 1985; Fatmi, 1974; G.D. Johnson *et al.*, 1982; N.M. Johnson *et al.*, 1985).

word for canyon. The dry badlands and canyons around Kaulial also have yielded important remains of fossil apes, including the most complete known skull remains of *Sivapithecus sivalensis* (see Fig. 1.2; Pilbeam, 1982; Pilbeam *et al.*, 1977a, 1977b, 1979, 1980; Ward & Pilbeam, 1983).

The two measured sequences of paleosols reported here from the Dhok Pathan Formation were in Kaulial Kas (Fig. 5.8). One sequence of paleosols (Figs 5.3 and 5.9) was measured 1.2 km west of Kaulial village, from the stratigraphic level of the *Sivapithecus sivalensis* skull to 56-m below, and has been magnetostratigraphically dated at 8.45 to 8.55 million years old (Tauxe & Opdyke, 1982; corrected to time scale of Berggren, Kent, & Van Couvering, 1985). This is stratigraphically below the most productive level for fossil apes (U-sandstone and associated claystones of Badgley, 1982, 1986; Badgley & Behrensmeyer, 1980). A second sequence (Figs 5.4 and 5.10) was measured 1.5-km southeast of Kaulial village, and has been magnetostratigraphically dated at 7.7 to 7.8 million years. By this time there was a faunal overturn including the incoming of the large antelope *Selenoportax lydekkeri* (Barry, Lindsay, & Jacobs, 1982; Barry *et al.*, 1985). Ape fossils were very rare by this time: the youngest one in Kaulial Kas is from the paleochannel sandstone above the upper measured section (locality 442 of Fig. 5.7) and is an abraded edentulous mandible (Pilbeam *et al.*, 1980). Apes became locally extinct along with this faunal overturn in both Pakistan and nearby northern India (G. D. Johnson *et al.*, 1983). These sequences of paleosols in Kaulial Kas may reveal paleoenvironments of these early apes and other creatures, as well as paleoenvironmental changes coincident with their local extinction.

5.1.1.1 Murree Formation. The oldest rock of the Khaur area cropping out in the core of the Khaur Anticline is the Murree Formation: well-cemented, purple-gray and greenish-gray sandstones, with some intraformational conglomerate and interbedded dark-red and purple claystone. This weather-resistant formation also forms strike ridges north of Khaur, in the Murree Cantonment, for which the formation is named (Cotter, 1933). A type section has been designated (by Fatmi, 1974) in a strike ridge north of the village of Dhok Maiki (Fig. 5.1). The Murree Formation unconformably overlies marine Eocene sediments of the Kohat, Kuldana, and Char Gali Formations and the Sakesar Limestone (listed in order of geological age: Shah, 1977). It is overlain conformably by the Kamlial Formation. The Murree Formation reaches a thickness of 3030 m in the northern Potwar Plateau, and thins southward to 180 m in the northern Salt Range and west to 9 m at Banda Daud Shah in the Kohat region (Shah, 1977).

The formation as a whole is poorly fossiliferous, but silicified wood and bones of fish, turtles, crocodiles and mammals have been found (Fatmi, 1974), along with marine foraminifera and shark teeth probably derived from unconformably underlying Eocene marine rocks (Shah, 1977). The remains of rhinoceros, chalicothere, elephant and pig are compatible with an early Miocene age (Barry & Cheema, 1984). This age determination is supported by magnetostratigraphic studies of the formation near Chinji village, where it is some 16 to 18.3 million years old (N. M. Johnson *et al.*, 1985).

Claystones of the Murree Formation show many features of paleosols. Sandstones in the formation have conspicuous

Fig. 5.6 (to right). A geological map and cross section of the area around Khaur, Pakistan (data from Barry, Behrensmeyer, & Monaghan, 1980; Cotter, 1933; Gill, 1952; Tauxe & Opdyke, 1982). By the interpretation of Cotter the lower Dhok Pathan Formation of Barry and colleagues was mapped as Nagri Formation.

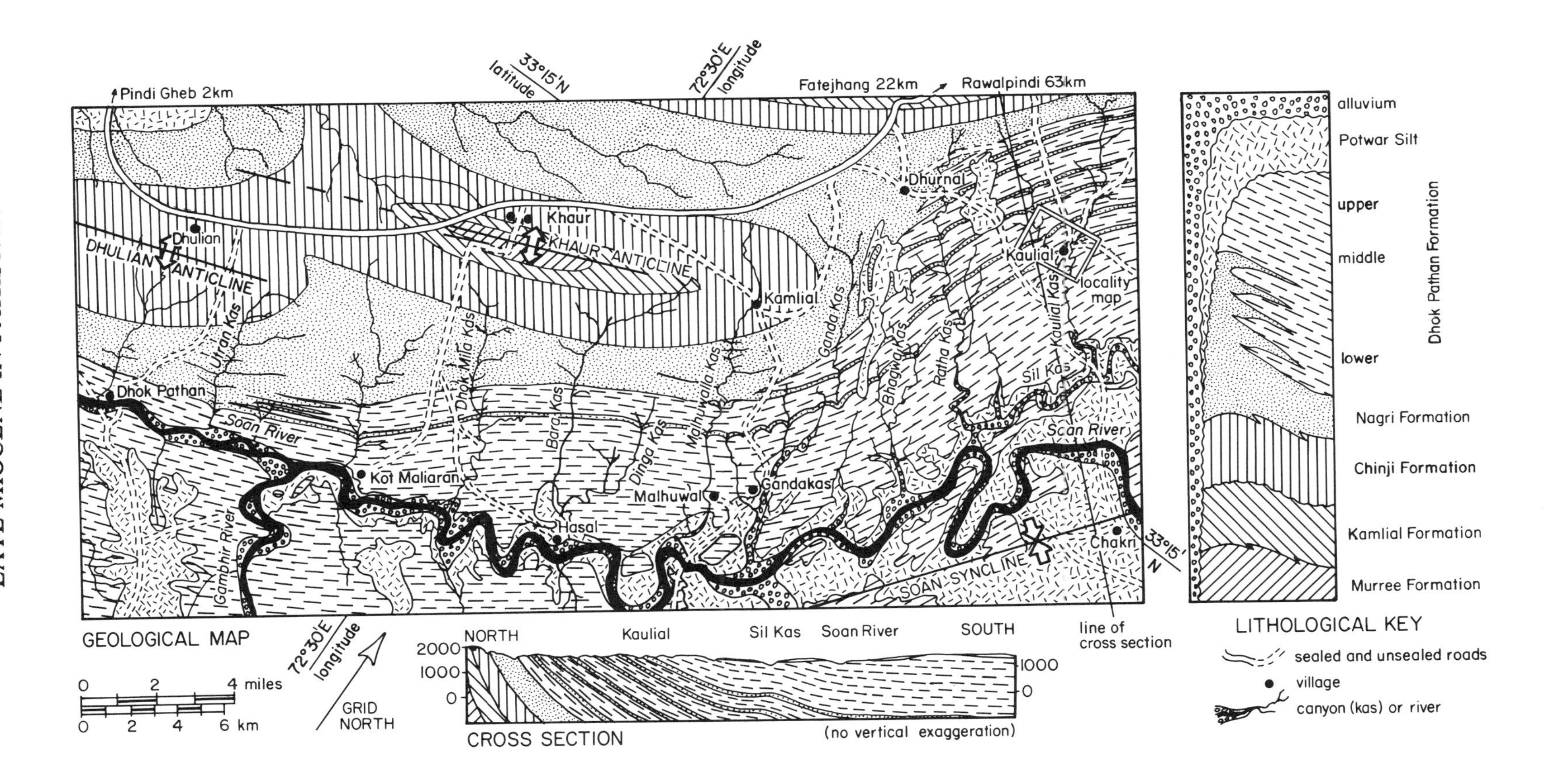
Pindi Gheb 2km
latitude 33°15'N
72°30'E longitude
Fatejhang 22km
Rawalpindi 63km
Dhulian
DHULIAN ANTICLINE
Khaur
KHAUR ANTICLINE
Dhurnal
Kaulial
locality map
Kamlial
Utran Kas
Dhok Mila Kas
Bara Kas
Dinga Kas
Malhuwalla Kas
Ganda Kas
Bhagwa Kas
Ratha Kas
Kaulial Kas
Sil Kas
Dhok Pathan
Soan River
Kot Maliaran
Malhuwal
Gandakas
Hasal
Gambhir River
Soan River
SOAN SYNCLINE
Chakri
33°15' N
GEOLOGICAL MAP
72°30'E longitude
GRID NORTH
0 2 4 miles
0 2 4 6 km
NORTH
Kaulial
Sil Kas
Soan River
SOUTH
line of cross section
2000
1000
0
1000
0
CROSS SECTION
(no vertical exaggeration)
alluvium
Potwar Silt
upper
middle
lower
Dhok Pathan Formation
Nagri Formation
Chinji Formation
Kamlial Formation
Murree Formation
LITHOLOGICAL KEY
sealed and unsealed roads
village
canyon (kas) or river

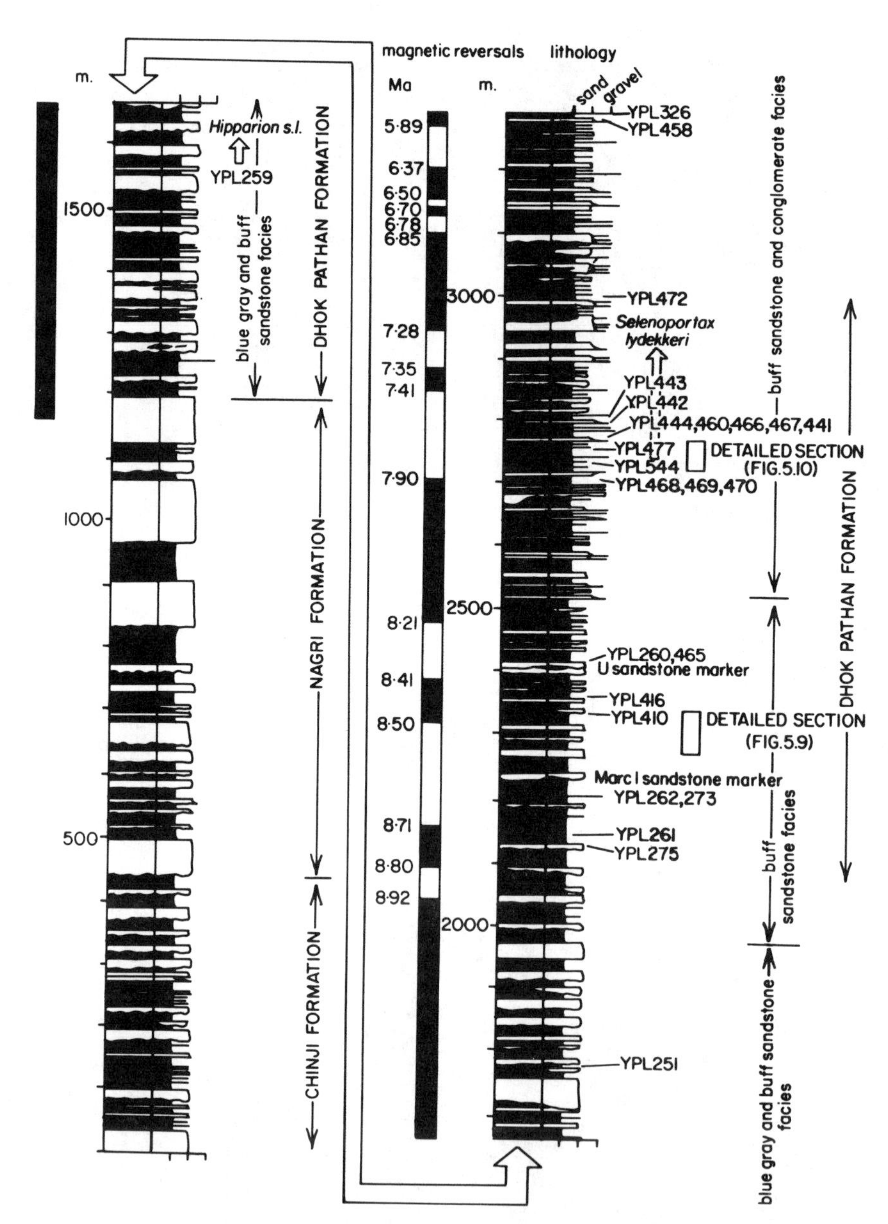

Fig. 5.7. A measured section of sedimentary rocks in Kaulial Kas, near Khaur, Pakistan (section measured by Marc Monaghan and published with magnetostratigraphy by Tauxe & Opdyke, 1982; here adjusted to time scale of Berggren, Kent, & Van Couvering, 1985; fossil localities YPL documented by Barry, Lindsay, & Jacobs, 1982).

cross bedding and other features of fluvial paleochannels. Their paleocurrent directions (N. M. Johnson *et al.*, 1985) and heavy mineral suites (Chaudhri, 1984) are evidence for a source area to the west and north in forerunners of the Hindu Kush and Himalayan ranges. These were modest ranges surrounded with foothills bearing good soil cover, unlike the barren hills and high glaciated peaks of the present, because sandstones of the Murree Formation contain largely weather-resistant

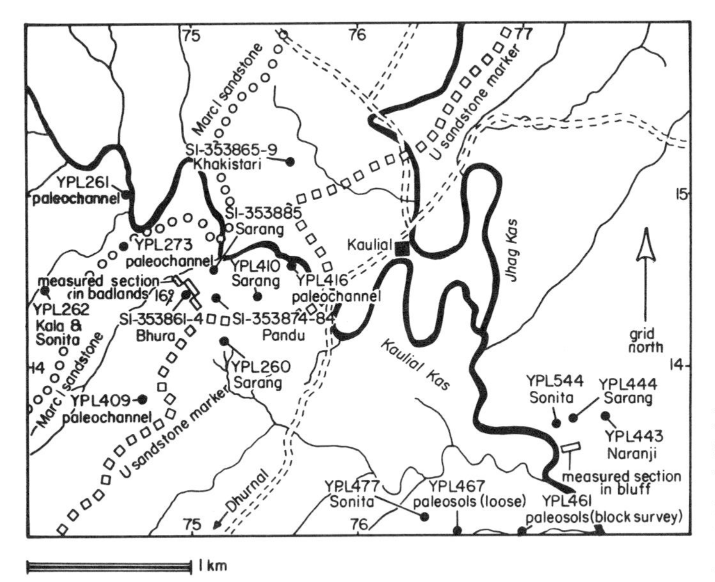

Fig. 5.8. (to left) Fossil localities, stratigraphic marker horizons and the location of measured sections near Kaulial village, Potwar Plateau, Pakistan (map and grid based on Attock sheet 43 C/11, 1:50,000).

mineral grains such as quartz, zircon, and tourmaline (Chaudhri, 1975; Krynine, 1937). Streams that deposited the Murree Formation probably flowed into the long axial sub-Himalayan trough now drained by the Ganges. During early Miocene time this trough extended into Pakistan and was bounded to the south by the Sargodha basement ridge (Farah *et al.*, 1977; Seeber, Quittmeyer, & Armbruster, 1980; Yeats & Lawrence, 1984). This low, deeply weathered ridge of Precambrian rocks also may have provided some clayey and silty sediment to the Murree Formation, but most of the sand was derived from hill ranges to the north and west (Chaudhri, 1975).

5.1.1.2 Kamlial Formation. The contact between the Murree and overlying Kamlial Formation is a transitional one, marked by an increased proportion of claystone over sandstone (Cotter, 1933). The overall similarity of the two formations in lithology, cementation, and weathering behavior has been recognized by uniting them within the Rawalpindi Group, as distinct from the lighter colored and less weather-resistant overlying Chinji and other formations of the Siwalik Group (Fatmi, 1974). The Kamlial Formation consists of gray, purple, and brick-red sandstones, with purple and brick-red claystones and intraformational claystone breccia (Shah, 1977).

The formation crops out well in the elliptical strike ridge outlining the Khaur Anticline (Fig. 5.6). The eastern part of this ridge is the type section of the formation, where it is 580-m thick (Fatmi, 1974). The Kamlial Formation reaches a maximum thickness of 650 m in the Soan Gorge south of Rawalpindi, and thins to 180 m near Jhatla in the central Salt Range, 150 m near Pamal-Domeli in the eastern Salt Range, 60 m along the Ling River east of Rawalpindi and 25 m in the Surghar Range to the west where it lies directly on Eocene Sakesar Limestone (Shah, 1977).

Middle Miocene fossil mammals have been collected from many parts of the Kamlial Formation (Barry *et al.*, 1985), and in the Chinji area the formation has been magnetostratigraphically dated at about 14.3 to 16 million years old (N. M. Johnson *et al.*, 1985).

As for the very similar Murree Formation, the Kamlial Formation was probably deposited in a subsiding axial basin south of the Himalayan mountain front and

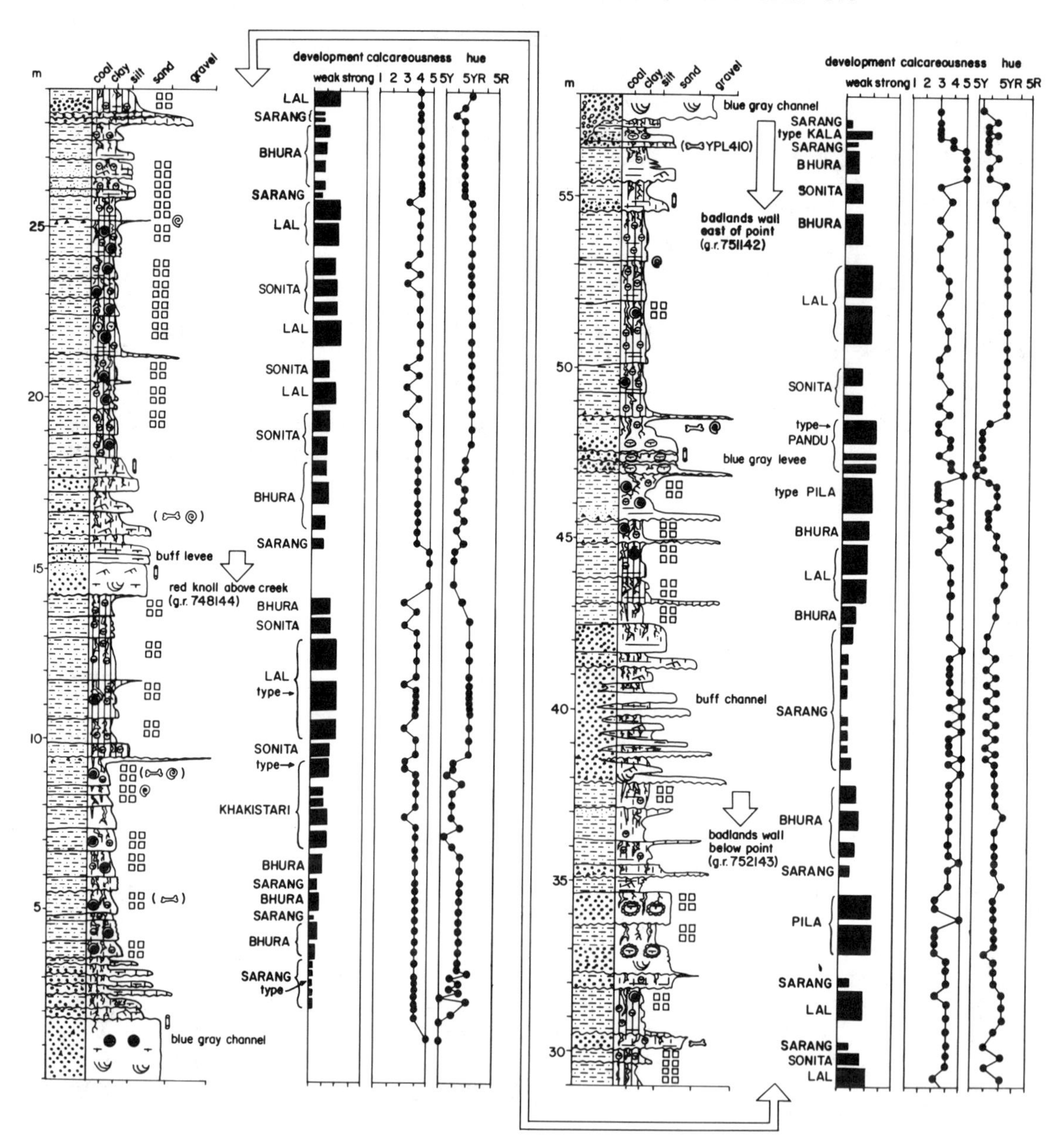

Fig. 5.9. A measured section of late Miocene (8.5 to 8.6 million years old) paleosols within the lower Dhok Pathan Formation, at a stratigraphic level laterally equivalent to the Nagri Formation, west of Kaulial village, Potwar Plateau, Pakistan. Development and calcareousness using scale of Retallack (1988a); hues from a Munsell color chart (1975). Lithological key as for Fig. 5.17.

north of basement hills of the Indian peninsula. Paleocurrents taken near Chinji village were all eastward (N. M. Johnson *et al.*, 1985). Sandstones of the Kamlial Formation are variable in composition, but include more feldspar and schist fragments than the underlying Murree Formation (Krynine, 1937). This indication of a source in less deeply weathered hills is compatible with its onlap of Eocene limestones in the Surghar Range to the west (Shah, 1977). Nevertheless, rates of sediment accumulation remained about the same for Murree and Kamlial Formations (N. M. Johnson *et al.*, 1985). The Kamlial Formation was deposited in a basin that was expanding by onlap onto highlands around the margin.

The red and purple clayey paleosols of the Kamlial and Chinji Formations include prominent calcareous nodules (Pilgrim, 1913), which are especially obvious and large compared with those of geologically younger paleosols reported in more detail here. My own casual observations of paleosols within the Kamlial Formation were made along the main trunk road 17-km north of Khaur, on the ridge 1-km southeast of Khaur, and in roadcuts within the oilfield 1.5 km-south of Khaur. These red clayey paleosols had abundant drab-haloed root traces and calcareous nodules up to 10-cm across. The nodules were avoided by burrows and drab-haloed root traces, and also were found as clasts in conglomerates of paleochannels, so were original features of soils. Most of the paleosols had moderately developed calcareous nodular horizons (stage II of Gile, Peterson, & Grossman, 1966), but one of the paleosols along the main trunk road north of Khaur had a laterally continuous calcareous horizon (K horizon or stage III). Considering what is known about the conditions for formation of such calcareous horizons in soils of Pakistan (Ahmad, Ryan, & Paeth, 1977) and elsewhere (Gile, Hawley, & Grossman, 1980), the Kamlial Formation formed in a subhumid to semiarid floodplain in which soil formation was uninterrupted for periods of tens of thousands to hundreds of thousands of years. The conspicuous greenish-gray (5GY6/1) surface horizons and root traces

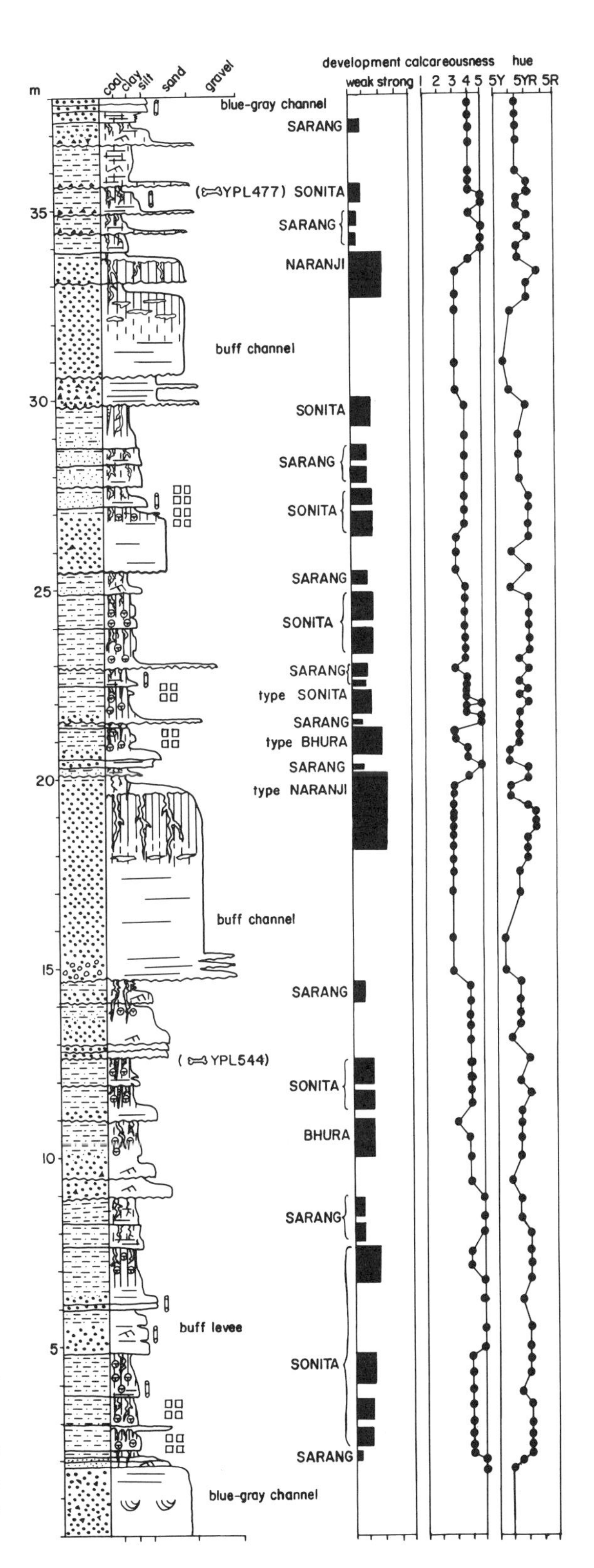

Fig. 5.10 (to right). A measured section of late Miocene (7.7 to 7.8 million years old) paleosols within the main part of the Dhok Pathan Formation, southeast of Kaulial village, Potwar Plateau, Pakistan. Constructed using same conventions as for Fig. 5.9.

in these paleosols, and their dark reddish-brown hues (2.5YR3/4), are similar to those seen in other sequences of paleosols thought to have formed under well-drained lowland forests, and then suffered dehydration of ferric hydroxides and local reduction spotting around organic matter during burial (Retallack, 1976, 1983, 1991).

5.1.1.3 Chinji Formation. Brick-red claystones and siltstones with subordinate thickness of gray sandstones characterize the Chinji Formation, which usually underpins rolling modern topography, unlike the strike ridges of indurated Kamlial and Murree Formations.

The Chinji Formation overlies the Kamlial Formation with a gradational contact in the far western Potwar Plateau (Hussain *et al.*, 1979), but sharply and conformably throughout most of the plateau. Where its upper contact has been observed, the Chinji Formation passes gradationally upward into the Nagri Formation. The type section of the Chinji Formation is south of Chinji village (Fig. 5.1), which is located on a strike ridge of the Nagri Formation (N. M. Johnson *et al.*, 1985). In this part of the southern Potwar Plateau the Chinji Formation is 700 m thick. It thickens to 1150 m near Khaur to the north and to 1800 m in the Shin Ghar Range to the west (Shah, 1977). It thus has the geometry of a clastic wedge, unlike the basin-fill geometry of the underlying Rawalpindi Group sediments.

The Chinji Formation yields middle to late Miocene mammalian fossils (Barry *et al.*, 1985). In the Chinji area, it has been magnetostratigraphically dated as 14.3 to 10.8 million years old (N.M. Johnson *et al.*, 1985).

The Chinji Formation represents a marked change in fluvial depositional style. The Kamlial and Murree Formations have abundant sandstones (65% of thickness of measured sections), mainly in sheetlike units, whereas the Chinji Formation has less sandstone (25%) in beds that include both sheets and ribbons (N. M. Johnson *et al.*, 1985). Deeply weathered rock fragments are abundant in both Chinji and Kamlial Formations, but the Chinji Formation has more recognizable metamorphic rock fragments and minerals such as staurolite (Chaudhri, 1984; Krynine, 1937). Paleocurrent directions remain easterly in the Chinji Formation and also are more consistent in direction than in the Kamlial Formation (Sheikh & Shah, 1984). Rates of sediment accumulation determined paleomagnetically, remained about the same in the Chinji and Kamlial Formations (N. M. Johnson *et al.*, 1985).

Paleosols in the Chinji Formation are brighter red (higher Munsell chroma) and have calcareous nodules less developed than in the Kamlial Formation. My own casual observations of these paleosols in the creek bed 3-km south of Khaur confirm published descriptions of Chinji paleosols elsewhere (G. D. Johnson *et al.*, 1981, profile CH-2-1; Pilgrim, 1913). The best-developed profile south of Khaur had a dusky-red (10R3/3) surface horizon (10 cm) with abundant drab-haloed (dark greenish-gray, 5BG4/1) root traces, penetrating a subsurface horizon of dark reddish-brown (5YR3/4) claystone. Scattered calcareous nodules up to 1 cm in diameter form an indistinct horizon about 50-cm below the surface. These features of the paleosols may reflect a well-vegetated basin in a subhumid to semiarid climate as before, but a basin that was better drained now that irregularities on pre-Miocene basement had been filled with sediments of the underlying Murree and Kamlial Formations. The weaker development of Chinji paleosols compared with those of the Kamlial Formation may reflect a slight quickening of long-term rates of sediment accumulation. The rising metamorphic ranges of the future Hindu Kush and Himalaya were becoming an increasingly important source of sediment, but their coarse-grained alluvial fans were still distant from the axial drainage basin of the Chinji Formation exposed near Chinji and Khaur.

5.1.1.4 Nagri Formation. Thick greenish gray, medium to coarse grained, cross-bedded sandstone characterizes the Nagri Formation. There are also buff colored sandstones, intraformational conglomerates, and reddish brown, gray, and yellow claystones. Both the underlying Chinji and overlying Dhok Pathan Formations are more clayey and red and have transitional contacts with the Nagri Formation. Sandstones are found in all these formations, but in the Nagri Formation they comprise at least 65 percent of its thickness (Gill, 1952).

The type section is near the village complex of Sethi Nagri in the southern Potwar Plateau (Fatmi, 1974), where the Nagri Formation is 200-m thick (N. M. Johnson *et al.*, 1985). It is much thicker northward in the area around Khaur and Dhurnal, where it is 1500 m (Shah, 1977). It also thickens westward from only 500 m in the southeastern Salt Range near Jalalpur to 2150 m in the Shin Ghar Range of the Kohat region (Shah, 1977). The southeastward thinning of this sandy clastic wedge can be seen especially well in the area around Khaur (as interpreted by Barry, Lindsay, & Jacobs, 1982, rather than Cotter, 1933), where the sandstones of the upper Nagri Formation pass laterally into the mixed claystones and sandstones of the Dhok Pathan Formation.

The Nagri Formation contains fossil wood, as well as a variety of late Miocene fossil mammals, including the oldest hipparionine horses in the Indian subcontinent (Barry, Lindsay, & Jacobs, 1982; Barry *et al.*, 1985). A volcanic ash bed in the Nagri Formation near the villages of Bhilamur and Qadirpur, between Nagri and Chinji in the southern Potwar Plateau, contains zircons riddled with fission tracks indicating an age of 9.46 ± 0.59 million years (G. D. Johnson *et al.*, 1982). This is an important datum for magnetostratigraphic correlation of the formation in its type area to ocean-floor lavas 10.8 to 8.5 million years old (N. M. Johnson *et al.*, 1985). The Nagri Formation is of comparable age in the area west of Khaur, but the transition between the Nagri and Dhok Pathan Formations to the east (as exposed in Kaulial Kas, Fig. 5.6) may have taken place by about 9 million years ago, because of interdigitation of the two formations here.

Most sandstones of the Nagri Formation are distinctly different from older Miocene sandstones of this region. They are a greenish to bluish gray color, and dominated by rock fragments of schist (Behrensmeyer & Tauxe, 1982; Krynine, 1937). They also contain the oldest kyanite (Chaudhri, 1984) and blue-green hornblende in the sequence (N. M. Johnson *et al.*, 1985). The appearance of these sandstones also coincides with a dramatic increase in the rate of sediment accumulation, from about 12 cm/1000 years for the Murree-Chinji interval to 30 cm/1000 years for the Nagri and Dhok Pathan Formations, as indicated by magnetostratigraphic studies (N. M. Johnson *et al.*, 1985). The sandstones were shed from a major uplift of the central metamorphic and ultramafic axis of the Himalaya and Hindu Kush ranges, then extensively glaciated and contributing appreciable quantities of little weathered sediments to antecedent streams penetrating the rising mountain front. Paleocurrents of tongues of upper Nagri Formation within the lower Dhok Pathan Formation in the Khaur area (Behrensmeyer & Tauxe, 1982) are easterly into an extended drainage of the Ganges River, as found earlier in the Miocene.

In addition to these blue-gray sandstones, there also are buff-colored sandstones, which contain a greater proportion of reddened and otherwise weathered sand grains (Krynine, 1937). These two different kinds of sandstones are found also in the overlying and laterally equivalent Dhok Pathan Formation, where blue-gray sheet sandstones from large braided streams have easterly paleocurrents, but buff ribbon sandstones of smaller sinuous streams have southeasterly paleocurrents (Behrensmeyer & Tauxe, 1982). The buff sandstones may represent drainage of Himalayan foothills with a deeply weathered soil mantle, unlike the fresh alluvium of the blue-gray sandstones derived from the glaciated alpine core of the Himalaya. Similar differences in the composition of modern alluvium can be seen in the Kosi River along the Himalayan foothills, compared with the Ganges River, which is a major antecedent stream penetrating deeply into the Himalaya (Sinha, Sahay, & Prasad, 1964).

There also are two differently colored suites of paleosols in the Nagri Formation (G. D. Johnson *et al.*, 1981). However their color probably does not reflect provenance, considering chemical analyses reported here of paleosols in the lower Dhok Pathan Formation where it interfingers with the Nagri Formation (Figs 2.22 and 2.23). Olive-gray (5YR3/2) and dark reddish-brown (10YR3/6) paleosols formed on clayey and silty parent material chemically like buff sandstones and some dark reddish brown (10YR3/6) paleosols on material chemically like the blue gray sandstones. Both the drab and reddish paleosols have profile forms, drab haloed root traces, and calcareous nodules similar to those found in paleosols in the Chinji Formation. Broadly similar vegetation, climate, and time for formation can be inferred. The drab-colored paleosols probably formed on mixed parent materials in waterlogged lowlands of the major axial drainage reaching well into the core of the Himalaya. These paleosols were similar to what are locally called terai and dhankhar soils of the modern Ganges River (B. D. Deshpande, Fehrenbacher, & Beavers, 1971; B. D. Deshpande, Fehrenbacher, & Ray, 1971; Gupta, Agarwal, & Mehrotra, 1957). In contrast, reddish paleosols of streams with sources in the forested sub-Himalayan foothills were similar to bhabar soils of broad alluvial fans sloping up from the Ganges River toward the Himalaya (Murthy *et al.*, 1982).

5.1.1.5 Dhok Pathan Formation. At its type locality north of Dhok Pathan village, 19-km southwest of Khaur (Fig. 5.6), this formation is alternating sandstones and claystones in fairly equal proportions (Gill, 1952). The sandstones are light to pinkish-gray or buff and the claystones are brown to reddish-brown, unlike the blue-gray sandstones and dark-red claystones of the Nagri Formation.

Further west in Kaulial Kas (Fig. 5.6) this lighter and browner lithology is found to the south of Kaulial village, stratigraphically above a marker bed of blue-gray sandstone ("U-sandstone" of Behrensmeyer & Tauxe, 1982), which can be regarded as the uppermost tongue of the Nagri Formation (Cotter, 1933). The upper measured section of paleosols (Fig. 5.10) is in this portion of what can be regarded as typical Dhok Pathan Formation. Northwest of Kaulial village and down section there is a considerable thickness (1350 m) of the Dhok Pathan Formation passing transitionally down into the Nagri Formation. These interbedded strata of buff sandstones and olive, red, and brown claystones pass laterally to the west into the Nagri Formation, which penetrates them as tongues of thick blue-gray sandstones (Fig. 5.6). The other measured section of paleosols reported here (Fig. 5.9) is in this part of the formation.

The upper Dhok Pathan Formation includes thin conglomerates with clasts of gneisses, quartzites, and Eocene sediments. A broken pebble of quartzite from such conglomerates near Haritalyangar, India, has been unconvincingly interpreted as a stone tool (K. N. Prasad, 1982). The upper contact of the formation with the Soan Formation is placed below thick conglomerates of this kind, well developed in the western Potwar Plateau (Gill, 1952). These conglomerates and gravelly sandstones do not crop out in the Khaur area, where the entire 884 m of the exposed Dhok Pathan

Formation has been tilted to form an angular unconformity with the overlying Potwar Silt.

The Dhok Pathan Formation is thickest in the northern Potwar Plateau near Khair-e-Murat (1830 m) and in the west near Kohat (1680 m), thinning to 900 m near Kallar Kahar in the southern Salt Range and 330 m near Pamal Domeli in the southeastern Salt Range (Fatmi, 1974; Shah, 1977). It retained the form of a clastic wedge, seen also for the Chinji and Nagri Formations.

The Dhok Pathan Formation is the most fossiliferous rock unit in the Khaur area (Badgley, 1982, 1986; Badgley & Behrensmeyer, 1980; Pilbeam *et al.*, 1977a, 1977b, 1979, 1980), and includes the upper part of the interval zone of *Hipparion s.l.* and the zone of *Selenoportax lydekkeri* (of Barry, Lindsay, & Jacobs, 1982), both late Miocene. Near Khaur and Chinji the Dhok Pathan Formation has been magnetostratigraphically dated at 8.3 to 5.3 million years old, but in the zone of interfingering with the Nagri Formation in Kaulial Kas it is as old as 9 million years (N. M. Johnson *et al.*, 1985; Tauxe & Opdyke, 1982; corrected to time scale of Berggren, Kent, & Van Couvering, 1985).

As already discussed, the lower Dhok Pathan Formation in Kaulial Kas was deposited in two separate river systems like the Nagri Formation, with which it is laterally equivalent. There was a waterlogged lowland (dhankhar) of an axial drainage system of large braided streams reaching well into the Himalayan mountains, as well as a series of well drained alluvial fans (bhabhar) coursed by tributary, loosely sinuous streams of forested hill ranges (Behrensmeyer & Tauxe, 1982; Retallack, 1985). Such a system of drainage found still in the soils (Murthy *et al.*, 1982) and sediments (V. R. Baker, 1986; Sinha, Sahay, & Prasad, 1964) of the Gangetic Plains of Uttar Pradesh, India, persisted into the main portion of the Dhok Pathan Formation above the highest tongue of Nagri sandstone. Here the axial drainage is represented by light gray to pink pebbly sandstones and the foothills drainage by yellowish-brown ribbon sandstones. The lighter and warmer color (higher Munsell value and redder Munsell hue) of these sandstones compared with those lower in the Dhok Pathan Formation may reflect generally better drainage, along with increased stream energy and greater redeposition of earlier deposited alluvium in both stream systems.

Paleosols in the main part of the Dhok Pathan Formation are less varied in color and character than those in the interval of interdigitation with the Nagri Formation. Gone are the drab-colored, dark grayish brown (2.5Y4/2) paleosols and thick, bright, reddish brown (5YR4/5) paleosols of the lower Dhok Pathan Formation and Nagri Formation. Instead there are thinner, less developed, and less varied paleosols of reddish-brown (5YR4/4) to dark brown (7.5YR4/4). These observations confirm the idea that sedimentation rates remained high, as shown also by magnetostratigraphy (G. D. Johnson, Raynolds, & Burbank, 1986; G. D. Johnson *et al.*, 1982). Changes observed in paleosols may also reflect a lowering regional base level, perhaps due to climatic drying, detected also by petrographic studies (Krynine, 1937). These and other aspects of the paleoenvironment of the Dhok Pathan Formation will be discussed in more detail after thorough consideration of paleosols in the sections measured in Kaulial Kas.

5.1.1.6 Potwar Silt. Unconformably overlying an erosional landscape on the dipping Siwalik Group in the Khaur area is a blanket of yellow siltstones. The Potwar Silt was first recognized (by de Terra & de Chardin, 1936) as a mappable unit in the area of the Soan River and Kas Dovac, 11-km south of Rawalpindi, where the silts are 61-m thick (Gill, 1952).

Only 2-km south of Rawalpindi, the Potwar Silt can be seen to be younger than undeformed Siwalik sediments dated magnetostratigraphically and radiometrically using fission tracks on zircons from volcanic ash at 1.9 million years old. These sediments and ashes unconformably overlie tilted Siwalik sediments dated magnetostratigraphically at 2.1 million years old (G. D. Johnson, Raynolds, & Burbank, 1986; Rendell, Hailwood, & Dennell, 1987). This deformation at about 2 million years ago created the Soan Syncline and the Potwar Plateau (Burbank, Raynolds, & Johnson, 1986).

The unconformity between the Potwar Silt and the Siwalik Group has yielded a variety of stone tools, thought to be 400,000 to 700,000 years old. The bulk of the overlying Potwar Silt with its distinctive Middle Paleolithic artifacts (Late Soan industry) is probably younger than this, mainly late Pleistocene or about 40,000 years old (Allchin & Allchin, 1982).

Imbricated conglomerates of locally derived rocks are extensive at the base of the Potwar Silt, and occasionally within it. Most of the silt is massive or indistinctly bedded. There are some paleosols with dark-brown surface (A) horizons and with calcareous nodules forming a subsurface horzon (Bk). These paleosols are similar to surface soils on the silt, and may be taken as evidence of a similar dry wooded grassland to that predating human abuse of these landscapes. Fossil snails and mammal bones also have been found, although these are rare and not especially useful indicators of age or paleoenvironment (de Terra & Paterson, 1939). The Potwar Silt can be regarded as an eolian deposit that accumulated during periods of Pleistocene glacial advance.

5.2 Description and Interpretation of the Paleosols

A variety of paleosols were recognized in the two measured sections within the late Miocene Dhok Pathan Formation in Kaulial Kas (Figs 5.9 and 5.10). Their names were derived from Urdu adjectives that describe obvious field features of the paleosols. In many cases these were words for colors, which were recorded on freshly excavated rock as well as on the naturally dry, weathered badlands slopes to facilitate recognition of further examples on the badlands exposures. In other respects, the approach and methods used for these descriptions are similar to those used for Kenyan paleosols described and interpreted in the last chapter. Point counting of thin sections of the Pakistani paleosols was performed by G. J. Retallack (specimens SI-353773 to 794), R. Goodfellow (SI-353785 to 794), and S. Cook (others). Chemical analyses, including major and trace elements using ICP, loss on ignition and Walkley-Black organic carbon, were done by C. Cool (University of Washington, Seattle), and bulk density from paraffin-coated clods by G. J. Retallack. Once again, the paleosol series are listed in the stratigraphic order in which they were encountered.

5.2.1 Sarang Series (new name)

5.2.1.1 Diagnosis. Thin, very weakly developed paleosols, with root traces and burrows penetrating clearly interbedded reddish-brown (5YR) claystone and light yellowish-brown (10YR) sandstone and siltstone.

5.2.1.2 Derivation. Sarang is an Urdu word for spotted or variegated (Platts, 1960), in reference to the colorful alternation of relict bedding only slightly mixed by burrows and root traces.

5.2.1.3 Description. The top of the type Sarang clay paleosol (Figs 5.11 and 5.12A) is in the low red knoll 80-cm stratigraphically above the top of the paleochannel sandstone forming a low dry waterfall in the creek bed (grid reference 748144 on Attock 1:50,000 sheet 43 C/11), 1.2-km west of Kaulial village, Potwar Plateau, northern Pakistan (Figs 5.6 and 5.8). This is in the late Miocene, lower Dhok Pathan

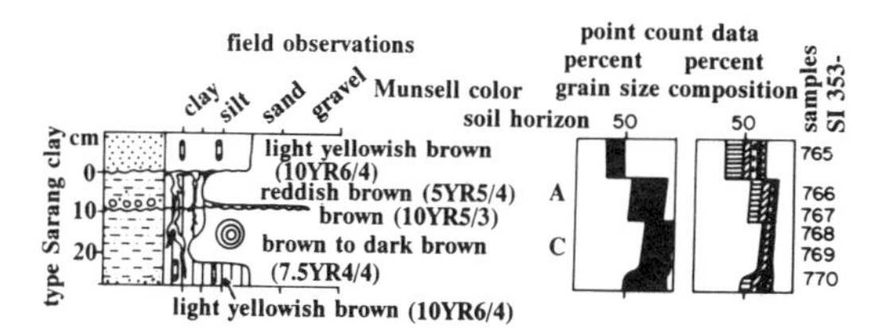

Fig. 5.11. Measured section, Munsell colors, soil horizons, grain size, and mineral composition of the type Sarang clay paleosol in the lower measured section of the Dhok Pathan Formation, west of Kaulial village, Pakistan. This section is at 2.5 to 2.9 m in Fig. 5.9. Lithological key as for Fig. 5.17.

Formation, here dated paleomagnetically at about 8.5 million years old (Tauxe & Opdyke, 1982; corrected to time scale of Berggren, Kent, & Van Couvering, 1985), near the top of the interval where it interfingers with the Nagri Formation to the west (at 2274 m in Fig. 5.7 and 2.3 m in Fig. 5.9). At this point these strata dip at 16 degrees toward the southeast (azimuth 134 degrees magnetic).

+13 cm; C horizon of overlying paleosol (unnamed Sarang Series); medium-grained sandstone; light yellowish-brown (10YR6/4), with an upper portion (2 cm) light greenish-gray (5GY7/1); weathers light brown (7.5YR6/4); distinct planar and wavy bedding; common burrows, 3 and 7-8 mm in diameter, filled with brown (7.5YR4/4) silty claystone from overlying layers; strongly calcareous; intertextic silasepic in thin section; common quartz, feldspar, limestone fragments, and clayey matrix; the matrix is possibly in part deeply weathered and crushed metamorphic rock fragments, also present in small amounts in less weathered form; abrupt, smooth contact to

0 cm; A horizon; silty claystone; reddish-brown (5YR5/4); weathers light brown (7.5YR6/4); common red (10YR4/6) and yellow (10YR7/6) clayey root traces; prominent (4-7-mm diameter), light greenish-gray (5GY7/1) metagranotubules after roots and burrows, filled with medium-grained sandstone from above, and with a light greenish-gray (5G7/1) halo extending 1-3 mm into the matrix; few, 8-9-mm diameter, brown (10YR5/3) to yellowish-brown (10YR5/6) ferric concretions; strongly calcareous; agglomeroplasmic insepic in thin section, with local clasts of crystic limestone and indistinct, small, irregular ferric mottles; abundant clay, with common quartz, calcite, and feldspar; some mica and rock fragments; abrupt, smooth contact to

-9 cm; C horizon; silty claystone, brown (10YR5/3); weathers light brown (7.5YR6/4); this is a thin layer, perhaps an erosional lag deposit, of abundant ferric concretions like those in the overlying layer; this layer also is penetrated by root traces and drab-haloed metagranotubules from above; strongly calcareous; agglomeroplasmic insepic in thin section of matrix, isotic in concretions and nodules, intertextic silasepic within metagranotubules; abrupt, wavy contact to

-10 cm; C horizon; silty claystone; brown to dark brown (7.5YR4/4); weathers light brown (7.5YR6/4); abundant yellowish-brown (10YR5/6) root mottles, as well as metagranotubules from overlying layers; strongly calcareous; porphyroskelic insepic in thin section of matrix; common quartz, claystone, and limestone grains; some feldspar and mica; gradual, wavy contact to

-15 cm; C horizon; silty claystone; brown to dark brown (7.5YR4/4); weathers light brown (7.5YR6/4); abundant fine (1-mm), root traces of light gray (5Y7/2); few brown (10YR5/3) and yellowish-brown (10YR5/6) ferric concretions; coarse-granular and coarse subangular-blocky peds, defined by silans and slickensided clay skins (argillans); also found was a large (7 cm in diameter) spherical structure of sand with irregular concentric laminae of claystone, similar to *Termitichnus,* the fungus gardens of mound-building termites; strongly calcareous; porphyroskelic insepic in thin section; mainly clay with common quartz and feldspar, little mica; few calcareous rhizoconcretions, with a central tube of sparry calcite and a marginal zone of micrite (calciasepic diffusion neocalcan); clear, smooth contact to

-23 cm; C horizon; medium-grained clayey sandstone; light yellowish-brown (10YR6/4); weathers light brown (7.5YR6/4); common brown (7.5YR4/4) fine root traces; few large (15 mm in diameter) root traces with central tube (2 mm in diameter) of light greenish-gray (5GY7/1) sandstone, flanked by narrow (2-mm) halo of greenish-gray (5G7/1) silty claystone and then a thin (1-mm) rind of brown (7.5YR4/4) claystone; few scattered, small (2-3-mm), calcareous nodules; indistinct relict bedding; strongly calcareous; porphyroskelic to agglomeroplasmic, unistrial insepic in thin section of matrix; intertextic silasepic in tubules and calciasepic in nodules; abundant clay; common quartz and feldspar; some calcite and mica; clear, smooth contact to.

-28 cm; A horizon of underlying paleosol (unnamed Sarang Series); silty claystone; brown (7.5YR4/4); weathers light brown (7.5YR6/4); common large, greenish-gray (5GY7/1) drab-haloed root traces.

5.2.1.4 Further examples. Numerous other Sarang Series paleosols were excavated (24 of them in the lower measured section of Fig. 5.9 and 14 of them in the upper measured section of Fig. 5.10). The Sarang clay eroded-phase paleosol (Fig. 5.19; at 57.2 m in Fig. 5.9) is immediately beneath a thick blue-gray paleochannel sandstone. The Sarang clay (Fig. 5.15; at 34.8 m in Fig. 5.10), Sarang clay cumulic-surface phase (Fig. 5.13, at 22.4 m in Fig. 5.10) and Sarang clay-loam (Fig. 5.19, at 56.6 m in Fig. 5.9) also are very weakly developed profiles with prominent relict bedding. The most pronounced soil development seen in Sarang profiles are micritic rhizoconcretions that are sparse in the type Sarang clay but more abundant in the Sarang silty-clay rhizoconcretionary-variant (Fig. 5.13: at 21.5 m in Fig. 5.10), and the Sarang clay rhizoconcretionary-variant (Fig. 5.13, at 22.9 m in Fig. 5.10). Also a little better developed than usual for these profiles is the Sarang silty-clay massive-surface phase (Fig. 5.20, at 20.4 m in Fig. 5.10), which lacks both calcareous and drab-haloed root traces but has strong-brown (7.5YR5/6) clayey root traces in a matrix of dark reddish-brown (7.5YR6/6). Similar Miocene paleosols, a little more red than these examples, have also been described from Siwalik sediments near Jhelum, Pakistan (profile MK-12-1A of G. D. Johnson *et al.*, 1981; and profile type Rl of Visser & Johnson, 1978).

5.2.1.5 Alteration after burial. Considering the deep burial of these paleosols as already discussed, compaction could have reduced clayey parts of the profiles to 58 percent of their original thickness in the lower measured section and 60 percent in the upper section, while sandy parts of the profile could have been crushed to 70 percent and 75 percent respectively. These estimates are unlikely maximal compactions, considering the calcareous nature of these profiles and slight deformation of trace fossils (*Skolithus, Termitichnus*, and root traces). Also likely is local drab discoloration due to burial gleization around relict organic matter and reddening due to dehydration of ferric hydroxides after burial. There may also have been a loss of organic matter on burial, but there probably was not much originally, considering the sparse root traces, lack of well developed soil structure, and evident activity of termites.

5.2.1.6 Reconstructed soil. Sarang paleosols were little more than rooted and burrowed sediments of interbedded brown sandstone and red claystone. The original soils

Fig. 5.12. Field appearance of the very weakly developed type Sarang clay paleosol, with its clear relict bedding and root traces (A), compared to the moderately developed type Lal clay with its massive to blocky structure (B), both in the lower measured section of the Dhok Pathan Formation, west of Kaulial village, Pakistan. Hammers for scale.

probably did not include drab colors, and may have been brown (10YR) near the weakly humified surface, over an orange to brownish red (7.5YR) subsurface horizon. They are very calcareous, with limestone rock fragments and redeposited soil nodules, and so were probably alkaline in pH, with high cation exchange capacity and base saturation. Their reddish hue, abundant near-opaque oxides, burrows like those of termites, and deeply penetrating root traces are evidence of good drainage and generally oxidizing conditions.

5.2.1.7 Classification. These very weakly developed paleosols would be identified as sedimentary beds were it not for their root traces and burrows. They can be identified as Fluvents (of Soil Survey Staff, 1975), Calcaric Fluvisols (of Food and Agriculture Organization, 1974), Alluvial Soils (of Stace *et al.*, 1968), Uf1.11 (of Northcote, 1974) and Dundee dundee (of MacVicar *et al.*, 1977). Comparable surface soils are widespread in streamside locations within the Indo-Gangetic alluvium of India and Pakistan (map units Jc42-2/3a, Jc45-2a, Jc50-2a, Jc52-2a, Jc55-2ab of Food and Agriculture Organization, 1977b) and within streamside alluvium on granitic areas of Kruger National Park, South Africa (Botha, 1985; Gertenbach, 1983; Venter, 1986).

5.2.1.8 Paleoclimate. Paleosols of such weak development cannot usually be taken as good evidence of paleoclimate. However, burrows and spherical structures interpreted as trace fossils of termites are indications of a warm climate. In addition, root traces and burrows of these paleosols sometimes penetrate deeply into cross-bedded sandstones of underlying paleochannels. This is an indication of a severe dry season, when streams were well below the levels of exceptional floods that deposited the beds of alluvium within these paleosols. Even at this time in the late Miocene, monsoonal circulation may have been established over India and Pakistan and their great northern mountains.

5.2.1.9 Ancient vegetation. Sarang profiles show abundant relict bedding and scattered large and small root traces, as would be expected in soils supporting vegetation early in the ecological succession toward woodland. This probably colonized river margin locations frequently disturbed by flooding, as indicated by the common association of Sarang paleosols with fluvial paleochannel and levee facies.

Comparable surface soils commonly support early successional vegetation (Food and Agriculture Organization, 1977b; Gertenbach, 1983; Soil Survey Staff, 1975; Stace *et al.*, 1968; Venter, 1986). In the present Indo-Gangetic Plains, such vegetation is dominated by khair (*Acacia catechu*) and sissu (*Dalbergia sissoo*), with sissu saplings tending to dominate the bushier khair and a ground cover of other shrubs and grasses (vegetation unit 5/IS2 of Champion & Seth, 1968). There is no fossil record of either of these taxa in Pakistan, although a fossil fruit identified as *Dalbergia sissoo* has been found in Lower Siwalik rocks at Balugoloa, near Jawalamukhi in Himachal Pradesh, India (Lakhanpal & Dayal, 1966). Fossil pollen of *Acacia* is as old as Eocene in Africa, as outlined for Lwanda Series paleosols in Kenya, but this genus may not have entered India until Pliocene biogeographic exchange with Africa (Mani, 1974c).

5.2.1.10 Former animal life. Sarang profiles are the most prolifically fossiliferous paleosols in the lower Dhok Pathan Formation, where bone also is locally abundant in associated paleochannel and levee facies (Badgley, 1982, 1986; Badgley & Behrensmeyer, 1980; Behrensmeyer, 1987, 1988; Pilbeam *et al.*, 1980).

The locality (Yale University-Geological Survey of Pakistan site 410, at grid reference 754144 on Attock 1:50,000 map 43 C/11) for a partial skull of *Sivapithecus sivalensis* (Fig. 1.2) was found in an unnamed Sarang Series paleosol within the deep gully south of the main watercourse of Kaulial Kas, about 600-m north of the lower measured section (Fig. 5.9; in which it correlates to the Sarang profile at 56.5 m). Only bones of the face and lower jaws were found here, and they appear to have suffered carnivore damage (Pilbeam, 1982; Ward & Pilbeam, 1983).

Fossils were more abundant and better exposed in a sequence of two unnamed Sarang paleosols at a slightly higher stratigraphic level (at 2420 m in Fig. 5.7: near marker O of Barry, Lindsay, & Jacobs, 1982; called the "U-level" by Badgley, 1982, 1986) in a low knoll and shallow gullies atop a sandstone dip slope, 1-km south of the lower measured section (Fig. 5.9). This is a well known fossil locality (Yale-Geological Survey of Pakistan number 260, at grid reference 752142). Badgley (1982) has listed the following forms: apes (*Sivapithecus sivalensis*, 4 teeth; and *S. punjabicus*, 6 teeth, maxilla, 3 mandibles), hyaena (*Palhyaena sivalensis*, mandible), civet (*Viverra chinjiensis*, mandible), proboscidean (*Stegolophodon* sp., 6 teeth, 2 tusks, metapodial and 1 shaft; as well as indeterminate proboscidean tooth and tusk fragments, vertebra and radius), three-toed horse ("*Cormohipparion*" or ?*Hipparion*, 3 teeth), rhinoceros (3 skulls, 4 teeth, maxilla, mandible, 4 vertebrae, scapula, 2 metapodials and 3 phalanges), anthracothere (podial, astragalus), chevrotain (femur), chowsinghalike antelope (*Elachistoceras khauristanensis*, 2 teeth, tibia, podial), unidentified antelope (tooth, femur, radius, ulna, calcaneum), and other unidentified mammals (134 fragments). Badgley interpreted this assemblage as a local carnivore or scavenger accumulation that had been little affected by fluvial transport.

Sarang Series paleosols in the lower Dhok Pathan Formation also contain a variety of trace fossils. These include large (up to 7 cm in diameter), concentrically layered, sand spheres, similar to the ichnogenus *Termitichnus* (of Bown, 1982), connected to pinnately branched, narrow (3-4-mm) galleries (Fig. 2.7). Some of these galleries are filled with darker clay than the paleosol matrix and include ovoid pellets or meniscuslike internal banding (spreiten of Seilacher, 1964). In thin section and under the scanning electron microscope (Fig. 2.29) the *Termitichnus* spheres consist of a complex network of broadly concentric micritic partitions, enclosing vermiform hollows filled with sparry calcite and occasional grains of red claystone, quartz, and feldspar. The whole structure is very similar to a fungus garden of a young nest of the mound-building termite *Macrotermes*. Similar Pliocene termite nests have been found in Tanzania ("unique ovoid" of Sands, 1987), where they are also associated with pinnately branched galleries, locally filled with pellets. Mound-building fungus gardeners such as *Macrotermes*, *Odontotermes* and *Trinervitermes* remain common in India today (Sen-Sarma, 1974). Termites have a fossil record extending back at least to the Cretaceous (Burnham, 1978) and complex nests have been found in rocks that are at least as old as Oligocene (Bown, 1982), but possibly as old as late Eocene (Berggren, Kent, & Van Couvering, 1985).

Sarang paleosols in the lower Dhok Pathan Formation also contain sand-filled burrows referable to the ichnogenera *Edaphichnium*, *Macanopsis*, and *Skolithus*. *Edaphichnium* is thought to be an earthworm burrow, partly filled with fecal pellets (Bown & Kraus, 1983). *Macanopsis* and *Skolithus* are burrows of a generalized kind that could have been made by a variety of insects and spiders (Ratcliffe & Fagerstrom, 1980).

Within the upper measured section of the main part of the Dhok Pathan Formation (Fig. 5.10), Sarang paleosols were not found to be so fossiliferous as in the lower measured section. One locality (444 of Yale-G.S.P. at grid reference 773137) at a stratigraphic level above the uppermost sandstone of the measured section (above

Fig. 5.10; 2780 m in Fig. 5.7; or reference U of Barry, Lindsay, & Jacobs, 1982) yielded a skull and axis of the large antelope *Selenoportax lydekkeri*. Also noted in Sarang paleosols in the upper measured section were trace fossils of *Edaphichnium*, *Macanopsis* and *Skolithus*.

5.2.1.11 Paleotopographic setting. The close association of Sarang paleosols with paleochannel and levee facies is indicative of a stream margin location. Nevertheless, there are no indications of gleization from the deeply penetrating root traces and burrows and from the low ferrous iron/ferric iron ratios (Figs 5.13, 5.19, and 5.20). They probably formed on well-drained point bars and levees of streams, inundated only during extensive flooding. Such a paleotopographic setting also is compatible with the lateral variation in thickness of these paleosols, including shallow swalelike features (especially prominent at 38-41 m in section of Fig. 5.9, and 35-37 m in Fig. 5.10, and at fossil locality 260 of Pilbeam *et al.*, 1980). Fossiliferous Sarang paleosols also have been noted in other parts of the Siwalik Group within fine-grained sedimentary fills of abandoned paleochannels (Behrensmeyer, 1987, 1988).

5.2.1.12 Parent material. Sarang paleosols have been so little altered by soil formation that they can be considered close to parent material for associated paleosols of other kinds. Most Sarang paleosols consist of alluvium derived from already weathered soils in drainage basins in the forested foothills of the Himalayas (buff drainage of Behrensmeyer & Tauxe, 1982), but some formed on alluvium rich in little-weathered rock fragments of schist from the axial drainage that extended deeply into the high Himalaya (blue-gray drainage). The principal sand grains in both drainages were quartz, feldspar, muscovite, and rock fragments of limestone, schist, and shale (Krynine, 1937). The main clay minerals from the metamorphic source regions are chlorite and illite, whereas kaolinite may be derived from bauxitic paleosols of Eocene age and smectite from preexisting soils. Minor amounts of smectite and chlorite may have formed during soil formation in Sarang paleosols, as has already been discussed (see Fig. 2.32).

5.2.1.13 Time for formation. Sarang paleosols represent little more than a few seasons of rooting and burrowing into fresh alluvium. They are comparable to Shahdara Series soils less than 100 years old flanking the Ravi River near Lahore, Pakistan (Ahmad, Ryan, & Paeth, 1977). Sarang profiles lack the calcareous nodules formed in place and the clay coatings seen in soils as young as 400 years near the dry bed of the Ghaggar River in northern India (Courty & Féderoff, 1985). Comparison with numerous very weakly developed paleosols in other parts of the world (summarized by Birkeland, 1984) also supports the idea that these paleosols represent only a few tens to hundreds of years.

5.2.2 Bhura Series (new name)

5.2.2.1 Diagnosis. Thin paleosols with a dark brown (7.5YR), clayey surface (A) horizon, distinct peds, cutans, and drab-haloed root traces, over a subsurface (Bk) horizon with a weakly developed aggregation of fine (2-3-mm) calcareous nodules and some persistent relict bedding.

5.2.2.2 Derivation. Bhura is Urdu for brown (Platts, 1960), in reference to the color of the surface horizon of these paleosols.

5.2.2.3 Description. The type Bhura clay paleosol (Fig. 5.13) is the dark brown band (below 21.3 m in Fig. 5.10, or 2749 m in Fig. 5.7), above the prominent red paleosol (type Naranji clay) developed on the lower buff paleochannel in the large bluff (grid

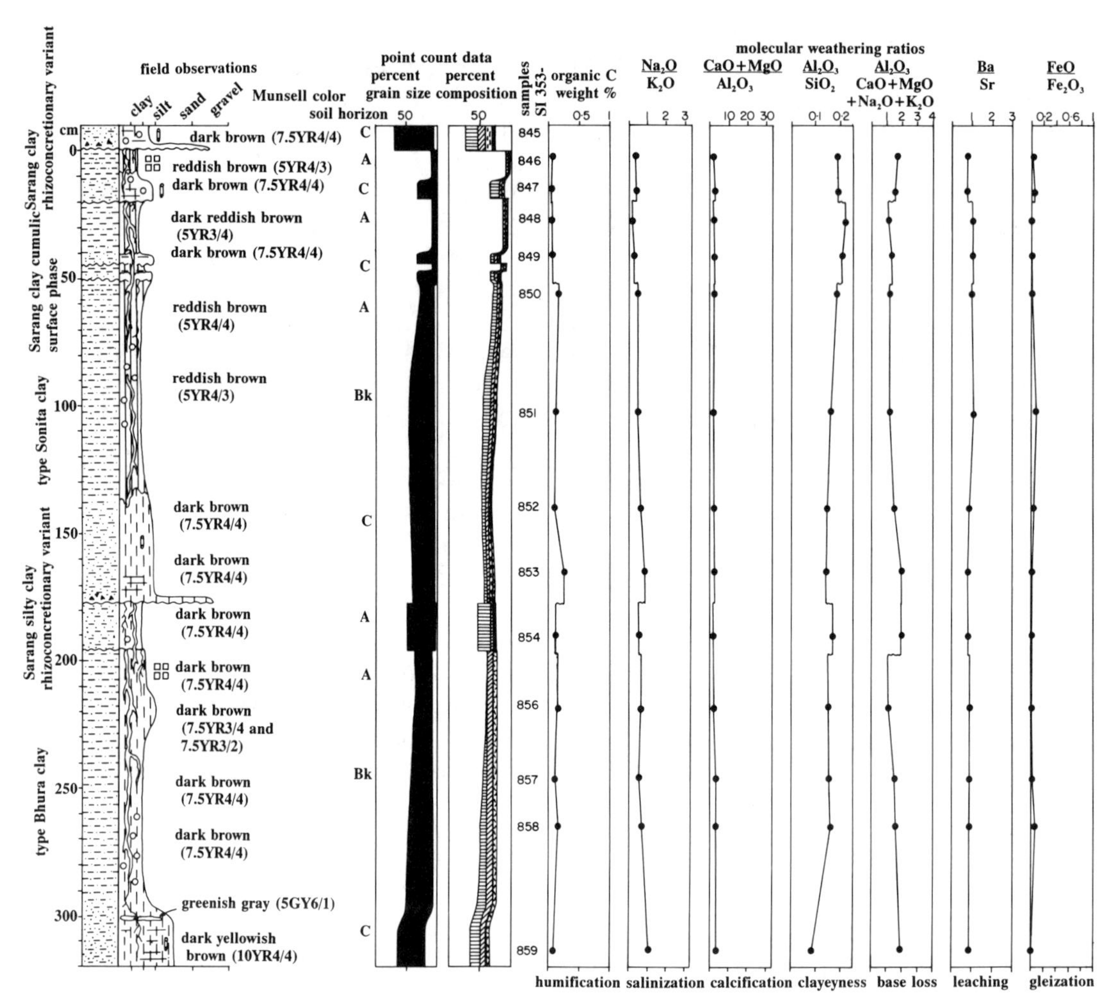

Fig. 5.13. Measured section, Munsell colors, soil horizons, grain size, mineral composition, organic content and selected molecular weathering ratios of type Bhura clay, Sarang silty-clay rhizoconcretionary-variant, type Sonita clay, Sarang clay cumulic-surface phase and Sarang clay rhizoconcretionary-variant paleosols, in the upper measured section in the Dhok Pathan Formation, southeast of Kaulial village, Pakistan. This section is at 20.3 to 23.5 m in Fig. 5.10. Lithological key as for Fig. 5.17.

reference 773135) northeast of the main watercourse of Kaulial Kas, 1.5-km southeast of Kaulial village, Potwar Plateau, Pakistan (Figs 5.6 and 5.8). This is in the late Miocene, Dhok Pathan Formation, above the uppermost tongue of Nagri sandstone, and is dated paleomagnetically at 7.8 million years (Tauxe & Opdyke, 1982; corrected to the time scale of Berggren, Kent, & Van Couvering, 1985).

+18 cm; C horizon of overlying paleosol (Sarang silty-clay rhizoconcretionary-variant); silty claystone; dark brown (7.5YR4/4); weathers light brown (7.5YR6/4); common greenish-gray (5GY6/1) metagranotubules with greenish-gray (5G5/1) haloes, after roots; few yellowish-brown (10YR5/4) calcareous rhizoconcretions, strong-brown (7.5YR5/6) resorted calcareous nodules and very dark brown (7.5YR2/2) films of iron-manganese (mangans); indistinct coarse granular structure, defined by reddish-brown (5YR4/4) impersistent clay skins (argillans); prominent relict bedding, especially toward base of unit; very strongly calcareous; porphyroskelic skelmosepic in thin section; grains mainly quartz, feldspar, limestone, schist, chert, and micritic nodules; abrupt, smooth contact to

0 cm; A horizon; silty claystone; dark brown (7.5YR4/4); weathers brown (10YR5/3); common gray

(5Y5/1) metagranotubules, with greenish gray (5GY5/1) calcareous sheaths (diffusion calcans), after root traces, these whole rhizoconcretionary structures are up to 13 mm in diameter; few small (2-4-mm) chips of black (5Y2.5/1) charcoal; coarse-granular and thick-platy peds, defined by dark brown (7.5YR4/4) slickensided clay skins (argillans); few scattered strong-brown (7.5YR4/6) calcareous nodules; moderately calcareous in matrix; porphyroskelic skelmosepic in thin section; common quartz; few feldspar, opaque oxides, schist, limestone, and muscovite grains; gradual, wavy contact to

-20 cm; A horizon; silty claystone; dark brown (7.5YR3/4 and 7.5YR3/2); weathers brown (10YR5/3); common greenish-gray (5GY6/1) metagranotubules, with greenish-gray (5GY6/1) diffusion calcans and in places an outer rim (mangan) of black (10YR5/6), after root traces, all together up to 10 mm in diameter; this horizon has a distinctive bioturbation of wisps and hollow pear-shaped outlines up to 6-mm thick and 3-cm across of dark brown (7.5YR3/2) claystone, separating and enclosing siltstone of yellowish-brown (10YR5/6), perhaps the remains of pupal cells of scarabaeid beetles; few strong-brown (7.5YR5/6), irregular, calcareous nodules; moderately calcareous in matrix; porphyroskelic skelmosepic in thin section; common quartz; few feldspar, opaque oxide, schist, limestone, and muscovite grains; gradual, wavy contact to

-33 cm ; A horizon; silty claystone; dark brown (7.5YR4/4); weathers brown (7.5YR5/4); common light gray (5Y7/1) metagranotubules, with greenish-gray (5GY6/1) calcareous sheath (diffusion calcan), after root traces ; very-coarse granular peds defined by reddish-brown (5YR4/4) clay skins (argillans); few fine (less than 1-mm) micritic nodules; strongly calcareous in matrix; porphyroskelic skelmosepic in thin section; common quartz; few chert, limestone, feldspar, opaque oxide, biotite, and muscovite grains; diffuse, smooth contact to

-54 cm; Bk horizon; silty claystone; dark brown (7.5YR4/4); weathers brown (7.5YR5/4); common light greenish-gray (5GY7/1) metagranotubules with greenish-gray (5GY6/1) diffusion calcans, after root traces, still up to 10 mm in diameter; common strong brown (7.5YR5/6) calcareous nodules up to 6 mm in diameter; indistinct relict bedding; strongly calcareous; intertextic unistrial insepic in thin section of matrix, porphyroskelic calciasepic in nodules; sand grains mainly quartz, with some feldspar, schist, limestone, ferruginized claystone, and opaque oxides; diffuse, smooth contact to

-96 cm; C horizon; siltstone; dark yellowish-brown (10YR4/4); weathers brown (7.5YR5/4); prominent relict bedding; patchy subhorizontal mottles of greenish-gray (5GY6/1) extend along some bedding planes in the upper part of this horizon; few narrow (6 mm in diameter), reddish-brown (5YR4/4), clay-filled burrows that branch in a dichotomous and pinnate manner; very strongly calcareous; granular unistrial calciasepic in thin section, well cemented with sparry calcite; common quartz; few grains of schist, limestone, feldspar, and muscovite; few fine calcite crystal tubes with micritic diffusion calcans, after fine root traces; abrupt, smooth contact to

-131 cm; C horizon; clayey siltstone; dark yellowish-brown (10YR4/4); weathers brown (7.5YR5/4); prominent relict bedding; few reddish-brown (5YR4/4) clay-filled burrows, as in overlying horizon; very strongly calcareous; granular unistrial calciasepic, as in overlying horizon; abrupt, smooth contact to

-152 cm; A horizon of underlying paleosol (Sarang silty clay massive surface phase); silty claystone; dark reddish-brown (5YR3/4); weathers reddish-yellow (7.5YR6/6); indistinct relict bedding; common strong-brown (7.5YR5/6) calcareous nodules, with sharp boundaries suggesting that they are resorted; few yellowish-brown (10YR5/4), clayey, fine root traces; strongly calcareous; porphyroskelic mosepic in clayey matrix, calciasepic in nodules; common quartz; few muscovite, feldspar, schist, and limestone grains.

5.2.2.4 Further examples. Bhura paleosols were found in both measured sections of paleosols in the lower and middle Dhok Pathan Formation in Kaulial Kas (19 of them in Fig. 5.9 and 2 in Fig. 5.10). The Bhura clay ferric-concretionary variant (Fig. 5.17; at 45.4 m in Fig. 5.9) has concretions more abundant than the calcareous nodules characteristic of Bhura paleosols. Like many other Bhura profiles, this one is not as well developed or thick as the type Bhura clay paleosol. Bhura paleosols commonly show some relict bedding, but can be distinguished from Sarang paleosols by their comparatively better development: incipient soil structure extending down at least 10 cm (after compaction) from the surface, and an incipient horizon of calcareous nodules (Bk) showing signs that they formed in place. Sonita Series paleosols have a profile form similar to Bhura profiles, but have a red hue (5YR) rather than brown (7.5YR) in their freshly excavated surface horizons.

5.2.2.5 Alteration after burial. The estimates for compaction already outlined for Sarang Series paleosols probably also affected Bhura Series paleosols. The clayey surface of the type Bhura clay, for example, may originally have been at a depth of 90 cm rather than its present 54 cm. This must be regarded as a maximum, because compaction to this extent is not evident from mildly deformed clastic dikes, trace fossils, and calcareous rhizoconcretions in these paleosols. Reddening during burial also is likely, although this was probably not as significant as for associated Sonita and Lal paleosols, because the color of Bhura paleosols would have been darkened by even more organic matter than they have now (Fig. 5.13), which is less than found in comparable surface soils. Bhura paleosols also may have been locally discolored around calcareous rhizoconcretions by burial gleization, because there is little other evidence for gleization during soil formation in their chemical or mineralogical composition (Fig. 5.13). Some cementation during burial may account for the sparry calcite filling calcareous rhizoconcretions, most prominent in lower parts of these profiles. Other carbonates of small nodules and slender rhizoconcretions are micritic and show replacive textures typical of pedogenic carbonate. Potash is slightly enriched compared with soda in these paleosols (Fig. 2.23). They may have been affected by limited illitization, which left much smectite unaltered and illite near the surface poorly crystallized (Table 3.1).

5.2.2.6 Reconstructed soil. Bhura paleosols can be imagined as a thin, clayey, brown (probably 7.5YR-10YR), moderately organic surface (A) horizon with incipient granular structure, over an horizon (Bk) with some relict bedding and small, scattered, calcareous nodules. Their calcareous parent material, and indications from replacive nodules and rhizoconcretions that they were becoming more calcareous, are compatible with alkaline pH, and high cation exchange capacity and base saturation. There is no evidence of very high pH or salinity, considering the up-profile decline in soda/potash ratios (Fig. 5.13) and lack of domed columnar peds or other structures of sodic or saline soils (Northcote & Skene, 1972). The overall reddish hue, calcareous nodules, deeply penetrating root traces, and low ferrous iron/ferric iron ratios are evidence of free drainage and oxidizing Eh. Nevertheless, these were clayey soils that may have been prone to puddling after rain. This may be the reason for local iron-manganese stain around rhizoconcretions.

5.2.2.7 Classification. The subsurface horizon (between 20 and 33 cm) of clay shells in the type Bhura clay is not regarded as a pedogenically significant horizon, because it was of limited lateral extent, and for reasons outlined in due course is interpreted as a local aggregation of scarabaeid beetle pupal cells. All the Bhura paleosols appear to be weakly developed with some surficial clay formation, decalcification, ferruginization, and accumulation of organic matter, and an incipient development of a subsurface horizon of calcareous nodules, as in Inceptisols (of Soil Survey Staff, 1975). Ustochrepts and Ustropepts are likely alternatives. The former is especially favored by evidence of climatic seasonality outlined in the next section. Comparable Fluventic Ustochrepts are widespread now on the Indo-Gangetic Plains in a zone between Sangrut and Delhi in India (Sehgal & Sys, 1970). Similar surface soils in their color, structure, tubules, nodules, textural profile, and micromorphology are the Mahupur Series of West Bengal (Kooistra, 1982; Murthy *et al.*, 1982). These soils differ from Bhura paleosols in their deeper (67 cm) calcareous nodules and less clayey texture (only up to 31% clay). A good part of these differences can be blamed on the burial compaction of the Bhura clay and on systematic overestimation of clay in thin sections compared with mechanical analysis of the modern soil (Murphy & Kemp, 1984).

In the Food and Agriculture Organization (1974) classification, Bhura paleosols are developed to the extent of Cambisols, and were calcareous enough to qualify as Calcic Cambisols. Comparable surface soils have been mapped over wide areas of the Indo-Gangetic Plains in Uttar Pradesh (map units Bk39-2a and Bk40-2a of Food and Agriculture Organization, 1977b).

Among Australian soils, Bhura paleosols are most like Brown Clays (of Stace *et al.*, 1968) and Ug5.31 (of Northcote, 1974). The most similar South African profiles (of MacVicar *et al.*, 1977) are Oakleaf mutale. Similar but slightly less clayey profiles (Oakleaf limpopo) are found on alluvial levees between Dundee dundee profiles (similar to Sarang Series paleosols) by streams and Valsrivier lindley profiles (generally similar to Lal Series paleosols) within floodplains in Kruger National Park, South Africa (Botha, 1985; Gertenbach, 1983; Venter, 1986).

5.2.2.8 Paleoclimate. Comparable surface soils of Punjab, in northern India, with diffuse calcareous horizons at a depth of about 90 cm, as reconstructed for Bhura paleosols, are found in regions receiving mean annual rainfall of 500 to 840 mm (Seghal, Sys, & Bhumbla, 1968). Similar estimates can also be gained by comparison with climosequences of calcareous soils on the Serengeti Plains of Tanzania (de Wit, 1978), the Mojave Desert of California (Arkley, 1963), and the Great Plains of the North American interior (Jenny, 1941).

Evidence for a monsoonal, wet-dry seasonality includes the co-occurrence and occasional intergrowth of calcareous nodules and ferric concretions (as in Fig. 2.12), which is also a feature of surface soils in monsoonal India (Courty & Féderoff, 1985; Sehgal & Stoops, 1972). Seasonality of rainfall may also explain the diffuse nature of the calcareous, subsurface (Bk) horizon, the sand-filled clastic dikes seen in the surface of some of these paleosols, and the local iron-manganese stain of rhizoconcretions.

The abundant burrows and root traces and modest accumulation of organic matter and soil structure in these paleosols are compatible with a warm climate, but cannot be regarded as compelling evidence.

5.2.2.9 Ancient vegetation. The large drab-haloed root traces, modest organic content, and soil structure of these paleosols are evidence of forest vegetation. Their degree of development and the association of almost all Bhura paleosols with paleochannel and levee facies are evidence that they supported streamside forest.

Generally similar soils in Uttar Pradesh, India (Food and Agriculture Organization, 1977b; Kooistra, 1982; Murthy *et al.*, 1982), as well as in South Africa (Gertenbach, 1983; MacVicar *et al.*, 1977), Australia (Stace *et al.*, 1968) and the United States (Soil Survey Staff, 1975) support grassy forest and woodland, wooded grassland, and shrubland. Bhura paleosols lack the abundant fine root traces and granular structure of grassland soils. They also are more deeply weathered and have larger root traces than soils of desert shrubland.

No plant fossils were found in Bhura paleosols, so the floristic nature of its vegetation is not known. In much of northern India today, riparian woodland and forest (vegetation unit 4E/RSI of Champion & Seth, 1968) is dominated by trees of *Terminalia arjuna* and *Lagerstroemia speciosa*, but includes a variety of other trees, shrubs, vines, and grasses. Similar vegetation is possible for Bhura paleosols, because *Terminalia* leaf fossils have been found at Balugoloa, near Jawalamukhi in Himachal Pradesh (Lakhanpal & Dayal, 1966), and fossil wood similar to that of *Lagerstroemia* from Kalagarh, Uttar Pradesh (U. Prakash, 1981), both in fluvial deposits of the Lower Siwalik Group of early Miocene age.

5.2.2.10 Former animal life. Two Bhura paleosols in the lower measured section (Fig. 5.9) within the Dhok Pathan Formation yielded animal fossils. One of these (at 5 m in Fig. 5.9) contained only fragments of tortoise carapace. Another paleosol (at 16.5 m in Fig. 5.9) in the badlands slope 200-m south of the section line (at grid reference 751143) yielded the following fossils: astragalus of small, chowsinghalike antelope (*Elachistoceras khauristanensis*), lower molar of large three-toed horse ("*Cormohipparion*" sp.), and opercula of ampullariid pond snail (*Pila prisca*).

The snail opercula are dense, calcareous, large (13-15-mm wide by 20-25-mm long) and thick (2-3-mm), and this may be why they were preserved in preference to the shells of the snails (Retallack, 1984a). They commonly have their adapical margin broken, perhaps by birds or mammals feeding on them. *Pila prisca* is widespread in the Siwalik Group (Prashad, 1924; Wadia,1928), and allied forms also are known from Miocene rocks of East Africa and China (Newton, 1914; Wenz, 1938). The living apple snail (*Pila globosa*) of Indian ponds and paddyfields is an aquatic pulmonate that can breathe air by means of a siphon if caught in muddy or slimy water. During the dry season, they close the opening of the shell with their robust operculum, and aestivate (Hawkins, 1986). Dry-season aestivation is likely also for the Miocene snails of India, because the fossil opercula show two or three strongly marked growth rings (Prashad, 1924).

Bhura paleosols examined in the upper measured section yielded no bone, but the type Bhura clay paleosol did contain very poorly preserved, pear-shaped clay shells about 3 cm in basal diameter. These are very similar to the ichnofossil genus *Coprinsphaera* (of Sauer, 1955), especially specimens from Argentine Oligocene paleosols (Frenguelli, 1938a, 1938b, 1939). These South American trace fossils have been interpreted as brood cells of dung beetles, like those made by South American scarabaeids, such as *Phanaeus*, *Coprophanaeus*, *Sulcophanaeus*, *Oxysternon*, *Bolbites*, and *Dichotomius*. In these beetles, the clay shell is fashioned around both the egg and the food mass of dung, and serves to protect the developing larva as it feeds on the dung until pupation. This distinctive nesting behavior is less common among living Indian dung beetles than South American ones, but Indian *Heliocopris bucephalus* does line its dung boli with packed soil (Arrow, 1931). The fossil *Coprinsphaera* of the type Bhura clay may have been at depths within the soil of 33 to 55 cm before burial compaction, whereas living *Heliocopris* form much larger cells as much as 120 cm down in the soil (Halffter & Edmonds, 1982). Increases in the size, depth, and elaborateness of dung beetle nests have been prominent themes in the Tertiary evolution of dung beetle nesting behavior (Retallack, 1990b).

5.2.2.11 Paleotopographic setting. Within the lower measured section in the lower Dhok Pathan Formation (Fig. 5.9), every paleochannel is intimately associated with Sarang Series paleosols, which are in turn flanked above and below by Bhura paleosols. Other kinds of paleosols are more removed from paleochannels. These relationships in a vertical section can be translated into lateral relationships on ancient landscapes using "Walther's facies rule", and by this line of reasoning Bhura paleosols may have formed along streamsides in locations a little less prone to disturbance than Sarang Series paleosols.

In the upper measured section within the main part of the Dhok Pathan Formation (Fig. 5.10), the type Bhura clay is above a paleochannel, but the other Bhura paleosol is not. By 7.8 million years ago, then, these brown, moderately well-humified soils were a less prominent part of the soilscape, and did not form a distinct streamside gallery forest as they did 8.5 million years ago.

All of the Bhura paleosols have deeply penetrating root traces and burrows and are highly oxidized. Some iron-manganese stain around root traces may be an indication of wet-season puddling, but they generally were well drained and not so prone to seasonal waterlogging as associated Kala, Pila, Khakistari, or Pandu paleosols.

5.2.2.12 Parent material. The type Bhura clay contains common quartz, schist, and limestone fragments, like that of the alluvium of a major stream draining the high Himalaya ("blue gray sandstones" of Behrensmeyer & Tauxe, 1982). In addition, the overall chemical composition of the type Bhura clay (Fig. 2.23), including that of trace elements whose abundance in rocks and soils is well understood (Ba, Co, Cr, Ni, Sr, Zn in Aubert & Pinta, 1977; Kabata-Pendias & Pendias, 1984), is also schistlike and not especially deeply weathered. This is surprising because the brown color of the type Bhura clay is much more like that of alluvium derived from the soil-covered foothills of the Himalayan ranges, such as that of the underlying type Naranji clay and its parent material (a "buff sandstone" of Behrensmeyer & Tauxe, 1982). Probably this paleochannel was no longer a significant source of alluvium after it was abandoned to allow formation of the type Naranji clay. This is not to imply that all Bhura paleosols had a parent material of little weathered alluvium from major Himalayan antecedent streams. Other Bhura paleosols closely associated with paleochannels of foothills drainage may prove to have formed on that kind of alluvium, when analyzed petrographically and chemically.

5.2.2.13 Time for formation. Bhura series paleosols show a degree of development of calcareous nodules and clay accumulation near the surface comparable with soils of northern India some 400 to 4500 years old (Courty & Féderoff, 1985). These pedogenic features are intermediate in development between those of surface soils of the Ravi River floodplain near Lahore, Pakistan, that are some 2300 years old (Sultanpur Series of Ahmad, Ryan, & Paeth, 1977) and some 10,000 to 20,000 years old (their Bhalwal Series). Such estimates of a few thousand years for the development of Bhura paleosols also can be gained by comparison with weakly developed soils (in the strict sense of Retallack, 1988a) in other parts of the world (Birkeland, 1984).

5.2.3 Khakistari Series (new name)

5.2.3.1 Diagnosis. Very clayey, brown to gray (10YR-2.5Y) paleosols, with surface brecciation and clastic dikes, and subsurface ferric concretions and mottles (Bg), and relict bedding.

5.2.3.2 Derivation. Khakistari is Urdu for ash-colored (Platts, 1960), in reference to the light grayish color of these paleosols.

5.2.3.3 Description. The type Khakistari clay paleosol (Fig. 5.14) is the uppermost band of similar drab paleosols in the low red knoll above a dry tributary of Kaulial Kas (grid reference 748144), 1.2-km west of Kaulial village, Potwar Plateau, Pakistan (Figs 5.6 and 5.8). This is in a tongue of late Miocene, Nagri Formation, within the lower Dhok Pathan Formation (2299 m in Fig. 5.7, 9.4 m in Fig. 5.9), paleomagnetically dated at about 8.5 million years (Tauxe & Opdyke, 1982; corrected to time scale of Berggren, Kent, & Van Couvering, 1985).

+14 cm; Co horizon of overlying paleosol (Sonita clay ferric-concretionary variant); silty claystone; dark reddish-brown (5YR3/4); weathers reddish-brown (5YR5/4); common small (3-4-mm) ferric concretions; few small (1-2-mm) calcareous nodules; prominent relict bedding; indistinct coarse angular-blocky peds, picked out by impersistent slickensided clay skins (argillans); very calcareous; porphyroskelic skelmosepic in thin section; common spherical micropeds of near-opaque ferruginized clay; common quartz, feldspar, mica, and rock fragments; few metagranotubules, with agglomeroplasmic insepic internal fabric; micritic nodules have a calciasepic microfabric; clear, smooth contact to

0 cm; A horizon; claystone breccia; dark brown (10YR3/3) claystone clasts in a dark reddish-brown (5YR3/4) silty claystone matrix; this appears to be the disrupted surface of the paleosol filled with reddish silty-claystone infiltrating down from above; weathers reddish-brown (5YR5/4); common light greenish-gray (5G7/1), drab-haloed root traces; few, near-vertical, light gray (2.5Y7/2) clastic dikes (silans), up to 8-mm wide; moderately calcareous; porphyroskelic bimasepic in thin section of both crack-filling material from above and less ferruginized clasts; common quartz; few grains of feldspar, limestone, and schist; clear, irregular contact to

-8 cm; A horizon; claystone; brown (10YR5/3); weathers light brown (7.5YR6/4); common large (up to 3

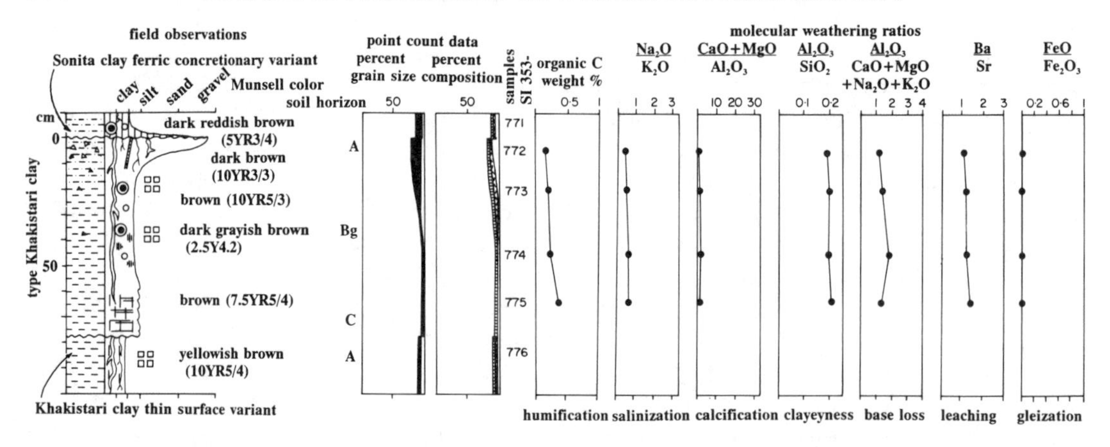

Fig. 5.14. Measured section, Munsell colors, soil horizons, grain size, mineral composition, organic content, and selected molecular weathering ratios of the Khakistari thin-surface phase and type Khakistari clay paleosols, in the lower measured section of the Dhok Pathan Formation, west of Kaulial village, Pakistan. This section is at 8.5 to 9.6 m in Fig. 5.9. Lithological key as for Fig. 5.17.

cm in diameter) black (5Y2.5/1) root traces (Fig. 3.1), with a wide halo of greenish-gray (5GY6/1) and an additional narrow halo of dark greenish-gray (5G4/1); common angular clasts up to 2 cm long of yellowish-red (5YR4/6) claystone; slickensided clay skins (argillans) of reddish-brown (5YR4/4) define coarse angular-blocky peds; moderately calcareous; porphyroskelic bimasepic in thin section, very clayey with scattered stress argillans; few micritic calcareous rhizoconcretions; some quartz and red claystone clasts; few ferric rhizoconcretions; diffuse, irregular contact to

-34 cm; Bg horizon; claystone; dark grayish-brown (2.5Y4/2); weathers light yellowish-brown (10YR6/4); common drab-haloed root traces, up to 8 mm in diameter of black (5Y2.5/1), with a wide halo of greenish-gray (5GY6/1) and an outer halo of dark reddish-brown (5YR2.5/2); common irregular mottles, in part picking out relict beds, of dark reddish-brown (5YR3/3) and yellowish-brown (10YR5/4); distinct, coarse angular-blocky peds, defined by slickensided clay skins (argillans); few small (2-mm) ferric concretions and very small (less than 1-mm) calcareous nodules; strongly calcareous; porphyroskelic insepic in thin section; few ortho-isotubules with diffusion sesquan, after burrows; some sparry calcite crystal tubes, with micritic sheath (diffusion calcan), after fine root traces; diffuse, irregular contact to

-66 cm; C horizon; claystone; brown (7.5YR5/4); weathering light yellowish-brown (10YR6/4); few greenish-gray (5G6/1) drab-haloed root traces; common mottles of yellowish-brown (10YR5/4) and reddish-brown (5YR5/3), picking out conspicuous relict bedding; few small (4-mm) calcareous nodules and ferric concretions; strongly calcareous; porphyroskelic insepic in thin section; few fine ferruginous and calcareous rhizoconcretions, as in overlying horizon; clear, smooth contact to

-78 cm; A horizon of underlying paleosol (Khakistari clay thin-surface phase); claystone; yellowish-brown (10YR5/4); weathers light yellowish-brown (10YR6/4); common greenish-gray (5G6/1) drab-haloed root traces; distinct, coarse angular-blocky peds, defined by reddish-brown (5YR4/4) slickensided clay skins (argillans); strongly calcareous; porphyroskelic bimasepic in thin section; common quartz, red claystone, microcline, and opaque oxides; few calcite crystal tubes and ortho-isotubules with diffusion sesquans, after roots; few fragments of land snail.

5.2.3.4 Further examples. Khakistari Series paleosols were seen in a single band of 5 superimposed paleosols in the lower measured section of the lower Dhok Pathan Formation (from 5.6 to 9.4 m in Fig. 5.9). The two profiles below the type Khakistari clay are thinner and have more relict bedding than usual for these paleosols, and one of these is called the Khakistari clay thin-surface phase (Fig. 5.14; at 8.6 m in Fig. 5.9). Several Khakistari paleosols also were seen at a comparable stratigraphic level in Ganda Kas, 8 km to the west, within the trench sampled for a paleomagnetic transition by Lisa Tauxe (Tauxe & Badgley, 1988). Khakistari and Pandu Series paleosols are an important part of the "gray silt" lithofacies mapped throughout the Khaur region (by Behrensmeyer & Tauxe, 1982) as tongues of Nagri Formation within the lower Dhok Pathan Formation. Similar paleosols also have been reported from Siwalik sediments around Jhelum, Pakistan (profiles KK4 and MK-3-1B of G. D. Johnson *et al.*, 1981; and type O of Visser & Johnson, 1978).

5.2.3.5 Alteration after burial. Such clayey paleosols could have been compacted to as much as 58 percent from an original thickness above the C horizon of as much as 114 cm. Some deformation of clastic dikes and root traces was noticed, but not to this extent, which can be regarded as a maximum likely compaction. Reddish mottles in these paleosols may have been brown originally with ferric hydroxides, rather than hematite formed during burial. Such reddening would not have altered the dominantly gray color of these paleosols, which itself may be due in part to burial gleization. Some loss of organic matter during burial also is likely, and is especially suggested by declining abundance of organic carbon toward the top of the type Khakistari clay paleosol (Fig. 5.14). The Khakistari clay is too clayey for pervasive cementation, and too smectitic and consistent in potash composition for illitization to have been profound.

5.2.3.6 Reconstructed soil. Khakistari paleosols can be reconstructed as thick (up to 114 cm), sticky, gray clays, cracked and copiously rooted near the surface (A) horizon, and strongly mottled in a subsurface clayey (Bg) horizon, over a stiff, gray, calcareous clay with relict bedding. The surface is a little decalcified, but the whole profile has remained calcareous and probably had an alkaline pH and high base saturation and cation exchange capacity. There is no evidence for salinization from the up-profile decline in the soda/potash ratio (Fig. 5.14). The prominent ferric mottles, and complex zones of iron-manganese stain around drab-haloed root traces with relict organic matter (Fig. 3.1), as well as a peak in ratios of ferrous iron/ferric iron in the subsurface (Bg) horizon, can be taken as evidence of gleization and low Eh for a good part of the year. Nevertheless, wide cracks and deeply penetrating root traces are an indication that the soil also dried out on a regular basis.

5.2.3.7 Classification. Considering the distinctive features of these paleosols related above to seasonal waterlogging, Khakistari paleosols are best classified as Vertic Haplaquepts (of Soil Survey Staff, 1975) and as Calcaric Gleysols (of Food and Agriculture Organization, 1974). Similar surface soils are found in low-lying areas near the confluence of the Ganges and Brahmaputra Rivers, of eastern India and Bangladesh (map units Gc9-3a and Jc50-2a of Food and Agriculture Organization, 1977b). Such soils also are found in the mangrove region of the Brahmaputra Delta, but there is no evidence in Khakistari paleosols of the peat, pyrite, burrows, and brackish-adapted molluscs found in marine-influenced soils. The Khakistari clay is broadly similar to described soils on Ganges alluvium near Meerut, Uttar Pradesh (Shankaranaryana & Hirekirur, 1972) and on the Hooghly floodplain of West Bengal (Kanagarh Series of Murthy *et al.*, 1982; Kooistra, 1982).

There is no clear equivalent to Khakistari paleosols in the Australian classification (Stace *et al.*, 1968), in which they can be included within the large and heterogeneous group of Grey Clays. In the Northcote (1974) key, these paleosols are Ug5.24. Among South African soils (MacVicar *et al.*, 1977), they are like Rensburg rensburg profiles, which are widespread on basaltic colluvium and alluvium in northern Kruger National Park (Botha, 1985; Gertenbach, 1983; Venter, 1986).

5.2.3.8 Paleoclimate. Khakistari paleosols have a paradoxical mix of features. On the one hand, they are drab colored overall with prominent red mottles and ferric concretions. These features and their root traces with relict organic matter and rinds of iron manganese (quasimangans) beyond the drab haloes are all indications of a waterlogged soil. On the other hand, they have surficial cracks and brecciation, abundant slickensides, and deeply penetrating root traces as indications of good drainage. This conflicting mix of features is understandable as a reflection of paleoclimate with very pronounced wet and dry seasons, as is common in Vertisols. Khakistari paleosols do show some vertic features, but lack the undulating deformation of surface horizons (mukkara structure of Paton, 1974) or intersecting fans of slickensides (lentil peds of Krishna & Perumal, 1948) found in Vertisols. Perhaps if soil development had continued for longer without interruption by flooding, Khakistari paleosols may have developed into Vertisols.

The large root traces and snail shells found in Khakistari paleosols are compatible with a climate that was warm, and certainly not frigid. Their modest decalcification is indicative of subhumid to semiarid range of rainfall. Khakistari paleosols are not developed sufficiently to be regarded as reliable indicators of these aspects of paleoclimate.

5.2.3.9 Ancient vegetation. Considering the evidence for seasonality already discussed, as well as the abundant large root traces up to 3 cm in diameter, Khakistari paleosols probably supported seasonally wet lowland forest. This is the kind of vegetation found on comparable surface soils (Calcaric Gleysols) near the confluence of the Brahmaputra and Ganges Rivers in eastern India and Bangladesh (units Gc9-3a and Jc50-a of Food and Agriculture Organization, 1977b). Comparable soils (Vertic Haplaquepts) in the Mississippi Delta plain of the United States support deciduous woodland (Soil Survey Staff, 1975). Australian Grey Clays include a variety of soils not especially like Khakistari paleosols, and support grassy woodland and wooded grassland (Stace *et al.*, 1968).

Khakistari paleosols are not as black and are more deeply rooted than soils supporting tropical seasonal swamp forest (vegetation unit 4D/SS1 and SS2 of Champion & Seth, 1968) or tropical freshwater swamp forest (their unit 4C/FS1 and FS2) in the wet lowlands of eastern India and Bangladesh. In view of the closely associated red and formerly well drained paleosols (Bhura, Sonita and Lal Series) and evidence of seasonal drying of Khakistari paleosols, they are more likely to have formed within seasonally wet lowlands (terai or dhankar) near sub-Himalayan alluvial fans (bhabar). In these regions of India today, vegetation type somewhat similar to that envisaged for the Khakistari clay is moist tarai sal forest (unit 3C/C2c of Champion & Seth, 1968), which is characterized by the trees, *Shorea robusta*, *Trewia nudiflora*, and *Syzygium cumini*, and a small climbing palm (rattan cane, *Calamus tenuis*).

Fossil wood comparable to either *Shorea*, *Parashorea*, or *Pentacme* has been found in Lower Siwalik fluvial sediments near Kanagarh, Uttar Pradesh (Awasthi, 1982; Trivedi & Ahuja, 1979b). Fossil leaves similar to those of living *Syzygium claviflorum* have been found in sediments of about the same age in Nepal (M. Prasad & Prakash, 1984), in a fossil assemblage similar to Indian tropical swamp forest (unit 4D/SS5 of Champion & Seth, 1968). Sal (*Shorea robusta*) and allied plants avoid highly calcareous and clayey soils (Puri, 1950) and so are unlikely to have grown in Khakistari paleosols, for which the Nepali fossil assemblage (of M. Prasad & Prakash, 1984) is more reasonable ecologically.

5.2.3.10 Former animal life. No bones were seen within these paleosols within the measured section (Fig. 5.9), but this unit of gray claystones was traced 1-km northeast across Kaulial Kas to an isolated knoll of fossiliferous Khakistari paleosols (at grid reference 756152). This yielded a labial fragment of a large molar with the distinctive silky-appearing enamel of large giraffe (cf. *Brahmatherium megacephalum*), thick enamel fragments of unworn molar cusps of a gomphothere elephant

(*Choerolophodon corrugatus*, in the sense of Tassy, 1983) and numerous thick opercula of apple snails (*Pila prisca*, of Prashad, 1924; Wenz, 1938). These aquatic snails that can aestivate during the dry season were much more common in Khakistari paleosols than in Bhura paleosols. Seasonal growth rings are prominent also in these opercula, as well as in thin sections made from the gomphothere tooth enamel, where banding is similar to that documented for mastodons (by D. C. Fisher, 1984).

5.2.3.11 Paleotopographic setting. Khakistari paleosols were lowland soils associated with the axial, or proto-Gangetic River drainage ("blue gray sandstones"), as can be seen from their mapped distribution in the Khaur area (in "blue gray siltstones" of Behrensmeyer & Tauxe, 1982). Waterlogging is apparent from their overall gray color, strongly colored mottles, ferric concretions, and preservation of remnant organic matter within some root mottles. Khakistari paleosols also have more smectite than illite, unlike associated Lal clay paleosols (Fig. 2.32). Comparable differences in clay composition can be seen in hydromorphic gley (dhankar) soils compared with associated better drained soils in Uttar Pradesh, India (Gupta, 1961; Gupta, Agarwal, & Mehrotra, 1957). Their clayey and calcareous composition would have made them prone to seasonal puddling (Buehrer, Martin, & Parks, 1939). Nevertheless, Khakistari paleosols lack peaty surface horizons, and nodules of siderite or pyrite found in permanently waterlogged soils. Evidence of seasonal drying is provided by surface cracks and deeply penetrating root traces. Khakistari paleosols may have formed on clayey fill of abandoned channels or broad swales in scroll bars and levees, as seen now on the modern floodplain of the Indus River in Pakistan (Holmes & Western, 1969).

5.2.3.12 Parent material. Khakistari paleosols formed on calcareous clayey alluvium from a proto-Gangetic River system ("blue gray sandstones" of Behrensmeyer & Tauxe, 1982) that penetrated deeply into the metamorphic core of the Himalaya. This can be seen from rare fragments of little-weathered schist in the type Khakistari clay. A similar provenance is indicated also by its chemical composition (Fig. 2.22) similar to that of the type Bhura and Lal clays: lower in silica and richer in alumina, potash, and magnesia than Pandu, Pila, and Kala paleosols that developed on alluvium of the foothills drainage ("buff sandstones" of Behrensmeyer & Tauxe, 1982).

5.2.3.13 Time for formation. The type Khakistari clay shows only weak development (in the strict sense of Retallack, 1988a), with considerable destruction of relict bedding near the surface but only the beginnings of a calcic horizon. It can be compared with the surface soils of known age already cited for the Bhura Series, and similarly represents only a few thousand years of soil development. Other Khakistari paleosols are very weakly developed, like Sarang Series paleosols, and may represent only a few hundred years.

5.2.4 Sonita Series (new name)

5.2.4.1 Diagnosis. Thin paleosols with a red (5YR) clayey surface (A) horizon with distinct soil structure, over a subsurface (Bk) horizon with some relict bedding and weakly developed, small (2-3-mm), scattered calcareous nodules.

5.2.4.2 Derivation. Sonita is Urdu for red, crimson, and purple (Platts, 1960).

5.2.4.3 Description. The type Sonita clay paleosol (Fig. 5.13) is separated from the type Bhura clay paleosol below by a thin profile (Sarang silty-clay rhizoconcretionary-variant) in the large bluff (at grid reference 773135), 1.5-km southeast of Kaulial village, Potwar Plateau, Pakistan (Figs. 5.6 and 5.8). this is in the main part of the late Miocene, Dhok Pathan Formation (at 2750 m in Fig.5.7, and at 22.2 m in Fig. 5.10), above the uppermost tongue of Nagri sandstone, and is dated paleomagnetically at about 7.8 million years old (Tauxe & Opdyke, 1982; corrected to the time scale of Berggren, Kent, & Van Couvering, 1985).

+7cm; C horizon of overlying paleosol (Sarang clay cumulic-surface phase); silty claystone; dark brown (7.5YR4/4); weathers reddish-brown (5YR4/4); these two relict beds of siltstone with an intervening bed of shale are broken into very thick platy peds, 3-4-cm across, by clay skins (illuviation argillans) and narrow (2-3 cm in diameter), yellowish-brown (10YR5/8), calcareous rhizoconcretions, with central metagranotubules of very dark grayish brown (10YR3/2) siltstone; strongly calcareous; intertextic insepic in thin section of matrix, with porphyroskelic calciasepic rhizoconcretions and porphyroskelic insepic to mosepic papules (claystone clasts); common quartz; some grains of mica, feldspar, and rock fragments of limestone and schist; abrupt, broken contact to

0 cm; A horizon; silty claystone; reddish-brown (5YR4/4); weathering reddish-brown (5YR5/4); common fine (2-3 mm in diameter), light gray (5Y7/1) metagranotubules, with a 5-6-mm wide greenish-gray (5GY6/1 and 5GY5/1) halo and an outermost narrow (1-mm) dark reddish-brown (5YR3/3) rind (diffusion mangan), after root traces; few large (1-2 cm in diameter) calcareous rhizoconcretions; some indistinct septate nodular structures, generally similar to *Termitichnus*; indistinct coarse-granular peds; strongly calcareous in matrix; porphyroskelic skelmosepic in thin section of both matrix and drab haloes to metagranotubules, but calciasepic in micritic nodules; common quartz and feldspar; few grains of mica, opaque oxides, schist, and limestone; diffuse, irregular contact to

-20 cm; Bk horizon; silty claystone; reddish-brown (5YR4/3); weathers brown (7.5YR5/4); common small (6-7-mm), reddish-yellow (7.5YR6/6), irregular calcareous nodules and rhizoconcretions; few gray (5Y6/1) metagranotubules with greenish-gray (5GY6/1) haloes, after roots; few clay galls (papules) and clay skins (illuviation argillans) of reddish-brown (5YR4/4) claystone; strongly calcareous; agglomeroplasmic skelmosepic in thin section, porphyroskelic calciasepic in rhizoconcretions and nodules; common quartz; few rock fragments, feldspar, and mica; diffuse, smooth contact to

-90 cm; BC horizon; clayey siltstone; dark brown (7.5YR4/4); weathers brown (7.5YR5/4); distinct relict bedding; few gray (5Y5/1) metagranotubules, with reddish-brown (5YR4/4) iron stain (sesquiargillans), after burrows; few calcareous rhizoconcretions and nodules as in overlying horizon; strongly calcareous; intertextic crystic in thin section due to pervasive calcite cement; agglomeroplasmic skelmosepic in metagranotubules and porphyroskelic calciasepic in calcareous rhizoconcretions and nodules; some rhizoconcretions have a central crystal tube of sparry calcite; common quartz; some feldspar, opaque oxides, and mica; clear, smooth contact to

-99 cm; C horizon; siltstone; dark brown (7.5YR4/4); weathers light brown (7.5YR6/4); distinct relict bedding; few reddish-brown (5YR4/4) clay galls (papules), burrows, and root traces; the basal 1 cm of this horizon has angular claystone clasts up to 2-cm across, as well as irregular stratiform greenish-gray (5GY6/6) mottles; very strongly calcareous; unistrial intertextic calciasepic in thin section; few crystal tubes of sparry calcite and meta-isotubules of porphyroskelic insepic claystone, after roots and burrows; common quartz; some feldspar, mica, and opaque oxides; clear, wavy contact to

-115 cm; A horizon of underlying paleosol (Sarang silty-clay rhizoconcretionary-variant); dark brown (7.5YR4/4); weathers light brown (7.5YR6/4); common greenish-gray (5GY6/1) metagranotubules with a greenish-gray (5G5/1) halo, after roots; few yellowish-brown (10YR5/4) calcareous rhizoconcretions; few strong-brown (7.5YR5/6) resorted calcareous nodules with local rims (mangans) of very dark brown (7.5YR7/2); indistinct coarse granular peds defined by reddish-brown (5YR4/4) impersistent argillans; prominent relict bedding, especially toward base of horizon; very strongly calcareous; porphyroskelic skelmosepic in thin section; grains mainly quartz, feldspar, limestone, schist, and chert.

5.2.4.4 Further examples. Sonita profiles were common in both measured sections in the Dhok Pathan Formation (12 profiles in Fig. 5.9 and 13 in Fig. 5.10). These profiles are more red than brown Bhura paleosols but are developed to about the same extent. Their moderate development is intermediate between that of Sarang and Lal Series paleosols, and they vary in thickness and degree of expression of relict bedding, rhizoconcretions, and nodules. The type Sonita clay is well developed for these paleosols. On the other hand, the Sonita silty-clay in the upper measured section within the main part of the Dhok Pathan Formation (Fig. 5.15, at 35.3 m in Fig. 5.10) is thin with relict bedding and short clastic dikes filled with sandstone. This represents the least degree of development accepted for a Sonita paleosol before it would be regarded as Sarang Series paleosol. Two additional unnamed Sonita paleosols were seen at a fossil

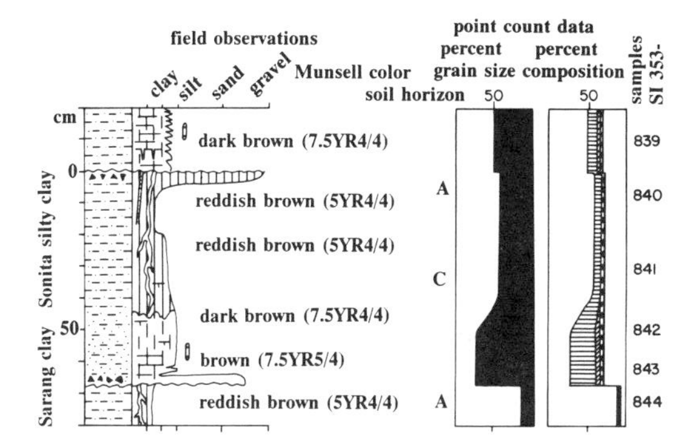

Fig. 5.15. Measured section, Munsell colors, soil horizons, grain size, and mineral composition of the Sarang clay and Sonita silty clay paleosol, in the upper measured section of the Dhok Pathan Formation, southeast of Kaulial village, Pakistan. This section is at 34.8 to 35.8 m in Fig. 5.10. Lithological key as for Fig. 5.17.

locality (Yale-GSP 262, excavation levels 3 and 4 of Badgley, 1982), in the long tributary creek of Kaulial Kas, 2.1-km west of Kaulial village (at grid reference 741144) stratigraphically below both measured sections reported here, within the lower Dhok Pathan Formation where it is laterally equivalent to the Nagri Formation (2219 m in Fig. 5.7). Another unnamed Sonita paleosol with prominent clastic dikes was seen in the bank by the stream bed inside the bend of Kaulial Kas 0.5-km south of Kaulial village (grid reference 764142) at a stratigraphic level between the two measured sections of paleosols in the Dhok Pathan Formation (2606 m in Fig. 5.7). There may be additional Sonita or Lal paleosols in the Siwalik sediments near Jhelum, Pakistan (profiles MK-6-1A, MK-6-3A and MK-9-2 of G. D. Johnson *et al.*, 1981; and type Rm of Visser & Johnson, 1978).

5.2.4.5 Alteration after burial. As already discussed for Sarang series paleosols, clayey parts of Sonita paleosols in the lower measured section of paleosols (Fig. 5.9) may have been compacted to as much as 58 percent of their former thickness, and those of the upper section to as much as 60 percent. By these estimates the type Sonita clay would have had a Bk horizon at a depth of about 33 cm.

Organic matter remaining in the type Sonita clay is only slightly less than in the type Bhura clay, despite the better soil structure of the Bhura clay. Both profiles probably lost most of their organic matter. Both also are likely to have been mildly illitized and reddened by dehydration of ferric hydroxides during burial. Reddening would have affected Bhura paleosols less than Sonita paleosols, which could have had an orange brown hue (close to 7.5YR, and with moderately high chroma). Drab-haloed root traces locally discoloring Sonita paleosols may be due to burial gleization. A small amount of sparry calcite in root holes is probably a burial cement, but most of the carbonate in replacive rhizoconcretions, nodules, and as dispersed micrite appears to be pedogenic.

5.2.4.6 Reconstructed soil. Sonita paleosols can be imagined with clayey, red surface (A) horizons about 30-cm thick and only weakly humified, over an horizon (Bk) with diffuse small calcareous nodules. Relict bedding is prominent in the parent material, and is distinct also in the lower horizon (Bk). There has been some limited decalcification near the surface and accumulation of carbonate in the subsurface, but the parent material already was very calcareous. Thus pH probably was alkaline and base saturation and cation exchange capacity high. There is no indication from soda/potash ratios (Fig. 5.13) or soil structures of salinization. Sonita paleosols are among the most ferruginized paleosols in the Dhok Pathan Formation, and probably were well drained and dry for most of the year.

5.2.4.7 Classification. Sonita paleosols are similar to Bhura paleosols in most taxonomically significant features. Sonita paleosols do have more common surface cracks, and are best identified as Vertic Ustochrepts (of Soil Survey Staff, 1975). As intergrades between Fluvents (like

Sarang Series) and Haplustalfs (like Lal Series), Sonita paleosols also are similar to surface soils of the Ustochrept region of Punjab, India, that have been called "Alfic Ustochrepts" (Sehgal & Stoops, 1972), but this soil subgroup was not included in the comprehensive soil taxonomy (Soil Survey Staff, 1975). In other classifications, Sonita paleosols are Calcic Cambisols (of Food and Agriculture Organization, 1974), Red Clays (of Stace *et al.*., 1968), Ug 5.36 (of Northcote, 1974), and Oakleaf makulek (MacVicar *et al.*, 1977), for reasons similar to those given for Bhura paleosols.

5.2.4.8 Paleoclimate. The type Sonita clay may have had a diffuse zone of calcareous nodules at a reconstructed depth of 33 cm. In the present alluvial plains of Punjab, India, soils with carbonate nodules at this depth are found in regions receiving 320 to 590 mm mean annual rainfall (Sehgal, Sys, & Bhumbla, 1968). This estimate is similar to that gained by comparison with calcareous soils in other parts of the world (Arkley, 1963; de Wit, 1978; Jenny, 1941). A semiarid to subhumid climate also is indicated by the dominance of smectite in these paleosols (see Fig. 2.32) and soils of subhumid to semiarid climate, unlike the illite and chlorite that dominates soils in wetter regions of Indo-Pakistan (Razzaq & Herbillon, 1979). These indications of dry climate should be viewed with caution considering the great contribution of parent material to clays and carbonates of these weakly developed paleosols. Closely associated moderately developed Lal paleosols are better paleoclimatic indicators.

A seasonal paleoclimate for Sonita paleosols is likely, considering their clastic dikes, calcareous rhizoconcretions, and the occurrence in them of both ferric concretions and calcareous nodules. Spherical micropeds and burrows like those of termites are evidence of a warm climate. These also are climatically sensitive features better expressed in associated Lal paleosols.

5.2.4.9 Ancient vegetation. The abundant stout, drab-haloed root traces and profile form of Sonita paleosols are evidence of dry woodland or forest vegetation. Probably they formed under vegetation intermediate in ecological succession between that envisaged for Sarang and Lal Series, but on drier better-drained sites more thoroughly scavenged of organic matter by termites than Bhura Series paleosols.

Comparable surface soils in North America (Soil Survey Staff, 1975), Australia (Stace *et al.*, 1968), and South Africa (Gertenbach, 1983; MacVicar, *et al.*, 1977) support grassy woodland, bushland, wooded grassland, and shrubland. The drier and open kinds of vegetation are unlikely for Sonita paleosols considering their copious stout root traces and lack of granular ped structure typical of grasslands, or of vesicular structure and evaporites typical of desert soils. Grassy woodland or forest is also suggested by comparison with very similar surface soils within the Indo-Gangetic alluvial plains of India (map units Bk39-2a and Bk40-2a of Food and Agriculture Organization, 1977b). These are well within the present region of sal (*Shorea robusta*) forests. As already discussed for the Khakistari Series, sal and related plants do not and probably did not grow in soils as calcareous or clayey as the Sonita clay, even though they have a fossil record in India back to early Miocene time. The modern vegetation most similar to that envisaged for Sonita Series is dry plains tropical deciduous forest (5B/C1b of Champion & Seth, 1968), which includes such trees as *Terminalia tomentosa* and *Diospyros tomentosa.* This vegetation extends in drier sites eastward into the areas with moist deciduous forests likely for associated Lal Series paleosols of the lower Dhok Pathan Formation. Permineralized wood similar to that of living *Shorea*, *Terminalia*, and *Diospyros* has been recovered from early Miocene Siwalik sediments near Kalagarh, Uttar Pradesh, India (Awasthi, 1982; U. Prakash, 1978, 1981; U.Prakash & Prasad, 1984;

Trivedi & Ahuja, 1979a, 1979b), but other elements of this diverse Indian fossil flora indicate affinities with wet evergreen dipterocarp forest (unit 1B/CI of Champion & Seth, 1968).

5.2.4.10 Former animal life. None of these paleosols in the lower measured section reported here (Fig. 5.9) were found to be fossiliferous, but fossils have been reported from Sonita paleosols lower within the Dhok Pathan Formation (at Yale-G.S.P. locality 262, grid reference 741144, at 2606 m in Fig. 5.7). The level in the excavation here with iron-manganese nodules (level 2 of Badgley, 1982) is a Kala Series paleosol, but the other levels (3 and 4) are Sonita paleosols, and have yielded the following: hyaenid (ulna), proboscidean (indeterminate teeth and tusk), gomphothere (2 incomplete juvenile teeth), three-toed horse ("*Cormohipparion*" or *?Hipparion*, tooth), rhinoceros (tooth, scapula), pig (*Propotamochoerus hysudricus*, juvenile tooth, tibia), giraffe (humerus), antelope (*Miotragocerus punjabicus*, podial), gazelle (*Gazella*, mandible, radius, metapodial, and phalange) and 15 indeterminate bones.

In the main part of the Dhok Pathan Formation fossils were found in a Sonita paleosol on a small spur within the gullies (Yale-G.S.P. locality 544, at grid reference 775137), 200 m north of the upper measured section (at a level of 12.6 m in Fig. 5.10). The following specimens (G.S.P. numbers 16244, 16248, 16249, 16251) were found by David Archibald and me: tortoise (scrap), proboscidean (tusk), three-toed horse ("*Cormohipparion*" sp., 4 teeth), rhinoceros (*Chilotherium intermedium*, tooth), giraffe (partial tooth), and antelope (*Selenoportax lydekkeri*, 2 teeth). Another possible fossiliferous Sonita paleosol is the "red silt outcrop" at a stratigraphic level equivalent to the upper measured section (35.5 m in Fig. 5.10) 1 km to the southwest (Yale-G.S.P. locality 477, at grid reference 764128). This locality yielded remains of bamboo rat (*Kanisamys sivalensis*), anthracothere, and large giraffe (Barry, Lindsay, & Jacobs, 1982).

Sonita paleosols commonly contain pinnately branching, narrow (3-4-mm) burrows, similar to those in Sarang Series paleosols thought to be the work of termites. Also common are larger (11-15 mm in diameter), near-vertical, sand-filled burrows, referable to the ichnogenus *Skolithus*, like burrows made by a variety of insects and spiders (Ratcliffe & Fagerstrom, 1980).

5.2.4.11 Paleotopographic setting. Sonita paleosols have some diffuse thin haloes of iron-manganese (mangans) around drab-haloed root traces, which could be interpreted as evidence of surface water gleization during wet season puddling of these clayey soils. However, ferrous iron/ferric iron ratios in the Sonita clay are among the lowest found in the Pakistani paleosols examined, and the paleosols have common deeply penetrating rhizoconcretions, so probably formed on well drained parts of the floodplain. Sonita Series paleosols are not so intimately associated with paleochannels as Sarang and Bhura paleosols, but like them also appear to have been part of the large alluvial fans flanking Himalayan foothills rather than the lowland axial drainage. Relict bedding, occasional cumulic horizons, and weak development of Sonita paleosols are indications that they formed on young land surfaces within reach of major floods.

5.2.4.12 Parent material. The type Sonita clay, like the underlying type Bhura clay paleosol, has a petrographic and chemical composition that indicates formation on little-weathered alluvium of the proto-Gangetic River system. Other Sonita paleosols, however, are associated with paleochannels that drained the soil-covered foothills of Himalayan ranges, and they may prove to have formed on their alluvium.

5.2.4.13 Time for formation. Sonita

paleosols are intermediate in development between Sarang and Lal Series. They are very similar to Bhura Series in features significant as indicators of time for formation, which was probably also only a few thousand years.

5.2.5 Lal Series (Retallack, 1985)

5.2.5.1 Diagnosis. Thick (1 m or so), red (5YR) clayey paleosols, with a subsurface (Bt) horizon enriched in clay above an horizon (Bk) with scattered fine nodules of carbonate.

5.2.5.2 Derivation. Lal is Urdu for red (Platts, 1960).

5.2.5.3 Description. The type Lal clay paleosol (Fig. 5.16) is in a low red knoll above a dry tributary of Kaulial Kas (grid reference 748144), 1.2-km west of Kaulial village, Potwar Plateau, Pakistan (Figs 5.6 and 5.8). This is in the lower part of the Dhok Pathan Formation (2290 m in Fig. 5.7, 11.7 m in Fig. 5.9), where it interfingers with the Nagri Formation, at a level paleomagnetically dated at about 8.5 million years old (Tauxe & Opdyke, 1982; corrected to time scale of Berggren, Kent, & Van Couvering, 1985).

+30 cm; C horizon of overlying paleosol (unnamed Lal Series); silty claystone; dark reddish-brown (5YR3/4); weathers reddish-brown (5YR5/4); common, fine (1-2-mm), brownish-yellow (10YR6/6), calcareous rhizoconcretions and nodules; few, fine (1-3-mm), reddish-brown (5YR4/4), ferric concretions; strongly calcareous; porphyroskelic insepic to inundulic in ferruginized matrix, vomasepic within drab haloes to root traces, and micritic calciasepic within nodules that may show very faint dark banding (neosesquans and quasisesquans); few grains of quartz, feldspar, mica, and opaque oxides; rare spherical micropeds of inundulic ferruginized claystone; clear, smooth contact to

0 cm; A horizon; silty claystone; reddish-brown (5YR4/4); weathers reddish-brown (5YR5/4); common clastic dikes (up to 1-cm wide), burrows (2-3-mm diameter) and large root traces (up to 6 mm in diameter), all filled with light gray (5Y7/1) siltstone, and for root traces also with a drab halo of greenish-gray (5GY7/1); common yellowish-brown (10YR5/4) and brownish-yellow (10YR6/6), fine (1-2-mm), root traces; moderately calcareous; porphyroskelic insepic to inundulic in ferruginized matrix, insepic in metagranotubules, and vomasepic in drab haloes to root traces; few surprisingly large (coarse-sand) quartz grains; few feldspar, mica, and spherical micropeds of ferruginized claystone, these last with margins of varying distinctness; diffuse, irregular contact to

-8 cm; A horizon; silty claystone; dark reddish-brown (5YR3/4); weathers reddish-brown (5YR5/4); common, silt-filled, light greenish-gray (5GY7/1) root traces, up to 3 mm in diameter, with narrow (1-mm) haloes of yellowish-brown (10YR5/6) claystone; fine subangular blocky peds, defined by slickensided clay skins (illuviation argillans); common small (5-6-mm), reddish-brown (5YR4/4), ferric concretions and few, small (3 mm), calcareous nodules; strongly calcareous; porphyroskelic inundulic in ferruginized matrix and vomasepic in root traces; mineral content and other microstructures as in above horizon; diffuse, wavy contact to

-27 cm; Bt horizon; claystone; dark reddish-brown (5YR3/4); weathers reddish-brown (5YR5/4); few root traces filled with light gray (5Y7/1) siltstone, and with a narrow (1-mm) halo of light greenish-gray (5GY7/1) claystone; common, fine (1-2-mm) root traces of dark grayish-brown (10YR3/2) with halo of yellowish-brown (10YR5/4); fine, subangular blocky peds, defined by locally slickensided clay skins (argillans); few small (4-6 mm in diameter), light brown (7.5YR6/4), calcareous nodules; few reddish-brown (5YR4/4) ferric concretions; strongly calcareous; porphyroskelic inundulic in ferruginized matrix, vomasepic in root traces; few quartz, feldspar, and mica; common spherical micropeds of ferruginized claystone; diffuse, wavy contact to

-52 cm; Bk horizon; silty claystone; dark reddish-brown (5YR3/4); weathers reddish-brown (5YR5/4); common fine (1-mm in diameter) root traces of greenish-gray (5GY6/1) claystone; indistinct peds, with scattered slickensided clay skins (argillans); common small (4-6 mm in diameter) light brown (7.5YR6/4) calcareous nodules; few reddish-brown (5YR4/4) ferric concretions; distinct relict bedding in places, and few strata-concordant, 1-2-cm thick mottles of greenish-gray (5GY6/1), with a narrow marginal rim (sesquan) of brown (10YR5/3); strongly calcareous, even away from nodules; porphyroskelic inundulic in ferruginized matrix, vomasepic in root traces and calciasepic in micritic nodules; few quartz, feldspar, and mica; rare intertextic, insepic metagranotubules after burrows; clear, wavy contact to

-68 cm; 2Bk horizon; claystone; dark reddish-brown (5YR3/4); weathers reddish-brown (5YR5/4); few, fine (1-2-mm) root traces of light greenish-gray (5GY7/1); indistinct relict beds, with scattered slickensided clay skins (argillans); few very small (1-mm) calcareous nodules and ferric concretions; few and scattered clasts of red claystone up to 2-cm across; strongly calcareous; porphyroskelic inundulic in ferruginized matrix, vomasepic in root traces, and calciasepic in micritic nodules; few grains of quartz, feldspar, and mica; diffuse, smooth contact to

-84 cm; 2Co horizon; silty claystone; dark reddish-

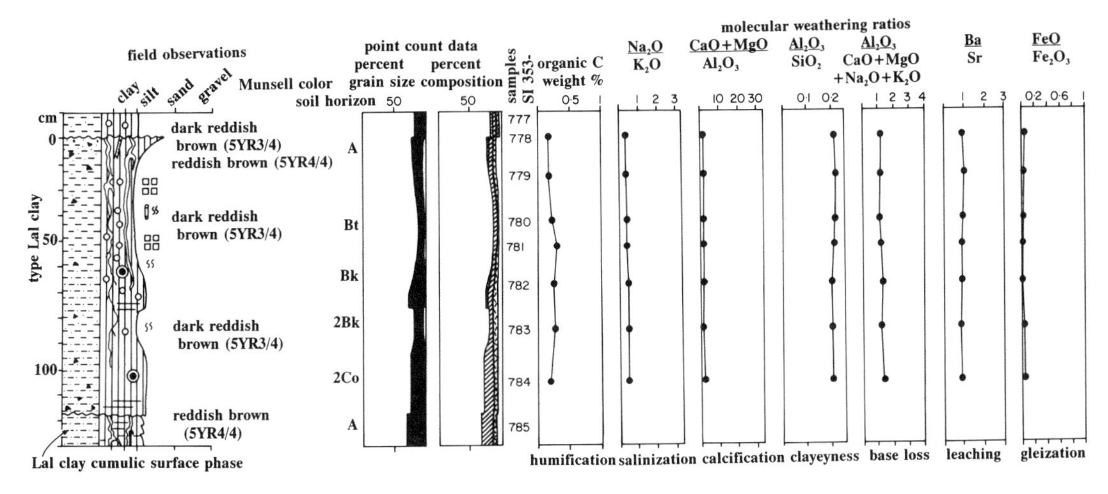

Fig. 5.16. Measured section, Munsell colors, soil horizons, grain size, mineral composition, organic content, and selected molecular weathering ratios of the Lal clay cumulic-surface phase and type Lal clay paleosols, in the lower measured section of Dhok Pathan Formation, west of Kaulial village, Pakistan. This section is at 11.6 to 13 m in Fig. 5.9. Lithological key as for Fig. 5.17.

brown (5YR3/4); weathers reddish-brown (5YR5/4); common fine (1-2-mm) root traces of olive (5Y5/3), greenish-gray (5G6/1), dark reddish-brown (5YR3/2), and pink (7.5YR7/4); indistinct relict bedding; a few very small (1 mm in diameter) ferric concretions, but no calcareous nodules were seen; this layer appears to be the silty base of an upward-fining unit below that of the main part of the paleosol; strongly calcareous; porphyroskelic inundulic in ferruginized matrix, and vomasepic in root traces; few grains of quartz, feldspar, and mica; clear, smooth contact to

-107 cm; A horizon of underlying paleosol (Lal clay cumulic surface phase); silty claystone; reddish-brown (5YR4/4); weathers reddish-brown (5YR5/4); common, fine (1-mm), dark greenish-gray (5BG4/1) root traces, with surrounding narrow (2-mm) halos of greenish-gray (5GY5/1), some of them with an additional narrow (2-mm), dark outer rim (sesquan) of dark reddish-brown (5YR3/4); distinct relict bedding of graded beds, 2-4-cm thick, with dark reddish-brown (5YR3/4) sandy bases and dark reddish-brown (5YR4/4) clayey tops; medium blocky peds, defined by slickensided clay skins (argillans); strongly calcareous; porphyroskelic inundulic in ferruginized matrix, vomasepic in root traces; few quartz, feldspar, and mica grains.

5.2.5.4 Further examples. Lal paleosols were seen only in the lower part of the Dhok Pathan Formation, and not in the main part of the formation higher than the uppermost tongue of Nagri sandstone. The type Lal clay is the best developed of 15 similar profiles seen in the lower measured section of paleosols (Fig. 5.9). The other Lal paleosols are thinner or less nodular, and some such as the Lal clay cumulic surface phase (Fig. 5.16; at 10.5 m in Fig. 5.9) had beds of silt and clay added to their surface. Some of the red paleosols described from Siwalik sediments near Jhelum, Pakistan, and already compared with Sonita Series paleosols, may prove on closer examination to be Lal paleosols.

5.2.5.5 Alteration after burial. Like associated Sarang paleosols, the type Lal clay may have been compacted to 58 percent of its former thickness, so that the depth of its calcareous horizon was more like 90 cm. This estimate does not seem so unreasonable for Lal paleosols as for Sarang and Sonita Series, because Lal paleosols are more clayey, are less calcareous, and have contorted clastic dikes that are not traceable for depths of more than 12 cm from the surface. Loss of organic matter and burial dehydration of ferric hydroxides during burial may have altered the color of these paleosols from a more brownish-red (closer to 7.5YR) original color. Burial gleization of root traces also is likely, but some of these have marginal dark stains (diffusion sesquans) that could be interpreted as evidence of surface-water gley of these clayey paleosols. The type Lal clay is dominantly smectitic, but mild illitization is evident from depth functions of potash and soda (Fig. 2.22) and illite crystallinity (Fig. 2.32, Table 3.1). No clear evidence of burial cementation was observed in thin section.

5.2.5.6 Reconstructed soil. The Lal clay can be imagined as a thick, uniformly clayey, reddish-brown soil. Its cracked and rooted surface (A) horizon graded down into a clayey subsurface (Bt) horizon with a few calcareous nodules that reached peak abundance in an horizon (Bk) below that (90 cm or so). Its surficially decalcified and smectitic composition is evidence of a moderately alkaline pH and high cation exchange capacity and base saturation. The highly oxidized nature of this paleosol (very low ferrous iron/ferric iron ratio: Fig. 5.16) and its calcareous nodules and deeply penetrating root traces are indications of good drainage. Nevertheless, there are ferric concretions and also diffusion sesquans on drab-haloed root traces that may indicate some surface water gleization, as would be expected in a soil so clayey.

5.2.5.7 Classification. These massive red clayey paleosols have the outward appearance of Oxisols, but their base status is much greater than in such soils. There is a discernable argillic horizon (in the sense of Soil Survey Staff, 1975) in the textural data, but these paleosols are so clayey overall that this is not reflected in their chemical composition, especially the alumina/silica ratios (Fig. 5.16). These profiles thus qualify as Alfisols in the U.S. taxonomy, or more precisely as Haplustalfs. Such soils are widespread on alluvial fans flanking the Himalayan foothills in the region from Gurdaspur to Chandigarh in Punjab, India (Razzaq & Herbillon, 1979; Sehgal & Sys, 1970), but also occur locally at lower elevations within the belt of Ustochrept soils similar to Sonita and Bhura Series paleosols.

In the Food and Agriculture Organization (1974) classification, Lal paleosols are best identified as Calcic Luvisols, also mapped as widespread on the Indo-Gangetic alluvial plains in the soil map of South Asia (Food and Agriculture Organization, 1977b, units Lo49-1ab, Lo49-1b, Lo49-2a).

Uniform, calcareous, clayey red soils in Australia (Stace *et al.*, 1968) include Calcareous Red Earths and Euchrozems. The latter is most likely for Lal paleosols considering their spherical micropeds and very clayey texture. In the Northcote (1974) key, Lal paleosols are Gn3.13. In the South African classification (MacVicar *et al.*, 1977), they are similar to Valsrivier marienthal. Brown and less clayey Valsrivier profiles are found on floodplains in northern and central Kruger National Park, South Africa (Botha, 1985; Gertenbach, 1983; Venter, 1986).

5.2.5.8 Paleoclimate. On the modern Indo-Gangetic Plains of Punjab, India, soils with calcareous horizons 90 cm from the surface are found in regions receiving some 500-840 mm mean annual rainfall (Sehgal, Sys, & Bhumbla, 1968), similar to estimates from comparison with soils in other parts of the world (Arkley, 1963; de Wit, 1978; Jenny, 1941). The subhumid end of this range would be compatible with the mixture of kaolinite, illite, chlorite, and smectite in the type Lal clay, even in the very fine or pedogenic fraction (Fig. 2.32), in comparison with clays of soils on similar parent materials in such climates in Pakistan (Razzaq & Herbillon, 1979) and India (Gupta, 1968).

Seasonality of paleoclimate is in evidence from the deep oxidation, burrowing, and root penetration that may reflect a dry season, versus ferric concretions and marginal sesquans to drab-haloed root traces possibly produced during wet-season puddling of these very clayey soils. There also are intergrowths of calcareous nodules, concentric zones of iron stain within calcareous nodules (see Fig. 2.11), and a diffuse scatter of calcareous nodules, as in surface soils of monsoonal regions of India (Courty & Féderoff, 1985; Sehgal & Stoops, 1972).

Spherical micropeds like fecal and oral pellets of termites, and abundant burrows like their galleries, are evidence of a warm paleoclimate. Also compatible with tropical conditions are the deep red color and thorough weathering of Lal paleosols, which are unlikely to be due also to prolonged weathering considering relict bedding within the calcareous horizon (Bk), but could in part have been inherited with parent material.

5.2.5.9 Ancient vegetation. These are the thickest and best developed paleosols with large root traces in the measured sections, and so they are likely to have supported old-growth forest communities and to be representative of the vegetation of large areas of the well-drained floodplain remote from streams. Considering the various paleoclimatic indicators of a seasonally dry tropical climate already discussed for Lal paleosols, they probably supported deciduous tropical woodland or dry forest.

Comparable North American soils (Soil Survey Staff, 1975) support deciduous forest and wooded grassland, and in Australia (Stace *et al.*, 1968) they are under open grassy woodland. Unlike associated Pila and Kala paleosols, Lal profiles show none of the fine structure or the planar lower limit of rooting found in grassland soils.

Dry woodland and forest also is more likely by comparison with similar surface soils on the alluvial plain of the Ganges River in Uttar Pradesh, India (Food and Agriculture Organization, 1977b). This is well within the region of sal (*Shorea robusta*) forests. Although Lal paleosols are among the least calcareous in the measured section, they may still have been too calcareous for sal (Puri, 1950). An edaphically reasonable analog in the modern vegetation of India is west Gangetic moist mixed deciduous forest (unit 3C/C3a of Champion & Seth, 1968). Compared with dry plains deciduous forest (envisaged here for Sonita Series paleosols), moist deciduous forests of the Indo-Gangetic Plains have many of the same tree species, but have overall greater diversity and a lusher appearance because of their vines and epiphytes. A likely ancient example of a west Gangetic moist mixed deciduous forest is the early Miocene fossil flora from Siwalik sediments at Balugoloa, near Jawalamukhi, in Himachal Pradesh, India; which includes *Dipterocarpus turbinatus*, *Smilax* sp., *Fissistigma senii*, *Berchemia balugoloensis*, *Zizyphus sivalicus*, *Dalbergia sissoo*, Papilionaceae indet., *Lagerstroemia* sp., *Mallotus* sp., and *Ficus precunea* (Awasthi, 1982; Lakhanpal & Guleria, 1986).

5.2.5.10 Former animal life. The surficial breccias of two Lal paleosols in the lower measured section within the lower Dhok Pathan Formation (at 25 and 53 m in Fig. 5.9) yielded small, poorly preserved shell fragments and opercula of ampullariid pond snails, similar to *Pila prisca* locally abundant in the Khakistari Series. Compared with associated paleosols, Lal Series are among the least calcareous, and this may explain the rarity of shell and bone in them (Retallack, 1984a).

Lal paleosols have abundant spherical micropeds associated with narrow (3-4 mm in diameter) branching systems of burrows, similar to trace fossils in the Sarang Series attributed to termites. No subterranean nests of *Termitichnus* were found. Lal paleosols do contain irregular pockets of anomalously sandy and calcareous sediment forming a local matrix to claystone breccia in the very top of the profile. These may represent the remains of termite mounds, and a deliberate search for undisturbed mounds could prove rewarding. Termite mounds commonly are more alkaline and calcareous than surrounding soil (B. Jones and Pemberton, 1987; Lee & Wood, 1970), and this could have allowed local preservation of snail opercula in the surface breccia of a weakly calcareous Lal paleosol.

5.2.5.11 Paleotopographic setting. All the Lal paleosols were found within floodplain sequences removed from deposits of paleochannels and levees. These profiles also show the greatest differentiation of a clayey subsurface (Bt) horizon of all the Pakistani paleosols studied. These are indications that they formed on old parts of the floodplain distant from streams. There are some ferric concretions and thin diffusion sesquans associated with root traces that may reflect surface-water gleization. Wet-season puddling would in any case be expected in soils as clayey as these within a flat alluvial plain. There is no evidence of groundwater-gleization, and there are some reasons to discount it considering the highly ferruginized nature of the paleosol and its deeply penetrating root traces. Although well drained for most of the year, these paleosols probably did not form on terraces very elevated above river level. Some of the profiles have cumulic surface horizons. The type Lal clay includes within its surficial clastic dikes some quartz grains larger than those found in the profile or in overlying alluvium. These grains also are larger than usual for eolian dust, and may have been introduced by an exceptionally powerful flood.

5.2.5.12 Parent material. The type Lal clay has few sand-sized grains remaining to indicate its provenance. Its chemical composition is rich in alumina and magnesia and low in silica compared to associated paleosols (Fig. 2.22); compatible with a source terrain of little-weathered limestone and schist, rather than soils, or felsic or ultramafic rocks. Surprisingly, then, its parent material may have been floodborne silty alluvium of the proto-Gangetic River system that reached well into the high Himalaya ("blue gray sandstones" of Behrensmeyer & Tauxe, 1982), rather than that of alluvial fans flanking the soil-covered foothills of Himalayan ranges ("buff sandstones"). Other Lal paleosols could have formed on alluvium of the foothills drainage, but additional chemical studies would be needed to prove this. Their red color and poverty of easily weathered grains may have more to do with Miocene soil formation than the provenance of their parent material.

5.2.5.13 Time for formation. Lal Series paleosols have moderately developed argillic (Bt) and calcic (Bk) horizons, but not quite to the extent seen in surface soils (Bhalwal Series) some 10,000 to 20,000 years old on the alluvial plains of northern Pakistan (Ahmad, Ryan, & Paeth, 1977). Among soils of the Merced River in the San Joaquin Valley of California (Harden, 1982), the type Lal clay is developed to about the extent of soils on the 10,000-year-old upper Modesto surface. Among soils of the Las Cruces area, New Mexico (Gile, Hawley, & Grossman, 1980), the degree of development of both argillic and calcic horizons is comparable to soils on the 8,000-15,000-year-old Isaack's Ranch surface, although these soils are more calcareous and less clayey overall. Each Lal paleosol probably represents a few thousands to a few tens of thousands of years of formation.

5.2.6 *Pila Series (new name)*

5.2.6.1 Diagnosis. Thick, silty, yellowish-brown (10YR to 7.5YR) paleosols, with granular surface (A) horizons, and weakly developed, only moderately calcareous, clayey (Bt) horizons.

5.2.6.2 Derivation. Pila is Urdu for yellow or pale (Platts, 1960). This weathered color distinguishes these paleosols in outcrop from brown and red associated profiles.

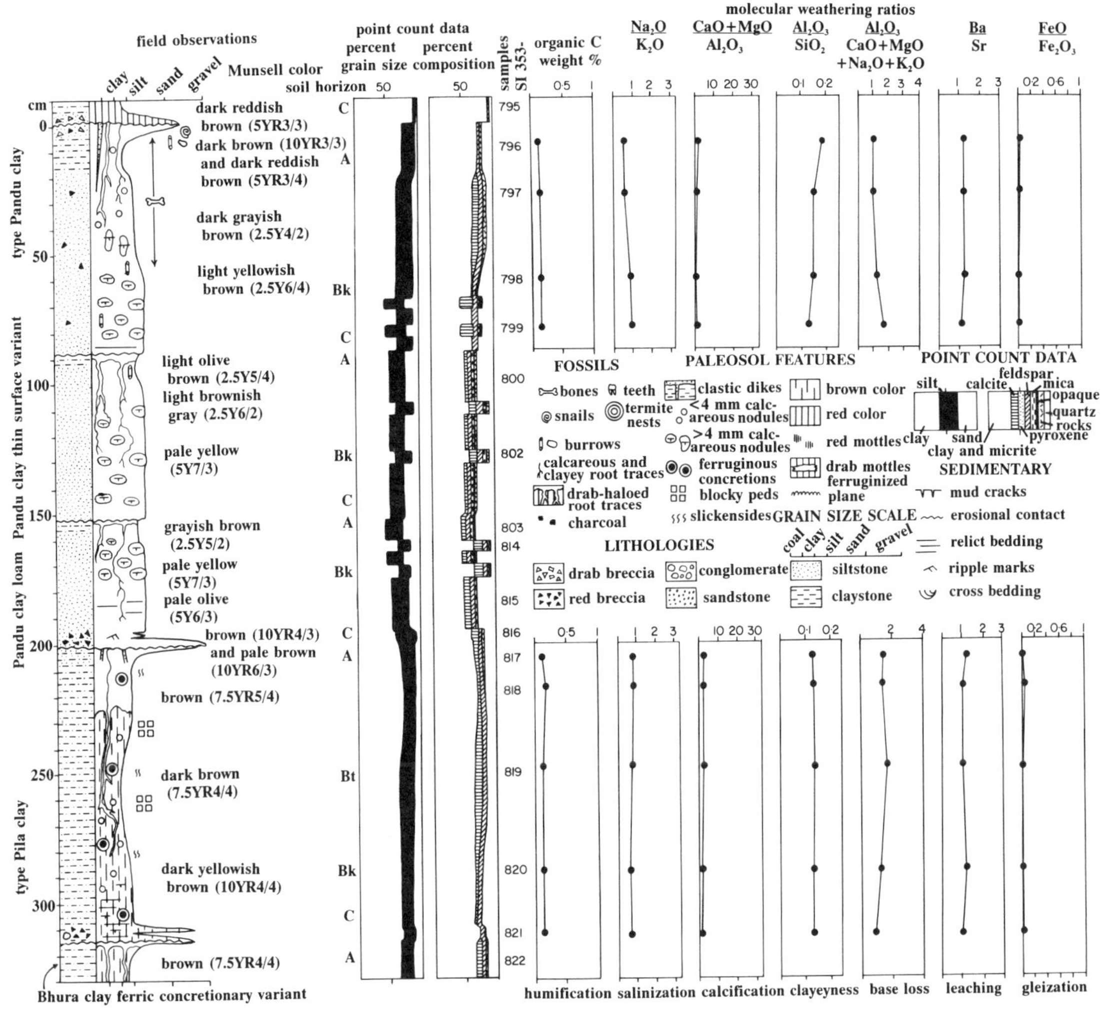

Fig. 5.17. Measured section, Munsell colors, soil horizons, grain size, mineral composition, organic content, and selected molecular weathering ratios of the type Pila clay, Pandu clay-loam, Pandu clay thin-surface phase and type Pandu clay paleosols, in the lower measured section of the Dhok Pathan Formation, west of Kaulial village, Pakistan. This section is at 45.2 to 48.8 m in Fig. 5.9.

5.2.6.3 Description. The type Pila clay (Fig. 5.17) is immediately below a prominent band of Pandu Series paleosols (at 2522 m in Fig. 5.7, and 46.9 m in Fig. 5.9) in the southern part of a small amphitheater of badlands below a promontory capped with Potwar Silt, about 200-m south of a tributary of Kaulial Kas (grid reference 751142), 1.2-km west of Kaulial village, Potwar Plateau, Pakistan (Figs 5.6 and 5.8). This is in the lower Dhok Pathan Formation, immediately below the "middle gray silts," which is a tongue of the laterally equivalent Nagri Formation (as mapped by Behrensmeyer & Tauxe, 1982). This is late Miocene in age and about 8.5 million years old (Tauxe & Opdyke, 1982; corrected to the time scale of Berggren, Kent, & Van Couvering, 1985).

+27 cm; C horizon of overlying paleosol (Pandu clay-loam); fine-grained sandstone; pale olive (5Y6/3); weathers light brownish-gray (2.5Y6/2); distinct relict bedding, including breccia beds with clasts up to 6-mm across of dark reddish-brown (5YR3/3) silty claystone; few light gray (N2) strata-concordant mottles, especially in lower 10 cm of this horizon; few medium (7 mm in diameter) burrows filled with dark reddish-brown (5YR3/3) claystone; very strongly calcareous; porphyroskelic insepic and calciasepic in thin section; few calcite crystal tubes, some with micritic crystic sheaths (neocalcans), after root traces; common quartz; few grains of feldspar, mica, and schist; clear, wavy contact to

+4 cm; C horizon of overlying paleosol (Pandu clay-loam); claystone breccia of dark reddish-brown (5YR3/4) clasts in a sandy matrix of brown (10YR4/3) and pale brown (10YR6/3); weathers light brownish-gray (2.5Y6/2) overall; clasts are up to 4-cm long, but most are tabular and 5-6-mm long; some clasts are curved as if they were once flakes between thin mudcracks; very strongly calcareous; agglomeroplasmic insepic in thin section; few sparry calcite tubes with micritic crystic sheaths (neocalcans), after root traces; common quartz; few grains of feldspar, mica, schist, and opaque oxides; gradual, irregular (where penetrating cracks below) contact to

0 cm; A horizon; silty claystone; brown (7.5YR5/4); weathers yellow (10YR7/6); common root traces, up to 4 mm in diameter, of brownish-yellow (10YR6/6) silty claystone, with drab haloes of greenish-gray (5GY5/1) and gray (5Y5/1) claystone; rare, very pale brown (10YR7/4), small (2-mm) calcareous nodules and small (4-mm) reddish-brown (5YR3/3) ferric concretions; coarse-granular peds, defined by locally red (2.5YR4/6) and slickensided clay skins (argillans); moderately calcareous; porphyroskelic skelmosepic in thin section of matrix, calciasepic in micritic nodules intergrown with opaque concretions; few metagranotubules after burrows; common quartz; few grains of feldspar, schist, mica, and opaque oxides; diffuse, wavy contact to

-26 cm; Bt horizon; silty claystone; dark brown (7.5YR4/4); weathers yellow (10YR7/6); common, copiously branched root traces up to 5 mm in diameter of greenish-gray (5G6/1) silty claystone, with drab haloes up to 12-mm wide of greenish-gray (5G6/1); few burrows up to 15 mm in diameter of greenish-gray (5G6/1) siltstone; this horizon has the greatest abundance of ferric concretions, 7-8 mm in diameter, of reddish-brown (5YR3/3) with wispy areas of surrounding iron stain (ferrans) of dark red (2.5YR3/6) and local patches with admixed manganese oxides of dark reddish-brown (2.5YR2.5/4); few small (2-3-mm) reddish-yellow (7.5YR7/6) calcareous nodules; very-coarse angular-blocky peds, defined by locally slickensided clay skins (argillans); moderately calcareous; porphyroskelic skelmosepic in thin section of matrix, agglomeroplasmic mosepic to intertextic in metagranotubules, calciasepic in nodules and opaque in concretions; few sparry calcite crystal tubes with calciasepic micritic sheaths (neocalcans) after root traces; diffuse, wavy contact to

-69 cm; Bk horizon; silty claystone; dark yellowish-brown (10YR4/4); weathers yellow (10YR7/6); fine (2-3-mm) very pale brown (10YR7/4) calcareous nodules are more common in this than in any other horizon; common greenish-gray (5G5/1), copiously branched, drab-haloed root traces; few dark reddish-brown (2.5Y2.5/4) and dark red (2.5YR3/6) ferric concretions; indistinct relict bedding; few scattered slickensided clay skins (argillans); moderately calcareous; porphyroskelic skelmosepic in thin section; few complexly intergrown calciasepic calcareous nodules and opaque ferric concretions; few calcite crystal tubes with micritic calciasepic sheaths (neocalcans) after root traces; common quartz; some feldspar, schist, and opaque oxides; diffuse, wavy contact to

-96 cm; C horizon; silty claystone; dark yellowish-brown (10YR4/4); weathers yellow (10YR7/6); prominent relict bedding, in places with strata-concordant greenish-gray (5G5/1) mottles; few, fine (1-2-mm in diameter) greenish-gray (5G5/1), drab-haloed root traces; near the base of this horizon are two relict beds of claystone breccia, including ferric concretions and calcareous nodules up to 15-mm across; moderately calcareous; porphyroskelic skelmosepic in thin section, with scattered calciasepic micritic nodules; common quartz and feldspar, few mica, schist, limestone, and opaque oxides; clear, wavy contact to

-116 cm; A horizon of underlying paleosol (Bhura clay ferric-concretionary variant); brown (7.5YR4/4); weathers yellow (10YR7/6) because of material washed down from above; common metagranotubules of light gray (5Y7/1), with haloes of greenish-gray (5GY6/1), after root traces and burrows; abundant medium-sized (6-8-mm in diameter) dark reddish-brown (2.5YR2.5/4) and dark red (2.5YR3/6) ferric concretions; few reddish-yellow (7.5YR7/6) calcareous nodules; indistinct very-coarse granular structure, defined by scattered slickensided clay skins (argillans); moderately calcareous; porphyroskelic skelmasepic in thin section, with complex aggregations of micritic calciasepic nodules within ferruginized concretionary shells, perhaps after termite nests.

5.2.6.4 Further examples. Pila paleosols are rare: only two additional profiles were seen within the lower measured section of the lower Dhok Pathan Formation. The Pila clay nodular-variant (Fig. 5.18; at 2390 m in Fig. 5.7, and 33.7 m in Fig. 5.9) has more prominent relict bedding, including very sandy subsurface beds compared with the type Pila clay. The nodular-variant paleosol also has lumpy, tubular to ellipsoidal, light brown (7.5YR6/4) calcareous nodules, up to 5 cm in diameter, with a distinctive outer rind (sesquiargillan) of red (2.5YR4/8) and yellowish-red (5YR5/6). As discussed under animal life, this carbonate may have been locally precipitated in termite nests and galleries, because the surrounding clayey matrix is only moderately calcareous (in the strict sense of Retallack,

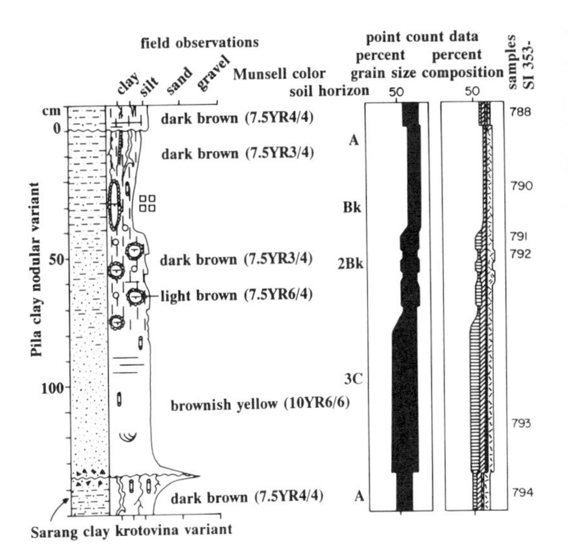

Fig. 5.18. Measured section, Munsell colors, soil horizons, grain size, and mineral composition of the Pila clay nodular variant paleosol, in the lower measured section of Dhok Pathan Formation, west of Kaulial village, Pakistan. This section is at 32.2 to 33.8 m in Fig. 5.9. Lithological key as for Fig. 5.17.

1988a). Paleosols generally similar to Pila Series have been reported from Siwalik Group sediments near Jhelum, Pakistan (profile Mk-3-1B of G.D. Johnson *et al.*, 1981; and type Yb of Visser & Johnson, 1978). Yellowish paleosols with somewhat similar profile form become much more common in Siwalik sediments younger than about 6 million years old (A. K. Behrensmeyer, personal communication, 1989).

5.2.6.5 Alteration after burial. Pila paleosols were affected by compaction of clayey portions of the profile to 58 percent of their thickness, so that calcareous horizons of the type Pila clay may have been at depths of 118 cm and of the Pila clay nodular-variant at 69 cm. These do not seen unreasonable estimates considering the impersistence of clastic dikes and the degree of contortion of copiously branched root traces in these paleosols. Also likely is a substantial loss of organic matter during burial, because organic carbon is much less abundant (Fig. 5.17) than would be expected considering their granular structure and abundant root traces. Burial reddening could also have affected these paleosols, but they were probably only a little browner (10YR) than their present color (7.5YR), because their high oxidation (low ferrous iron/ferric iron ratio) and deeply penetrating root traces indicate a well-drained original soil. Apart from local sparry calcite in root traces deep in the profile, there is little indication of cementation during burial. Nor is there bulk chemical evidence of illitization (Fig. 2.22).

5.2.6.6 Reconstructed soil. Pila paleosols can be imagined as moderately developed soils with brown, granular surface (A) over clay-enriched darker brown subsurface (Bt) horizon and below that (69-118 cm down), a culmination in abundance of calcareous nodules (Bk horizon) over clayey calcareous alluvium. They show some chemical evidence of salinization in their up-profile increase in soda/potash ratios, unlike that found in associated paleosols (Fig. 5.17). They could not have been strongly salinized considering the abundance of root traces and low values of the soda/potash ratio as well as the lack of structures such as domed-columnar peds (Northcote & Skene, 1972) found in sodic and saline soils. This together with their calcareous composition, is compatible with an alkaline pH and a high cation exchange capacity and base saturation with a substantial sodium content. Pila paleosols may also have been seasonally gleyed more profoundly than envisaged for Sonita and Lal Series paleosols but not to the extent of Khakistari paleosols. This is especially indicated by the abundance of ferric concretions and brown color of these paleosols, as well as their up-profile increase in ferrous iron/ferric iron ratios (Fig. 5.17). Some wet-season surface-water gley also is compatible with the chemical indications of mild salinization already discussed.

5.2.6.7 Classification. The combination of granular structure of a well-drained moderately decalcified soil and evidence of mild seasonal gleization and salinization is best expressed by identification of Pila paleosols as Aquic Haplustolls (of Soil Survey Staff, 1975). They are more calcareous than surface soils of Uttar Pradesh, India, identified as Aquic Hapludolls (by S. B. Deshpande, Fehrenbacher, & Beavers, 1971; S. B. Deshpande, Fehrenbacher, & Ray, 1971; Sehgal & Sys, 1970) and show much less evidence of salinization and gleization than Calcic Aeric Halaquepts of Punjab, India (Sehgal & Sys, 1970). Pila paleosols are similar to Haplustolls of the alluvial fan (bhabar) of the Gola River, in Uttar Pradesh, India (described by Hurelbrink & Fehrenbacher, 1970), and especially to soils of the Basiaram Series (Kooistra, 1982; Murthy *et al.*, 1982). These surface soils include the unusual combination of ferric concretions and carbonate nodules, and also have relict bedding deep within the profile. They are, however, much less clayey than Pila paleosols, a difference that may in part be due to overestimation of clay common when working with petrographic thin sections (Murphy & Kemp, 1984).

In the Food and Agriculture Organization (1974) classification, soils most like Pila Series paleosols are Calcaric Phaeozems. These were not mapped in India on the Food and Agriculture Organization (1977b) soil map, which does include less calcareous Haplic Phaeozems in Uttar Pradesh (map unit Je75-2a) and more salinized Orthic Solonetz in Uttar Pradesh and Punjab (map unit Lo5-2a). These FAO maps were made with the assumption that mollic soils were rare in tropical regions with high evapotranspiration and termite activity. This assumption has since proved incorrect for soils of the terai of Uttar Pradesh (S. B. Deshpande, Fehrenbacher, & Beavers,1971; S. B. Deshpande, Fehrenbacher, & Ray, 1971), and needs to be reexamined for other areas of Indo-Pakistan within the terms of the FAO classification.

Among Australian soils (Stace *et al.*, 1968), Pila paleosols are best classified with Wiesenboden and Ug6.3 of Northcote (1974). Pila paleosols are less calcareous than Solonized Brown Soils, but have more carbonate and a less developed clayey subsurface horizon than Non-calcic Brown Soils and Brown Earths.

In the South African classification (MacVicar *et al.*, 1977), Pila paleosols resemble Bonheim bonheim found, for example, in seasonally waterlogged abandoned stream beds (dambos) in areas of shale and basalt bedrock in Kruger National Park (Botha, 1985; Gertenbach, 1983; Venter, 1986).

5.2.6.8 Paleoclimate. The two Pila paleosols examined in detail are mildly salinized, are decalcified compared to their parent material and originally may have had calcareous nodular horizons at depths of 69 and 118 cm, as in modern soils in Punjab, India, receiving 440-740 and 610-980 mm mean annual rainfall, for each paleosol respectively (Sehgal & Stoops, 1972; Sehgal, Sys, & Bhumbla, 1968). Similar estimates can be gained by comparison with calcareous soils in other parts of the world (Arkley, 1963; de Wit, 1978; Jenny, 1941).

Intergrown and co-occurring calcareous nodules and ferric concretions in Pila paleosols are evidence of a strongly seasonal climate, as for other Pakistani paleosols studied, and for surface soils of Punjab, India (Courty & Féderoff, 1985; Sehgal & Stoops, 1972). Other evidence of a marked dry season includes clastic dikes in the surface of the paleosols, very deep penetration of some root traces, abundant slickensides and strong ferruginization of the exterior of calcareous nodules, including nodular termite nests.

Warm temperatures are indicated by burrows and nests attributed to termites in Pila paleosols. Geochemical evidence of mild salinization also is compatible with high evapotranspiration in a warm climate.

5.2.6.9 Ancient vegetation. The abundant fine and common stout root traces and granular peds in the surface horizon, together with features already discussed as evidence of mild seasonal gleization are most like those of soils under moist lowland, grassy woodland, or wooded grassland. Pila paleosols would have been less prone to waterlogging than the type Kala clay, which also has some features of grassland soils. Pila paleosols also are more decalcified and better structured, and they may have supported a luxuriant growth of tall grasses as well as scattered trees.

In North America, similar surface soils support prairie (Soil Survey Staff, 1977), and in Australia, a distinctive sclerophyll woodland called mallee scrub (Stace *et al.*, 1968). In Uttar Pradesh, India, generally similar soils are found under grassland and wooded grassland within abandoned watercourses and in low-lying depressions (terai) near the base of alluvial fans (bhabars) from the Himalayan foothills (S. B. Deshpande, Fehrenbacher, & Beavers, 1971; S. B. Deshpande, Fehrenbacher, & Ray, 1971; Hurelbrink & Fehrenbacher, 1970; Murthy *et al.*, 1982; Sehgal & Sys, 1970). Pila paleosols were not so salinized, dark, and clayey as soils now supporting wooded grassland of babul (*Acacia arabica*), nor so profoundly waterlogged as soils of seasonally dry swamp forests and allied vegetation envisaged for Khakistari and Pandu paleosols. The modern vegetation type most likely for soils like Pila paleosols is "moist sal savanna" (unit 3C/DSI of Champion & Seth, 1968). Trees of this modern vegetation include *Shorea robusta* (sal), *Careya arborea*, *Emblica officionalis*, and *Wrightia tomentosa*. Their grass cover is thick and tall (up to 1.8 m) with *Imperata cylindrica*, *Themeda arundina*, and *Cymbopogon nardus*.

Few of these plants have a Miocene fossil record in India. Fossil leaves of wide-leaved tall grasses have been found in early Miocene Siwalik sediments near Sudnatti in Jammu-Kashmir (B. Sahni, 1964) and near Jawalamukhi in Himachal Pradesh (A. K. Mathur, 1978), but none of these remains are preserved sufficiently well for identification to generic or even subfamily level. Also found in early Miocene Siwalik sediments is permineralized wood similar to that of *Shorea* or the allied genera *Parashorea* and *Pentacme* from near Kalagarh, Uttar Pradesh (Awasthi, 1982; Trivedi & Ahuja, 1979b) and to *Careya arborea* from near Nalagarh, Himachal Pradesh (U. Prakash, 1979).

5.2.6.10 Former animal life. No bone or shell was found in Pila paleosols. This is to be expected considering the small outcrop area of these paleosols and their less calcareous composition, compared with other kinds of paleosols reported here. Vegetation envisaged for Kala Series was generally similar to that of Pila paleosols, and these two kinds of paleosols may have shared some mammal species.

Nevertheless, Pila paleosols have a variety of burrows not found in Kala paleosols. Most of these are similar to remains attributed to termites in Sarang Series paleosols. These include narrow (3-4-mm) galleries and sand-sized spherical micropeds that may be fecal or oral pellets. Some of the lumpy, near-spherical calcareous nodules with red clay skins in the Pila clay nodular-variant are similar to degraded fungus combs of the living termite *Odontotermes* (Fletcher, 1914). Elongate near-vertical nodules may be the remains of stalks of germinated mushrooms, sometimes found emerging from these combs. Even within weakly calcareous soils, calcite can selectively precipitate in fungi (B. Jones & Pemberton, 1987) and fungus combs of termites (Lee & Wood, 1971).

Also found in Pila paleosols were near vertical, unbranched, sand-filled burrows, 10 to 15 mm in diameter, referable to the trace fossil genus *Skolithus*. These generalized kinds of burrows are made by a variety of insects and spiders (Ratcliffe & Fagerstrom, 1980).

5.2.6.11 Paleotopographic setting. The Pila

clay nodular-variant and the unnamed Pila paleosol above it are developed on sandstones with wedge-shaped geometry and cross-bedding, here interpreted as proximal crevasse splay deposits from a stream of the former foothills drainage system ("buff sandstones" of Behrensmeyer & Tauxe, 1982). The type Pila clay, on the other hand, is immediately beneath tabular sandstones interpreted as a levee deposit, which can be mapped laterally into paleo-channels of the proto-Gangetic axial drainage of the high Himalaya ("blue-gray sandstones" of Behrensmeyer & Tauxe, 1982). The increased ferrous iron/ferric iron ratio, high soda/potash ratio, overall brown color, and abundant ferric concretions of the type Pila clay can be taken as indications of a propensity for seasonal waterlogging, though this was mild in comparison with that for Khakistari and Pandu paleosols. Pila paleosols probably formed in low-lying abandoned channels, chutes within the levee, or other depressions near the transition zone between alluvial fans from the Himalayan foothills and the more consistently swampy lowlands of the proto-Gangetic River.

5.2.6.12 Parent material. Although the type Pila clay immediately underlies little weathered blue-gray sandstone, its subsurface horizons (Bk and C) show deeply weathered quartzofeldspathic alluvium probably derived from foothills drainage ("buff sandstones" of Behrensmeyer & Tauxe, 1982). Sandy subsurface horizons of the other two Pila paleosols are petrographically similar. In chemical composition also (Fig. 2.22), the type Pila clay is less like Khakistari and Lal paleosols than Kala and Pandu Series, in showing a felsic addition to the overall schistlike composition of this alluvium.

5.2.6.13 Time for formation. Neither the clayey nor calcareous subsurface horizons of Pila clay paleosols qualify as argillic or calcic horizons (in the strict sense of Soil Survey Staff, 1975), and the Pila clay nodular-variant and overlying unnamed Pila paleosol show clear evidence of relict bedding. Pila paleosols are thus developed to about the extent of surface soils 2300 years old along the Ravi River, near Lahore, Pakistan (Ahmad, Ryan, & Paeth, 1977) and up to 5000 years old along the dry bed of the Ghaggar River, India (Courty & Féderoff, 1985). Their development is thus intermediate between that for Sonita and Bhura Series paleosols on the one hand and Lal Series on the other, and represents several thousand years.

5.2.7 *Pandu Series (new name)*

5.2.7.1 Diagnosis. Thin, silty to sandy, brown to gray (10YR to 2.5Y) paleosols, with a clayey surface (A) horizon, and shallow horizon (Bk) of calcareous nodules.

5.2.7.2 Derivation. Pandu is Urdu for yellowish-white light-colored soil, or a mixture of clay and sand (Platts, 1960). These are all good descriptions of these paleosols.

5.2.7.3 Description. The type Pandu clay (Fig. 5.17) is in a badlands slope just west from a topographic bench of calcareous siltstone that includes several of these paleosols, 400-m southwest of the main watercourse of Kaulial Kas (grid reference 751142), and 1.2-km west of Kaulial village, Potwar Plateau, Pakistan (Figs. 5.6 and 5.8). This paleosol is on the top of this bench forming siltstone (2324 m in Fig. 5.7; and 48.6 m in Fig. 5.9), which is mapped (by Behrensmeyer & Tauxe, 1982) as the "middle gray silts" tongue of the Nagri Formation, here interfingering with that part of the late Miocene Dhok Pathan Formation some 8.5 million years old (Tauxe & Opdyke, 1982; corrected to the time scale of Berggren, Kent, & Van Couvering, 1985).

+15 cm; C horizon of overlying paleosol (unnamed

Sonita Series); silty claystone; dark reddish-brown (5YR3/3); weathers light reddish-brown (5YR6/4); indistinct relict bedding; few thin beds of greenish-gray (5G5/1) siltstone; few slickensided clay skins (argillans) defining indistinct, very coarse, subangular-blocky peds; common fine (less than 4 mm in diameter) greenish-gray (5G5/1) calcareous rhizoconcretions; few small (3-mm), brownish-yellow (10YR6/6), calcareous nodules; strongly calcareous, even away from nodules; porphyroskelic insepic in thin section; few calcite crystal tubes, after fine root traces; few angular clasts of ferruginized claystone; clear, irregular contact to

0 cm; A horizon; claystone breccia; dark brown (10YR3/3) in abundant clasts, with conspicuous silty claystone matrix of dark reddish-brown (5YR3/4); the clasts are 1.5 to 2 cm in size and appear to be ripped up peds of the paleosol infiltrated by brown clay from above; weathers brown (10YR5/3); common large (up to 4 cm in diameter) metagranotubules of dark greenish-gray (5GY4/1), medium-grained sandstone, after burrows and root traces; common fine root traces are yellowish-brown (10YR3/4) claystone; few slickensided clay skins (argillans) define indistinct medium-granular peds, similar to ripped up clasts higher in this horizon; few brownish-yellow (10YR6/6) calcareous nodules, up to 1 cm in diameter; moderately calcareous away from nodules; porphyroskelic skelmasepic in thin section; common quartz; few grains of mica, schist, and opaque oxides; few ferruginized spherical micropeds; few calciasepic nodules and near opaque ferruginous concretions; clear, smooth contact to

-17 cm; A horizon; siltstone; dark grayish-brown (2.5Y4/2); weathers brown (10YR5/3); common calcareous rhizoconcretions of greenish-gray (5GY5/1) and light gray (5Y7/1), with local black (5B2.5/1) mangans; few clay skins (illuviation sesquiargillans) continuing down from cracks in horizon above and defining coarse subangular-blocky peds; moderately calcareous; porphyroskelic skelmasepic in thin section of matrix, but agglomeroplasmic skelmasepic in metagranotubules at center of calciasepic micritic sheaths (diffusion neocalcans) of calcareous rhizoconcretions; few woody fragments; few calciasepic micritic calcareous nodules and strongly ferruginized spherical micropeds; gradual, smooth contact to

-54 cm; Bk horizon; sandy siltstone; light yellowish-brown (2.5Y6/4); weathers pale yellow (2.5Y7/4) and light gray (5Y7/1); common ellipsoidal to bulbously irregular calcareous nodules, up to 3 cm in diameter; few brownish-yellow (10YR6/6) clay-filled, fine root traces; few near-vertical, 6 to 7 mm in diameter, dark reddish-brown (5YR3/4) and gray (5Y5/1) metagranotubules, after burrows; indistinct relict bedding; strongly calcareous in matrix as well as nodules; agglomeroplasmic mosepic with abundant calciasepic micritic carbonate in nodules, cement, and rock fragments; few ferric concretions and woody fragments; clear, wavy contact to

-89 cm; A horizon of underlying paleosol (Pandu clay thin-surface phase); clayey siltstone; light olive-brown (2.5Y5/4); weathers light yellowish-brown (2.5Y6/4); common light gray (5Y6/2) metagranotubules and drab-haloed root traces; few dark reddish-brown (5YR3/4) clay skins (illuviation sesquiargillans); few root traces of yellowish-brown (10YR5/4) claystone; moderately calcareous; granular skelmosepic in thin section of matrix, but porphyroskelic and agglomeroplasmic insepic in metagranotubules after burrows; few angular clasts of strongly ferruginized claystone; common quartz and feldspar; few grains of schist, mica, limestone, and opaque oxides.

5.2.7.4 Further examples. Only three Pandu paleosols were seen, all in the "middle gray silts" tongue of the Nagri Formation in the lower Dhok Pathan Formation of the lower measured section (Fig. 5.9). The type Pandu clay is the uppermost profile, above the Pandu clay thin-surface phase paleosol (Fig. 5.17; at 47.6 m in Fig. 5.9), which in turn is above the Pandu clay-loam (at 47.2 m). Other Pandu paleosols also were seen in the section excavated for paleomagnetic samples (and reported by Tauxe & Badgley, 1988), in nearby Ganda Kas, again in the "middle gray silts" (of Behrensmeyer & Tauxe, 1982). Broadly similar paleosols also have been reported from Miocene Siwalik sediments near Jhelum, Pakistan (type Sp of Visser & Johnson, 1978), and may be widespread in the Siwalik Group of Nepal (Munthe *et al.*, 1983).

5.2.7.5 Alteration after burial. Compaction of clayey parts of the type Pandu clay to 58 percent of their thickness, and of sandy parts to 70 percent, as outlined for Sarang paleosols, would have meant that its calcareous nodules were at a depth of 93 cm, rather than their present 54 cm. This compaction left thick calcareous rhizoconcretions and burrows in silty subsurface horizons relatively undeformed (Fig. 2.1), but destroyed the continuity of clastic dikes in the clayey surface horizon. Organic carbon values are low and decline up-section (Fig. 5.17). Much of this was probably lost after burial, but not before encouraging the gray color of these paleosols due to burial gleization. Cementation with carbonate during burial is possible for some of the abundant carbonate low in these profiles, but the nodules and rhizoconcretions have micritic, replacive, and other pedogenic microtextures. Mild illitization may also be indicated by chemical trends in alkali oxides (Fig. 2.22).

5.2.7.6 Reconstructed soil. The type Pandu clay can be envisaged as a moderately developed soil, with a brown clayey surface (A) horizon over a gray silty subsurface (Bt) horizon, studded with calcareous nodules and rhizoconcretions. These calcareous soils were probably alkaline in pH, with high cation exchange capacity and base saturation. There is no geochemical or structural evidence for salinization. The drab color of the profile, moderate values of the ferrous iron/ferric iron ratio (Fig. 5.17), preserved wood fragments, and local films of iron-manganese (mangans) are evidence of waterlogging, as well as burial gleization. Nevertheless, the profile was probably dry for most of the year, to allow cracking, root and burrow penetration, and formation of calcareous rhizoconcretions and nodules.

5.2.7.7 Classification. Pandu paleosols have calcareous nodules like Aridisols, and these can appear prominent in modern outcrop (Fig. 2.10). In fresh excavations, however, the nodules are not distinct from calcareous matrix, and Pandu paleosols have abundant root traces, woody fragments, and signs of gleization due to waterlogging unusual in Aridisols. Pandu paleosols are best identified as Typic Haplaquepts (of Soil Survey Staff, 1975) or Calcaric Gleysols (of Food and Agriculture Organization, 1974). Pandu Series are silty to sandy textured paleosols, in some respects similar to Khakistari Series, with which I have seen them associated in Ganda Kas within the "gray silts" facies of the Khaur area (Behrensmeyer & Tauxe, 1982). Such textural variety is also found in the surface soils of the modern alluvial lowlands (dhankar) of the Indo-Gangetic Plains. Itwa Series soils of Uttar Pradesh, India (Kooistra, 1982; Murthy *et al.*, 1982), show overall similarity to Pandu paleosols, although these surface soils are less clayey and have a more pronounced clayey subsurface (Bt) horizon.

Among Australian soils (Stace *et al.*, 1968), Pandu paleosols, like Khakistari Series, are best included within the heterogeneous group of Grey Clays. They are better developed than Calcareous Sands, but less developed than Grey Brown Calcareous Soils, which usually are on limestone bedrock. In the Northcote (1974) key, Pandu paleosols are Ug6.5, and in the South African binomial system (MacVicar *et al.*, 1977), Katspruit killarney.

5.2.7.8 Paleoclimate. The calcareous nodules of the type Pandu clay may have been as much as 93-cm deep, as in soils of Punjab receiving 510-850 mm mean annual rainfall (Sehgal *et al.*, 1968). Similar semiarid to subhumid climate also is indicated by comparison with calcareous soils elsewhere in the world (Arkley, 1963; de Wit, 1978; Jenny, 1941). An even drier paleoclimate is indicated by the depth to nodules in the other two Pandu paleosols, but their surface horizon has been either poorly preserved or eroded away, so that they cannot be trusted as reliable guides to former rainfall.

Seasonality of climate is in evidence from the copious surficial cracking of the paleosol and banding within calcareous rhizoconcretions and ferric concretions. A warm climate would be compatible with the diversity of burrows, including traces attributed to termites.

5.2.7.9 Ancient vegetation. The large drab-haloed root traces of Pandu paleosols, and the abundant woody fragments seen in thin section, are evidence of woodland or forest vegetation. This was a forest capable of withstanding seasonal waterlogging, but not so wet as swamp forests with tabular root systems in peaty surface horizons. It also endured marked dry seasons, as already discussed, so may have been deciduous and open. It was not, however, so open that it supported much grass, which would have created a granular and organic surface horizon, more like that seen in Kala and Pila Series paleosols.

In North America, soils of this kind support mainly meadows, although some were under forest before clearance (Soil Survey Staff, 1975). In Australia, superficially similar soils are mainly under wooded grassland and grassy woodland (Stace *et al.*, 1968). Generally similar soils in India and Bangladesh, already discussed for the Khakistari Series, are found within the alluvial lowlands of the Ganges and Brahmaputra Rivers all the way from Uttar Pradesh out into the mangrove swamps of the delta margin in the Bay of Bengal (Food and Agriculture Organization, 1977b). The most similar surface soils are on sandy, lowland alluvium of Uttar Pradesh and support western light plains alluvial sal (unit 3C/C2dii of Champion & Seth, 1968). This is an open forest with some grassy ground cover. Its main trees are *Shorea robusta* (sal), *Terminalia bellerica*, *Lagerstroemia parviflora*, and *Mallotus philippensis*. All of these genera of dicots were present in India during Miocene time. Permineralized wood similar to that of *Terminalia mannii* and to either *Shorea*, *Parashorea*, or *Pentacme* has been found in lower Siwalik sediments near Kalagarh, Uttar Pradesh (Awasthi, 1982; Trivedi & Ahuja, 1979b). Fossil leaves referred to *Lagerstroemia* sp. and *Mallotus* spp. have been reported from lower Siwalik sediments near Jawalamuhki, Himachal Pradesh (Lakhanpal & Dayal, 1966; A. K. Mathur, 1978).

5.2.7.10 Former animal life. Fossils from the band of three Pandu paleosols studied in detail here (Fig. 5.17) are most easily collected from the dipslope bench west of the measured section for several hundred meters (grid reference 751143). The most common fossils were molars of three-toed horse (Smithsonian specimen F353874), all greater than 25-mm wide, and thus probably "*Cormohipparion*" *theobaldi* or "*C.*" *perimense* (Bernor & Hussain, 1985). Also found was a bone (left scaphoid; specimen F353875) of a ruminant about the size of a domestic sheep or goat. Of the fossil antelope known from this stratigraphic level (Badgley, 1982), this is likely to have been *Selenoportax vexillarius*. Much tortoise scrap also was found (specimens F343877 to F343880).

In the same bench exposures, a variety of trace fossils can be collected, in many cases well preserved by calcite cementation. Most distinctive are burrows 12 to 14 mm in diameter, partly filled with ellipsoidal pellets 2 to 4 mm in size (Fig. 2.2, far right). These are most like the ichnogenus *Edaphichnium* (Bown & Kraus, 1983; Freytet & Plaziat, 1982), which, unlike the somewhat similar *Ophiomorpha, Oligichnus* and *Granularia*, is not observed to branch. There are some narrow side branches on the Pakistani traces, but petrographic study revealed these to be lateral rootlets of drab-haloed root traces invading the burrows. *Edaphichnium* is thought to be the burrow or entrance tube of large tropical earthworms (Bown & Kraus, 1983). In India, as in other countries, earthworm casts are commonly more calcareous than surrounding soil (Joshi & Kelkar, 1952), and this may explain in part the excellent preservation of these structures in Pandu paleosols.

Also found on the bench of Pandu paleosols were large (4 cm in diameter), sand-filled, inclined burrows, ending in a bulbous terminal chamber, as in the ichnogenus *Macanopsis* (Häntzschel, 1975). Such burrows may be created by crabs, including freshwater crabs such as *Potamon emphysetum* found fossilized in Siwalik sediments of India (G. D. Johnson, 1977). However, the fossil burrows in Pandu paleosols extend only to depths of 17 cm, which would not have been deep enough to reach the water table during dry season. It is more likely that they were excavated by large beetles, spiders, or small mammals, known to produce similar burrows (Ratcliffe & Fagerstrom, 1980; Voorhies, 1975).

Common narrow burrows and large concentrically banded sandstone spheres of *Termitichnus* (Smithsonian specimen F353881), similar to galleries and fungus gardens of termites, are identical in Pandu paleosols to those already discussed for Sarang series.

5.2.7.11 Paleotopographic setting. Pandu paleosols formed in a sequence of graded beds, 50-90-cm thick, fining upwards from planar-bedded to cross-bedded siltstone to silty claystone. These "gray silts" have been mapped as a thin wedge westward into paleochannels of the Nagri Formation (by Behrensmeyer & Tauxe, 1982). They are here interpreted as a crevasse splay deposit of the distal levee of the proto-Gangetic River system ("blue gray sandstones" of Behrensmeyer & Tauxe, 1982). A lowland setting also is compatible with the varied lines of evidence for seasonal waterlogging already discussed, but in each of these features Pandu paleosols were not as severely affected by waterlogging as otherwise similar Khakistari paleosols. Pandu paleosols probably formed on elevated, sandy, and silty parts of levee systems that were well drained for a good part of the year. Khakistari paleosols, on the other hand, formed on clayey fill of abandoned channels, chutes, and floodplain lakes, further removed from active channels. Such low lying clayey and slightly more elevated sandy soils are now seen in the modern channel belt of the Indus River (Holmes & Western, 1969).

5.2.7.12 Parent material. Pandu paleosols formed in siltstones and sandstones with quartz, feldspar, chert, muscovite, and opaque oxides, but few weathered clasts of schist and limestone. Prior weathering of the parent material of the type Pandu clay is indicated by its chemical composition, particularly high silica, low alumina, and moderate magnesia content, which allies it with Pila and Kala Series rather than Lal and Khakistari paleosols. Such alluvium was typical for the drainage of the foothill streams to the Himalayan ranges ("buff sandstones" of Behrensmeyer & Tauxe, 1982). This is surprising because the gray silts containing Pandu paleosols can be mapped laterally into little-weathered blue gray sandstones of the proto-Gangetic River system that reached into the high Himalaya. The distal crevasse splay envisaged as a parent deposit for the Pandu paleosols may have included a substantial portion of redeposited levee sediment from streams of the foothills drainage. The gray color of these paleosols, like the converse situation of the red but chemically schistlike Lal clay, is an indication that color may reflect soil and burial history, rather than the provenance of parent material.

5.2.7.13 Time for formation. Pandu paleosols have moderately developed calcareous horizons (stage II of Gile, Peterson, & Grossman, 1966), like the Bhalwal Series soils, some 10,000 to 20,000 years old near Lahore, Pakistan (Ahmad, Ryan, & Paeth, 1977). In the well known chronosequence of calcareous soils near Las Cruces, New Mexico, U.S.A., such development is seen after 25,000 to 75,000 years (Gile, Hawley, & Grossman, 1980). These estimates of a few tens of thousands of years are the longest for Pakistani paleosols of the two measured sections in the Dhok Pathan Formation, but are comparable with times estimated for undescribed paleosols reported here from the Chinji and Kamlial Formations near Khaur, Pakistan, and from the Tut, Chogo, and Onuria paleosols of Kenya.

5.2.8 Kala Series (new name)

5.2.8.1 Diagnosis. Thin silty claystone paleosols, with brown (10YR), granular-structured surface (A) horizons, with prominent black specks after charcoal and iron-manganese nodules (Bg), over a nodular calcareous subsurface (Bk) horizon.

5.2.8.2 Derivation. Kala is Urdu for black, dark color, soot, and smoke (Platts, 1960), in reference to the black speckled appearance of surface horizons of these paleosols.

5.2.8.3 Description. The type Kala clay paleosol (Fig. 5.19) is in the badlands slope 20-cm below the prominent blue-gray sandstone (at 56.9 m in Fig. 5.9, and 2332 m in Fig. 5.7) capping the red beds about 600-m southwest of Kaulial Kas (grid reference 751142), and 1.2-km west of Kaulial village, near Khaur, Potwar Plateau, Pakistan (Figs 5.6 and 5.8). This is in the late Miocene lower Dhok Pathan Formation below a sandstone tongue of the laterally equivalent Nagri Formation, and is here paleomagnetically dated at about 8.5 million years old (Tauxe & Opdyke, 1982; corrected to time scale of Berggren, Kent, & Van Couvering, 1985).

+7 cm; C horizon of overlying paleosol (Sarang clay eroded-phase); siltstone; brown (10YR4/3); weathers strong-brown (7.5YR5/6); conspicuous laminae of medium-grained sandstone are greenish-gray (5GY6/1) and light greenish-gray (5GY7/1); few linguoid ripple marks are defined by impersistent flaser beds of dark reddish-gray (5YR4/2) claystone; few medium (13 mm in diameter) burrows filled with dark-brown (7.5YR4/4) claystone from above; few fine, light yellowish-brown (10YR6/4) root traces; moderately calcareous; granular silasepic in thin section; common quartz; few schist, limestone, feldspar, mica, and opaque oxides; abrupt, wavy boundary to

0 cm; A horizon; claystone; dark brown (10YR3/3); weathers yellowish-brown (10YR5/4); common fine (1-2-mm), dark greenish-gray (5G4/1) and yellowish-brown (10YR5/4) root traces; indistinct granular and platy peds defined by slickensides and patchy black (5B2.5/1) films of iron-manganese (mangans); few, dark brown (10YR3/3) ferric concretions, up to 10 mm in diameter; moderately calcareous; porphyroskelic clinobimasepic in thin section; few ortho-isotubules with iron-stained rims (diffusion ferrans), after fine roots; common quartz and strongly ferruginized clay clasts; few clasts of schist, limestone, mica, and feldspar; gradual, smooth contact to

-10 cm; A horizon; silty claystone; dark brown (7.5YR3/4); weathers yellowish-brown (10YR5/4); common greenish-gray (5GY5/1) and olive (5Y5/3), fine (1-2-mm), root traces; indistinct granular and platy peds, defined by reddish-brown (2.5YR4/4) slickensided clay skins (sesquiargillans); common black (5BG3/1) specks (3-4-mm) of charcoal; common dark brown (7.5YR3/4) ferric concretions, as in overlying horizon; one distinct lamina of greenish-gray (5GY6/1) siltstone, 3-mm thick, was traced as a relict bed, uninterrupted for 10-cm laterally; few, narrow (2-mm) clastic dikes, filled with sandy material from above; moderately calcareous; porphyroskelic bimasepic in thin section; few agglomeroplasmic insepic metagranotubules, after root traces; few calciasepic nodules; few ferruginized claystone clasts; common quartz; few grains of schist and opaque oxides; clear, smooth contact to

-21 cm; Bg horizon; siltstone; very dark grayish-brown (10YR3/2); weathers brown (10YR5/3); common fine (1-2-mm), yellowish-brown (10YR5/6) root traces, some of which show a distinct yellowish-brown (10YR4/4) central fill and a narrow (2-mm) halo of greenish-gray (5G5/1) claystone, as well as an outer dark rim (diffusion mangan) of very dark brown (10YR2/2); indistinct platy peds defined by dark brown (10YR2/2) and brownish-yellow (10YR6/6) clay skins (argillans); common small (2-mm) brownish-yellow (10YR6/6) calcareous nodules, small (3-4-mm), angular, near-cubic, black charcoal fragments, and small (2-5-mm), irregular, black (5B2.5/1) iron-manganese nodules; strongly calcareous; agglomeroplasmic mosepic in thin section of matrix, but porphyroskelic calciasepic in calcareous nodules and opaque iron-manganese nodules; common charcoal and mildly coalified, elongate wood fragments; common quartz; few grains of schist, feldspar, mica, and opaque oxides; clear, smooth contact to

-30 cm; Bg horizon; silty fine-grained sandstone; brownish-yellow (10YR6/6); weathers yellowish-brown (10YR5/4); few very fine (less than 1 mm in diameter), strong-brown (7.5YR5/8) root traces; few strata-concordant, irregular, light greenish gray (5GY7/1) mottles, 6-9-mm thick; common small (2-mm), brownish-yellow (10YR6/6) calcareous nodules; strongly calcareous, even away from nodules; intertextic insepic in thin section of matrix, but calciasepic in pervasive cement and in nodules; common quartz and charcoal; few grains of schist, muscovite, and opaque oxides; few calcite crystal tubes after fine root traces; gradual, smooth contact to

-70 cm; Bk horizon; silty fine-grained sandstone; brownish-yellow (10YR6/6); weathers yellowish-brown (10YR5/4); common reddish-yellow (7.5YR6/6) calcareous nodules, up to 3 cm in diameter, arranged into three separate diffuse horizons; distinct relict bedding; very strongly calcareous, even away from nodules; intertextic crystic in thin section; common quartz and feldspar, few schist, muscovite, and opaque oxides; few calcite crystal tubes with micritic sheaths (neocalcans), after root traces; clear, smooth contact to

-91 cm; A horizon of underlying paleosol (Sarang clay-loam); dark yellowish-brown (10YR4/4); weathers brown (10YR5/3); common small (3-4 mm in diameter), greenish-gray (5GY6/1), drab-haloed root traces and burrows; prominent relict beds of light yellowish brown (10YR6/4), dark yellowish-brown (10YR4/4) laminae, 0.5-1-mm thick, and strata-concordant greenish-gray (5GY6/1) mottles, 3-4-mm thick; very strongly calcareous; intertextic insepic in thin section; common quartz and woody debris; some schist, microcline, muscovite, and opaque grains.

5.2.8.4 Further examples. Only the type Kala clay was found in the measured sect-

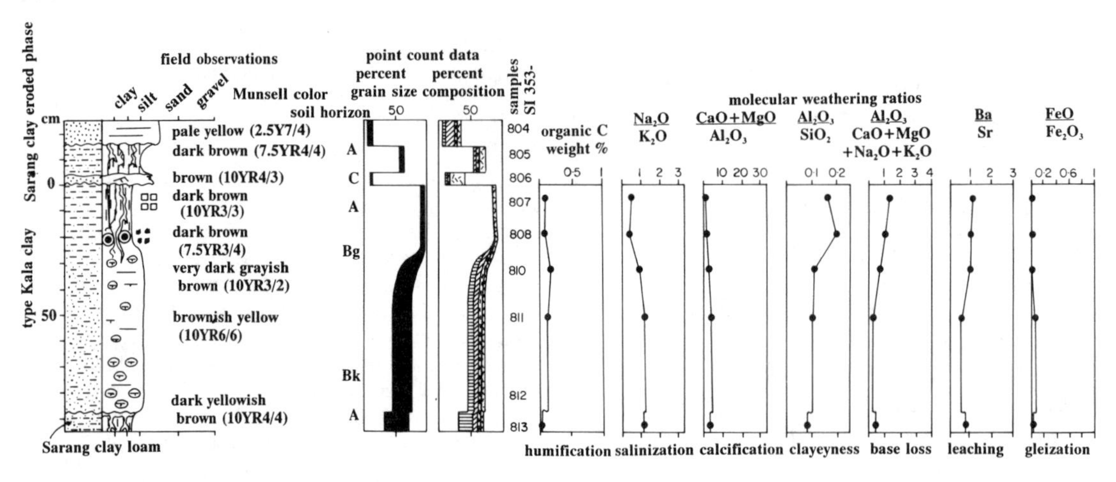

Fig. 5.19. Measured section, Munsell colors, soil horizons, grain size, mineral composition, organic content, and selected molecular weathering ratios of the Sarang clay-loam, type Kala clay and Sarang clay eroded-phase paleosols, in the lower measured section of the Dhok Pathan Formation, west of Kaulial village, Pakistan. This section is at 56.4 to 57.6 m in Fig. 5.9. Lithological key as for Fig. 5.17.

ions, but another unnamed paleosol of this kind was seen lower within the late Miocene lower Dhok Pathan Formation, at a fossil site (Yale-G.S.P. 262, excavation level 2 of Badgley, 1982; at grid reference 741144, and 2606 m in Fig. 5.7), in a tributary of the main watercourse of Kaulial Kas, 2.1 km west of Kaulial village, Pakistan (Figs 5.6 and 5.8). This other Kala paleosol has a prominent subsurface (Bg) horizon darkened with charcoal and iron-manganese nodules, but is less silty that the type Kala clay and has a less distinct transition downward into silty parent material.

5.2.8.5 Alteration after burial. Cementation soon after burial of the silty lower part of the type Kala clay may have resisted severe compaction during deep burial, which was probably not quite to 70 percent of original thickness estimated for sandy beds in this part of the lower Dhok Pathan Formation. The upper clayey part of the profile, in contrast, could have been compacted to 58 percent, and the top of the calcareous nodular (Bk) horizon would have been at about 108 cm. Loss of organic matter early during burial is quite likely, especially considering the up-profile decline of organic carbon (Fig. 5.19). Burial gleization of drab-haloed root traces and reddening of mottles may have occurred. However, the dark rims of iron-manganese around some drab-haloed root traces are evidence of open-system alteration with translocation of metals, as in surface water gleization, rather than the closed-system reduction of iron-bearing minerals typical of burial environments (Retallack, 1983, 1990a). Bright mottles also are typical of waterlogged soils, and in this paleosol are not extremely red (only 7.5YR). Thus the effects of burial gley and reddening were minor and added to original features of soils that probably were seasonally waterlogged. Mild illitization is indicated by depth functions of alkali oxides (Fig. 2.22).

5.2.8.6 Reconstructed soil. The type Kala clay can be imagined as a moderately developed soil with a thin granular to platy structured surface (A) horizon and abundant fine root traces, over a mottled subsurface (Bg) horizon rich in iron-manganese nodules and charcoal, and an horizon (Bk) of calcareous nodules about 108-cm below the surface. The surface horizon may well have been brown with organic matter since lost, but the subsurface (Bg) horizon was probably nearly as colorfully mottled in black, gray, and orange as the paleosol is now, because of seasonal waterlogging. The profile has been decalcified at the surface, where pH may have been near neutral, but further down the profile it was alkaline, with high cation exchange capacity and base saturation. There may have been some heavy metal toxicity at this level, to judge from the abundant iron-manganese nodules. This, together with seasonally anaerobic conditions, may have suppressed microbial decay of charcoal and woody debris at that level. Soda/potash ratios are moderately high at depth (Fig. 5.19), but decline toward the surface, so that salinization was not a problem.

5.2.8.7 Classification. The type Kala clay has some of the appearance of Mollisols, but fails to qualify because of its coarse structure and shallow surface horizon. The bright mottles, iron-manganese nodules, and relict bedding of this paleosol are more like those of Aquepts (of Soil Survey Staff, 1975). Many of these soils in India today are sodic to saline (Sehgal & Sys, 1970), but the type Kala clay has neither the high soda/potash ratios nor domed columnar peds of such soils. Considering its brown color and degree of oxidation, decalcification of its surface, and evidence for strong seasonality of paleoclimate in this and associated paleosols, the best identification of this profile in the U.S. taxonomy is Aquic Eutrochrept.

Within the Food and Agriculture Organization (1974) classification, the type Kala clay has some features of waterlogging, similar to Gleysols like the Pandu and Khakistari Series. However, these mottles and iron-manganese nodules must be weighed against deep penetration of root traces and chemical indications of oxidation (ferrous iron/ferric iron ratio), leaching (barium/strontium and alumina/bases ratios) and decalcification (alkaline earths/alumina ratios). This paleosol is thus best identified with Gleyic Cambisols, which are widely associated with Eutric Cambisols in the alluvial plain of tributaries of the Indus River in northern Pakistan (map units Be70-2/3a and Be71-2/3a of Food and Agriculture Organization, 1977b) and with Calcaric Gleysols and Eutric Fluvisols in the lower reaches of the Ganges and Damodar Rivers in India and Bangladesh (map units Gc9-3a and Je71-2a).

Comparable surface soils in Australia (Stace *et al.*. 1968) are Wiesenboden. These are usually gray to black, but some have brown to red subsoils like the Kala clay. In the Northcote (1974) key it is Ug6.3.

In the South African system (MacVicar *et al.*, 1977), the Kala clay is most like Willowbrook chinyika, which is found in abandoned channels with grassy vegetation (dambos) in areas of basaltic bedrock in northern Kruger National Park (Botha, 1985; Gertenbach, 1983; Venter, 1986).

5.2.8.8 Paleoclimate. Surface soils of Punjab, India, with calcareous nodular horizons at depths of 108 cm are found in regions receiving about 590-820 mm mean annual rainfall (Sehgal, Sys, & Bhumbla, 1968). Similar estimates can be gained by comparison with soils in other parts of the world (Arkley, 1963; de Wit, 1978; Jenny, 1941).

Evidence for seasonality of rainfall in this paleosol includes the co-occurrence of ferric concretions and iron-manganese nodules that are indications of waterlogging, along with deeply penetrating root traces and calcareous nodules from a dry season. Woody fragments and charcoal are more conspicuous in this than in other paleosols studied, and may indicate common fires during the dry season.

No compelling evidence of paleotemperature was seen, but warm climate is compatible with the degree of bioturbation and weathering of the surface horizon.

5.2.8.9 Ancient vegetation. The type Kala clay has abundant fine root traces in the surface, but also has occasional stout root traces and much woody debris and charcoal. As already discussed, it probably was waterlogged for a part of the year, but also dried out to allow deep root penetration. It probably supported vegetation dominated by grasses or sedges, and scattered trees.

Comparable surface soils in North America are largely under meadow, but some are forested (Soil Survey Staff, 1975). In Australia such soils support grassland and wooded grassland in areas of footslopes where seepage concentrates (Stace *et al.*, 1968). In South Africa such soils support grasslands of seasonally wet watercourses, or dambos (Botha, 1985; Gertenbach, 1983), although in Africa generally, dambos more often have thicker, darker, and more clayey soils (Vertisols of Food and Agriculture Organization, 1977a; M. M. Cole, 1986). Comparable soils in India and Pakistan (Food and Agriculture Organization, 1977b) are of limited local extent in dry stream beds and other depressions, but are widely scattered through the Indo-Gangetic Plains. They are flooded during the monsoon, but water table retreats to depths of 2 to 4 m during the dry season.

In India such soils commonly support low alluvial wooded grassland (unit 3/ISI of Champion & Seth, 1968), with the following trees: *Salmalia malabarica*, *Adina cordifolia*, and *Albizzia procera*. Grasses are dense and tall, sometimes 4-5-m high, and dominated by *Saccharum procerum* and *Phragmites maximum*. Fossil grasses of this broad-leaved vegetative form have been found in early Miocene Siwalik rocks at several localities in India, as already discussed for Pila Series paleosols. Permineralized wood similar to that of living *Albizzia lebbek* has been found in early Miocene lower Siwalik sediments near Nalagarh, Himachal Pradesh, India (U. Prakash, 1975).

5.2.8.10 Former animal life. No burrows were seen in the type Kala clay, possibly because it was seasonally waterlogged. No bones were seen in it either, but fossils have been collected from another unnamed Kala paleosol lower in the late Miocene Dhok Pathan Formation (Yale-G.S.P. locality 262, excavation level 2 of Badgley, 1982; at grid reference 743145 and 2606 m in Fig. 5.7). From there in the badlands speckled with iron-manganese and charcoal I have seen abundant crocodile scutes (*Crocodylus palaeindicus*) and foot bones of a large three-toed horse ("*Cormohipparion*" *theobaldi* or "*C*" *perimense*). From the excavation, Badgley (1982) recovered the following fossils: big cat (cf. *Felis*, tooth), proboscidean (tusk), gomphothere (indeterminate tooth), three-toed horse (tooth), rhinoceros (metapodial), pig (5 teeth, mandible, metapodial), chevrotain (patella), gazelle (*Gazella*, horncore, podial, phalange), and 28 indeterminate bone fragments. She has interpreted this assemblage as a local accumulation formed by predators and scavengers, not appreciably transported by streams.

5.2.8.11 Paleotopographic setting. The type Kala clay is not far below the base of a large paleochannel of blue gray sandstone of the former proto-Gangetic drainage of the high Himalaya. Its chemical depth function is erratic and its composition is intermediate between these little-weathered sediments and the more deeply weathered alluvium of the foothills drainage ("buff sandstones" of Behrensmeyer & Tauxe, 1982). Considering the various lines of evidence for slight gleization already discussed, it probably formed in a local, seasonally inundated, abandoned channel, chute, or lake in the region where the main lowland drainage was onlapped by alluvial fans of streams draining the foothills.

The other Kala paleosol lower in the Dhok Pathan Formation (at locality 262) is associated with a buff sandstone paleochannel (Behrensmeyer & Tauxe, 1982) and may have formed in a local depression near a former stream draining soil-covered Himalayan foothills.

5.2.8.12 Parent material. Both Kala paleosols formed on silty alluvium of quartz, feldspar, schist, and limestone. Only the type Kala clay was evaluated by laboratory studies, and it is chemically intermediate between buff sandstones of foothills drainage likely as parent materials for Pila Series, and blue-gray sandstones with little weathered schist derived from the high Himalaya, which are likely parent materials for Khakistari and Lal Series paleosols. There is a marked difference in chemical composition between its clayey surface and silty subsurface horizons, especially in titania, zirconium, gold, and chromium (Fig. 2.22), which usually are stable in soils. The silty subsurface is more felsic in composition, like alluvium of the foothills drainage, whereas the clayey surface is chemically more schistlike as is alluvium delivered by major streams from the high Himalaya.

5.2.8.13 Time for formation. The type Kala clay has a profile form similar in features significant as indicators of time for formation to Pandu paleosols. Relict bedding is a little more prominent and the calcareous nodules less pronounced, so that development of the type Kala clay can be considered just short of moderate (in the strict sense of Retallack, 1988a). This paleosol is much better developed than Sonita and Bhura paleosols, which show peds and calcareous nodule development similar to the other Kala paleosol lower in the Dhok Pathan Formation (at locality 262). Considering the various surface soils of known age already compared with these other Miocene paleosols, Kala paleosols probably represent soil formation for the order of 10,000 years.

5.2.9 Naranji Series (new name)

5.2.9.1 Diagnosis. Thick, sandy, red (5YR-2.5YR) paleosols, with clayey surface (A) horizon and a clay-enriched sandy subsurface (Bt) horizon, only moderately calcareous.

5.2.9.2 Derivation. Naranji is Urdu for orange-colored (Platts, 1960), a color especially prominent in lower horizons of these paleosols in thick-yellow paleochannel sandstones.

5.2.9.3 Description. The type Naranji clay paleosol (Figs 5.20 and 5.21) is the stratigraphically lower paleosol (at 20 m in Fig. 5.10 or 2748 m in Fig. 5.7) of two similar paleosols developed on thick (3-5-m) sandstone paleochannels, in the prominent bluff east of the main watercourse of Kaulial Kas (at grid reference 773135), 1.5 km southeast of Kaulial village, Potwar Plateau, Pakistan (Figs 5.6 and 5.8). This is in the main part of the late Miocene Dhok Pathan Formation, above the uppermost tongue of Nagri sandstone, and is here paleomagnetically dated as 7.8 million years old (Tauxe & Opdyke, 1982; corrected to the time scale of Berggren, Kent, & Van Couvering, 1985).

+10 cm; C horizon of overlying paleosol (Sarang silty-clay massive-surface phase); siltstone; brown (7.5YR5/4); weathers reddish-yellow (7.5YR7/6); conspicuous relict bedding; common burrows, mostly 3 to 4 mm in diameter, but occasionally 13 mm in diameter, filled with dark reddish-brown (5YR3/4) claystone from overlying layers; few, narrow (2-3-mm) clastic dikes of siltstone; just above the base of this horizon is a thin (1-cm), strata-concordant mottle of greenish-gray (5G6/1); strongly calcareous; intertextic insepic to calciasepic in thin section; few porphyroskelic calciasepic rhizoconcretions, with a central crystal tube of sparry calcite, and marginal iron staining (diffusion sesquan); common quartz, feldspar, and opaque oxides; abrupt, smooth contact to

0 cm; A horizon; silty claystone; dark reddish-brown (5YR3/4); weathers reddish-yellow (7.5YR7/6); common yellowish-brown (10YR5/4), light greenish-gray (5GY7/1), and dark greenish-gray (5GY4/1) calcareous rhizoconcretions; common small (1-2-mm), black (5B2.5/1) specks of charcoal; few, narrow (3-mm) clastic dikes of yellowish-brown (10YR5/4) siltstone; few strong-brown (7.5YR5/6) calcareous nodules, up to 10 mm in diameter, and with sharp margins as if redeposited; distinct relict bedding; strongly calcareous; porphyroskelic insepic in thin section; few micritic calcareous rhizoconcretions with ferruginized woody remains; few agglomeroplasmic mosepic metagranotubules, after burrows; scattered quartz, feldspar, schist, and opaque oxides; gradual, smooth contact to

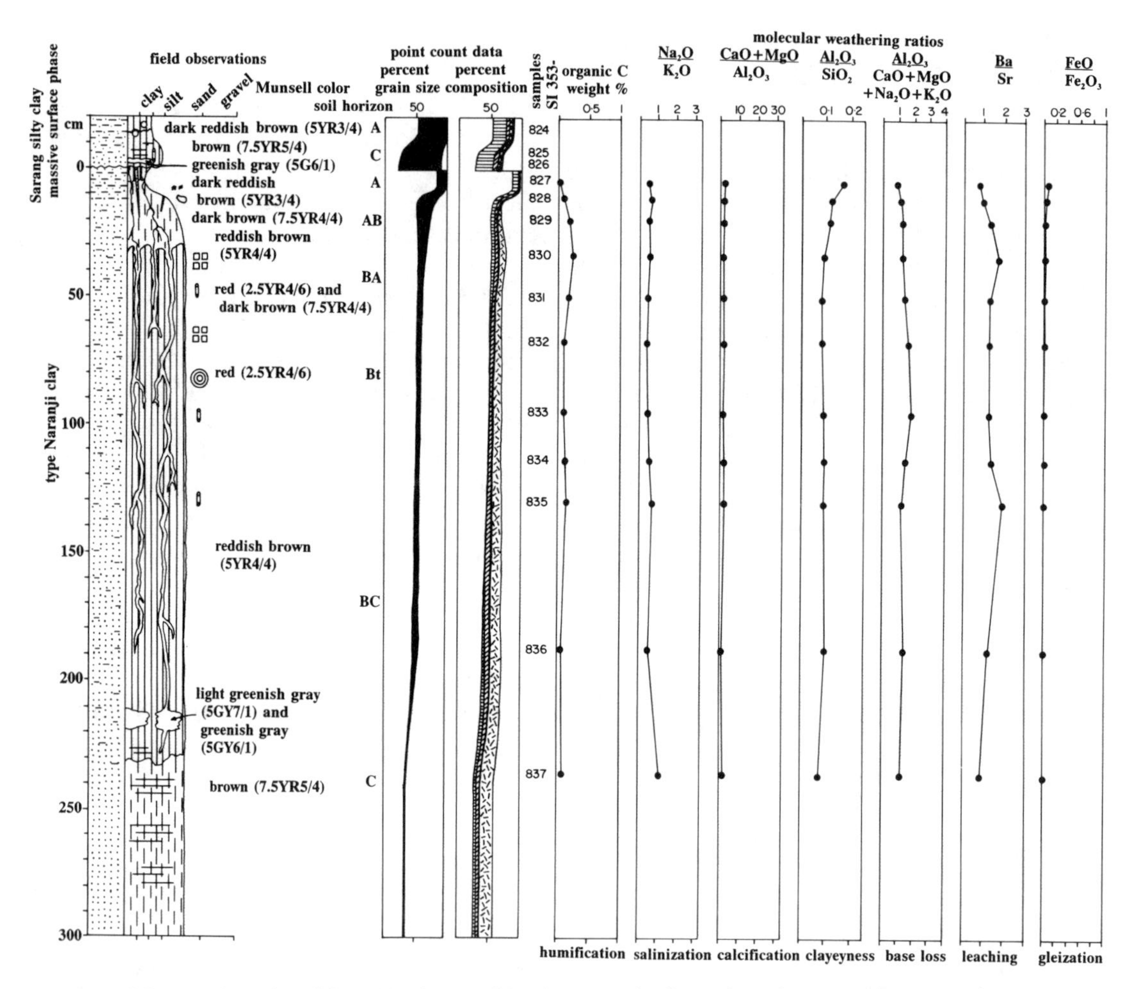

Fig. 5.20. Measured section, Munsell colors, soil horizons, grain size, mineral composition, organic content and selected molecular weathering ratios of the type Naranji clay and Sarang silty-clay massive-surface phase paleosols, in the upper measured section of the Dhok Pathan Formation, southeast of Kaulial village, Pakistan. This section is at 17.1 to 20.3 m in Fig. 5.10. Lithological key as for Fig. 5.17.

-10 cm; A horizon; clayey medium-grained sandstone; dark brown (7.5YR4/4); weathers light brown (7.5YR6/4); common large (up to 3 cm in diameter), drab-haloed, calcareous rhizoconcretions, filled with light yellowish-brown (10YR6/4) siltstone, with a thin dark grayish-brown (10YR4/3) clay skin (argillan) or a halo of greenish-gray (5GY6/1) claystone in the largest examples; very coarse angular-blocky peds defined by narrow (3-mm) clastic dikes (skeletans) of yellowish-brown (10YR5/4) siltstone; few coarse sand-sized grains of brown (10YR5/3) and pink (10YR7/4); few black (5BG4/1) charcoal specks, as in overlying horizon; moderately calcareous; porphyroskelic to agglomeroplasmic skelmasepic in thin section; few micritic calciasepic calcareous rhizoconcretions with a central tube of mosepic agglomeroplasmic clayey siltstone; few ferruginized woody and opaque charcoal fragments; common quartz; few grains of feldspar, schist, limestone, and opaque oxides; gradual, wavy contact to

-31 cm; AB horizon; coarse-grained sandstone; reddish brown (5YR4/4); weathers brown (7.5YR5/4); common large (up to 3 cm in diameter), drab-haloed calcareous rhizoconcretions, of greenish-gray (5GY5/1), with an outer dark rim (mangan) of dark reddish-brown (5YR3/2); few large (up to 3 cm in diameter) metagranotubules of greenish-gray (5GY6/1) sandstone, after burrows; few angular claystone clasts up to 3-cm long, lying along indistinct relict bedding planes; moderately calcareous; agglomeroplasmic skelmosepic in thin section, with few porphyroskelic calciasepic micritic nodules; common quartz; few grains of feldspar, muscovite, and opaque grains; gradual, wavy contact to

-43 cm; BA horizon; coarse-grained sandstone; mottled red (2.5YR4/6) and dark brown (7.5YR4/4); weathers yellowish-red (5YR5/6); common, large (up to 3 cm in diameter), drab-haloed, calcareous rhizoconcretions (Fig. 2.3) of light gray (5Y7/1) and greenish-gray (5GY6/1), weathering to yellowish-red (5YR5/6) and reddish-yellow (7.5YR6/6); very coarse angular-blocky peds, defined by indistinct dark-brown (7.5YR4/4) iron-stain (diffusion sesquans); moderately calcareous; agglomeroplasmic skelmasepic in thin section, but agglomeroplasmic calciasepic in calcareous rhizoconcretions; few opaque ferric concretions; common quartz and schist; few grains of feldspar, muscovite, and opaque oxides; gradual, wavy contact to

-56 cm; Bt horizon; coarse-grained sandstone; red (2.5YR4/6); weathers yellowish-red (5YR5/8); common large (8-11 mm in diameter), copiously branched, drab-haloed, calcareous rhizoconcretions, with a central strand of brownish-yellow (10YR6/6) siltstone, a broad (3-5-mm) halo of light greenish-gray (5GY7/1) micritic carbonate and often also an outermost thin (1-mm) stain (sesquan) of reddish-brown (5YR5/3) claystone and sesquioxides; common meta-isotubules, 8 to 9 mm in diameter, dark brown (7.5YR4/4) silty claystone, after burrows; moderately calcareous; agglomeroplasmic skelmasepic in matrix, but calciasepic with concentric iron stain (quasisesquans) in rhizoconcretions; common quartz and schist; few volcanic rock fragments, feldspar, and opaque oxides; diffuse, wavy contact to

-109 cm; BC horizon; coarse-grained sandstone; reddish-brown (5YR4/4); weathers yellow (10YR7/6); common large drab-haloed calcareous rhizoconcretions of light greenish-gray (5GY7/1) and greenish-gray (5GY6/1), extending down from overlying horizon; prominent distinct mottles of yellowish-brown (10YR5/4) and dark reddish-brown (5YR3/4); few meta-isotubules, up to 6 mm in diameter, of reddish-brown (5YR3/4) claystone, after burrows; near the base of this horizon (at -210 to -220 cm) are strata-concordant, light greenish-gray (5GY7/1) and greenish-gray (5GY6/1) mottles; relict bedding also becomes obvious toward the base of this horizon; moderately calcareous; agglomeroplasmic skelmosepic in thin section of matrix, but rhizoconcretions have calciasepic micritic sheath (neocalcan) and concentric zones of iron stain (diffusion neosesquans); few, irregular, small, opaque nodules of iron-manganese; common quartz and schist; few volcanic rock fragments, feldspar, muscovite, and opaque grains; diffuse, irregular contact to

-217 cm; C horizon; coarse-grained sandstone of paleochannel; brown (7.5YR5/4); weathers pale brown (10YR6/3); prominent relict cross-bedding; moderately calcareous; intertextic insepic in thin section; few agglomeroplasmic calciasepic micritic calcareous rhizoconcretions, with calcite crystal tubes in their center, as in overlying paleosol; many grains have a thin weathering rind (sesquiargillan); grains include quartz, schist, siltstone, mafic volcanics, opaque grains, pyroxene, biotite, muscovite, and micritic limestone.

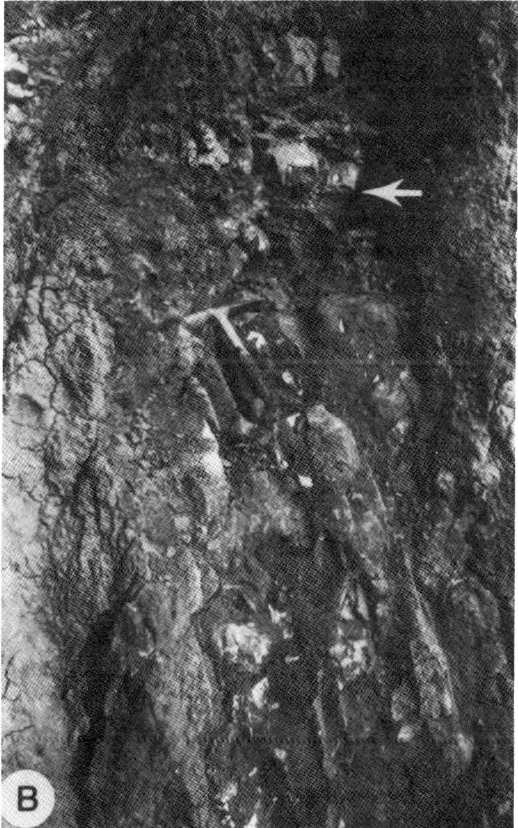

Fig. 5.21. Outcrop (A) and excavation (B) of the type Naranji clay paleosol, showing sharp contact and bedded overlying sequence (above arrow), contrasted with gradational lower contacts deep into paleochannel sands below. Human figure (A) and hammer (B) for scale.

5.2.9.4 Further examples. Only two Naranji paleosols were identified, both of them formed on prominent yellow paleochannel sandstones in the same bluff east of the main watercourse of Kaulial Kas, 1.5-km southeast of Kaulial village (Figs

5.6 and 5.8). The second example is stratigraphically higher (33 m in Fig. 5.10) than the type Naranji clay, but also in the late Miocene (7.8 million years old) Dhok Pathan Formation. There may be another example at a slightly higher stratigraphic level in the Dhok Pathan Formation (2810 m in Fig. 5.7), some 7.5 million years old, at a fossil locality (Yale-G.S.P. 443, at grid reference 775137), 300-m northeast of the upper measured section, also 1.5-km southeast of Kaulial village, Pakistan (Figs 5.6 and 5.8).

5.2.9.5 Alteration after burial. The uppermost horizon of the type Naranji clay may have undergone considerable compaction, but this can be disregarded considering its prominent relict bedding, which is evidence that this was a late cumulative addition to the profile. Compaction of the sandy main portion of this paleosol to 75 percent of its former thickness, as estimated from consideration of likely burial depth, should be regarded as a maximum value. This also is suggested by observations in thin section, compared with sandstones of known burial depth (J. M. Taylor, 1950). In the Naranji clay paleosol, quartz grains show mainly tangential but very few long contacts, but there is considerable local deformation of grains of claystone and weathered schist. Considerable losses of organic matter during burial are likely considering the low analytical values for organic carbon (Fig. 5.20), but the shape of the depth function for carbon is what would be expected in a thick soil with a cumulic surface horizon. Naranji paleosols may have been reddened to their present color (2.5YR in places) by burial dehydration of ferric hydroxides. They probably were originally orange to reddish-brown (7.5YR), because they have neither the structure nor textures of dark brown, drab, or black soils, such as gleyed soils, Mollisols, or Vertisols. There probably was some burial gleization of calcareous rhizoconcretions (see Fig. 2.3), but local dark rims may indicate a contribution of slight surface-water gleization, as in Kala Series paleosols. Local fill of the central root hollow of calcareous rhizoconcretions with sparry calcite was the only observed burial cementation within the profile. There is similar cementation of the sandy parent material of this paleosol, but the present outcrop is friable and leached of carbonate to depths of about 30 cm. There is no discernible chemical evidence of illitization (Fig. 2.23), but these paleosols may have been mildly affected like associated profiles.

5.2.9.6 Reconstructed soil. The type Naranji clay can be imagined as a thick sandy soil with a thin, brown, clayey surface (A) horizon, partly cumulic in nature and with weakly developed platy peds. Below that, transitional horizons (AB, BA) of sandy texture, riddled with large root traces, passed down into an orange to red horizon (Bt) of clay enrichment with crude blocky structure. Below that again was another gradual transition (horizon BC) into cross-bedded sand of a paleochannel. The Naranji clay contains numerous calcareous rhizoconcretions but is only moderately calcareous in the sandy matrix. It probably had neutral to alkaline pH and moderately high cation exchange capacity and base saturation. There was no problem with salinization, to judge from the low soda/potash ratios (Fig. 5.20). The high oxidation, deeply penetrating root traces, and subsurface clayey horizon are evidence of free drainage. However, there are thin dark stains of iron-manganese around some root traces, and charcoal and woody material near the surface, perhaps because of slight and short-lived, seasonal, surface-water gleization as envisaged for the type Lal clay. This may have been a problem during the last phases of formation of this profile, when it was covered by weakly permeable bedded clays.

5.2.9.7 Classification. The broad bulge of

subsurface clay enrichment with its clay skins (Fig. 5.20) qualifies as an argillic horizon (of Soil Survey Staff, 1975). This, together with evidence of high base status from calcareous rhizoconcretions in the type Naranji clay, are indications that it was a Psammentic Haplustalf. Similar surface soils on the alluvial plains of Punjab, India, have been identified (by Sehgal & Stoops, 1972) as "Alfic Ustochrepts," a name not accepted in the U.S. taxonomy (Soil Survey Staff, 1975). These Punjab soils were envisioned as intergrades between Ustochrepts (like Sonita and Bhura Series paleosols) and Haplustalfs (like Lal paleosols here). A Nabha Series surface soil of Punjab, described and identified as an Udic Ustochrept (by Kooistra, 1982; Murthy *et al.*, 1982), is similar in some respects to the type Naranji clay. Its degree of clay enrichment is less pronounced and its overall clay content is less than the paleosol. On the other hand, a soil profile from near Sadhpur village, India, identified as a Udic Haplustalf (by Tomar, 1987) has a more pronounced clay bulge than the type Naranji clay.

Naranji paleosols are sandy, as in Arenosols of the Food and Agriculture Organization (1974) classification, but they developed on alluvium, which is excluded for Arenosols. More likely they were Luvisols, not so strongly colored originally as Chromic Luvisols, and closest to Orthic Luvisols. Such soils are widespread in Indo-Gangetic alluvium of Punjab, Uttar Pradesh and Bihar, India (Food and Agriculture Organization, 1977b, map units Lo5-2a, Lo34-2a, Lo35-2a, and Lo44-1b).

Among Australian soils (Stace *et al.*, 1968), they are best grouped with Red Earths, although that category includes many very ancient relict soils on bedrock that are quite different from Naranji paleosols. In the Northcote (1974) key for Australian soils they are Gn2.45.

In the South African system (MacVicar *et al.*, 1977), Naranji paleosols are most like Valsrivier lilydale. Generally similar, though redder and less clayey, soils are found on floodplains in regions of granitic gneiss basement in southern Kruger National Park, South Africa (Botha, 1985; Gertenbach, 1983; Venter, 1986).

5.2.9.8 Paleoclimate. Naranji paleosols are moderately developed, with calcareous rhizoconcretions, but show no calcareous nodular horizon, as in regions of transition between noncalcareous and calcareous soils (the pedocal/pedalfer transition of Marbut, 1935). In the climatic transect from the dry plains of Punjab up into increasingly humid Himalayan foothills, this zone of transition receives about 750-1000-mm mean annual rainfall (Sehgal, Sys, & Bhumbla, 1968). In the southern United States, such weakly calcareous soils are found in regions receiving more than 600 mm of rain but beyond the 1000-mm isohyet soils are more deeply decalcified than the type Naranji clay paleosol (Birkeland, 1984).

Seasonality of waterlogging could be an explanation for the thin dark haloes found around some calcareous rhizoconcretions. More convincing evidence is the internal banding of these micritic rhizoconcretionary sheaths and the clastic dikes seen in the surface of these paleosols.

A warm climate is indicated by the diversity of burrows in Naranji paleosols, especially those similar to termite remains. However, the single possible example of a fungus garden of termites is very poorly preserved and probably was a clast in the sandy parent material of the paleosol.

5.2.9.9 Ancient vegetation. The stout, deeply penetrating, calcareous rhizoconcretions and deep subsurface horizon (Bt) of clay enrichment in the Naranji clay paleosol are both indications of seasonally dry forest or woodland vegetation. An impression of a forest of great stature might be gained from the depth of this paleosol, but this depth is more likely a consequence of its freely drained, coarse-grained parent material. This thick paleochannel could have been a source of groundwater during the dry season, so that its forest or woodland may have been a little lusher than that of thinner associated clayey paleosols.

Broadly comparable Indian, African, and North American soils support deciduous dry forest and wooded grassland (Food and Agriculture Organization, 1977a, 1977b; Gertenbach, 1983; Soil Survey Staff, 1975). In Australia, such soils are covered with sclerophyll forest and grassy woodland (Stace *et al.*, 1968). In the alluvial plains of northern India, streamsides and stream beds support locally distinctive vegetation, such as wooded grassland of palms (*Phoenix sylvestris*; unit 5/E8a of Champion & Seth, 1968) and swamp forest of *Syzygium cumini* and *Barringtonia acutangula* (unit 4D/5S3). The pattern of stout fossil root traces in Naranji paleosols is unlike that of either grasslands or palms. These paleosols also were better drained and less affected by salinization than swamps or wooded grasslands of palms. A more likely modern analog is dry tropical riverain forest (unit 5/ISI of Champion & Seth, 1968). This is dominated by *Terminalia arjuna* and is taller than surrounding dry deciduous forests, from which it differs also in having a more shrubby and evergreen bushy understory, but few grasses and vines. Fossil *Terminalia* leaves and wood have been found in early Miocene Lower Siwalik sediments near Koilabas, Nepal (Tripathi & Tiwari, 1983) and Kalagarh, Uttar Pradesh, India (U. Prakash, 1981).

5.2.9.10 Former animal life. No fossil bones were found in Naranji paleosols in the measured section, but one of these paleosols is a fossil locality (Yale-G.S.P. 443; Fig. 5.8) at a higher stratigraphic level within the Dhok Pathan Formation (2810 m in Fig. 5.7) some 7.5 million years old (Tauxe & Opdyke, 1982; corrected to time scale of Berggren, Kent, & Van Couvering, 1985). Recorded from this locality (by Barry, Lindsay, & Jacobs, 1982) are tortoise carapace, large giraffe, chevrotain (*Dorcatherium majus*), and large antelope (*Selenoportax lydekkeri*).

The type Naranji clay paleosol contains sand-filled greenish-gray burrows up to 3 cm in diameter, which extend to depths of 40 cm, where they end in a simple chamber tilted a little more toward the horizontal than the steeply inclined entrance burrow. These burrows are very similar to burrows referred to the ichnogenus *Macanopsis* in Pandu paleosols, and as discussed there, probably were dens of large beetles or spiders.

Found at a depth of 90 cm in the type Naranji clay was a calcareous concentrically banded sphere similar to the ichnogenus *Termitichnus*. As discussed for Sarang Series paleosols, this may have been a fungus garden of a termite. However, the specimen seen in the type Naranji clay was too poorly preserved to collect, was unusually small (only 2 cm in diameter), and showed internal bands truncated by the outer surface. Probably it was eroded out of a streamside soil and deposited with the parent material of the type Naranji clay.

5.2.9.11 Paleotopographic setting. All the Naranji paleosols seen were developed on top of sandstone paleochannels, which presumably were abandoned channels like those seen now on sub-Himalayan alluvial fans (V. R. Baker, 1986) and along the Indus River (Holmes & Western, 1969). These abandoned channelways remain as local sinuous depressions in alluvial landscapes. Although low lying, Naranji paleosols were well drained for most of the year, as evident from their highly ferruginized claystones and deeply penetrating calcareous rhizoconcretions. This, together with the already weathered nature of their parent sandstones, is evidence that they formed on alluvial fans (bhabar) of the foothills drainage of the Himalayan ranges ("buff sandstones" of Behrensmeyer & Tauxe, 1982), rather than in a major lowland drainage system like the present

Indus or Ganges Rivers.

5.2.9.12 Parent material. In general, the thickest soils commonly form on the coarsest grained parent materials (Retallack, 1990a), and such parent material control may be a partial explanation for the unusual thickness of Naranji paleosols compared with associated paleosols on siltstone and claystone. This also compromises the use of soil thickness as an indication of tropical paleoclimate or a long duration of weathering, relationships that can be documented in surface soils (Birkeland, 1984).

Because of the thickness of sandstone beneath the Naranji paleosols and their coarse grain size, their parent material can be assessed in some detail. Quartz is the most conspicuous component. It is present as both clear grains with minor inclusions probably from plutonic sources and as strained composite grains from gneisses or other metamorphic rocks. A few grains had adhering quartz overgrowths, indicating derivation from quartzites, which are common in the Paleozoic sequence of the Himalayas. Next in abundance were fragments of schist and limestone. Paleozoic schists also are widespread in the Himalaya, as are Mesozoic to Eocene limestones (Wadia, 1975). Less common were fragments of mafic volcanics, siltstone, pyroxene, biotite, and opaque oxides. Even well below the type Naranji clay paleosol these sand grains showed thin weathering rinds (diffusion sesquans), indicating that they were derived from preexisting soils on foothills of the Himalayan ranges ("buff sandstones" Behrensmeyer & Tauxe, 1982), rather than from major streams into the little-weathered high Himalaya ("blue gray sandstone"). Such a provenance also is indicated by the chemical composition of these sandstones, higher in silica, lower in alumina and magnesia, and in general less schistlike than blue-gray sandstones or closely associated Sonita and Bhura Series paleosols (Fig. 2.23).

5.2.9.13 Time for formation. Traces of relict bedding persist throughout the type Naranji clay, and its clayey subsurface (Bt) horizon barely qualifies as argillic. It is a little better developed than Sultanpur Series surface soils, some 2300 years old, yet short of development seen in Bhalwal Series soils 10,000 to 20,000 years old along the Ravi River, near Lahore, Pakistan (Ahmad, Ryan, & Paeth, 1977). Its subsurface clay enrichment is comparable to that seen on Organ I alluvial fan surfaces some 2200 to 4600 years old near the Las Cruces, New Mexico (Gile, Hawley, & Grossman, 1980), and on the post-Modesto II and upper Modesto river terrace surfaces along the Merced River in the San Joaquin valley of California (Harden, 1982). Thus, formation over several thousands of years is most likely for Naranji paleosols.

5.3 Summary and prospects for Miocene paleosols from Indo-Pakistan

Each of the Miocene paleosols described from Pakistan has been identified within several classifications for surface soils (Table 5.1) and has particular paleoenvironmental implications (Table 5.2). Their relationships and distribution across ancient landscapes also are worthy of attention as clues to the way in which these alluvial sediments were put together. To explore these relationships the paleosols from each stratigraphic level have been plotted according to hue, as a crude guide to former drainage and humification, versus development, as a guide to time for formation (Fig. 5.22). From these relationships former soilscapes of the Indo-Gangetic Plains can be reconstructed for two successive times during the late Miocene.

At the lower stratigraphic level studied in detail, in that part of the lower Dhok Pathan Formation some 8.5 million years old, there are drab-colored paleosols on clayey (Khakistari Series) and sandy (Pan-

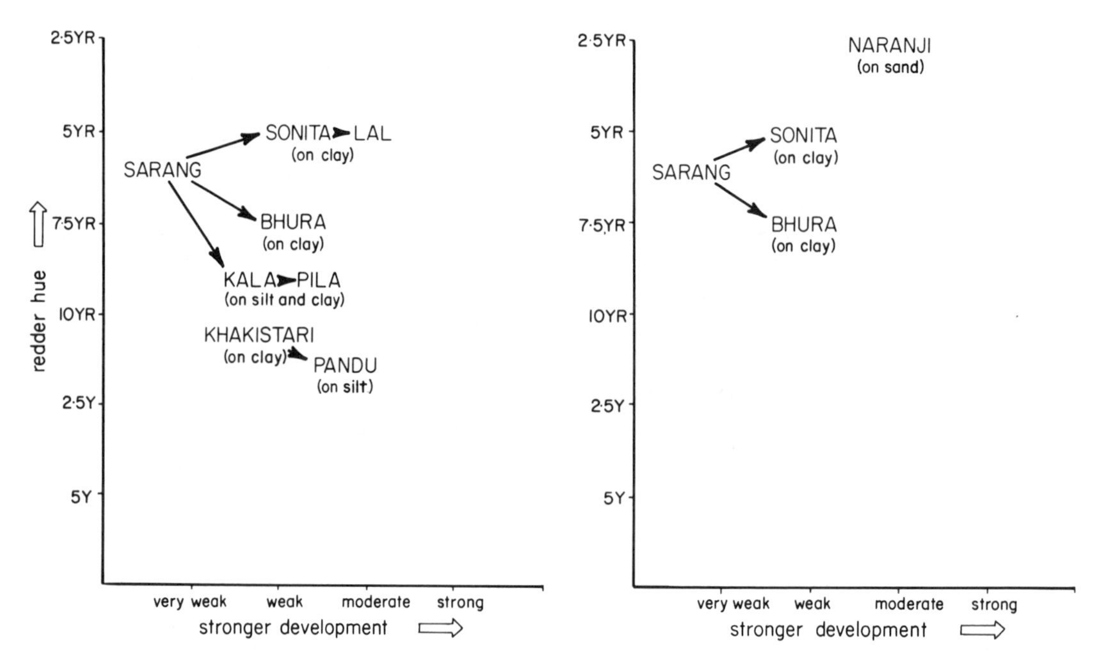

Fig. 5.22. Interpretive relationships between Pakistani paleosols from both the lower measured section of the Dhok Pathan Formation some 8.5 to 8.6 million years old (to left) and from the upper measured section some 7.7 to 7.8 million years old (to right), in terms of former degree of drainage and humification (red versus drab hue) and duration of soil formation (weak versus strong development) and the likely direction of plant ecological succession and of soil development (arrows).

du Series) alluvium of the wet lowlands of a major river draining the Himalayan interior. These paleosols were like dhankar and terai soils of the modern Indo-Gangetic Plains, which support seasonally waterlogged forest. These drab-colored paleosols are distinct from better-drained red and brown paleosols, which were like the bhabar soils of alluvial fans now draining Himalayan foothills in northern India and Pakistan. The colors of the paleosols reflect the degree to which they were waterlogged during formation and during early burial, and not necessarily the provenance of the soil parent materials, because many of the red and brown paleosols are chemically more like the little-weathered alluvium of the major Himalayan drainages, which would have been a copious source of sediment in this extensive subsiding axial trough south of the uplifted Himalaya. Conversely, some of the drab colored paleosols are chemically more like the foothills alluvial fans, which included a variety of different kinds of well-drained soils. The least developed paleosols of alluvial fans (Sarang Series) show few soil features beyond root and burrow penetration, and reveal the nature of this red (5YR), already-weathered alluvium from soil-covered slopes of foothills. Other red paleosols (Sonita toward Lal Series) show progressive alteration of relict bedding by root traces and burrows, along with surface decalcification and subsurface accumulation of clay and carbonate nodules, and formed under seasonally dry forests away from streams. A few paleosols (only three profiles of Pila and Kala Series out of 80 paleosols in the section) are brown (10YR) and show some evidence of slight seasonal waterlogging. These paleosols of local low lying channelways and other depressions have abundant fine root traces in a clayey surface (A) horizon with a subsurface

TABLE 5.1. Identification in modern soil classifications of Miocene paleosols series of northern Pakistan

Paleosol Series	Urdu Meaning	Diagnosis	U.S. Taxonomy	FAO Map	Australian Handbook	North-cote Key	South African System
Bhura	brown	thin, clayey, and dark reddish-brown (7.5YR), with subsurface calcareous (Bk) horizon	Fluventic Ustochrept	Calcic Cambisol	Brown Clay	Ug5.31	Oakleaf mutale
Kala	dark	thin, silty, and brown (10YR), with subsurface iron-manganese (Bg) and calcareous (Bk) horizons	Aquic Eutrochrept	Gleyic Cambisol	Wiesen-boden	Ug6.3	Willow-brook chinyika
Khaki-stari	ash colored	thin, clayey, uniform, and brown to gray (10YR-2.5Y), with subsurface ferric concretions and mottles (Bg)	Vertic Haplaquept	Calcaric Gleysol	Gray Clay	Ug5.24	Rensburg rensburg
Lal	red	thick, clayey, uniform, and red (5YR), with subsurface clayey (Bt) and calcareous (Bk) horizons	Haplustalf	Calcic Luvisol	Euchrozem	Gn3.13	Valsrivier marienthal
Naranji	orange	thick, sandy, moderately calcareous, and red (5YR-2.5YR), with subsurface clayey (Bt) horizon	Psammentic Haplustalf	Orthic Luvisol	Red Earth	Gn2.45	Valsrivier lilydale
Pandu	yellowish white	thin, silty, brown to gray (10YR-2.5YR), with sub-surface calcareous (Bk) horizon	Typic Haplaquept	Calcaric Gleysol	Gray Clay	Ug6.5	Katspruit killarney
Pila	yellow	thick, silty, moderately calcareous, and yellow (10YR-7.5YR), with subsurface weakly clayey (Bt) and calcareous (Bk) horizons	Aquic Haplustoll	Calcaric Phaeozem	Wiesen-boden	Ug6.3	Bonheim bonheim
Sarang	varie-gated	bedded sand and clay with root traces, brown to red (10YR-5YR)	Fluvent	Calcaric Fluvisol	Alluvial Soil	Uf1.11	Dundee dundee
Sonita	crimson	thin, silty, clayey, reddish brown (5YR), with subsurface calcareous (Bk) horizon	Vertic Ustochrept	Calcic Cambisol	Red Clay	Ug5.36	Oakleaf makulek

TABLE 5.2. Paleoenvironmental interpretation of Miocene paleosols of northern Pakistan

Paleosol Series	Locality	Age (Ma)	Paleoclimate	Ancient Vegetation
Bhura	upper and lower sections (Kaulial east and west)	8.5-7.8	500-840-mm annual rainfall, seasonally dry, tropical	riparian fringing forest
Kala	lower section (Kaulial west and YPL262/2)	8.6-8.5	590-820-mm annual rainfall, seasonally dry	seasonally wet, wooded grassland
Khakistari	lower section (Kaulial west)	8.5	seasonally dry	seasonally wet forest
Lal	lower section (Kaulial west)	8.5	500-840-mm annual rainfall, seasonally dry, tropical	dry tropical closed forest
Naranji	upper section (Kaulial east)	7.8-7.5	750-1000-mm annual rainfall, seasonally dry	dry tropical riverain forest
Pandu	lower section (Kaulial west)	8.5	510-850-mm annual rainfall or drier, seasonally dry, tropical	seasonally wet tropical forest
Pila	lower section (Kaulial west)	8.5	440-980-mm annual rainfall, seasonally dry, tropical	lowland alluvial wooded grassland
Sarang	upper section (Kaulial east) and lower section (Kaulial west and YPL260, 410)	8.5-7.8	seasonally dry, tropical	colonizing streamside woodland
Sonita	upper section (Kaulial east and YPL477,544) and lower section (Kaulial west and YPL262/3)	8.6-7.8	320-590-mm annual rainfall, seasonally dry, tropical	dry tropical open forest

Former Animal Life	Paleotopographic Setting	Parent Material	Time for Formation
three-toed horse ("*Cormohipparion*" spp.), chowsinghalike antelope (*Elachistoceras khauristanensis*), tortoise scute, apple snail (*Pila prisca*) (lower section): dung beetle nest (upper section)	levee of foothills drainage	clayey, quartzofeldspathic alluvium	400 to 5000 years
three-toed horse ("*Cormohipparion*" spp.), gazelle (*Gazella* sp.), chevrotain, elephant, pig, cat (*Felis* sp.), crocodile (*Crocodylus palaeindicus*) (all from YPL-262/2)	streamside swale of foothills stream	quartzofeldspathic silty and clayey alluvium	5000 to 15,000 years
early elephant (*Choerolophodon corrugatus*), giant giraffe (cf. *Brahmatherium megacephalum*), apple snail (*Pila prisca*) (all lower section)	floodprone streamside swales of proto-Gangetic River	clayey quartzofeldspathic alluvium	5000 years or less
apple snail (*Pila prisca*), termite burrows and pellets (all lower section)	floodplain of foothills drainage	clayey quartzofeldspathic alluvium	5000 to 15,000 years
early chevrotain (*Dorcatherium majus*), nilgailike antelope (*Selenoportax lydekkeri*), termite fungus garden, insect or spider burrow (all upper section)	abandoned sandy channel bed of foothills drainage	sandy quartzofeldspathic alluvium	2000 to 15,000 years
three-toed horse ("*Cormohipparion*" spp.), nilgailike antelope (cf. *Selenoportax vexillarius*), tortoise scrap, termite fungus garden, earthworm and insect or spider burrows (all lower section)	sandy levee of proto-Gangetic River	sandy quartzofeldspathic alluvium	10,000 to 25,000 years
termite burrows and pellets, insect or spider burrow (all lower section)	streamside swale of foothills drainage	quartzofeldspathic silty and clayey alluvium	2000 to 6000 years
apes (*Sivapithecus sivalensis, S. punjabicus*), hyaena (*Palhyaena sivalensis*), civet (*Viverra chinjiensis*), early elephant (*Stegolophodon* sp.), horse, rhino, chevrotain, chowsinghalike antelope (*Elachistoceras khauristanensis*), termite fungus garden, earthworm and insect or spider burrows (lower section, YPL-260 & 410): nilgailike antelope (*Selenoportax lydekkeri*), earthworm and insect or spider burrows (upper section & YPL444)	levee and point bar of seasonally dry streams	quartzofeldspathic clayey to sandy alluvium	400 years or less
elephant, horse, rhino, pig (*Propotamochoerus hysudricus*), antelope (*Miotragocerus punjabicus*), gazelle (*Gazella* sp.), insect or spider burrows (lower section, YPL262/3): tortoise, bamboo rat (*Kanisamys sivalensis*), anthracothere, three-toed horse ("*Cormohipparion*" spp.), nilgailike antelope (*Selenoportax lydekkeri*), rhinoceros (*Chilotherium intermedium*), giraffe, elephant tusk, insect or spider burrow (upper section, YPL477,544)	levee and floodplain of seasonally dry streams	clayey quartzofeldspathic alluvium	400 to 5000 years

accumulation (Bk) of calcareous nodules, and probably supported small local areas of low alluvial wooded grassland. A final common kind of paleosol (Bhura Series) is intermediate in color (7.5YR) and character between dry forest and seasonally wet grassland paleosols. These paleosols are mostly associated with paleochannel and levee deposits (including Sarang Series paleosols) of the foothills drainage, and probably formed under streamside gallery forest.

For the upper stratigraphic level examined, within the main part of the Dhok Pathan Formation some 7.8 million years old, there are fewer kinds of paleosols, indicating a less diverse soilscape. No drab colored paleosols were seen, although the thick, little weathered sandstones of major antecedent streams of the Himalaya were still present. Nor were there any thick red clayey paleosols. The only two moderately developed profiles were on thick paleochannel sandstones. All the other paleosols were weakly developed, thin, and not very deeply weathered. These differences may reflect a drier climate, in which water table was seasonally deep in the major axial drainage as well as in the alluvial fans of Himalayan foothills. They may also reflect a slightly accelerated rate of sediment accumulation. Weakly developed red soils (Sarang Series) were widespread on freshly deposited alluvium of both the axial drainage and foothills alluvial fans. When uninterrupted by flooding these developed a clayey well structured surface and subsurface accumulation of calcareous nodules under seasonally dry woodland and forest (of red Sonita Series) or a slightly lusher, dry forest (of brown Bhura Series). Distinctive deep clayey red paleosols (Naranji Series) formed under low alluvial forest on the sands of abandoned channels of the foothills alluvial fans.

The suite of paleosols some 8.5 million years old matches best the soils in that part of the lowland plain of the Ganges River and alluvial fans of the Gandak and Ghaghara Rivers, around Gorakhpur, Uttar Pradesh, India (map unit Je75-2a of Food and Agriculture Organization, 1977b; Agarwal & Mukherji, 1951). Native vegetation in this area before human disturbance was probably moist Gangetic deciduous forest, with local patches of low alluvial wooded grassland and swamp woodland (Srivastava, 1976). Broadly similar vegetation is now preserved from human destruction in Dudhwa National Park, Uttar Pradesh (Hawkins, 1986). At Gorakhpur, the mean annual temperature is 25.7°C, ranging from 16.0°C for January and to a hot 29.4°C in July. Rainfall averages 1274 mm annually, with only 15 mm in January but 346 mm in July (Champion & Seth, 1968).

The suite of paleosols about 7.8 million years old compares best with soils further west in the alluvial plains of Uttar Pradesh, near Saharanpur and Roorkee (map unit BK40-2a of Food and Agriculture Organization, 1977b). Here the Yamuna and Ganges Rivers issue forth from the Himalayan ranges as major streams, but there are also alluvial fans of foothills streams such as the Kosi and Gola Rivers (S. B. Deshpande, Fehrenbacher, & Beavers, 1971; S. B. Deshpande, Fehrenbacher, & Ray, 1971; Hurelbrink & Fehrenbacher, 1970; Tomar, 1987). Native vegetation is dry deciduous tropical forest, with local riparian woodland and forest. The dry deciduous forests include most of the tree species of the moist deciduous forests further east, but lack vines and epiphytes (Champion & Seth, 1968). The dry deciduous forests also are an important zoogeographic boundary between Oriental-Asian faunas of forests to the east and the Mediterranean-Ethiopian open country faunas to the west (Mani, 1974c, 1974d). At Roorkee, mean annual temperature is 23.4°C, with a January mean of 13.4°C and 29.5°C for July. Mean annual rainfall is 1050 mm, with 42 mm in January and 312 mm in July (Champion & Seth, 1968).

These comparisons with surface soils are compromised by extensive human modification of soils and biota, which is effectively a kind of desertification (Mani, 1974b). Agriculture in these rich alluvial plains dates back at least 7000 years, and the use of fire dates back 40,000, perhaps 500,000 years (Allchin & Allchin, 1982). Tool-using human ancestors may have been in Pakistan as long ago as 2 million years (Rendell, Hailwood, & Dennell, 1987). There also have been great changes in the courses of the Ganges and Indus Rivers since Miocene time (see Fig. 5.2). Furthermore, deserts and very high glaciated ranges as a source of dust were not so widespread during late Miocene time. Each of these factors may explain why rainfall estimates from depth to calcic horizons of paleosols (Table 5.2) are less than those from general comparison of soilscapes. They may also be reasons why the surface soils are more silty and brown (10YR to 7.5YR) than the clayey, red (5YR) Miocene paleosols (Kooistra, 1982; Tomar, 1987). These differences also could be due to a systematic overestimation of clay in thin sections compared with mechanical analysis used for surface soils (Murphy & Kemp, 1984), to the burial dehydration of ferric hydroxide minerals (T. R. Walker, 1967), and to unusually severe compaction of the paleosols. There are differences between the paleosols and surface soils of the Indo-Gangetic Plains, but a residue of similarities has remained.

The distribution of paleosols in the two measured sections may reveal causes and durations of episodic sedimentation, as has been demonstrated in other sequences of paleosols (Retallack, 1986a). Of particular interest is the expansion and retreat of the sandy braid plain of the proto-Gangetic River compared with the alluvial fans of soil-covered foothills (Behrensmeyer & Tauxe, 1982). The measured sequences are an unusually complete temporal record of environmental changes considering the abundance of weakly developed paleosols and implied high rate of sediment accumulation, supported also by paleomagnetic data (Tauxe & Opdyke, 1982). Another indication is the evenly sinuous pattern of development of successive paleosols, best seen in the diagrammatic profile of the lower measured section (Fig. 5.9), less so in the upper section (Fig. 5.10). In contrast, sequences of paleosols containing large temporal breaks show stepwise patterns of development (Retallack, 1986a, 1990a). The time represented by soil formation can be calculated by assigning minimum times of development for each paleosol series of 50 years (Sarang Series), 2000 years (Bhura, Khakistari, Sonita, Kala), 3000 years (Pila), and 5000 years (Lal, Pandu, Naranji). This adds up to 164,200 years and an overall rate of sediment accumulation of 34 cm/1000 years for the lower measured 55 m of paleosols (excluding paleochannels) about 8.5 million years old, for which there are three cycles of expansion of the blue gray sandstones of the laterally equivalent Nagri Formation. The upper measured section of 29 m of paleosols 7.8 million years old and representing only one cycle of expansion and retreat of the major lowland drainage shows a total duration of soil formation of only 40,650 years, for an overall rate of sediment accumulation of 71 cm/1000 years. Paleomagnetic estimates (from Tauxe & Opdyke, 1982; adjusted to time scale of Berggren, Kent, & Van Couvering, 1985) for the lower part of the section in Kaulial Kas (between 8.92 and 8.21 million years ago) give long term rates of 61 cm/1000 year, which for 55 m would be 90,185 years. For the upper part of the section in Kaulial Kas (8.21-6.85 million years old) the average rate is 47 cm/1000 years, which for 29 m is 61,625 years. The paleomagnetic and paleosol estimates are of about the same order of magnitude. The difference may reflect uncertainty of the calculations, but also unsteadiness of sedimentation, which is especially apparent when paleomagnetic data are examined on fine time scales (Badgley & Tauxe, 1990; Badgley, Tauxe, & Bookstein, 1986; N. M. Johnson *et al.*, 1985). Such unsteadiness is more apparent from abrupt changes in paleosol development within the upper measured section (Fig. 5.10) than the lower one (Fig. 5.9). In both cases, however, the episodic expansion and retreat of the proto-Gangetic braid plain recurred at intervals of the order of 50,000 years.

Possible causes of the expansive phase of the braid plain in these episodes in theory (Schumm, 1977, 1981) could be (1) elevation of Himalayan sources of the proto-Gangetic River system; (2) change from clayey to more sandy, less stable floodplains of the proto-Gangetic River; (3) change to drier climate, which lessened the flow of foothills drainage more than that of glaciated alpine ranges; (4) sparser vegetation of lowland floodplains compared with vegetation of higher alluvial fans and foothills; (5) fall of regional water table in lowland axial drainage compared with water table of foothills alluvial fans; (6) episodic increase in discharge of major antecedent streams of the high Himalaya without increase in discharge of foothills streams; (7) expansion of upland drainage net of major streams of high Himalaya; (8) steepening of slope of lowland braid plain; (9) increased sediment supply of major high Himalayan drainage compared with supply of foothills streams; (10) change in channel behavior of the major axial drainage from narrow and sinuous to broadly braided; (11) decreased progradation of foothills alluvial fans over the axial lowlands trough.

Some of these changes (2, 10, and 11) are intrinsic to the behavior of fluvial systems, and are likely to have been significant only if there is no evidence of extrinsic factors, of which the most important are climate (3, 4, 5, 6, and 9) and tectonic activity (1, 6, 7, and 8). Although the Himalayan ranges continued to rise during late Miocene time, tectonic causes are not supported by paleosol evidence for lower rates of sediment accumulation after expansion of the lowland braid plain sandstones compared with afterward. Paleosols associated with this braid plain (Pandu and Khakistari paleosols) include the best developed calcareous horizons found, and are in general better developed than paleosols of the alluvial fans of foothills drainage (Sarang, Bhura, Sonita, Lal, Pila, and Kala Series). The braid plain paleosols are also less well drained than interbedded alluvial fan paleosols, which would not be expected if the braid plain had been episodically steepened or elevated by uplift. On the other hand, there is evidence for drier climate during expansion of the proto-Gangetic braid plain than during other times. As a group the braid plain paleosols (Khakistari and Pandu paleosols) are more calcareous and have a shallower calcareous horizon than the clayey and partially decalcified paleosols of the foothills alluvial fans (Sarang, Bhura, Lal, Pila, and Kala Series). A drier and warmer more evaporative climate would produce less weathering in soils, which may be eroded to sandy rather than clayey alluvium, as well as lower water table, sparser vegetation, and lower discharge of foothill streams compared with major Himalayan antecedent streams fed by enormous drainage basins including melting glaciers. Episodic drying and cooling was a prominent feature of Pleistocene paleoclimates (Hays, Imbrie, & Shackleton, 1976), and similar episodic paleoclimatic fluctuations have been discovered in lake deposits of Miocene (Barnosky, 1984), Triassic, and Jurassic age (Olsen, 1986). The episodic changes in the Miocene paleosols and paleochannels of Pakistan reported here have the duration of climatic changes cued to Milankovitch variation in the obliquity of the Earth's rotational axis. Longer and shorter sequences of paleosols should also be examined from this perspective, for evidence of climatic change on different time scales. The enormously thick Siwalik Group of Pakistan and India is potentially a much more detailed and long range record of Neogene climatic fluctuation than well known sequences of paleosols in loess of Czechoslovakia (Kukla, 1977).

Many thousands more paleosols remain

to be studied in the Siwalik Group of India and Pakistan. Only the most prominent of them have been mapped (by Behrensmeyer, 1987; Behrensmeyer & Tauxe, 1982) in studies of alluvial architecture of the Siwalik Group at selected localities in the Potwar Plateau, Pakistan, including sediments ranging in age from 7 to 13 million years. More profiles have been described from Siwalik sediments 2.3 to 10.9 million years old in the area around Jhelum, Pakistan (Visser & Johnson, 1978) and 7 to 7.5 million years old near Haritalyangar, India (G. D. Johnson *et al.*, 1981). These studies and the work reported here represent only a preliminary reconnaissance of Siwalik paleosols.

Apes, such as *Sivapithecus*, some 8 to 9 million years ago in India and Pakistan have generally been reconstructed as living in a mosaic of forest, woodland, and grassland, based on studies of sedimentary petrography (Krynine, 1937), paleosols (G. D. Johnson, 1977; G. D. Johnson *et al.*, 1981; Retallack, 1985), and the ecology of living mammals allied to the fossils and their adaptive features (Badgley & Behrensmeyer, 1980; Pickford, 1977; Vasishat, 1985). On the other hand, studies of Siwalik fossil wood and leaves (Awasthi, 1982; Lakhanpal & Guleria, 1986; U. Prakash & Prasad, 1984; M. Prasad & Prakash, 1984; K. N. Prasad, 1971; Tripathi & Tiwari, 1983; Vishnu-Mittre, 1984), of fungi (Phadthare, 1990), and of pollen and spores (Y. K. Mathur, 1984; H. P. Singh, 1982), have tended to indicate forested conditions throughout Miocene time. This is not a serious difference, because plant fossils have not been found in close association with the fossil mammals and plant remains are preserved best in waterlogged lowland soils and sediments where vegetation is more mesophytic than elsewhere (Retallack, 1984a). On the other hand, other analyses of the fossil mammalian fauna (by Andrews, 1983; Tatersall, 1969a, 1969b), have been taken as indications of seasonally dry woodland vegetation. Such a view is also compatible with studies of the carbon isotopic composition of calcareous nodules of paleosols in the Siwalik Group (by Quade, Cerling, & Bowman, 1989), which indicate vegetation mostly of trees (C_3 plants) until about 6 million years ago, when tropical grasses (C_4 plants) became more widespread.

This study shows that there are elements of truth in each interpretation, and that the nature of the mosaic of vegetation can now be understood in greater detail. When *Sivapithecus* was most common some 8.5 million years ago, seasonally dry, deciduous tropical forest and woodland were widespread (Sonita and Lal Series), with lesser areas under early successional woodland (Sarang), riparian woodland (Bhura), and seasonally wet lowland forest (Khakistari, Pandu). Wooded grassland was a very minor part of the mosaic (3 paleosols of Kala and Pila Series out of 80 in Fig. 5.9), in seasonally inundated depressions, rather than extensive dry interfluves. *Sivapithecus* has been reported only from paleosols of streamside early successional woodland.

By the time *Sivapithecus* disappeared in this area some 7 million years ago, a drier climate is indicated by paleosols, and change toward more uniform dry, deciduous tropical forest and woodland that may have lacked the diversity of vines, epiphytes, and understory plants found earlier. Unlike the short term (20,000-100,000-year) climatic fluctuation already discussed, this drying was part of a long term trend, as a lengthening rain shadow was cast across the northern alluvial plains of India and Pakistan by the rising ranges of the Himalaya and Hindu Kush. Additional drying and the appearance of widespread grasslands is not indicated by paleosols (Behrensmeyer, personal communication, 1988) or their carbonate carbon isotopes (Quade, Cerling, & Bowman, 1989) until 6 million years ago.

Miocene vegetation during accumulation of the Siwalik sediments now can be seen to have been very different from the mosaic of forest, woodland, and wooded grassland around East African volcanic highlands. Such modern grassland mosaic environments have been influential in the interpretation of Miocene faunas of India (Badgley & Behrensmeyer, 1980), but are in some ways misleading modern analogs. The Pakistani paleosols reported here have calcareous rhizoconcretions and other products of a much more strongly seasonal climate than any of the Miocene paleosols described from Kenya. Nor do those few Pakistani paleosols thought to have supported grasslands show the dark granular surface horizons (mollic epipedon) and freedom from waterlogging seen among paleosols at Fort Ternan, Kenya. Seasonally dry, deciduous, tropical woodland and forest, also called monsoon forest, is now widespread in both India (Champion & Seth, 1968) and Africa (White, 1983), and has been a neglected element in interpreting Miocene paleo-environments of Pakistan (Andrews, 1983) and elsewhere in the Old World Tropics (Andrews, 1990a; Solounias & Dawson-Saunders, 1988). It was, however, probably the original concept of their past vegetation, considering that these were the habitats of most of the modern mammalian skeletal material gathered for comparison with the earliest collections of Siwalik fossil mammals by Falconer (1868). It was Hugh Falconer also who argued that fossil apes of the Siwaliks were most closely allied to the living orang-utan, a conclusion only recently resurrected for these fossils after several decades when other views prevailed (Andrews & Cronin, 1982; Pilbeam, 1982).

6. RECONSTRUCTED LIFE AND LANDSCAPES OF THE OLD WORLD TROPICS

Illustrated reconstructions of the past can be a valuable conceptual focus. Our impressions of the prehistoric past are based less on illustrations of fossil bones and leaves than on the work of artists such as Charles Knight, Bruce Horsfall, Zdenek Burian, Jay Matternes, Frank Knight, and Peter Murray. Their restorations of the past not only are aesthetically pleasing but are in effect hypotheses about life and landscapes of the past. Such reconstructions of extinct mammals are testable from studies of functional morphology, bone histology, and biomechanics, to name just a few of the scientific approaches that have become popular in recent years. Reconstructions of past landscapes also are becoming testable hypotheses with the application of sedimentary facies models and the study of paleosols. In this spirit, this chapter offers illustrated reconstructions, and a concise summary of the paleoenvironments of the studied paleosols, as well as some general thoughts on tropical soil formation and the evolution of tropical grasslands and of early human ancestors from apes. This book will not be the last word on these topics, but it does introduce little exploited and potentially useful lines of scientific evidence.

Some of the reconstructions of extinct mammals given here are little modified from previously published illustrations (Fleagle, 1988; D. Lambert, 1987; M. Lambert, 1981; Osborn, 1936-1942; G. Paul for Stanley, 1986; Thenuis, 1972). In other cases the reconstructions have been made from published technical descriptions of fossil remains (listed in Appendices 8, 9, and 10), taking inspiration from general information on allied modern mammals, as illustrated by Jonathan Kingdon (1971, 1974a, 1974b, 1977, 1979). The living African and Indian plants, illustrated to give substance to their names, are from standard botanical manuals, with the exception of a fossil grass from Fort Ternan, reconstructed from an exceptionally complete specimen (Dugas, 1989; Retallack, Dugas, & Bestland, 1990). The landscape and vegetation reconstructions owe much to scientific study of south Asian fluvial geomorphology (V. R. Baker, 1986), East African volcanology (Dawson, 1962, 1964; Downie & Wilkinson, 1962; Hay, 1986, 1989; Mariano & Roeder, 1983), and regional phytogeographic monographs (Champion & Seth, 1968; Lind & Morrison, 1974; White, 1983).

The particular species of extinct mammal illustrated on a paleosol was in all cases found fossilized within them. The documentation of Kenyan fossil finds has been exemplary (Pickford, 1984, 1986a; Pickford & Andrews, 1981; Shipman, 1977, 1981; Shipman *et al.*, 1981). Large and representative collections are known from some paleosols, and these animals can be shown to have lived and died on these soils. Such collections can be a very useful guide to the paleoecology of Miocene mammals. A few comparable collections and excavations of fossils have been made in paleosols of the Siwalik Group of Pakistan (Badgley, 1982, 1986; Pilbeam *et al.*, 1980). This study also made use of unpublished information from fossil locality files (Behrensmeyer & Raza, 1984) and my own collections (now in the Smithsonian Institution and Geological Survey of Pakistan). Nevertheless, most of the Siwalik vertebrates continue to be collected from large areas ("survey blocks" of Barry, Jacobs, & Lindsay, 1982; Barry, Behrensmeyer, & Monaghan, 1980; Behrensmeyer, 1987), and could not be used for establishing specific fossil faunas for each kind of paleosol. Although bone is neither so diverse nor abundant in paleosols of levees and floodplains as in paleochannels of the Siwalik Group (Badgley, 1986; Behrensmeyer, 1987), it is common enough to be worth pursuing. My collecting for only 15 hours resulted in 19 identifiable bones and 7 snail fragments traceable to specific paleosols on knolls or benches. Much loose bone was seen and left because it was not traceable to a specific layer. While it is very likely that the animals died in these paleosols, these data fall short of the kind of paleoecological study that could be achieved by more dedicated attention to the exact location of a large number of fossils.

6.1 Early Miocene in Kenya

By early Miocene time, Africa had drifted north to begin its ongoing collision with Europe. In the place of the modern Mediterranean Sea and European Alps was an archipelago of islands separating the shallow Paratethys Ocean from remnants of the deeper Tethys Ocean to the south. Africa was thus making tenuous early connection with Eurasia after a long late Paleozoic and Mesozoic connection with the Gondwana supercontinent (Scotese, Gahagan, & Larson, 1989). Southwestern Kenya at this time was about 5 degrees of latitude south of the equator (Reilly *et al.*, 1976). There was not yet much sign of the present Rift Valleys of East Africa, but there was a good deal of local relief around a series of carbonatite-nephelinite stratovolcanoes, cinder cones, and associated faults (Bishop & Trendall, 1966). The ancient volcanic centers are now deeply eroded as the Gwasi Hills near Rusinga Island, Mt Elgon near Kitale, and Tinderet Peak near Koru, Songhor, and Fort Ternan, but they once were much more imposing volcanoes (B. H. Baker, 1987). It is around these that there is an exceptionally good fossil record of plants, animals, and soils.

Carbonatite and nephelinite tuffs are not geologically common materials, and processes of fossilization and soil formation on such unusually alkaline rocks are not well understood. The early Miocene carbonatite-melilitite tuffs of southwestern Kenya have preserved natural casts of soft bodied organisms such as grasses, flowers, insects, spiders and lizard's tongues (Estes, 1962; A. C. Hamilton, 1968; L. S. B. Leakey, 1952; Wilson & Taylor, 1964), probably because of the tendency of these tuffs to become cemented by early weathering products (Barker & Nixon, 1983; Hay, 1986). The extraordinary fossil record of Miocene fossil fruits, millipedes, snails, and bones is due in part to the unusually alkaline and calcareous sediments and soils associated with these chemically distinctive volcanic rocks (Bishop, 1968; Hay, 1986; Pickford, 1986b). Calcareous sediments and soils are widespread in dry climatic regions (Birkeland, 1984), but such paleoclimate should not be assumed for paleosols developed on intrinsically calcareous and alkaline parent materials. It also is likely that soils on such unique parent materials supported plants and animals that were in some ways distinct from those of surrounding regions, as has been noted, for example, for decidedly more xeromorphic modern vegetation around the modern syenite-carbonatite intrusive complex of Phalarborwa in South Africa (M. M. Cole, 1986; Frick, 1986).

Early Miocene paleosols from Koru and Songhor around the Tinderet volcanic center show varying degrees of development interpreted as evidence that initially hard-setting, highly calcareous, and alkaline carbonatite tuffs could be softened, decalcified, and neutralized with progressive soil formation. In advanced stages of formation these paleosols appear very similar to soils on nonvolcanic parent materials, such as calcareous beach sands (Muchena & Sombroek, 1981). Like surface soils on alkaline volcanic rocks in East Africa (Weilemaker & Wakatsuki, 1984), these moderately developed paleosols on mixed carbonatitic and nephelinitic tuffs do not show evidence of amorphous or glassy low-density materials characteristic of many modern volcanic soils. It is in the earliest stages of development that these paleosols in carbonatite and nephelinite tuffs are especially distinctive. Many of the very weakly developed paleosols have calcareous hardpans, around which fossil root traces are deflected. These hardpans commonly preserve relict bedding and a complex system of anastomosing subplanar cracks filled with sparry calcite. This distinctive microfabric reflecting substantial loss of volume and carbonate precipitation is without analog in modern soils as far as I am aware. It is here termed *pervasively displacive*, and interpreted as an extreme manifestation of a displacive fabric observed during the early weathering of carbonatite-melilitite tuffs from historic and Plio-Pleistocene eruptions of the active carbonatite-nephelinite volcano, Oldoinyo Lengai in Tanzania. These tuffs are erupted with a large proportion of highly unstable minerals such as gregoryite and nyerereite, which are quickly leached to form an early cement of zeolites and calcite (Hay, 1978, 1989). Continued volume reduction and cracking is produced by the leaching of zeolite and weathering to clay of aluminosilicates such as pyroxene and biotite. These distinctive hardpans of weakly developed soils and paleosols on carbonatite-melilitite tuff may appear superficially similar to subsurface calcareous horizons of moderately developed soils. Pedogenic carbonate on these unusual volcanic parent materials is, however, more like that found widely in calcareous soils: a very fine grained unbedded mass of carbonate that replaces the outside of grains in irregular embayments. This replacive microfabric is associated with local displacive fabric in pedogenic nodules, but this is distinct from the outlandishly displacive fabric of hardpans in little-weathered carbonatite-melilitite ash. These peculiarities of fossil preservation and soil formation on carbonatite and nephelinite volcanics must be considered before one interprets paleoclimate and former biota of these paleosols.

Early Miocene paleosols of nephelinite stratovolcanoes and carbonatite cinder cones around the former volcanic center of Tinderet are generally more clayey and less calcareous than their parent materials. These indications of a moist climate must be weighed against the presence in some of them of pedogenic nodules of the kind found in soils under climates no wetter than subhumid. The depth to the calcareous horizon in these paleosols, when corrected for burial compaction and compared with that in surface soils on carbonatite and nephelinite volcanic ashes of the Serengeti Plains of Tanzania (de Wit, 1978; Hay & Reeder, 1978; Jager, 1982), indicates a subhumid to semiarid paleoclimate, with mean annual rainfall in the range of 400 to 1200 mm. A similar rainfall regime also is compatible with the smectitic, rather than kaolinitic or evaporitic, composition of these paleosols, compared with soils on East African volcanics (de Wit, 1978; Mizota & Chapelle, 1988; Mizota, Kawasaki, & Wakatsuki, 1988). Some of the nodules have ferruginized zones, and there also are local clastic dikes indicative of climatic seasonality, but this was not nearly so marked as during the middle Miocene in Kenya or the late Miocene in Pakistan. Abundant burrows, and particularly structures interpreted as food storage chambers of termites, are evidence for a warm tropical climate, as would be expected in near-equatorial regions around the base of volcanoes.

The vegetation of these volcanic paleosols can be interpreted from their overall profile form and patterns of root traces, as well as by comparison with a variety of surface soils and their vegetation in Africa (Fig. 6.1). Alluvial lowlands around the volcanoes and cinder cones (now near Songhor) probably supported dry tropical forest on clayey terraces (Tut clay paleosol). Similar profiles (Choka clay) on carbonatite-nephelinite tuff also supported dry tropical

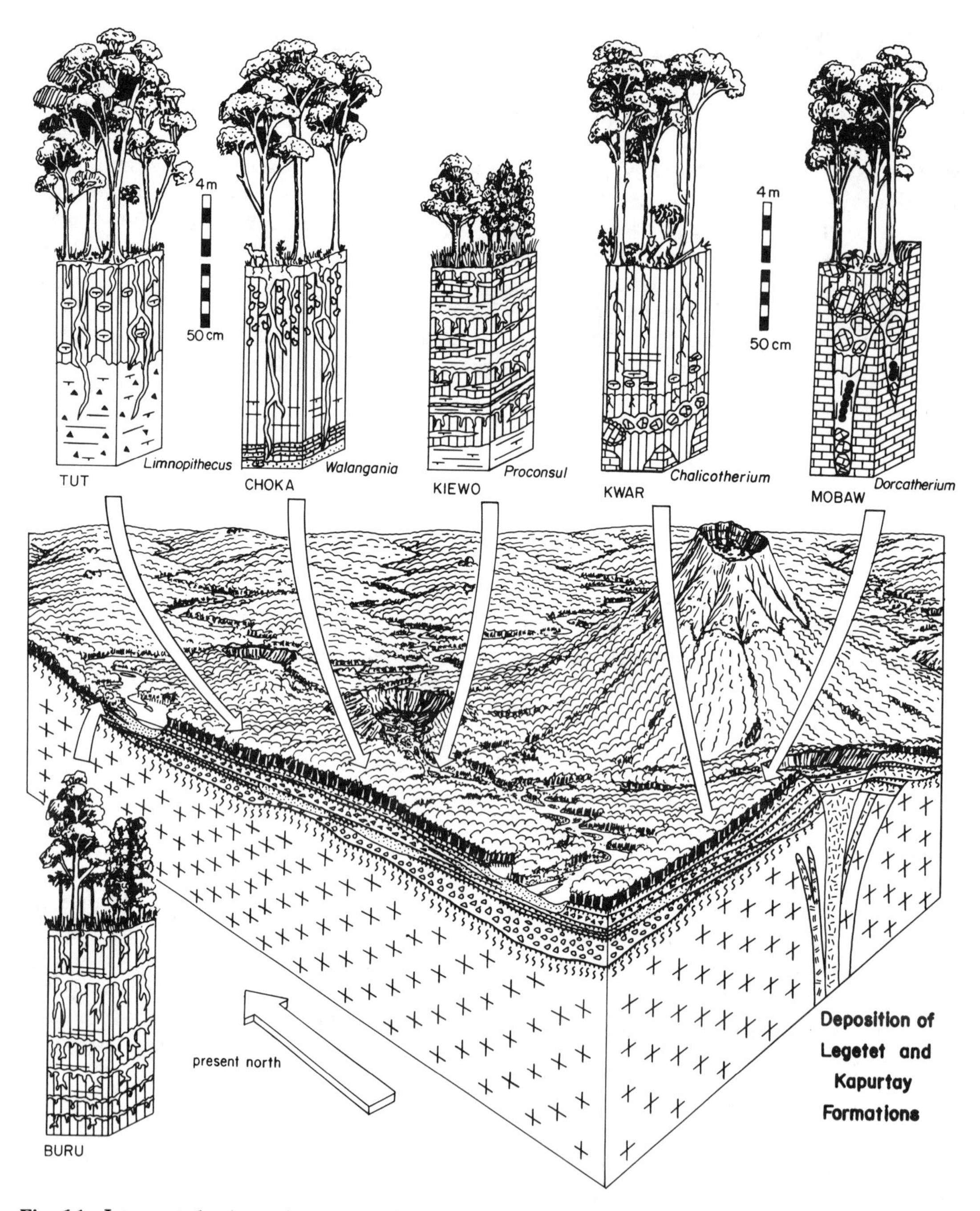

Fig. 6.1. Interpreted paleoenvironment, soils and vegetation in southwestern Kenya, during early Miocene time some 20 million years ago.

forest that may have been more open. Near-stream weakly calcareous, sandy, nephelinitic soils of areas prone to frequent flooding, with prominent relict bedding, may have been dense with pole trees early in the ecological succession to forest (Buru Series). On carbonatite tuffs, in contrast, soils (Kiewo Series) of colonizing forest were more alkaline, had shallow hardpans, and may have supported more open forest. Dry tropical forest further advanced in ecological succession may have covered weakly developed red paleosols on boulders of carbonatite welded tuff and associated paleokarst (Mobaw Series), as well as mixed carbonatite and nephelinite airfall ashes (Kwar Series) on the flanks of a small carbonatite cinder cone (now near Koru).

Few fossil remains of plants were preserved in these oxidized paleosols, but one of the upland ashy profiles (Kwar Series) did yield a fossil endocarp of hackberry (*Celtis* sp. cf. *C. mildbraedii*; Fig. 6.2). Hackberries are dominant trees of forests intermediate in ecological succession between colonizing regrowth and old growth forests of the Guineo-Congolian rain forest and peripheral drier kinds of tropical forest in East Africa (Lind & Morrison, 1974; White, 1983). Hackberry fruits are mineralized by calcite and silica while living (Yanovsky, Nelson, & Kingsbury, 1932). They would have been favored for preservation like bones and other fossil hard parts, even if only a minor component of the former vegetation.

Paleosols are evidence only of the general kind of vegetation, or plant formation. A better appreciation of the likely floristic affinities of the paleosols can be gained by considering fossil floras from other early Miocene rocks in East Africa. Like the paleosols, fossil plants indicate a variety of forest types in East Africa during early Miocene time. Guineo-Congolian rain forest and floristically similar dry peripheral forests of this humid basin are indicated by the presence of *Bersama* and *Cola* some 22 million years old from Bukwa, Uganda (Brock & MacDonald, 1969; A. C. Hamilton, 1968), and of *Antrocaryon* and *Entandrophragma* (Fig. 6.2) some 18 million years old on Mfangano Island, Lake Victoria (Collinson, 1983). Other wet-forest elements, perhaps from forerunners of Afromontane forests, are represented by fossil *Dipterocarpus* and *Cyathea* some 22 million years old from Mt Elgon, Uganda (Bancroft, 1932a, 1933; Bishop, Miller, & Fitch, 1969). Another fossil flora from near Mt Elgon in Uganda, 23 million years old, includes leaves similar to those of *Bauhinia*, *Berlinia*, *Cassia*, *Dalbergia*, *Olea*, *Maranthus (Parinari)*, *Pittosporum*, and *Terminalia* (Chaney, 1933), an assemblage most like that of Zambezian dry deciduous woodland of impoverished lateritic soils on Precambrian basement (of White, 1983). Another fossil flora some 19 million years old (J. A. Van Couvering, 1972) from near Precambrian basement west of Lake Mobutu (formerly Lake Albert) in Zaire, includes remains comparable with *Entandrophragma*, *Dichrostachys*, *Newtonia*, *Isoberlinia*, *Brachystegia*, and *Baphia* (Lakhanpal & Prakash, 1970), an assemblage most like Zambezian riparian woodland and miombo woodland (of White, 1983). Permineralized woods some 28 to 23 million years old in Ethiopia (Beauchamp, Lemoigne, & Petrescu, 1973; Lemoigne & Beauchamp, 1972; Lemoigne, Beauchamp, & Samuel, 1974; radiometrically dated by Justin-Visentin *et al.*, 1974; I. McDougall, Morton, & Williams, 1975; W.H. Morton *et al.*, 1979) are indications that dry woodlands of the Sudanian floristic realm also had differentiated by early Miocene time. Considered from the perspective of early Miocene vegetation of Africa, the paleosols of carbonatite-nephelinite ashes would initially have been highly alkaline, hard setting, and difficult soils from which to gain important nutrients such as phosphate, as already discussed. Their early successional vegetation may well have included plant species allied to those of surrounding Zambezian woodlands on impoverished lateritic soils. With further development, these alkaline volcanic soils became more clayey and less calcareous, and their vegetation may have included a greater variety of tropical forest trees. They can be envisaged as approaching the stature and diversity of dry forests, such

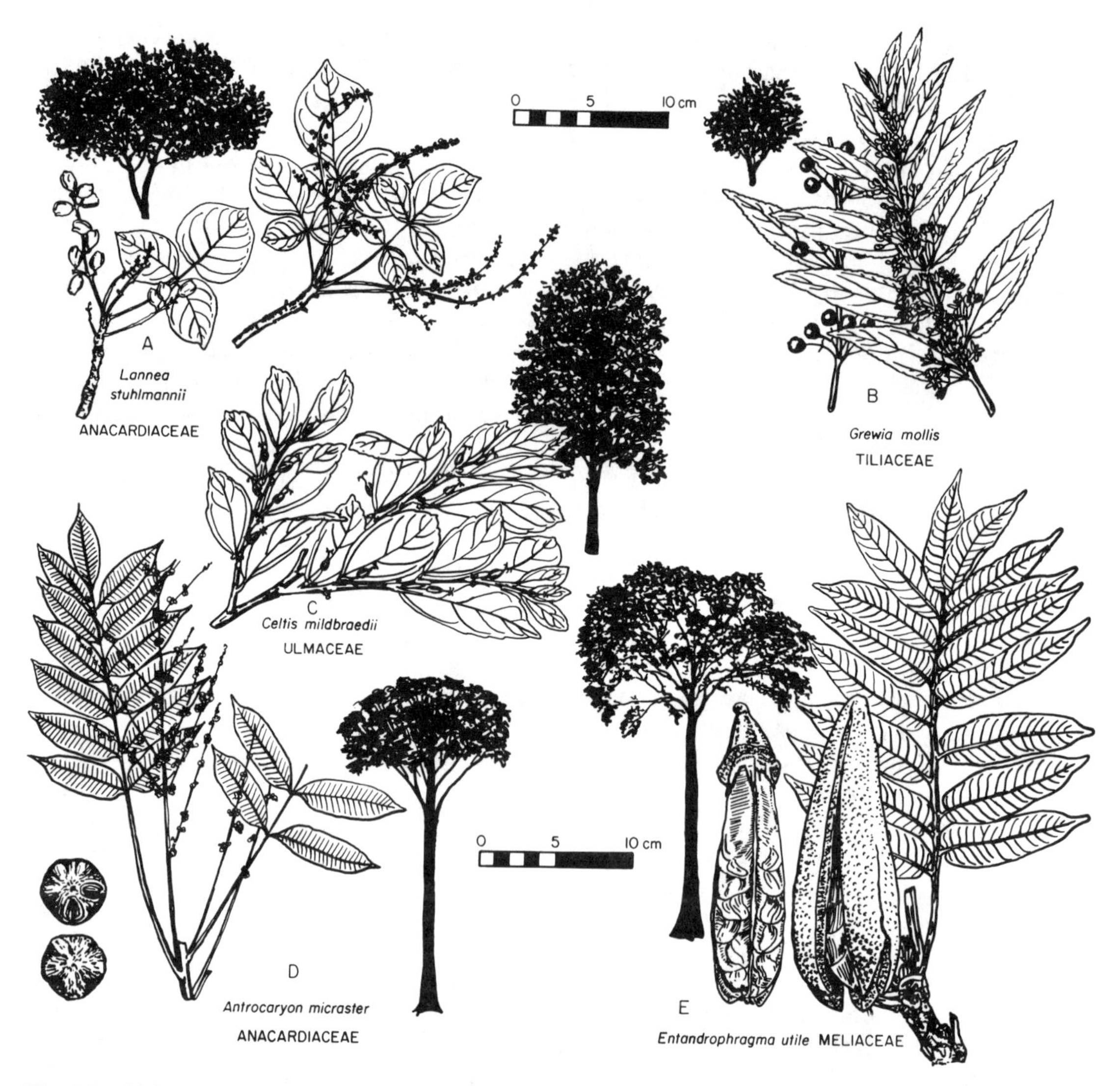

Fig. 6.2. Living African plants of genera found fossilized in early Miocene rocks of southwestern Kenya (illustrations and data from Burkill, 1985; Dale & Greenway, 1961; Harms, 1960; Irvine, 1961; Kokwaro, 1986; E. Palmer & Pitman, 1972; Voorhoeve, 1965).

as those of Kakamega in Kenya and Budongo in Uganda (Lind & Morrison, 1974). These early Miocene paleosols are not nearly so deeply weathered as surface soils under rain forest, nor so finely structured or shallow as soils under open grassy vegetation. These early Miocene forests of carbonatite-nephelinite volcanic centers may well be without an exact modern analog.

Unlike plant remains, fossil snails and mammals are abundant in early Miocene paleosols of the Tinderet volcanic center in Kenya (Fig. 6.3), and their various adaptive features show clear indications of forested conditions (Andrews, Lord, & Evans, 1979; Evans, Van Couvering, & Andrews, 1981). This is also indicated by the presence of extinct species of chevrotains (*Dorcatherium*), scaly-tailed flying squirrel (*Paranomalurus*), and bush babies (*Komba*, *Progalago*), as well as a

Fig. 6.3. Reconstruction of selected early Miocene mammals from southwestern Kenya [some illustrations redrawn from R. B. Horsfall for Scott, 1913 (scale); Fleagle, 1988 (A,C,D); D. Lambert, 1987 (B); Savage & Long, 1986 (I,S): others original and based on sources listed in Appendix 8].

diverse assemblage of tropical snails (Verdcourt, 1963). More open conditions are suggested by the presence of cane rats (*Diamantomys*), mole rats (*Bathyergoides*), and spring hares (*Megapedetes*). In the joint occurrence of such ecologically different taxa, this early Miocene fauna is distinct from modern mammalian faunas. Archaic deerlike ungulates (*Walangania*) and chevrotains (*Dorcatherium*) were common in many of these paleosols, all of which have a similar fauna overall. The open-country elements are more common in paleosols of colonizing forest (Kiewo Series) than in better developed paleosols (Tut, Mobaw, and Kwar Series). No fossils were found in weakly calcareous paleosols of colonizing forest (Buru Series), probably because such a chemical composition is unsuitable for the preservation of bone. Few fossils were found in highly calcareous carbonatitic paleosols of dry forest (Choka Series), perhaps in part because of the known toxic effects of fresh ash of this kind (Richard, 1942). A variety of fossil ape-monkeys have been found in these paleosols and their habitat preferences proved less marked than formerly suspected (Pickford, 1983). *Micropithecus clarki* was found only on upland carbonatite paleosols (Kwar and Mobaw Series), and *Dendropithecus macinnesi* and *Nyanzapithecus vancouveringorum* only in lowland mixed carbonatite-nephelinite paleosols (Kiewo, Choka, Tut). Other species (*Limnopithecus legetet*, *Rangwapithecus gordoni*, *Proconsul africanus*, and *P. major*) were widespread in each of the habitats, though most abundant in Kiewo paleosols, perhaps because this was an exceptionally good preservational environment and because these creatures were at greater risk of predation in such open streamside habitats.

Significantly younger within the early Miocene is the carbonatite-nephelinite volcanic center of Kisingiri, which includes a long sequence of fossiliferous paleosols mainly 17 million years old on Rusinga Island, Lake Victoria (Drake *et al.*, 1988). Detailed studies of these paleosols have been initiated (Bestland, 1990b; Thackray, 1989), and a preliminary interpretation of paleosols in nephelinitic alluvium and carbonatite tuffs of the main fossiliferous part of the Hiwegi Formation can be offered here. There are numerous very weakly developed paleosols, some very similar to Buru Series at Songhor, and these probably supported lowland colonizing forests along streamsides. Also common were red, moderately developed paleosols of several kinds, but broadly similar to Choka paleosols of Songhor, that may have supported dry woodland or forest. One of these examined in detail had nodules at 55 cm. Correcting for compaction under an overburden of about 2100 m (Le Bas, 1977), using standard compaction curves (Baldwin & Butler, 1985), gives an original depth of 96 cm, as found in surface soils in regions receiving 530-870-mm mean annual rainfall (de Wit, 1978). A few thin, brown, weakly calcareous paleosols also were seen, most like the Kala or Pila Series paleosols of Pakistan, and these may have supported local lowland alluvial wooded grassland, like the dambo vegetation of the Zambezian region of Africa (White, 1983). One of these had nodules at a depth of only 16 cm, which before compaction would have been 25 cm, as in surface soils receiving only 290-520-mm mean annual rainfall. There may have been some local grasslands in waterlogged depressions, but much of the landscape was forested.

These impressions of the former vegetation on Rusinga Island some 17 million years ago are compatible with what is known of its fossil flora (Chesters, 1957; Collinson, 1983), in which the following trees and shrubs were prominent: Annonaceae, *Cordia*, *Grewia*, *Lannea*, *Terminalia*, and *Zizyphus*. These genera are widespread in Africa today, but most like modern Zambezian dry woodlands (White, 1983). A similar view of vegetation at this time on Rusinga Island can be gained from the study of fossil snails, grasshoppers, and dung beetles, which include some assemblages like those of open vegetation (Paulian, 1976; Verdcourt, 1963; A. C. Walker & Pickford, 1983; J. M. Ritchie, personal communication, 1987).

Despite some opening up of vegetation, the fossil mammal fauna of Rusinga Island 17 million years ago is not very different from that of Koru and Songhor 20 million years ago. The two mammalian faunas differ only by 13.2 percent of their species (Pickford, 1981). The early Miocene large mammal fauna of Rusinga Island remained a forest-adapted assemblage of pigs, anthracotheres, archaic deerlike ungulates, chevrotains, rhinoceroses, hyraxes, gomphotheres, chalicotheres, creodonts, and ape-monkeys (Andrews, 1979; Evans,

Van Couvering, & Andrews, 1981), with some indications of local open habitats (Andrews & Van Couvering, 1975; A. C. Walker & Pickford, 1983). Habitat preferences of ape-monkeys more consistent than seen at Koru and Songhor, are apparent from the distribution of skulls and partly articulated specimens in paleosols on Rusinga Island. *Proconsul nyanzae* (in the sense of Kelley, 1986) was found consistently in paleosols of colonizing forest (at R1 site of 1931 mandible and of 1981 partly articulated limbs, at R106 site of 1948 skull, and at Kaswanga site; see Pickford, 1984, 1986a; A. C. Walker & Teaford, 1988). Partly articulated skeletons of three individuals of *Dendropithecus macinnesi* were found (at locality R3) in a red, carbonatite-rich paleosol that may have supported dry forest, similar to that of the Choka clay paleosol at Songhor.

During early Miocene time Africa was emerging from its biotic isolation of early Tertiary time, and there was some exchange of dry woodland mammals with the archipelago of islands to the north that would later become Eurasia. Archaic endemic African faunas were enriched by Eurasian forms, and elephants, among other less conspicuous mammals, infiltrated into Eurasia (Thomas, 1984). More profound faunal changes were to come to East Africa during middle Miocene time.

6.2 Middle Miocene in Kenya

By 14 million years ago Africa had moved a few degrees of latitude northward into Europe. This was also a time of low sea level (Haq, Hardenbol, & Vail, 1987), when the ancient Tethys Ocean was once again disrupted by a land bridge into Europe. The isolated part of Tethys was a forerunner of the Mediterranean Sea. Further north the shallow Paratethys Ocean, in the future area of the Black and Caspian Seas, continued to cover a large area of central Asia through Iran into the Indian Ocean. Fault systems of the future Red Sea and East African Rift Valleys were becoming active, but downdropping of these great troughs was not yet anywhere near its present extent. Associated with early faulting of the future western margin of the Gregory Rift in East Africa was the eruption of voluminous flood lavas of phonolite, now forming the Kericho and Uasin Gishu Plateaus of western Kenya, as well as much of the hilly country of the Aberdares and around Nairobi in central Kenya (Lippard, 1973a, 1973b). In western Kenya these plateau phonolites flowed around the older carbonatite-nephelinite stratovolcanoes, which continued to be active.

Middle Miocene paleosols near Fort Ternan (Figs 6.4 and 6.5), show many features attributable to a climate much drier than the climate that prevailed in Kenya during early Miocene time. Even a paleosol (Lwanda clay) developed on phonolite is smectitic and moderately calcareous. This, together with calcareous horizons very shallow within the paleosols (especially in Chogo and Onuria Series) and a variety of other similarities with surface soils of the Serengeti Plains of Tanzania (de Wit, 1978; Jager, 1982), is evidence of mean annual rainfall of some 250 to 550 mm. There is evidence also of seasonality of rainfall from cracks and slickensides in some of these paleosols (especially the type Rabuor clay). Seasonality probably was not nearly as pronounced as in the modern Serengeti Plains, where well-developed Vertisols are now found. Nor are there calcareous rhizoconcretions or other features of the likely monsoonal paleosols from late Miocene alluvial sediments in Pakistan. The paleosols at Fort Ternan do show abundant burrows and deep weathering of rock fragments, compatible with a warm tropical paleoclimate, but evidence for this aspect of paleoclimate from paleosols is less than compelling.

A pronounced paleoclimatic cooling and drying some 15 million years ago is seen in

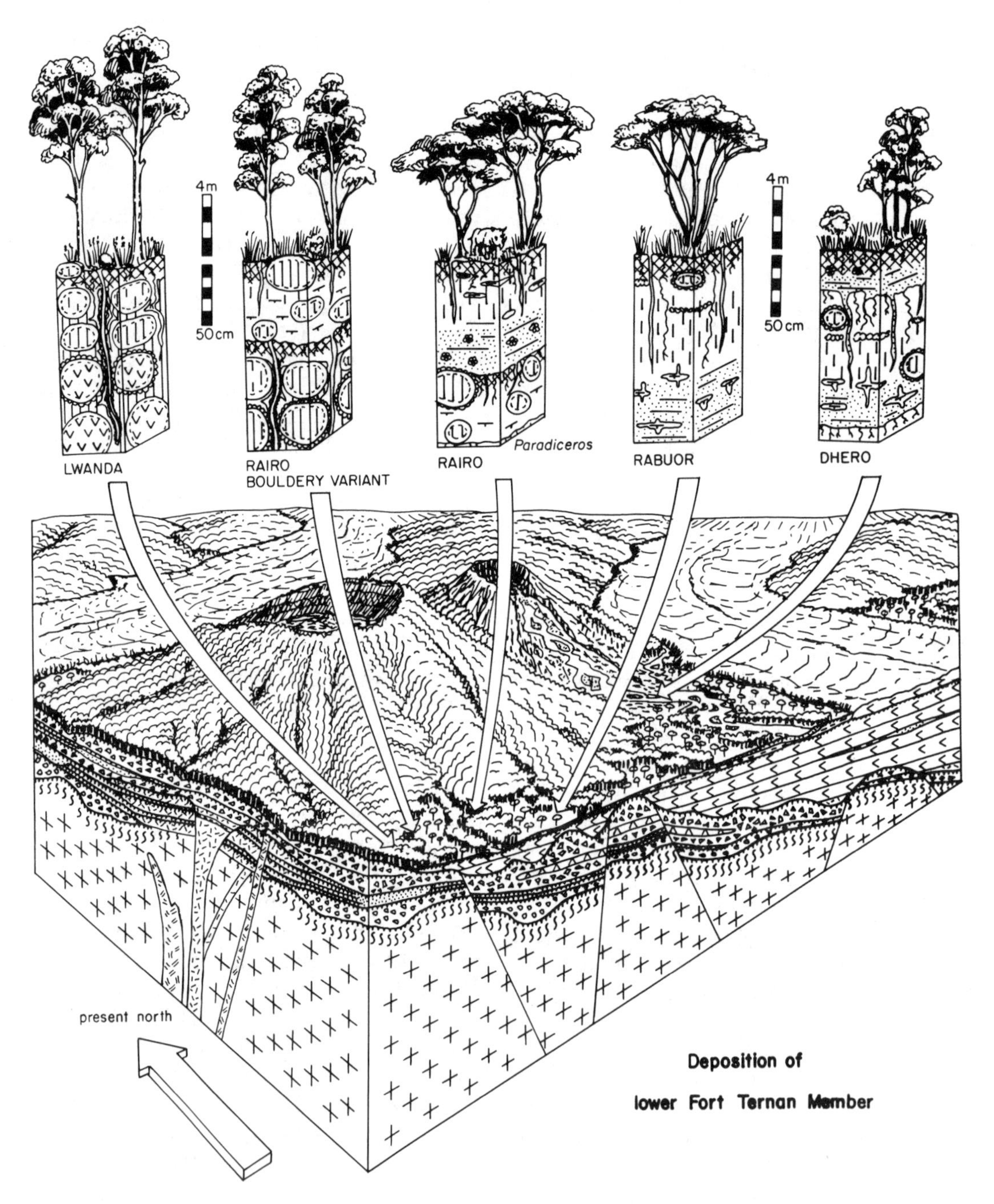

Fig. 6.4. Interpreted paleoenvironment, soils and vegetation of southwestern Kenya during the middle Miocene, some 15 million years ago.

a variety of Miocene paleoclimatic records from other parts of the world (Axelrod, 1989; Kennett, 1982; Rea, Leinen, & Janacek, 1985; Wolfe, 1981). Global climatic change may have been aided by atmospheric change from eruptions of flood basalt or from impact of large meteorites. The Ries and Steinheim Craters of West Germany were formed about 15 million years ago (Pohl *et al.*, 1977; Reiff,

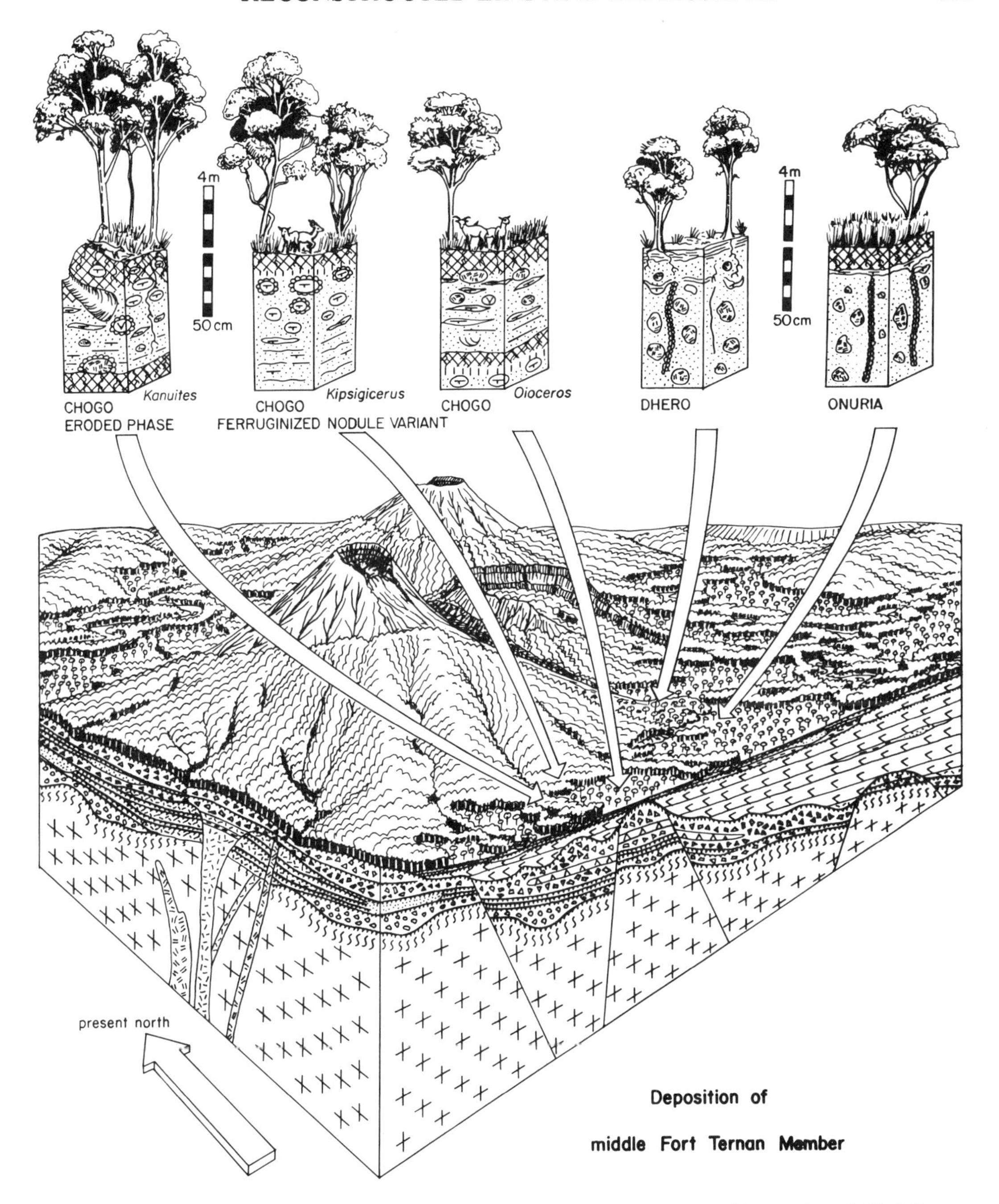

Fig. 6.5. Interpreted paleoenvironment, soils and vegetation of southwestern Kenya the middle Miocene, some 14 million years ago.

1977). The Columbia River Basalts of the northwestern U.S. are also of this age, and among the most voluminous outpourings of basalt in Earth history (Tolan, Beeson, & Vogt, 1984). At this time also, a major expansion of the Antarctic ice cap was encouraged by cooling of the north Atlantic Ocean following new connections with the Arctic Ocean across the subsiding Iceland-Faeroe ridge (Kennett, 1982). In

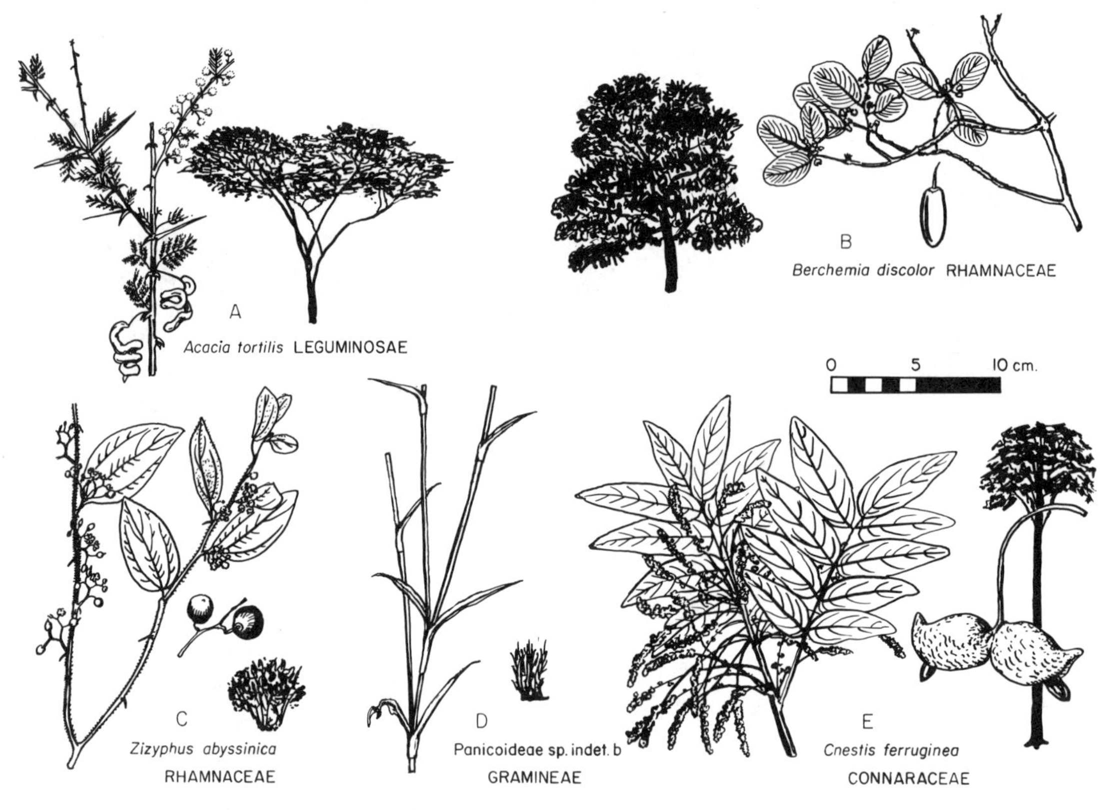

Fig. 6.6. Living African plants of genera found fossilized in the middle Miocene Fort Ternan beds, and a reconstruction of a fossil panicoid grass found there (illustrations and data from Burkill, 1985; Dugas, 1989; Eggeling, 1951; Hutchinson, Dalziel, & Keay, 1958; Johnston, 1972; E.Palmer & Pitman, 1972; Retallack, Dugas, & Bestland, 1990; Tree Society of South Africa, 1974).

East Africa this was a time of initial faulting of the western margin of the Gregory Rift and extrusion of voluminous plateau phonolites (B. H. Baker, 1987), so that there were local rain shadows and cool areas associated with elevation of the rift margin.

Along with climate, vegetation also had changed by middle Miocene time at Fort Ternan. A paleosol (Lwanda clay) developed on Baraget Phonolite some 15 million years old is thick and red, in some ways like forested soils, but has a finely structured clayey surface horizon and deep cracks, more like a soil of wooded grassland or grassy woodland. Colluvium and lahars of nephelinite and phonolite around volcanic footslopes at Fort Ternan include several very weakly developed paleosols (Rairo and Dhero Series). Judging from their relict bedding and stout root traces, these supported early successional woody vegetation. One paleosol on nephelinite-phonolite colluvium (type Rabuor clay) is thick, brown, clayey, and cracked, similar to soils now supporting wooded grassland and grassy woodland around phonolites in northern Serengeti National Park, Tanzania (Jager, 1982). Paleosols with thin, brown, granular-structured surface horizons on nephelinitic outwash sandstones (Chogo Series) and thick lahars (Onuria Series) are more like soils developed in carbonatite-nephelinite ash in the western Serengeti Plain, which support open grassland mosaic vegetation (de Wit, 1978; White, 1983). Some of these paleosols with an even boundary between surface clayey and

Fig.6.7. Reconstructions of selected middle Miocene mammals from the main fossil quarry at Fort Ternan National Monument, Kenya [illustrations redrawn from R. B. Horsfall for Scott, 1913 (scale); M. Lambert, 1981 (A,J); Thenius, 1972 (I): others original and based on sources listed in Appendix 9].

subsurface calcareous horizons and few stout root traces within the abundant fine ones (type Chogo clay and Onuria clay) may have supported open wooded grassland. This horizon boundary in other paleosols is disrupted by large root traces and the beginnings of a clayey subsurface horizon above the horizon of calcareous nodules (Chogo eroded-surface phase and Chogo ferruginized-nodule variant). These paleosols may have supported grassy woodland parts of the mosaic, in some cases (Chogo clay eroded-surface phase) flanking erosional gullies in the volcanic footslopes.

Former vegetation generally similar to modern Zambezian grassy woodlands and Somali-Masai *Acacia-Commiphora* wooded grasslands (of White, 1983) is indicated by fossils found in and on the paleosols. Fossil grasses and slender branches with stipular spines like those of living *Acacia* have been found in Chogo paleosols (Kenyan National Museum collections). A variety of tall panicoid and chloridoid fossil grasses and rare dicot leaf fragments are preserved in place of growth by nephelinitic sandstones overlying the type Onuria clay paleosol (Dugas, 1989; Retallack, Dugas, & Bestland, 1990). Fossil pollen grains isolated from the type Chogo clay are largely those of grasses (Bonnefille, 1984). Tree pollen found are largely of Afromontane affinities, and may have wafted in from much higher on the nearby volcano. Tree remains dominate a small assemblage of fossil fruits and seeds from the Chogo clay eroded phase paleosol (Shipman, 1977), including Annonaceae, Apocynaceae, *Cnestis*, Cucurbitaceae, Euphorbiaceae, Papilionaceae, *Berchemia*, *Zizyphus*, and *Celtis*. This assemblage of plants (Fig. 6.6) is most like that of modern lowland Zambezian riparian woodland (White, 1983). These fossil plants, like the paleosols at Fort Ternan, are evidence of a mosaic of woodland and wooded grassland.

Fossil mammals of Fort Ternan (Fig. 6.7) also reflect profound environmental change since the early Miocene, but this is less evident in fossil snail assemblages, which retained many woodland elements (Pickford, 1985). Most of these fossils were found in Chogo paleosols. The mammalian fauna includes large bulldozer browsers, such as gomphothere elephants (*Choerolophodon ngorora*) and rhinoceros (*Paradiceros mukirii*). There also were high browsers, including an ancestral giraffe (*Palaeotragus primaevus*). The dominance of antelope (particularly *Kipsigicerus labidotus* and *Oioceros tanyceras*) in this fauna compared with older Miocene faunas of East Africa also could be taken as an indication of open country, and some of the Fort Ternan antelope teeth do have microscopic striations similar to those found in animals that eat grass rather than leaves (Shipman *et al.*, 1981). However, the limb proportions and femoral head articulation of the fossil antelope from Fort Ternan are most like those of the living Indian nilgai (*Boselaphus tragocamelus*), which lives in thorn scrub and deciduous tropical forest (Gentry, 1970; Kappelman, 1991; Thomas, 1984). Other indications of woodland and forest vegetation include rare flying squirrels (*Paranomalurus* sp.), slow loris (Lorisinae indet.), and monkey-apes (cf. *Oreopithecus* sp., *Micropithecus* sp., *Proconsul* sp. cf. *P. africanus*). Also rare is the ape, *Kenyapithecus wickeri*, which has been documented only from the Chogo clay eroded phase paleosol, presumably formed under gallery woodland of a dry gully. Some preference for different parts of the mosaic also is shown by the common large mammals, but this is not nearly so marked as habitat preferences shown by modern faunas of grassland-woodland mosaics (Behrensmeyer, Western, & Dechant-Boaz, 1979).

The transition some 15 million years ago between mammalian assemblages like those of Koru, Songhor, and Rusinga Island, on the one hand, and Maboko and Fort Ternan, on the other hand, was one of the most profound faunal overturns in the fossil record of East African wildlife (a 77 percent difference from earlier faunas in analysis of Pickford, 1981). This faunal change was accomplished in part by immigration of mammals from Eurasia, where faunas gained a number of new African forms (Tassy, 1986; Thomas, 1985), including the first orang-utanlike apes (*Sivapithecus*) at Paşalar in Turkey (Andrews, 1990a). The magnitude and extent of these changes are indications that the underlying causes were more than just a local rain-shadow effect around a few East African volcanoes. Indeed, a popular explanation has been the invasion of Africa at this time by open-country, antelope-dominated faunas of eastern Asian origin (Gentry, 1970, 1987). This now seems unlikely, because paleogeographic connections to central Asia extended through high latitude, humid, forested lands (Steininger & Rögl, 1984). Reevaluation of Miocene mammalian faunas known from the Middle East, North Africa, and East Africa (Bernor, 1983) has led to the conclusion that much of the new open-country fauna of Fort Ternan evolved from ancestors that had been present in Africa at least since the faunal interchanges with Eurasia some 18 million years ago. The present region of the Sahara and Namibian deserts appears to have been a center for the evolution of these endemic African open-country faunas. The appearance of this fauna at Fort Ternan, Maboko, Nachola, and other localities in East Africa can be viewed as a geographic expansion of this dry climatic belt and its biota. Such a view also is supported by the degree of ecological integration of fossil plants and animals at Fort Ternan, especially in comparison with the tentative evolutionary experiment in an early grassland ecosystem revealed by studies of Oligocene paleosols in Badlands National Park, South Dakota (Retallack, 1983, 1988b, 1990a). The hypsodonty and striated microwear of some of the antelope teeth (Shipman *et al.*, 1981) may have been a response to grasses known from fossils to have been unusually heavily invested in phytoliths (Dugas, 1989; Retallack, Dugas, & Bestland, 1990). There were elephants and rhinos that could have destroyed trees and opened up the vegetation to grazers (Kortlandt, 1983), as well as sharp stipular spines on some of the trees (Kenyan National Museum collections).

The foregoing discussion should not be construed as implying that the fossil fauna of Fort Ternan was modern in all respects. Much has happened in the 14 million years separating this archaic grassland biota from the modern big game of East African grasslands. An especially significant event biostratigraphically was the immigration from Eurasia of hipparionine horses and other mammals some 9 million years ago (J. A. H. Van Couvering, 1980). There also has been an evolutionary diversification of antelopes, pigs, cercopithecid monkeys, and fissiped carnivores, along with an evolutionary decline in the diversity of elephants, apes, and chevrotains. Some archaic forms such as creodont carnivores and gelocid "deer" are now completely extinct. Along with these changes in taxonomic composition, the African fauna also has changed in adaptive grade. Increased hypsodonty of teeth and cursoriality of limb design are especially well seen in antelope and zebra. It is not surprising, then, that in comparison with modern faunas, the fossil assemblage from Fort Ternan is most like a woodland fauna (Evans, Van Couvering, & Andrews, 1981). Adaptively and taxonomically so it was, but fossil soils and plants presented here are evidence that it actually lived in an early version of grassland mosaic vegetation.

6.3 Late Miocene in Pakistan

The Indian subcontinent was part of the late Paleozoic and Mesozoic supercontinent of Gondwana. It drifted north to collide with Eurasia by Eocene time (Powell, 1979). By early Miocene time, underthrusting and crustal shortening had created forerunners of the Hindu Kush, Karakorum, and Himalayan ranges as a source for sediment of alluvial fans and vast river plains similar to the present lowlands of the Indus, Ganges, and Brahmaputra Rivers. During most of Miocene time, these great ranges were of alpine proportions. It is only since Pliocene time that these ranges and the Tibetan Plateau have been elevated to form the roof of the world (Dewey *et al.*, 1988).

The Indus and Ganges Rivers now penetrate deeply into this great wall of mountains as though their drainage predates uplift. During Miocene time, however, most rivers of northern Pakistan and India, including the present mountain headwaters of the Indus River, drained into an enlarged proto-Gangetic River, which flowed east into the Bay of Bengal (Raynolds, 1982). This main axial drainage was separated from an early forerunner of the Indus River emptying into the Indian Ocean by a line of hills, the Sargodha Ridge, of Precambrian basement on the northwestern margin of the Indian shield (Yeats & Lawrence, 1984). Since Pliocene time some 5 million years ago, overthrusting of the Potwar Plateau and foothill ranges has allowed progradation of alluvial fans over this basement ridge, and the Indus River has captured the former headwaters of the Ganges and its tributaries as far east as the Sutlej River. Like the ranges they drain, these streams transport mostly quartz, feldspar, and grains of schist and limestone. Their clays are largely illite, with some kaolinite and smectite. The modern Indo-Gangetic Plains are a superb natural laboratory for studies of soil formation, with their uniform alluvium over wide areas but varied climate from the western deserts to eastern dipterocarp jungles (Kooistra, 1982; Murthy *et al.*, 1982). The composition of the alluvial outwash and its depositional setting has been broadly similar for the past 20 million years. This thick pile of alluvial sediments of the Siwalik Group provides an outstanding record in fossil soils and mammals of paleoenvironmental changes since early Miocene time.

The assemblage of paleosols examined near Khaur, Pakistan, in that part of the Siwalik Group paleomagnetically dated (by Tauxe & Opdyke, 1982; corrected to time scale of Berggren, Kent, & Van Couvering, 1985) at some 8.5 million years old (Fig. 6.8) is similar overall to soilscapes now found much further east, for example, near Gorakhpur in Uttar Pradesh (Food and Agriculture Organization, 1977b). Judging from the depth of calcareous nodules, corrected for likely compaction, in moderately developed paleosols of former sub-Himalayan alluvial fans (Lal Series) compared with modern soils, mean annual rainfall was 500 to 750 mm. A slightly drier climate is indicated by the shallower calcareous nodules and more calcareous overall composition of paleosols in the former braid plain of the main proto-Gangetic River. The blue-gray paleochannels spread widely over the plains every 50,000 years or so, as estimated from times for formation of paleosols and from paleomagnetic data. This episodic expansion and retreat of the proto-Gangetic braid plain is not associated with the influx of coarse detritus or evidence of accelerated rates of sediment accumulation in paleosols that would be expected if it were cued to tectonic uplift. More likely, this episodic channel behavior was cued to climatic change due to Milankovitch orbital variations, as seen also in Miocene lake deposits elsewhere (Barnosky, 1984). Also seen in these Miocene paleosols was much evidence for a strongly seasonal climate: for example, co-occurrence of ferric concretions and calcareous nodules sometimes complexly intergrown, and calcareous rhizoconcretions deeply penetrating even in soils with evidence of seasonal waterlogging. Comparable features are common in Indian soils today (Courty & Féderoff, 1985; Sehgal & Stoops, 1972), and there are theoretical reasons to suspect that monsoonal circulation was strong over northern India and Pakistan throughout Miocene time (Parrish, Ziegler, & Scotese, 1982).

Fossil root traces and profile features of the paleosols 8.5 million years old near Khaur can be matched best among soils of the tropical moist deciduous forest zone of Uttar Pradesh and Bihar, India (Champion & Seth, 1968). This may have included early successional woodland (on Sarang Series with its prominent relict bedding), riparian forest (on brown Bhura Series), dry plains forest (on red Sonita Series), and a moist monsoon forest on the least disturbed parts of the floodplain (Lal Series with moderately developed argillic and calcic horizons). A few rare paleosols (1 Kala and 3 Pila Series out of 76 other profiles in one measured section) had thin, brown, granular-structured surface horizons, as well as mottles of a kind found now in seasonally waterlogged soils.

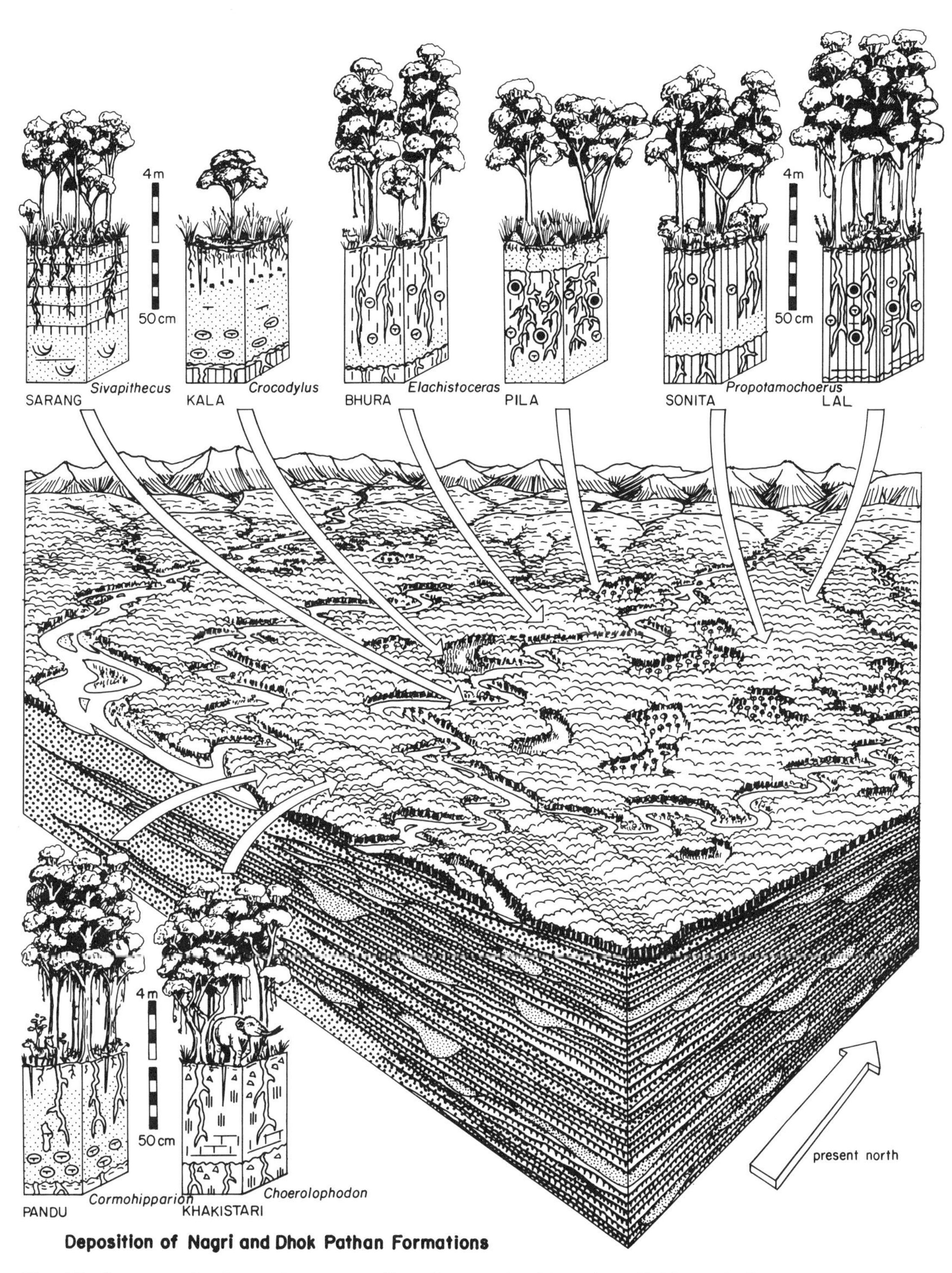

Fig. 6.8. Interpreted paleoenvironment, soils and vegetation in northern Pakistan, during the late Miocene some 8.5 to 8.6 million years ago.

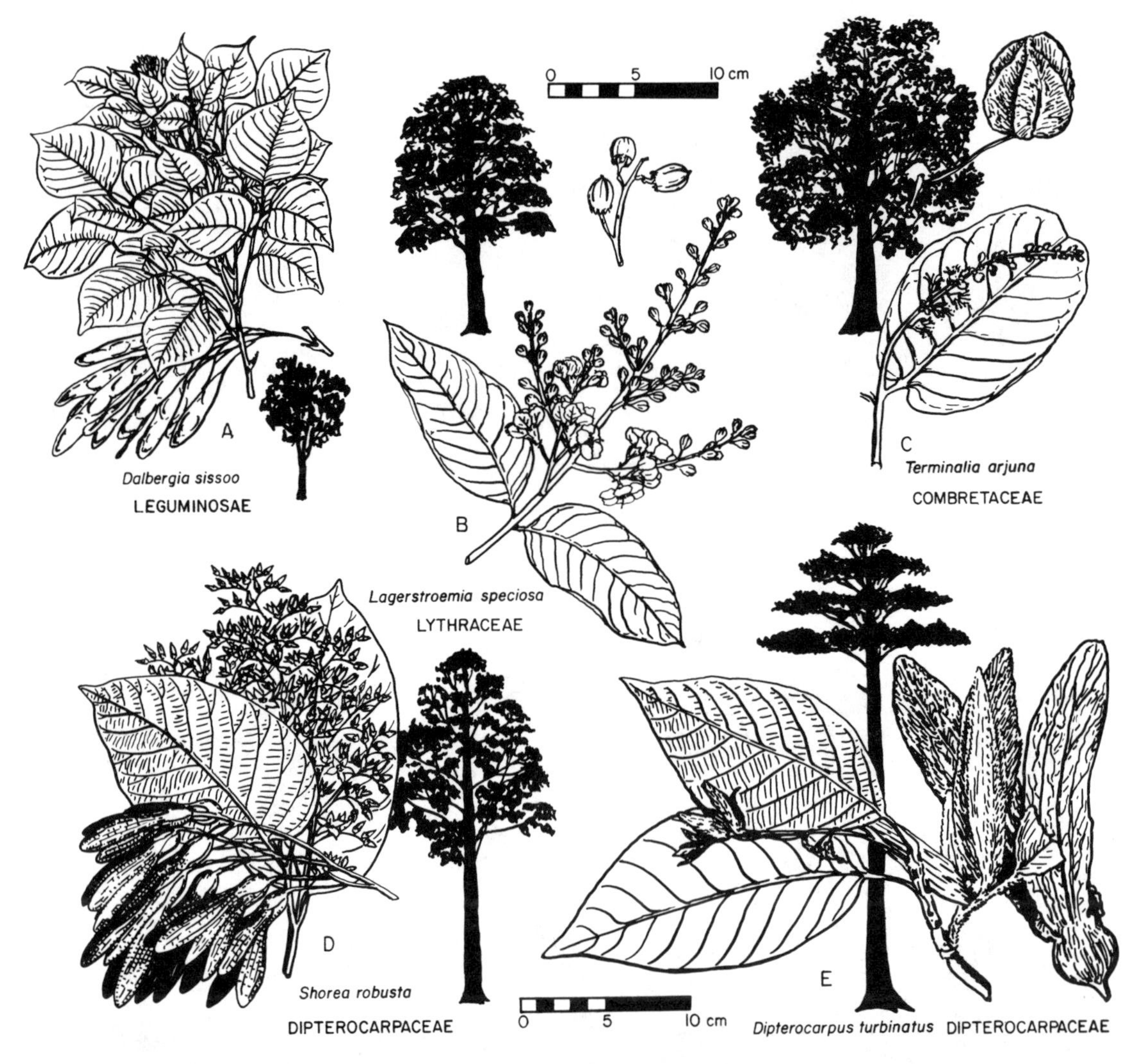

Fig. 6.9. Living Indian plants of genera found fossilized in the Miocene Siwalik Group of India (illustrations and data from Brandis, 1972; Champion & Seth, 1968; Gilg, 1960; Talbot, 1976).

These may have supported lowland alluvial wooded grassland in seasonally inundated abandoned channels (Kala) and clayey depressions (Pila). The foregoing paleosols and their vegetation were found in red and brown sediments of alluvial fans flanking soil-covered sub-Himalayan foothills. Two other kinds of gray paleosols may have supported seasonally swampy forests of sandy levees (Pandu Series) and clayey swales (Khakistari) within the braid plain of the proto-Gangetic River.

No fossil plants other than taxonomically indeterminate root traces and charcoal were found in these Pakistani paleosols, but there is evidence from fossil plants in Siwalik sediments in other areas that many of the plant associations now widespread on the Indo-Gangetic Plains had recognizable Miocene forerunners. For example, the early Miocene fossil flora closest to Pakistan, near Jawalamukhi, Himachal Pradesh, India (Awasthi, 1982; Lakhanpal & Guleria, 1986), has a taxonomic composition similar to that of modern moist Gangetic deciduous tropical

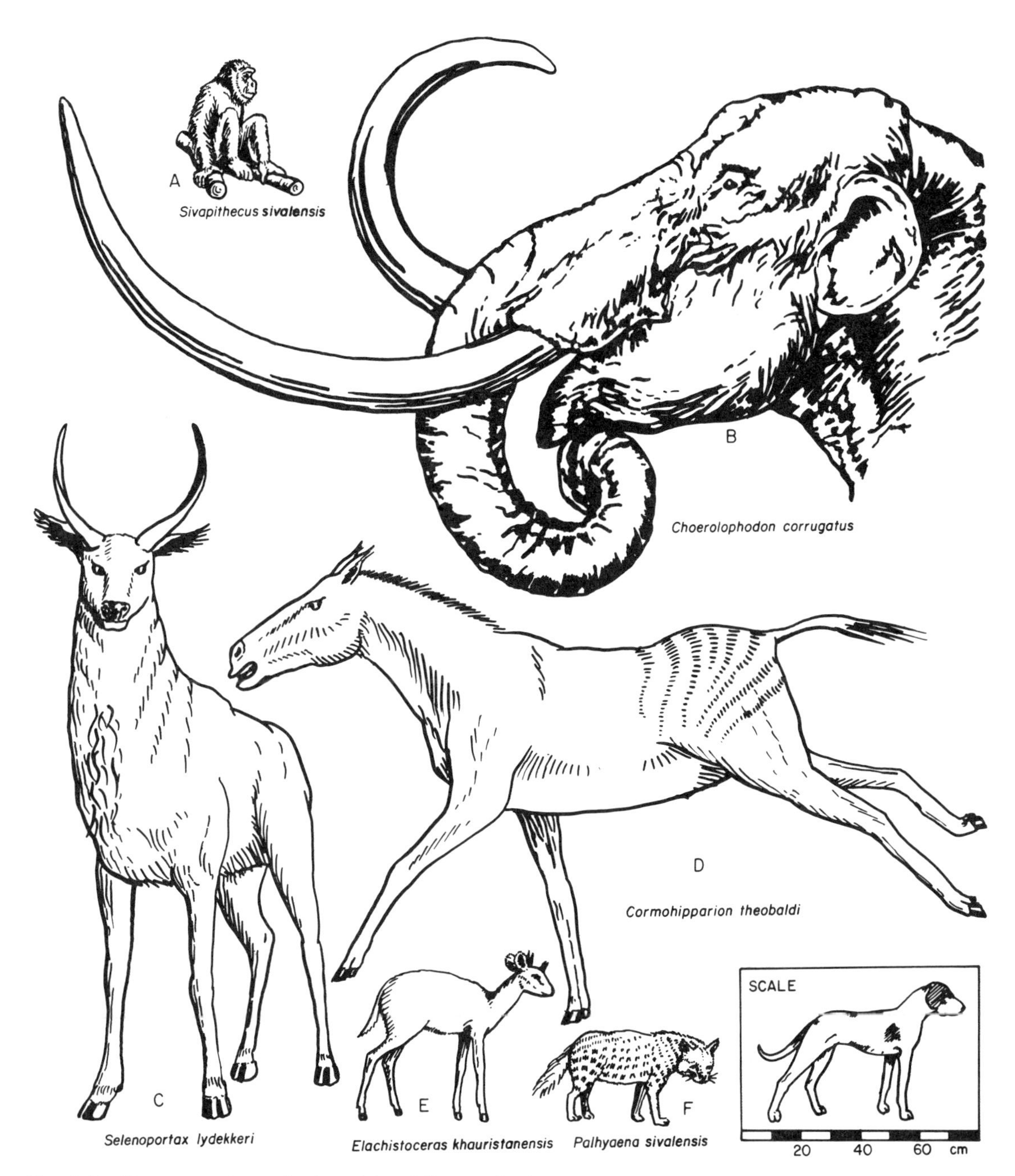

Fig. 6.10. Reconstructions of selected late Miocene fossil mammals from northern Pakistan [illustrations redrawn from R. B. Horsfall for Scott, 1913 (scale); Osborn, 1936-1942 (B); G. Paul for Stanley, 1986 (D); Savage & Long, 1986 (A): others original and based on sources listed for Appendix 10].

forest (of Champion & Seth, 1968). Other Indian fossil plant localities of the lower Siwalik Group, such as Nalagarh and Kalagarh, further east in Uttar Pradesh are of wetter forest types with a greater diversity and abundance of dipterocarps (Awasthi, 1982; G. D. Johnson, 1977). These various fossil floras do, however, include fossil relatives of plants now common in dry tropical forests (*Shorea robusta*) and associated stream margin scrub (*Dalbergia sissoo*), riparian woodland (*Lagerstroemia speciosa*), and seasonally dry swamp (*Terminalia arjuna*; Fig. 6.9). The fossil plants are also evidence that the ecotone between moist deciduous woodland and semievergreen forest in these alluvial lowlands was in Himachal Pradesh during early Miocene time, rather than further east in Bangladesh where it is today (Champion & Seth, 1968). Similarly, the late Miocene paleosols some 8.5 million years old from Pakistan reported here can be taken as evidence that moist deciduous tropical forest, which now extends not much west of the lowlands of Uttar Pradesh, was then as far west as Khaur, Pakistan.

Fossil mammalian faunas from Siwalik sediments near Khaur some 8.5 million years old (Fig. 6.10) were dominated by antelopes (*Miotragoceras*, *Elachistoceras*, *Selenoportax*) and horses ("*Cormohipparion*," ?*Hipparion*), which average 46 and 20 percent respectively of the assemblages in paleochannels where bone is locally abundant (Badgley, 1986). Bone also is locally common in nearstream paleosols (Sarang and Kala Series), but some was found in most of the different kinds of paleosols (with the exception of Lal and Pila Series). Although the data base of identified bones from paleosols could stand considerable improvement, some patterns of their distribution in the paleosols are worth mentioning if only to provoke future evaluation of their paleoecological significance. There is a surprising lack of habitat preference of large three-toed horses, "*Cormohipparion*" *theobaldi* and "*C.*" *perimense*, compared with that of other large mammals with equally robust teeth and bones. Their lack of habitat specificity within these tropical dry forests may have been related to their immigration from central Asia into the Indian subcontinent only a million years earlier (9.5 million years ago; Barry *et al.*, 1985). Crocodiles (*Crocodylus palaeindicus*) were seen in a paleosol thought to have formed in marshy wooded grassland (Kala Series). The small chowsinghalike antelope (*Elachistoceras khauristanensis*) was found in paleosols thought to have supported river margin colonizing (Sarang Series) and old growth forest (Bhura Series), whereas dry deciduous forest paleosols further from streams (Sonita and Pandu Series) yielded nilgailike antelopes (*Miotragocerus punjabicus*, *Selenoportax vexillarius*) and gazelles (*Gazella* sp.). Remains of large giraffe (cf. *Brahmatherium megacephalum*) and gomphothere elephant (*Choerolophodon corrugatus*) were found in a seasonally dry swamp forest paleosol of the proto-Gangetic braid plain (Khakistari Series), whereas another kind of elephant (*Stegolophodon* sp.) was found in streamside colonizing forest paleosols of the sub-Himalayan alluvial fans (Sarang Series). Remains of a large cat (*Felis* sp.) were found in a paleosol of marshy wooded grassland (Kala Series), and of early hyaena (*Palhyaena sivalensis*) and civet (*Viverra chinjiensis*) in a paleosol of streamside colonizing forest. Finally, the apes (*Sivapithecus sivalensis*, *S. punjabicus*) were found only in paleochannels and paleosols of streamside colonizing forest of the alluvial fans of the sub-Himalayan foothills (Sarang Series). Fossil apes are so rare that it is possible apes lived in a variety of dry forest paleosols from which few fossils of any kind have been found, but their absence in a large assemblage of bones (Badgley, 1982) from a marshy wooded grassland paleosol (Kala Series) may reflect avoidance of that habitat.

Some habitat specificity also is seen in trace fossils and snails in these late Miocene paleosols. Structures similar to termite galleries and fungus gardens were widespread in paleosols thought to be well drained for most of the year (Sarang, Sonita, Pila, and Lal Series), but were not seen in those with evidence of more prolonged wet-season waterlogging (Kala, Khakistari, and Pandu Series). Operculae of the seasonally aestivating pond snail, *Pila prisca*, were found in paleosols thought to have been near water or with a closed canopy vegetation (Bhura, Khakistari, and Lal Series), rather than in paleosols that were more distant from water or had open vegetation.

Another suite of paleosols at a higher stratigraphic level near Khaur, paleomagnetically dated at about 7.8 million years old, included many paleosols similar to those already discussed, although their diversity, depth, and degree of development were diminished (Fig. 6.11). No paleosols were found of the kinds at the lower level thought to have supported old growth dry forest, swamp forest, or low alluvial wooded grassland. These differences may record a shift to a drier and more monotonous tropical forest (on Sonita Series), with streamside colonizing forest (Sarang Series), riparian forest (Bhura Series), and local strips of tall riverain forest (Naranji Series) in abandoned channels of the sub-Himalayan alluvial fans. This sequence of paleosols in Pakistan 7.8 million years ago recorded a change from moist to dry deciduous tropical forest. Within the present Indo-Gangetic lowlands this ecotone is much further east, in Punjab and western Uttar Pradesh (Champion & Seth, 1968). It is a subtle ecotone botanically, because many of the same tree species are found in both kinds of forests. The ecotone is marked mainly by the lower diversity and lack of vines and epiphytes in the drier forests. This ecotone is now, however, an important biogeographic boundary for Oriental-Asian forest faunas to the east and Mediterranean-Ethiopian open-country faunas to the west (Mani, 1974c, 1974d; I. Prakash, 1974). By 7.8 million years ago in Pakistan, the rising Hindu Kush and Himalayan ranges were casting a rain shadow which was lengthening eastward. It had not yet, however, dried out to the extent that grasslands were widespread. Judging from the evidence of brown paleosols (Behrensmeyer, personal communication, 1989) and carbon isotopic composition of pedogenic calcareous nodules (Quade, Cerling, & Bowman, 1989), grasses did not become widespread until some 6 million years ago. The Miocene climatic drying of Pakistan was not so early nor severe as once thought (Krynine, 1937), but it had begun.

By 7.8 million years ago the mammalian fauna of the Potwar Plateau region of Pakistan suffered a subtle and protracted shift in composition that may be related to the drier climate and more monotonous dry tropical forest vegetation. A variety of rodents, chalicotheres, chevrotains, creodonts, and apes became locally extinct at about this time (Barry *et al.*, 1985). The last remains of apes *Sivapithecus sivalensis* and *S. punjabicus*, near Khaur in Pakistan have been paleomagnetically dated at about 7.5 million years old (Tauxe & Opdyke, 1982; corrected for time scale of Berggren, Kent, & Van Couvering, 1985). Near Haritalyangar in India, these orang-utanlike apes may have become extinct some 7.0 to 7.5 million years ago, but the giant ape *Gigantopithecus giganteus*, persisted in northern India until some 6.3 million years ago (G. D. Johnson *et al.*, 1983). These extinctions that reduced diversity of the mammalian fauna parallel the reduced diversity of paleosols seen in the upper measured section near Khaur, Pakistan. There also were some newly appearing species, particularly the large antelope *Selenoportax lydekkeri*. Fossil remains of this antelope show little preference for particular kinds of paleosol. They were found in all except Bhura paleosols and there were only two of these paleosols in the examined section. Also found in a well-preserved assemblage in paleosols (Sonita Series) of dry forest was three-toed horse ("*Cormohipparion*" spp.) and rhinoceros (*Chilotherium intermedium*). A variety of invertebrate trace fossils also were seen in this suite of paleosols some 7.8 million years old, including some evidence of fungus gardening termites and of clay shells, similar

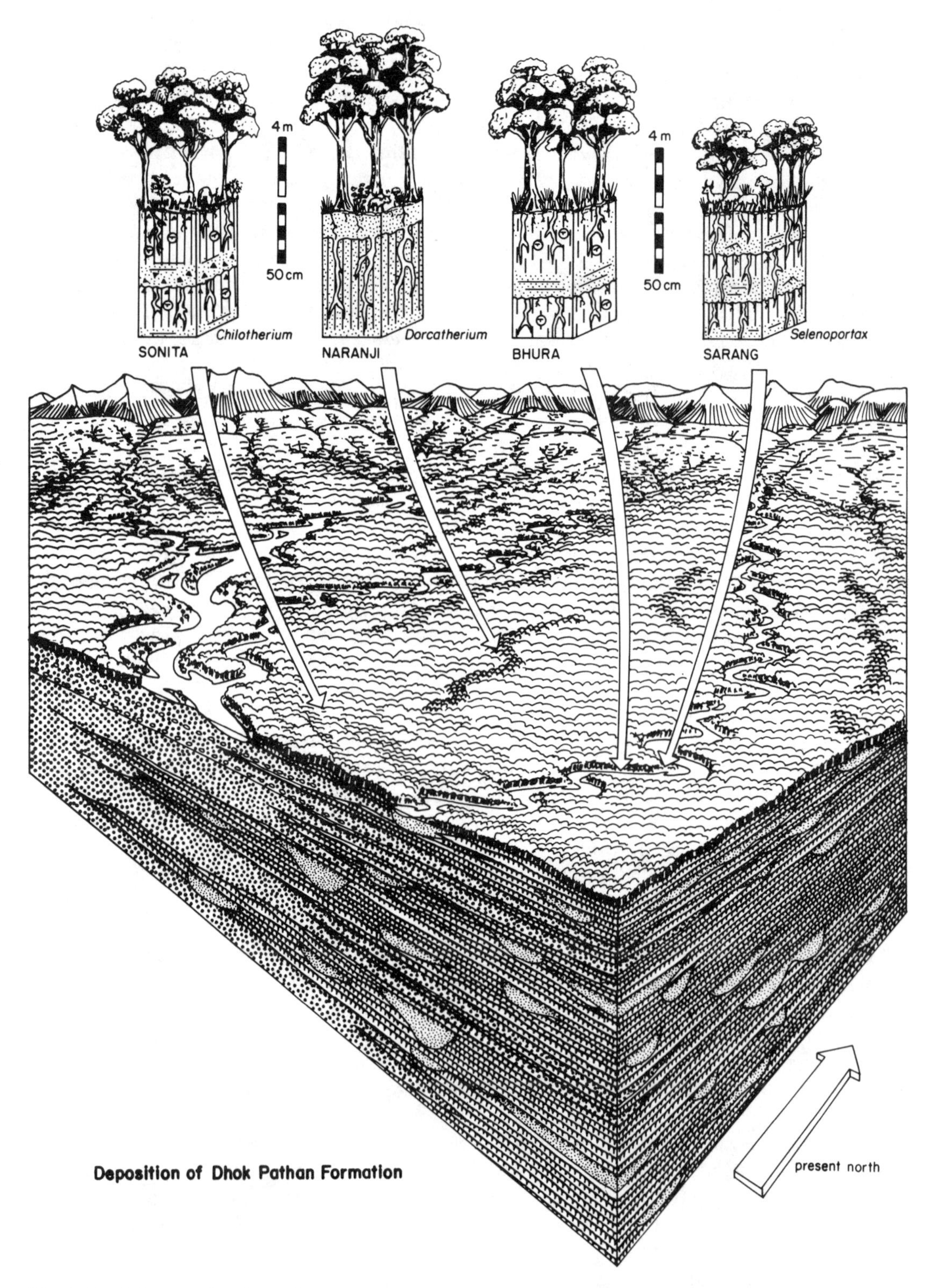

Fig. 6.11. Interpreted paleoenvironment, soils and vegetation of northern Pakistan, during late Miocene time, some 7.7 to 7.8 million years ago.

to the earthen lining in the brood cells of dung beetles, such as the living Indian *Heliocopris bucephalus*.

This floral and faunal shift due to climatic drying some 7 to 8 million years ago was a minor ecological adjustment in Siwalik faunas long isolated from other Asiatic and African faunas by the rising mountains of the Himalayan and other ranges (Barry *et al.*, 1985, 1986; Bernor *et al.*, 1988). The Siwalik Chronofauna can be traced back at least 15 million years to a time of global sea level and climatic change, as already discussed for Kenya. At this time an endemic Asiatic fauna of rhinoceroses and anthracotheres was enriched by the immigration from other parts of Asia and from Africa of antelope (?*Eotragus*), gomphothere elephants (*Deinotherium*, *Choerolophodon*), giraffes (*Giraffokeryx*), tragulids (*Dorcatherium*, *Dorcabune*), apes (*Dionysopithecus* sp. cf. *D. shuangouensis*) and cane rats (Thryonomyidae). The earliest orang-utanlike apes (*Sivapithecus*) appear at stratigraphic levels some 11.8 million years old. Another wave of immigration from other parts of Asia some 9.5 million years ago brought three-toed horses ("*Cormohipparion*," ?*Hipparion*), and some local extinctions, to create the fauna of the paleosols reported here. This was still very different from the modern fauna of the Indian subcontinent. The end of the distinctive Siwalik Chronofauna came when open-country mammals invaded Pakistan and western India from Africa and central and western Asia, beginning some 5 million years ago. This also was a time of pronounced global cooling, with expansion of the Antarctic ice cap, lowering of sea level, and desiccation of the Mediterranean Sea into a series of salt lakes (Kennett, 1982). The faunal overturn set in motion by these events was completed by about 3 million years ago, by which time the mammalian fauna of Indo-Pakistan was near-modern in general taxonomic composition. The most dramatic desertification of India has been accomplished by a successsion of human cultures over the past 7000 years (Allchin & Allchin, 1982).

6.4 What, If Anything, Is Characteristic of Tropical Soils?

A short answer to this question is, surprisingly, very little. During the pioneering days of soil science, the thin organic soils of temperate regions were contrasted with the thick red soils of tropical regions (Glinka, 1931). Such generalizations of soil geography are becoming increasingly difficult to maintain as more is learned of the diversity of tropical soils. For example, dark grassland soils have long been regarded as unique to temperate regions. They are shown that way in the Food and Agriculture Organization (1974, 1977a, 1977b) soil maps of the world, in which grassland soils of East Africa and India now known to be similar to those of Russia and North America (S. B. Deshpande, Fehrenbacher, & Beavers, 1971, S. B. Deshpande, Fehrenbacher, & Ray,1971; de Wit, 1978; Jager, 1982) were identified as weakly developed soils (Cambisols) rather than grassland soils (Chernozems).

As another example, the thick, red, deeply weathered, kaolinitic and lateritic soils long considered typical of tropical regions are now known in many temperate regions (McGowran, 1979; Paton & Williams, 1982; Singer & Nkedi-Kizza, 1980). In temperate regions such thick soils tend to be exhumed or relict profiles on land surfaces as old as Miocene. These could be urged as special circumstances that leave intact the generalization of thick strongly developed soils in the tropics. However, most of the thick red soils of tropical Africa and South America also can be shown to be relict paleosols on stable, Precambrian crystalline basement rocks and on land surfaces dating back to the Miocene or earlier (McFarlane, 1976). Many tropical soils were less affected by

Pleistocene glaciation than soils of temperate regions, and so are more ancient (Birkeland, 1984). The red color of soils also is related to their age. With time, brown ferric hydroxides such as goethite dehydrate to red ferric oxides such as hematite (Ruhe, 1969), a process that continues as soils are buried and dewatered (T. R. Walker, 1967). If such prominent soil features as profile thickness, redness, degree of weathering of clay, and organic matter content are related more to time for formation and vegetation of soils, then what is distinctive about tropical soils and paleosols?

It is small comfort to point out that soils and paleosols of warm climates lack permafrost features such as stone stripes, ice wedges, and sand wedges (Williams, 1986). There are, however, a number of potential indicators of warm temperate to tropical climate that could be used to infer such climatic conditions from paleosols. As suggested in the course of this research, these include: (1) abundance of burrowing organisms, especially those of termites or other creatures restricted to warm climates; (2) distinctive spherical micropeds of deeply weathered sand-size grains of claystone; (3) deeply and complexly sculpted paleokarst in calcareous bedrock; and (4) calcareous rhizoconcretions or joint occurrence of ferric concretions and calcareous nodules as evidence of intense seasonality of rainfall.

In general, life is more abundant and diverse in tropical than in temperate regions, and this may be reflected by extensive and deep bioturbation of tropical soils. However, many factors other than warmth might compromise such a generalization. The degree of bioturbation and diversity of burrows is, for example, adversely affected in waterlogged sites, in dry climates, and under some kinds of trees with phenolic chemical poisons. Thus the degree and abundance of bioturbation are insecure guides to warm paleoclimate in paleosols, compared with specific kinds of burrows excavated by animals known to be restricted to tropical regions. The complex mounds, subterranean fungus gardens and ramifying galleries of termites can be very distinctive (Bown, 1982; Sands, 1987; Tessier, 1959). If present at all, they usually are abundant. Termites today are almost exclusively tropical. "In no case are the climatic areas with cold winters and cool summers occupied by termites. The 49°F [8°C] annual isotherm line of both hemispheres encloses almost all native species" (Emerson, 1952, p. 172). The few termite species penetrating temperate regions are primitive forms that nest in wood, rather than in the ground. One limitation on the use of termite nests in paleosols as indicators of tropical paleoclimate is that fossils of termites and their borings in fossil wood are known only back to Cretaceous (Burnham, 1978; Rohr *et al.*, 1986). More may be learned of paleoclimate for paleosols when specific kinds of termite nests are compared with modern kinds of nests (Sands, 1987). Such a comparative approach may also be useful for evaluating the paleoclimatic significance of burrows of wasps, bees, and dung beetles, now being discovered in paleosols (Bown & Ratcliffe, 1988; Retallack, 1984b, 1990b; Ritchie, 1987).

Spherical micropeds of tropical soils are small (10-μm to 1-mm), near-spherical grains of strongly ferruginized and often kaolinitic or gibbsitic claystone (see Figs 2.28 and 2.29). There may be small quartz grains within them, but they contain no easily weathered minerals. They are hard and chemically inert. Many tropical soils consist almost entirely of such grains, and so seem sandy to the touch even though the main minerals are clays. Because of this stable porous structure, these soils may be excellent agricultural resources, provided they are copiously fertilized. Some spherical micropeds may be redeposited fragments from older tropical soils. They also are fashioned by the mouthparts of termites, or as fecal pellets of these and other soil invertebrates (Mermut, Arshad, & St Arnaud, 1984). Most fecal pellets in soils, however, have ovoid to tubular shapes, more organic matter, and a porous framework of easily weatherable minerals not seen in spherical micropeds (Retallack, 1990a). Spherical micropeds have been reported widely from tropical soils (Stoops, 1983). The geological range of spherical micropeds in paleosols remains little known: from my experience they can be traced back at least to middle Eocene time (Retallack, 1981; Smith, 1988).

The giant karst of Sichuan, celebrated in traditional Chinese watercolors, is a good example of the kind of rugged and intricate relief that can develop in limestone weathered under tropical conditions (Sweeting, 1973). In cool humid climates, karst relief is more subdued, with narrow dolines or grikes dissecting a level plateau of limestone. Numerous exceptions to this generalization now are known. Karst of so-called tropical form can be found in landscapes of steeply dipping limestone of highly variable purity or on very old landscapes (J. N. Jennings, 1985). A potentially valuable indicator of tropical paleoclimate is the unusually spongy and jagged karst, called black phytokarst. This is created by the activity of endolithic algae (Folk, Roberts, & Moore, 1973), but the climatic distribution of this form of karst is not well established. There are additional problems for interpreting buried karstlike relief, now known on carbonate rocks as old as Precambrian (Button & Tyler, 1981; Choquette & James, 1987: Schau & Henderson, 1983). Irregularly sculpted limestone also can result from dissolution below a soil mantle. One form of this cryptokarst is cavernous subsoil weathering, thought to indicate tropical conditions (V. P. Wright, 1982). An additional difficulty for ancient examples of karst is the problem of finding exposures large enough to reveal the entire relief of tropical paleokarst (Leary, 1981). There are thus many difficulties in evaluating the paleoclimatic significance of paleokarst, and these difficulties compromised interpretation of the paleokarst reported here on a Miocene carbonatite welded-tuff in Kenya (see Fig. 2.24). This particular failure can be blamed mainly on the weak development of this paleokarst, but I suspect that other studies of paleokarst will be equally inconclusive as guides to former climatic temperature.

A final possible line of evidence for tropical climatic conditions is the variety of indications in paleosols and soils of India and Pakistan of monsoonal climate. A strongly marked wet and dry season is indicated by the coexistence of deeply penetrating calcareous rhizoconcretions and drab soil with iron-manganese nodules (for example in Pandu Series paleosols). There also are paleosols (Pila Series) with both calcareous nodules and ferric concretions. It could be argued that such a mix of features represents a dry climate, for formation of calcareous rhizoconcretions and nodules, followed much later by wet climate, encouraging ferric concretions and iron-manganese nodules on the time scale of Milankovitch climatic change. However, none of these paleosols are so well developed as even the shortest, or 23,000-year Milankovitch cycle. This argument is also negated by the intergrowth of wet-season and dry-season features, both in the paleosols reported here (see Figs 2.1 and 2.12) and in surface soils of the Indo-Gangetic alluvial plains (Kooistra, 1982; Sehgal & Stoops, 1982). Monsoonal climates are not the only strongly seasonal climates, and it is possible that mixed paleoclimatic indicators may be found also in soils and paleosols of Mediterranean, cool temperate, and frigid climates. In these other seasonal climates, however, biological activity and physicochemical weathering are not so profound during the harsh season as in monsoonal climates. Monsoonal climatic features such as dispersed horizons of small, concentrically ferruginized, carbonate nodules are known in paleosols ranging back in age to Ordovician (Retallack, 1985).

In conclusion, tropical soils can be seen

to be more varied and less enigmatic than was apparent to the pioneers of soil science and their intellectual heirs working largely on postglacial loess and till soils of Europe and North America. Tropical soils are similar to those of warm temperate regions, and there are only a few features that can be used to interpret paleotemperature from paleosols. Considering the geographic restriction of Pleistocene glaciers and how short-lived was their expansion in the long run of geological time, tropical soils should be considered normal soils and play a greater role in studies of soil genesis than they have in the past. Miocene tropical paleosols like those described here have special value as windows on soil formation under conditions not greatly different from those found today, but without the confounding effects of human disturbance.

6.5 Expansion of Grasslands in the Old World Tropics

Paleosols offer a unique line of evidence for interpreting the evolution of vegetation such as grasslands. Unlike plant fossils, which can be transported considerable distances and are in any case seldom preserved in the dry, oxidizing habitats of many grasslands, paleosols are in the place they formed. Unlike vertebrate fossils, whose morphology evolved in response to paleoenvironmental pressures by means of a cumbersome process of selection of mutations within populations, paleosols have been relatively passive products of paleoenvironmental forces. The dark granular-structured surface horizons (mollic epipedons) of grassland soils, for example, reflect the dense mat of fine grass roots. The Miocene appearance of paleosols with the characteristic features of grassland soils in Kenya and Pakistan can now be seen as evidence additional to that of fossil plants and animals for understanding the origins of this major new kind of ecosystem in the Old World tropics.

The most ancient grasslands yet known from India and Pakistan are indicated by paleosols (Pila and Kala Series) known to be as old as 8.6 million years and thought to have supported seasonally waterlogged to marshy wooded grassland of abandoned channels and other local alluvial depressions. Preliminary studies of paleosols on Rusinga Island, Kenya, are now revealing that similar local dambo grasslands may have been present in Kenya as many as 17 million years ago. These kinds of grasslands are quite different from the well drained wooded grassland on rich volcanic soils envisaged at Fort Ternan 14 million years ago. This is currently the oldest evidence for grassland mosaic environments of the kind now so prominent in the national parks of East Africa (Retallack, Dugas, & Bestland, 1990).

The difference between these two kinds of grassland is not restricted to the various soil features of waterlogging, but is reflected also in the nature of their grasses. Seasonally waterlogged depressions, locally called "vlei" or "dambo" grasslands, in Africa today include sedges and arundinoid and bambusoid grasses, whereas dry grasslands like those of the Serengeti Plains of Tanzania are mainly panicoid and chloridoid grasses (White, 1983). Fortunately, a nephelinitic sandstone buried one of the grassland paleosols at Fort Ternan and preserved a suite of panicoid and chloridoid grasses (Retallack, Dugas, & Bestland, 1990). A tuff bed that preserved grasslike fossils some 23 million years old near Bukwa, Uganda, includes rhizomes with the characteristic form of sedges found along alkaline lake margins (A. C. Hamilton, 1968). Such local marshy grasslands or grassy woodlands may be an explanation for fossil birds more than 31 million years old related to forms now frequenting grasslands (Olson & Rassmussen, 1986). There is little evidence for extensive grasslands in the adaptive features or modern relatives of fossil mammals that old in Africa, or, for that matter, in the snail and mammal fauna of 14 million years ago at Fort Ternan (Evans, Van Couvering, & Andrews, 1981; Pickford, 1985). Nor is there much evidence for open grassland among late Miocene mammalian faunas of Pakistan (Badgley & Behrensmeyer, 1980). The high crowned teeth and elegant cursorial limbs of grassland mammals of the Old World presumably took some time to evolve under the selective pressure of open grassy vegetation.

Although Fort Ternan has furnished the oldest evidence of dry grasslands in the Old World tropics, it seems unlikely that grasslands originated there. The paleosols and fossil grasses at Fort Ternan are evidence of a tall grassland of a kind now promoted by large bulldozer herbivores such as elephants and rhinoceroses or by human disturbance (Kortlandt, 1983). The fossil grasses are diverse and copiously invested in silica phytoliths, as in modern grasses under considerable grazing pressure (Herrera, 1985; McNaughton & Tarrants, 1983). Some fossil antelope teeth show the striated microwear of grazers (Shipman *et al.*, 1981). There had been modest plant-animal coevolution, more than found in tentative early attempts at grassland ecosystems during Oligocene time in North America (Retallack, 1982, 1983, 1988b, 1990a). The middle Miocene fossil fauna of Fort Ternan, although adaptively most like modern faunas of woodland, was significantly better adapted to open vegetation than were early Miocene faunas of the region. Its appearance in East Africa some 15 million years ago was an abrupt faunal overturn, but it is likely that such faunas had evolved before that time in other drier areas of Africa, such as the present regions of the Saharan and Namibian deserts (Bernor, 1983). How much earlier, or whether, grasslands evolved in these regions may be discovered when sequences of paleosols there are examined with these questions in mind.

Since 15 million years ago, the grassland biome has expanded both its geographic and climatic range in the Old World tropics. Geographic expansion of grasslands can be seen both from studies of paleogeography (A. Sahni & Mitra, 1980; Scotese, Gahagan, & Larson, 1989; Steininger & Rögl, 1984), paleosols (herein; as well as by Bestland, 1990a, 1990b; Mutakyahwa, 1987; Thackray, 1989; Whybrow & McClure, 1981) and fossil plants (Awasthi, 1982; Axelrod, 1975, 1986; Bancroft, 1932a, 1932b, 1933; Beauchamp, Lemoigne, & Petrescu, 1973; Bonnefille, 1984; Chaney, 1933; Chesters, 1957; Collinson, 1983; A. C. Hamilton, 1968; Ilinskaya, 1988; Jacobs & Kabuye, 1987; Kong & Du, 1981; Kovar-Eder, 1988; Lakhanpal, 1966; Lakhanpal & Guleria, 1986; Lakhanpal & Prakash, 1970; Lemoigne & Beauchamp, 1972; Lemoigne, Beauchamp, & Samuel, 1974; Mai, 1981; Y. K. Mathur, 1984; M. Prasad & Prakash, 1984; U. Prakash & Prasad, 1984; Sah, 1967; Song & Liu, 1981; Song, Zheng, & Liu, 1984; Tripathi & Tewari, 1983; Xu, 1981; Yemane, Bonnefille, & Faure, 1985), compared with the modern distribution of vegetation over this region (Walter, 1973). The maps compiled here from this data base of published studies (Fig. 6.12) are still in many ways inadequate, but do refine previous attempts to map Miocene vegetation of this region (Axelrod & Raven, 1978; Lakhanpal, 1970; Wolfe, 1985). Especially well documented is the retreat of rain forest eastward through the Indo-Gangetic Plains into Bangladesh, ahead of the belt of moist monsoon forest, then dry monsoon forest, wooded grassland, and desert scrub. In the Khaur area of Pakistan, now supporting desert scrub, the paleosols reported here are evidence that the ecotone between moist and dry monsoon forest passed eastward between 7 and 8 million years ago. In other regions of the Old World tropics now desert, such as Arabia, Sahara, and Namibia, it is also likely that dry woodland and then wooded grassland appeared and then expanded their range. The most pronounced shifts in vegetation appear to have been cued to

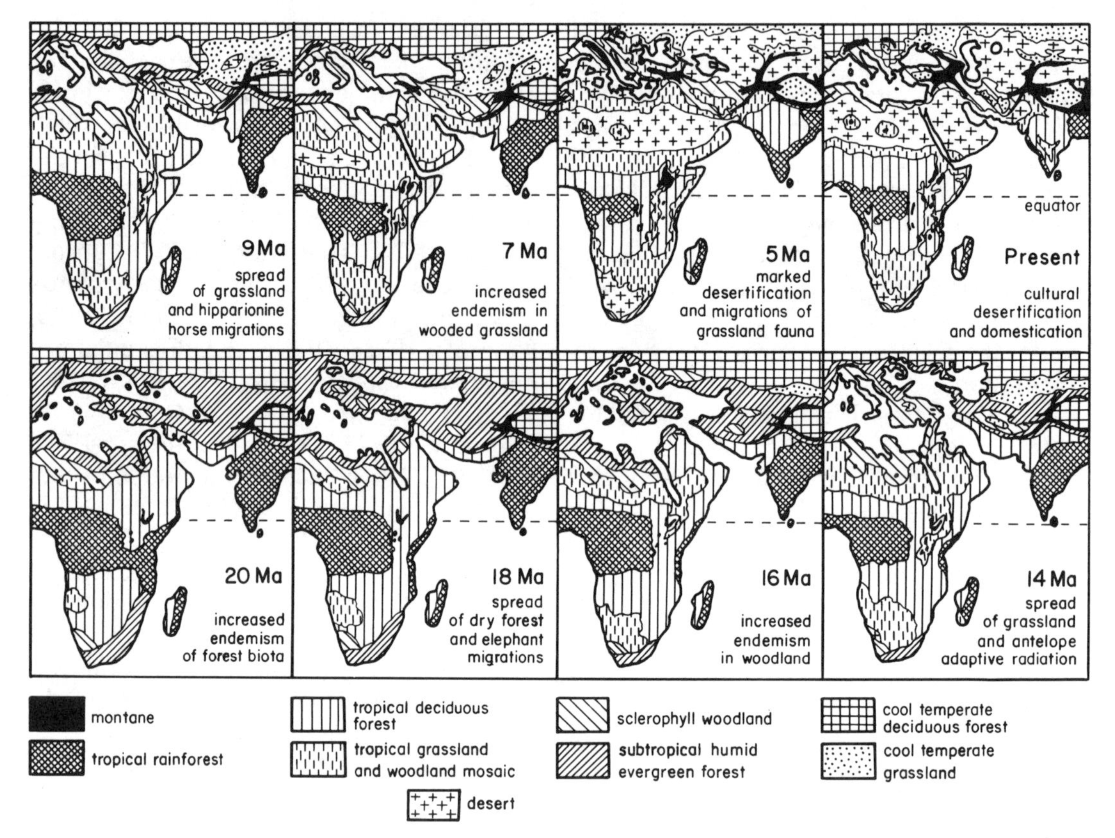

Fig. 6.12. Interpreted late Miocene to Pliocene and Recent phytogeography of the Old World tropics, showing expansion of tropical grassland and Mediterranean sclerophyll woodland at the expense of Tethyan laurel forests: based on evidence of shorelines (Steininger & Rögl, 1984), plate tectonics (Scotese, Gahagan, & Larson, 1989), modern plants (Walter, 1973), fossil plants (principally from Axelrod & Raven, 1978; Awasthi, 1982; Wolfe, 1985) and paleosols (herein as well as Bestland, 1990a, 1990b; Dugas, 1989; Mutakyahwa, 1987; Retallack, Dugas, & Bestland, 1990; Thackray, 1989).

global climatic changes 15, 10, and 5 million years ago (Kennett, 1982; Wolfe, 1981). These climatic oscillations were not always in the direction of drying. In Kenya, for example, there may have been expansion of rain forest 12 million years ago (Jacobs & Kabuye, 1987). Changes in vegetation over this interval were complex both in space and time, and may be considerably refined as additional sequences of paleosols come under scientific scrutiny.

Expansion of the climatic range of grasslands also is apparent from this study of paleosols. At Songhor, Koru, and Rusinga Island in Kenya, and near Khaur in Pakistan, paleosols contain root traces and other soil features of a variety of dry woodlands and forests, with some local seasonally waterlogged and early successional grassy vegetation. At Fort Ternan in Kenya there is evidence from fossil soils and grasses of wooded grassland and grassy woodland. The depth of calcareous nodules in all of these paleosols, corrected for burial compaction, indicates that mean annual rainfall at the grassland-woodland ecotone some 20 to 7.5 million years ago was about 400 mm (Fig. 6.13). Today in Africa this ecotone is at about 750 mm mean annual rainfall (Lind & Morrison, 1974; White, 1983). It may be that grasslands have expanded their climatic range from semiarid into subhumid regions during Pliocene to modern time.

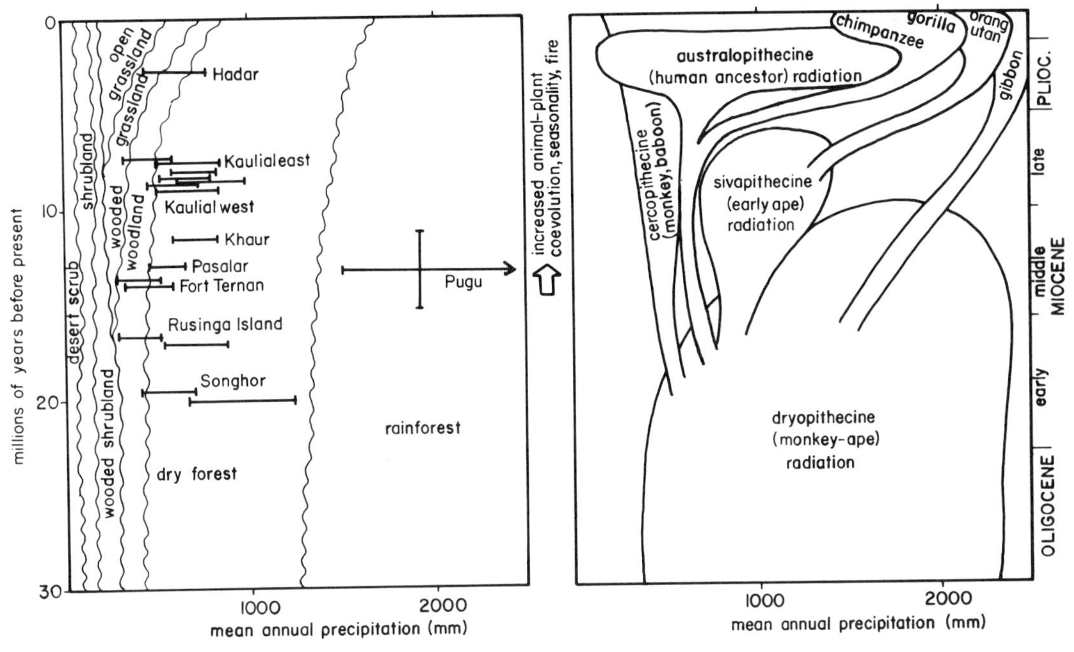

Fig. 6.13. An interpretative scenario for the changing Neogene climatic range of grasslands (left) and for the habitats of non-human primates (right) in the Old World tropics. Former rainfall is interpreted from depth to calcic horizon and vegetation from root traces and soil structure of paleosols (as for Fig. 6.12). Phylogenetic scheme for primates is after Andrews (1986); time scale from Berggren, Kent, and Van Couvering (1985).

Such evidence for geographic and climatic expansion of the grassland biome could clearly stand further refinement, but is by no means unexpected in theory. Grasslands are favored by dry unstable climates (Walter, 1973). During Pleistocene climatic fluctuations, grasslands expanded over enormous areas now under desert during wet phases, and expanded into areas now under woodland during dry phases, both in Africa (A. C. Hamilton, 1982; Van Zinderen-Bakker, 1976) and Indo-Pakistan (Allchin & Allchin, 1982). In addition, monsoonal circulation has been greatly strengthened by Plio-Pleistocene uplift of the Himalaya (Dewey *et al.*, 1988) and East African Rift System (B. H. Baker, 1987). Grasslands also are favored by nutrient rich soils, such as the volcanic ash soils of East Africa (Bell,1982) and calcareous alluvium of the Indo-Gangetic Plains (S. B. Deshpande, Fehrenbacher, & Beavers, 1971; S. B. Deshpande, Fehrenbacher, & Ray, 1971; Hurelbrink & Fehrenbacher, 1970), compared with nutrient-depleted lateritic soils that would have been more widespread earlier during Tertiary time on Precambrian basement of both India and Africa (McFarlane, 1976).

To these extrinsic factors can be added the continued coevolution of grasses and grazers (Bakker, 1983; Gabunia & Chochieva, 1982; Stebbins, 1981; Thomasson, 1985). The hidden rhizomes and meristems, telescoped leaves, abundant phytoliths, and small flowers and seeds of grasses enable them to withstand grazing better than many other kinds of plants. High crowned teeth and elongate limbs allow antelope and zebra to masticate dusty grasses rich in phytoliths and to flee predators over the open range. Elephants and grasshoppers also play a role in promoting grassland at the expense of woodland (Kortlandt, 1983; Owen-Smith,

1987). Grasslands also are promoted by humans, through land clearance, agricultural use of wheat and other grasses, and carelessness with fire (Johanessen, Davenport, & McWilliams, 1971; Vogl, 1974). Grasslands can thus be seen as a biological force that became more potent through time because of coevolution of their various elements.

6.6 Initial Steps in the Evolution of Human Ancestors from Apes

The view that the evolution of early human ancestors from apes was related to the transition from forest to wooded grassland has been popular since it was proposed by Charles Darwin (1871). On the other hand, arguments also have been advanced that human evolution from apes was not related to environmental change (Lovejoy, 1981). Paleosols have a role to play in this debate, as evidence for limitations of preservation and for former vegetation inhabited by the fossils of apes and early human ancestors they contain.

Many recent discoveries have served to refine and extend this debate, particularly with respect to the times when various human characteristics appeared. There now are indications that brain expansion clearly beyond the range of modern apes did not begin until the appearance of *Homo erectus* some 1.5 million years ago (Brown *et al.*, 1985; Tobias, 1981). There is evidence for the use of fire for cooking some 1.4 million years ago (Gowlett *et al.*, 1981), and stone tools of the Acheulian industry are almost this old. Crude pebble tools of the Oldowan tradition date back some 2.6 million years (Roche & Tiercelin, 1977). These tools were used by several species of early human ancestors that could walk erect as long ago as 3.7 million years (M. D. Leakey & Harris, 1987). These australopithecines remained in many ways apelike, especially in their cranial capacity and other features of their skulls. From these fossils back in geological time to Indian apes of 8 million years ago or African apes of 14 million years ago, fossils of any kind relevant to human ancestry are few and far between (Hill & Ward, 1988). Yet studies of the genetic similarity of modern humans and apes (Miyamoto, Slightom, & Goodman, 1987; Sarich & Wilson, 1967) reveal that the human evolutionary lineage diverged from that of the African apes (chimpanzees and gorillas) at about this time, some 5 to 10 million years ago. It now seems likely that australopithecines, so far known only in East and South Africa, evolved from generalized ape-monkeys such as *Proconsul* (Ward & Pilbeam, 1983), apes such as *Kenyapithecus* (Wolpoff, 1983) or *Ouranopithecus* (de Bonis *et al.*, 1990), or some unknown ancestor of *Sivapithecus* (Andrews, 1986). The Asian apes *Sivapithecus* and *Lufengpithecus*, now known from Turkish, Pakistani, and Chinese skulls from 15 to 7 million years old, already were more like the orang-utan (*Pongo*) than the chimpanzee (*Pan*) or gorilla (*Gorilla*), viewed as anatomically and genetically closer to australopithecines and humans (Andrews & Cronin, 1982; Fleagle, 1988; Pilbeam, 1982). These Asiatic ape fossils were once regarded as human ancestors, and that view is argued still in some quarters (Oxnard, 1987; Schwartz, 1984, 1987). Understanding the phylogeny and biogeography of human origins remains dependent on those all too rare fossils of creatures intermediate between apes and humans.

If the study of paleosols can help with these perennial questions of human origins it is in revealing preservational biases in the available record of ape and human ancestor fossils in paleosols. That many of these fossils were found in paleosols is now extensively documented (Badgley, 1986; Badgley for Pilbeam *et al.*, 1980; Behrensmeyer, 1987; Behrensmeyer for R. E. Leakey, Leakey, & Behrensmeyer, 1978; Bestland, 1990a; Bishop & Whyte, 1962; Hay, 1978; Pickford, 1984, 1986a; Pickford & Andrews, 1981; Retallack, 1986b, 1990a; Retallack, Dugas, & Bestland, 1990;

Shipman, 1977, 1981; Shipman *et al.*, 1981). The main destroyer of bones and teeth in soils is acidity of both soil solutions and rain water (Retallack, 1984a). Unpolluted rain water has a pH of 5.6, and may have been slightly more acidic in the geological past when atmospheric carbon dioxide levels were higher (Holland, 1984). Acidic soil solutions also are maintained by weak carbonic acid from the respiration of carbon dioxide by roots and soil animals (Brook, Folkoff, & Box, 1983), as well as organic acids and phenolic compounds leached from living and decaying plants (G. C. Fisher & Yam, 1984). Thus bone is seldom preserved in highly productive or acidic soils of humid climates, but can be abundant in soils of semiarid to subhumid regions. In general, then, there is little hope for a fossil record of apes, humans, or indeed any vertebrates from humid forest paleosols (Retallack, 1984a).

Some exceptions to this generalization have been recognized. Carbonatite-melilitite tuff, for example, is a highly alkaline and unusual kind of volcanic product that can harden like cement after rain (Barker & Nixon, 1983; Hay, 1978, 1989). It could be that the carbonatite-nephelinite stratovolcanoes of East Africa fortuitously preserved a record of early Miocene biota of wet forests (Bishop, 1968). By implication then, such an ecosystem could have been widespread in other parts of Africa and Asia where there did not happen to be such volcanoes. Some unusual features of paleosols on carbonatite, carbonatite-melilitite, and nephelinite tuffs have already been discussed (see Fig. 2.16), and a few of the paleosols on these materials are indeed spectacularly fossiliferous. Nevertheless, these highly alkaline volcanics as a parent material were less like other kinds of volcanics than calcareous beach sands (Weilemaker & Wakatsuki, 1984), and the paleosols are evidence that this material was prone to decalcification during soil formation. The Miocene paleosols of these Kenyan volcanoes include calcareous nodules and smectitic clay, suggestive of paleoclimate no wetter than subhumid. These paleosols compare best with soils of wooded grassland, grassy woodland, and dry forest, rather than wet forests. This also is suggested by the presence of such open-country animals as mole rats and spring hares, among a variety of forest-adapted mammals (Pickford & Andrews, 1981). Carbonatite tuff no longer seems to be an exceptional window into mammalian faunas in African rain forests of the past.

A second possible exceptional way to sample wet-forest fossil faunas is the preservation of bones and shells in stream deposits buffered by groundwater from the acidity of floodplain soils. This has been proposed as an explanation for the abundance and diversity of bones in sandstone paleochannels of the Siwalik Group of Pakistan (Behrensmeyer, 1987) and elsewhere (Retallack, 1983, 1988b; Retallack, Leahy, & Spoon, 1987). Unlike some of these other paleosol sequences, however, paleosols of the Siwalik sediments (especially Sarang, Sonita, Khakistari, and Kala Series) were locally rich in bone. Fossil collecting to date has focused on paleochannels, but paleosols also deserve attention. From paleosols also comes evidence that northern Pakistan some 7 to 8 million years ago was not under a humid jungle, but rather under subhumid to semiarid, seasonally dry forest and woodland. Thus a mammalian fossil record of humid forest ecosystems is not yet documented from Pakistan or India either, and this remains an irreducible gap in knowledge of ape and human evolution.

To the annihilation of the fossil record of mammals in humid forests can be added the tyranny of small numbers of early ape and human ancestor fossils. Over a century of search by many investigators worldwide has resulted in discovery of many hundreds of such fossils, and their demonstrated abundance in southwestern Kenya is especially impressive (Pickford, 1986b). Nevertheless, they were rare elements of the fossil faunas in which they were found. It is unusual for any particular paleosol to yield more than a handful of these remains. The Kiewo clay paleosols in the Songhor gullies, the Kwar clay near Koru in Kenya, and the Sarang paleosols near Kaulial in Pakistan have been very productive of fossil apes. Samples of ape fossils in paleosols cannot yet be regarded as statistically substantial evidence for their ecological preferences. The occurrence of rare fossils such as apes could just as well be controlled by random events of predation, drought, or burial as reflect their original ecology. It is only the common and robust elements of a fossil fauna of a paleosol that can be regarded as ecologically representative (Retallack, 1988b). For these reasons the place of ape fossils in the environmental reconstructions offered here should not be regarded as indications of former ecological preferences, but as documented occurrences.

Also of interest is how representative are paleosols of past landscapes. Numerous studies of paleosols, soils and soil erosion have shown that weakly developed paleosols are over-represented in fluvial and volcaniclastic sequences, compared to moderately and strongly developed soils on non-depositional landscapes (Bown & Beard, 1990; Cas & Wright, 1987; Retallack, 1983, 1986a, 1990a). In addition, soils formed under open vegetation are more likely to be eroded away than those under forest (Retallack, 1983, 1984a, 1986a, 1990a; Schumm, 1977). The moderately developed grassland paleosols at Fort Ternan have thus persisted against the odds, and their environment may have been more widespread than the paleosols. The odds of paleosol preservation in depositional systems are probably better than the odds of preservation of fossils in paleosols, considering that many paleosols lack obvious fossils. Although the paleosol record is biased in its own way, it is at least as good as the fossil record.

Interpretations of Miocene ape ecology will gain in precision as more fossils are found and placed within a paleoenvironmental setting based on fossil soils, snails, plants, and other mammals. Although data remain in many ways inadequate, let us reassess hypotheses of how human ancestors arose from apes from the new perspective provided by detailed studies of paleosols. After all, most of these hypotheses were first proposed on the basis of data even more meager than now available.

One set of hypotheses views the evolution of upright stance in early human ancestors as an adaptation to wielding and throwing weapons in an open grassy habitat (Ardrey, 1976; Dart, 1957; Washburn, 1967). Upright stance may also have been adaptive for spying enemies at a distance in long grass. These views promote a martial origin of early upright human ancestors, sometimes unfairly caricatured as "killer apes." This view had strong appeal after the horrors of two world wars earlier this century (Lewin, 1987).

A second set of hypotheses emphasizes new ecological opportunities in open grassy mosaics of vegetation. Erect stance may have evolved because hands were needed to manipulate smaller fruits and seeds of wooded grasslands compared with those of forests (Jolly, 1970). Perhaps apes stood erect to feed on small trees or bushes (Wrangham, 1980). Perhaps upright stance mitigated heat stress in open sunny tropical environments (Wheeler, 1984, 1985). Alternatively, walking on two legs may have been an energetically efficient way of searching large areas of open plains for carcasses to scavenge (Shipman, 1986b; Szalay, 1975; C. R. Taylor & Rowntree, 1973). Perhaps movement through a mosaic of woodland and grassland, as opposed to the monotony of forests, was needed to satisfy the varied demands for water, food, and shelter of larger than usual, ecologically versatile apes (Butzer, 1977; Foley, 1987; Foley & Lee, 1989). Exploitation of new habitats also is a feature of the hylobatian model of human origins in which arboreal small primates became diurnally bipedal in woodlands with less connected canopy (Tuttle, 1981). According to the aquatic model, erect stance evolved as part of an adaptation to gathering of shellfish and wading in the ocean (La Lumière, 1981; Morgan, 1972). Opportunistic views of human ancestors had strong appeal during the technological innovations of the 1950s (Lewin, 1987).

A final set of hypotheses emphasizes nurturing aspects of human nature. Perhaps erect stance was selected because hands were used to provision a home base needed to sustain infants that were larger and in greater need of socialization and instruction than usual for apes (Isaac, 1981; Lovejoy, 1981). Within the social environment of early ape troops, erect stance also would have been adaptive for making an impression during sexual or dominance displays (Livingstone, 1962), for threatening predators (Ravey, 1978), or for carrying infants and nest building materials (Hewes, 1961, 1964). Some of these views may owe inspiration to the women's liberation movement and ethnographic studies of hunter-gatherer societies during the 1960s (Lewin, 1987). They are all independent of a specific kind of paleoenvironment. It is the paleoenvironmental corollaries of these hypotheses that are potentially testable through the evidence of paleosols and their fossil apes and human ancestors.

From the varied paleosols reported in this book, it now can be seen that East African vegetation of 20 million years ago was a complex mosaic of dry forests of several kinds, including colonizing forests. There may also have been some local seasonally inundated wooded grasslands by 17 million years ago. This early Miocene mosaic supported a variety of early monkeylike apes including *Proconsul*, *Dendropithecus*, and *Limnopithecus*. Widespread dry wooded grassland appeared among the dry forests and woodlands by 14 million years ago. By that time few fossil apes are known (Harrison, 1989), and prominent among them are remains of *Kenyapithecus*, found mainly in woodland paleosols rather than those of wooded grassland. Some 8 million years ago in Pakistan, the orangutanlike ape *Sivapithecus* inhabited a region of seasonally dry, moist tropical forest, with local swamp forest, riparian forest, colonizing forest, and seasonally wet wooded grassland. The ape fossils have been found only in paleochannels and paleosols of streamsides that supported colonizing forest. The apes became locally extinct as climate dried out some 7 million years ago and vegetation changed to a more monotonous dry forest and woodland. From this study, the habitats of fossil apes can be seen to have included dry forest and woodland, which floristically may have been allied to the Guineo-Congolian, Sudanian, and Zambezian forests of Africa and the dry dipterocarp forests of Asia.

Other occurrences of fossil apes are evidence for only slightly broader habitat preferences. In eastern Saudi Arabia, fossil apes have been found with palm wood in deposits of a coastal lagoon (Andrews & Martin, 1987; Whybrow *et al.*, 1982). Miocene apes may also have lived in scattered remnants of the archaic, Tethyan, evergreen laurel forests that once extended from the Canary Islands through Spain and Europe into China (Axelrod, 1975; Begun, Moyá-Sola, & Kohler, 1990; Nagatoshi, 1987). Some penetrated higher and more northward than these forests into the climatically mild fringes of deciduous hardwood forests and swamps in Nepal, China, and France (Badgley *et al.*, 1988; Kortlandt, 1983; R. M. West, 1984).

Also of interest are the habitats of mammalian fossil faunas that do not contain apes (see Fig. 1.2). For example, fossil apes have not been found in northern Europe, the Americas, or Australia (Savage & Russell, 1983). Nor were they found in mangal paleosols of Saudi Arabia (Whybrow & McClure, 1981). Nor were they fossilized in paleosols of dry open vegetation in North or South Africa, despite the suitability of calcareous paleosols for the preservation of bone. Apes may have avoided the newly evolved grassland and sclerophyll vegetation, just as do their descendants: gibbons, orang-utans, chimpanzees and gorillas. No fossil apes or any other mammals have been found from ancient humid forests, probably because their soils do not preserve bone, as already discussed. From this perspective, the rarity of late Miocene ape fossils of any kind becomes more significant with progressive improvement of the fossil record of early and middle Miocene apes and of Plio-Pleistocene human ancestors (Andrews, 1986; M.D. Leakey & Harris, 1987; R.E. Leakey, Leakey, & Behrensmeyer, 1978; Pickford, 1986a; Pickford & Andrews, 1981). There is a modest fossil record of middle to late Miocene monkeys and baboons of open country (Andrews, 1981b; Fleagle, 1988; Pickford, 1986d). One would expect a better record of apes also if they had lived on such fossiliferous paleosols. Apes may well have retreated into humid forests, where they would not be preserved, during late Miocene climatic drying.

The idea that Miocene apes of the Old World were primarily arboreal frugivores of tropical forests and woodlands also is supported by a variety of other evidence; functional morphology of their limbs (Rose, 1986; Senut, 1988), microwear on their teeth (Covert & Kay, 1981; Teaford & Walker, 1984), and adaptive features of associated mammal faunas (Evans, Van Couvering, & Andrews, 1981). It may not seem a very surprising view when one considers the habitats of living apes (Andrews, 1982; D. A. Collins & McGrew, 1988), but it is a very different view from the one espoused only a few decades ago when *Sivapithecus* and *Kenyapithecus* were regarded as erect-walking tool-wielding human ancestors of open habitats (Lewin, 1987). On the contrary, it appears that grasslands, and probably also sclerophyll forests, were avoided by Miocene apes for some 10 million years as their preferred habitats in woodlands and forests became fragmented. Habitat fragmentation could have been significant for the evolution of forest-dwelling apes, but grasslands themselves may have been no more a selective pressure for their evolution than other habitats for which they were unsuited, such as beaches and high mountain tops.

This general line of thinking inclines me toward accepting a nurturing explanation for the origin of erect stance in early human ancestors, independent of paleoenvironmental change. Other reasons for thinking this way include the apparent habitat indifference and arboreal adaptations of the earliest well-known erect-walking australopithecines some 4 million years old (Retallack, 1990a; Susman, Stern, & Jungers, 1984). Erect stance can be considered initially an adaptation for the care of family among apes committed to woodland and forest. Erect stance also may have proved a preadaptation (exaptation) to later exploitation of unusual habitats for apes such as grassland mosaics, and for cooperative hunting using weapons. By that time, however, human ancestors may already have been such large animals with such complex culture that they had home ranges embracing several elements of the mosaic of vegetation (Foley, 1987).

This is not the view that I expected at the outset of these studies, knowing that paleosols provided a way of assessing local paleoenvironmental mosaics. My current satisfaction with this view is surely more emotional than analytical, as is so often the case with theories of human evolution (Lewin, 1987). Even though compatible with an array of data and theories, these hypotheses remain inadequately tested. Surely a more secure conclusion from the present study is that research on paleosols can now be added to a potent arsenal of scientific approaches to unraveling when, where, and how we evolved. Sir Winston Churchill (1942), who knew well both Kenya and Pakistan, offered some memorable lines in another context, which are a fitting conclusion to this study. "This is not the end. It is not even the beginning of the end. But it is, perhaps, the end of the beginning."

APPENDIX 1. Individual named Miocene paleosols in Pakistan and Kenya

Measured section	Strati-graphic Level (m)	Paleosol Number	Name
Koru	5.3	1	Mobaw bouldery clay
	6.5	2	Kwar clay calcareous stringer variant
	7.2	3	type Kwar clay
	7.6	4	type Mobaw silty clay loam
	8.0	5	Kwar silty clay
Songhor	2.2	11	Kiewo clay thin surface phase
	2.7	12	Kiewo clay single stringer variant
	2.9	13	type Kiewo clay
	4.2	14	type Choka clay
	6.0	15	type Tut clay
	6.7	19	Buru clay thin surface phase
	6.8	20	type Buru clay
Fort Ternan	7.8	22	type Lwanda clay
	8.3	23	Rairo bouldery clay
	9.3	24	type Rairo clay
	10.7	25	type Rabuor clay
	11.2	26	Dhero clay loam
	16.1	27	Chogo ferruginized nodule variant
	16.5	28	type Chogo clay and laterally equivalent Chogo clay eroded phase
	20.0	30	type Dhero clay
	20.9	31	type Onuria clay
	21.3	32	Dhero clay lapillistone variant
Kaulial west	2.4	35	type Sarang clay
	8.6	47	Khakistari clay thin surface phase
	9.4	48	type Khakistari clay
	9.8	49	Sonita clay ferric concretionary variant
	10.5	50	Lal clay cumulic surface phase
	11.7	51	type Lal clay
	32.2	80	Sarang clay krotovina variant
	33.7	81	Pila clay nodular variant
	45.4	98	Bhura clay ferric concretionary variant
	46.9	99	type Pila clay
	47.2	100	Pandu clay loam
	47.5	101	Pandu clay thin surface phase
	48.6	102	type Pandu clay
	56.6	110	Sarang clay loam
	57.0	111	type Kala clay
	57.3	112	Sarang clay eroded phase
Kaulial east	20.0	124	type Naranji clay
	20.4	125	Sarang silty clay massive surface phase
	21.3	126	type Bhura clay
	21.5	127	Sarang silty clay rhizoconcretionary variant
	22.2	128	type Sonita clay
	22.4	129	Sarang clay cumulic surface phase
	22.9	130	Sarang clay rhizoconcretionary variant
	34.8	141	Sarang clay
	35.3	142	Sonita silty clay

Note: Paleosol numbers are the stratigraphic order of paleosols in the measured sections (Figs 4.5, 4.10, 5.9, and 5.10).

APPENDIX 2. Munsell colors of selected Miocene paleosols from Pakistan and Kenya

Paleosol	Horizon	Specimen Number	Fresh Color	Minor Colors
Mobaw bouldery clay	A	KO-R96	2.5YR3/6	granules 2.5Y7/4,5Y7/3,5Y4/1,10YR8/1,5Y2/1; mangans 10YR2/1
	Bw	KO-R99	2.5YR4/6	granules 7.5YR6/6,10YR5/4,10YR2/1,10YR8/1
	clast	KO-R100	10YR5/6	phenocrysts 10YR8/1; diffusion ferrans 5YR3/3,5YR2.5/1; sesquans 5YR4/6
	C	not taken	5YR4/6	clasts 10YR7/8; diffusion ferrans 5YR3/3.5YR2.5/1
	2C	KO-R101	2.5YR4/6	granules 10YR7/6,2.5Y4/2,10YR8/1,10YR2/1
	2C	KO-R102	5YR5/8	boulders 10YR8/6; diffusion ferrans 10YR3/6,10YR2/1
	2Cg	KO-R103-4	N6/0,N8/0	paleokarst wall with ferran 10YR6/8
	C	KO-R105	mottled 10YR-6/8, 5YR3/4, and 5Y2/1	granules 10YR7/3
	nod.	KO-R106-7	5Y6/1	manganese nodules
	2Cg	KO-R108	7.5YR3/4	granules 10YR7/4,10YR3/2,10YR2/1,7.5YR5/8; mangans 10YR2/1
	R	KO-R109	10YR7/8	granules 10YR8/6,10YR7/1; mangans N3/0
welded tuff	-	KO-R110	10YR6/2	phenocrysts 10YR7/4,10YR7/6; ferrans 10YR7/8
Kwar clay calcareous stringer variant	A	KO-R90	2.5YR3/6	granules 10YR8/1,10YR7/6,10YR3/3,10YR2/1,5BG4/1
	Bt	KO-R91	2.5YR3/6	granules 10YR8/1,2.5Y6/1,10YR5/6; mangans 10YR2/1
	C	KO-R92	2.5YR4/6	tubules 5YR4/6 (specimen KO-R93); calcans 7.5YR6/4
	2C	KO-R94	2.5YR3/6	granules 7.5YR7/4,2.5Y5/4,2.5Y4/1,10YR2/1; tubules 5YR4/6
	2C	KO-R95	7.5YR6/4	granules 2.5YR3/4,5YR4/4,2.5Y7/6,5G4/1,10YR2/1; argillans 2.5YR3/6
type Kwar clay	A	KO-R87	5YR4/6	granules 10YR8/2,5Y5/2,10YR2/1; mangans 10YR2/1
	Bw	KO-R88	2.5YR3/6	granules 5YR5/4,5Y5/4,10YR8/2; mottles 2.5YR4/8
	C	KO-R89	2.5YR4/6	granules 5YR5/4,10YR4/3,10YR8/1,10YR2/1
type Mobaw silty clay loam	A	KO-R84	5YR3/3	granules 10R6/6,10YR8/1,10YR6/6,5Y5/4,5Y2/1; calcareous clasts 5Y8/2, with ferrans 2.5YR4/6; tubules 5YR4/3
	Bt	KO-R85	5YR3/3	bone 5YR8/1; snail 2.5YR4/8, with claystone 5Y7/2,5Y2/1
	C	KO-R86	2.5YR4/6	calcareous clasts 2.5Y7/2; mangans 5Y2/1; ferrans 5YR4/6, 2.5YR4/8
Kwar silty clay	A	KO-R80	7.5YR3/4	granules 10R4/8,7.5YR6/8,10YR5/6,5Y5/4,5Y2/1
	A	KO-R81	10R4/8	granules 7.5YR4/4,10YR8/2,10YR6/6,10YR5/4,5Y4/3; tubules 7.5R4/6
	C	K0-R82	2.5YR4/6	granules 10R4/4,10YR6/6,10YR7/8,10YR8/1,5Y4/1,5Y2/1
	C	KO-R83	2.5YR4/6	granules 10YR5/6,10YR6/3,10YR8/1,5YR5/4; mangans 5Y2/1; mottles 10YR7/2
siltst. 8.2 m	-	KO-R79	10YR7/4	ferrans 2.5YR4/4
clayst. 8.2 m	-	KO-R79	7.5YR3/4	mangans 5Y2/1
Kiewo	A	SO-R160	5YR4/4	tubules 10YR5/2; mammal tooth 2.5Y8/4
series unnamed #6	A	SO-R158	mottled 5YR-4/4,10YR6/3	
	C	SO-R159	5YR4/4	tubules 10YR6/3; dike 5Y8/3,10YR7/8
	C	SO-R161	5YR5/4	calcareous stringers 5YR6/4; tubules 10YR6/3, with sesquans 10YR5/3; dike 5Y8/3,10YR7/8
Kiewo	A	SO-R156	5YR4/4	drab root traces 2.5Y5/2
series unnamed #7	C	SO-R157	interbeds 5Y-R5/4,5YR6/4	tubules 10YR7/2, with sesquan 10YR6/2
Kiewo	A	SO-R154	5YR4/4	mangans 5YR2/1; root traces 2.5Y8/2, with halo 2.5Y7/2
series unnamed #8	C	SO-R155	7.5YR7/4	granules 5YR4/4; calcareous stringers 2.5Y8/1; mangans 10YR2/1; bone 10YR8/2; snail 2.5Y8/2
Kiewo	A	SO-R152	5YR3/4	tubules 2.5Y7/2, with sesquans 2.5Y6/2
series un-	C	SO-R153	5YR5/3	calcareous nodules 10YR6/3, with sesquans 5YR4/4; tubules

APPENDIX 2. continued

Paleosol	Hor-izon	Specimen Number	Fresh Color	Minor Colors
named #9				2.5Y4/2
Kiewo series un.#10	A	SO-R150	5YR3/4	tubules 2.5Y7/2, with sesquans 2.5Y6/2; mangans 10YR2/1
	C	SO-R151	5YR3/4	calcareous nodules 5YR7/3; mammal teeth 10YR8/6
Kiewo clay thin sur.ph.	A	SO-R148	5YR3/4	tubules 10YR5/2, with sesquans 10YR4/2; mangans 5YR2/1
	C	SO-R149	5YR5/4	beds 5YR6/4; tubules 2.5Y7/2, with sesquans 2.5Y6/2
Kiewo clay single stringer variant	A	SO-R144	5YR4/4	tubules 10YR7/3, with sesquans 10YR5/2
	C	SO-R145	interbeds 10YR5/4,5YR5/4	
	C	SO-R146	5YR4/4	mangans 5YR2/1; tubules 10YR6/3, with sesquans 10YR5/2
	C	SO-R147	interbeds 5YR5/4,5YR6/4	tubules 10YR7/2, with sesquans 10YR6/2
type Kiewo clay	A	SO-R142	5YR4/4	mottles 10YR7/2,10YR5/2
	C	SO-143	10YR7/2	mottles 10YR5/2; interbeds 5YR4/4; mangans 10YR4/2
type Choka clay	A	SO-134	5YR3/4	granules 5B4/1,5Y2/1; calcareous clasts 7.5YR6/4; tubules 2.5YR7/2, with halo 10YR4/2; tooth enamel 2.5Y7/6 and dentine 10YR6/6; bone 10YR8/4
	A	SO-R135	7.5YR6/4	granules 5YR3/4,7.5YR4/6,10YR4/4; tubules 10YR7/3,5YR4/4; calcareous stringers 5Y8/2, with manganese nodules 5Y2/1
	BA	SO-R136	mottled 5YR4/4,10YR5/2	granules 5YR6/4; tubules 5Y6/3,5Y7/3
	Bt	SO-R137	mottled 5YR4/4,10YR5/2	calcareous clasts 10YR5/2; mottles 10YR7/2; sesquans 10YR4/2
	Bt	SO-R138	5YR4/4	tubules 10YR6/2, with sesquans 10YR5/2; granules 5Y2/1
	C	SO-R139	mottled 5YR4/4,10YR5/2	granules 5Y6/4,5Y2/1,5Y8/2
	C	SO-R140	5YR4/4	mottles 10YR5/2; granules 5YR3/4,10YR7/8,5Y6/3,5Y2/1
	C	SO-R141	10YR6/2	granules 5Y2/1; clay skins 5YR6/4
type Tut clay	A	SO-R121	10YR4/2	granules 7.5YR5/4,5Y2.5/1; sesquans 5Y8/3
	A	SO-R122	7.5YR3/2	tubules 10YR4/2,10YR6/4, with sesquans 10YR4/2
	A	SO-R123	5YR3/4	tubules 10YR4/2,2.5Y8/4, with sesquans 10YR4/2
	nod.	SO-R124	5Y7/2	sesquan 10YR4/2, in matrix 5YR3/4; dike 5Y7/2
	Bt	SO-R125	5YR3/4	mangans 5YR2.5/1
	nod.	SO-126	5Y7/2	sesquan 10YR4/2, in matrix 5YR3/4
	Bk	SO-R127	5YR4/4	tubules 5Y7/2, with sesquans 10YR4/2
	Bk	SO-R128	5YR4/4	tubules 5Y7/2, with sesquans 10YR4/2; dike 5Y7/2
	BC	SO-R129	5YR6/4	
	C	SO-R130	5YR4/6	mottles 5YR6/4,5YR5/4; sesquans 7.5YR5/8, dike 5Y7/2,7.5YR5/6
	C	SO-R131	5YR4/6	tubules 5YR6/4, with sesquans 5YR5/4
	C	SO-R132	mottled 5YR3/4,10YR4/2, 10YR5/2	
	C	SO-R133	mottled 5YR3/4,10YR4/2	
Buru series unnamed #16	A	SO-R119	5YR3/4	tubules 2.5Y7/4, with sesquans 10YR3/2
	C	SO-R120	5YR6/4	tubules 2.5Y7/4, with halo 2.5Y7/4 and mangan 10YR6/2, 10YR2/1; dike 10YR7/3
Buru series un. #17	A	not taken	10YR3/2	
	C	not taken	7.5YR5/6	
Buru series unnamed #18	A	not taken	interbeds 5YR4/4,10YR3/2	
	C	not taken	7.5YR5/6	

APPENDIX 2. continued

Paleosol	Horizon	Specimen Number	Fresh Color	Minor Colors
Buru clay thin surface	A	SO-R118	interbeds 5Y-R4/4,10YR3/2	
phase	C	not taken	7.5YR5/6	
type Buru	A	SO-R116	5YR4/4	tubules 5Y7/2, with haloes 10YR5/2
clay	C	SO-R117	10YR3/2	
Buru un. #21	A	not taken	5YR4/4	tubules 5Y7/2, with haloes 10YR5/2
sandst. 7.5 m	-	SO-R115	2.5Y5/2	mottles 2.5Y6/4
type Lwanda	A	FT-R57	10YR3/3	granules 5YR3/4,10YR6/6,2.5Y7/6,5Y6/3,5G6/1; mangan 10YR2/1
clay	A	FT-R58	5Y8/2	granules 5Y8/1; boulder 5Y5/2, with phenocrysts 5YR8/1, 5YR7/2; ferrans 10YR4/6
	Bt	FT-R59	mottled 10R-3/2,10YR3/2	boulders 10YR3/2, with ferrans 10YR4/4 and phenocrysts 5YR8/1,5YR8/2, and marginal sesquan 10R3/2, including phenocrysts 10R8/1,10R6/2
	Bt	not taken	mottled 10Y-R5/6,7.5YR-3/2,2.5Y7/4	sesquan 5YR7/4
	C	FT-R60	2.5Y4/2	phenocrysts 2.5Y6/4,2.5Y8/2; sesquans 5YR7/6
	R	not taken	mottled 5Y-3/2,5Y2.5/2	phenocrysts 10YR7/2; glassy lava 10YR2/1,10YR4/4; sesquans 5YR5/2, with mangans 5YR2.5/1 and phenocrysts 7.5YR7/4
Rairo bouldery	A	FT-R54	10YR3/3	granules 10YR8/8,5Y7/6,5B4/1,5Y2/1; manganese nodules 10YR3/1; drab mottles 2.5Y5/2
clay	clast	FT-R55	10YR3/2	phonolite with phenocrysts 10YR8/1
	clast	not taken	5Y8/3	nephelinite with halo 5Y5/2
	C	FT-R56	10YR4/3	granules 7.5YR5/8,5Y8/3,5Y5/1,5Y2.5/1
	clast	not taken	2.5Y6/2	phonolite with phenocrysts 2.5Y8/2, mottles 2.5Y5/2
type Rairo clay	A	FT-R51	10YR4/2	granules 7.5YR4/6,5Y7/6,2.5Y7/6,5G4/1,5Y2/1; mottles 2.5Y6/2, with sesquans 10YR3/2; mangans 10YR2/1; calcareous stringers 10YR6/2
	A	FT-R52	10YR4/2	granules 10YR7/6,5YR5/2,5Y4/4,5B6/1,5Y2/1; calcareous stringers 10YR6/2
	C	FT-R53	10YR4/2	granules 7.5YR5/8,5Y7/4,2.5Y7/4,5GY4/1,5Y2/1; mangans 10YR3/2,10YR2/1
type Rabuor clay	A	FT-R45	10YR4/2	granules 2.5Y7/6,5Y6/4,5Y4/1; mottles 2.5Y6/2, with sesquans 10YR3/2; clay skins 7.5YR4/6
	clast	FT-R46	5YR4/3	phonolite, phenocrysts 5YR8/1,5YR8/2, sesquan 5YR4/4
	A	FT-R47	10YR4/3	granules 10YR7/4,7.5YR5/8,5GY6/1,5B4/1,5Y2/1; root traces 7.5YR4/6,7.5YR5/8; mangans 5YR2.5/1; clay skins 7.5YR4/6; mottles 5Y8/2, with haloes 5Y6/2
	Bt	FT-R48	10YR3/3	granules 5Y7/4,5Y7/6,10YR7/6,10YR2/1; root traces 7.5YR5/8
	clast	FT-R49	7.5YR3/2	phonolite with phenocrysts 7.5YR8/1,7.5YR8/2; weathering rind 10YR4/2, with phenocrysts 10YR8/1,10YR8/2; sesquans 2.5Y8/6
	C	FT-R50	10YR4/3	granules 7.5YR5/8,5Y6/4,5Y5/1,2.5Y7/6,5G5/1; calcareous stringers 10YR8/2,10YR7/2; bone N8/0, with sesquan 2.5Y8/6
Dhero clay loam	A	FT-R42	10YR4/2	granules 7.5YR5/6,10YR7/6,5GY6/1,5B4/1; root traces 5YR4/6; tubules 10YR6/4, with sesquans 10YR5/8 and outer ferran 5YR4/6; mangans 10R2.5/1
	C	FT-R43	2.5Y5/4	granules 7.5YR4/6,10YR8/2,2.5Y6/6,5Y6/4,5G5/1,5B4/1,5Y2/1
	clast	FT-R44	10YR8/2	carbonatite with sesquan 10YR4/2
tuff 11.4 m	-	FT-R41	10YR4/2	granules 10YR8/2,7.5YR5/6,2.5Y6/6,5BG5/1,5G5/2,5Y2/1
Chogo clay ferruginized	A	FT-R12	10YR4/2	granules 10YR6/6,5YR4/6,5GY6/1,5Y2.5/1; calcite crystal tubes 10YR8/2,10YR8/3
nodule	A	FT-R13	7.5YR4/4	granules 5YR3/4,10YR6/6,5Y7/6,5GY6/1

APPENDIX 2. continued

Paleosol	Hor-izon	Specimen Number	Fresh Color	Minor Colors
variant	nod.	FT-R14	10YR5/3	granules 10YR5/6,2.5Y5/4,N4/0,N2/0
(to east)	Bk	FT-R15	10YR4/2	granules 2.5YR4/6,7.5YR4/6,10YR8/6,5GY6/1,5B5/1,5Y2.5/1; bone 10YR8/4,10YR8/2, with sesquan 7.5YR4/6
	Bk	FT-R16	2.5Y5/2	granules 2.5Y5/2,7.5YR5/8,5Y6/6,5G5/1,5Y2.5/1; tubules 10YR5/3, with haloes (10YR7/1)
	nod.	FT-R17	10YR5/3	granules 10YR6/8,5YR5/4,5GY5/1,5Y2.5/1
	C	FT-R18	10YR4/1	granules 2.5YR8/4,2.5YR4/2,10YR3/4,5G5/2; calcareous cement 10YR8/2,7.5YR8/2
Chogo clay	A	FT-R35	10YR4/2	granules 10YR8/4.5YR5/8,7.5YR6/8,5Y4/3,5Y2/1
ferruginized nodule variant	A	FT-R36	10YR6/3	granules 10YR5/8,10YR7/6,5BG5/1,5Y2/1; sesquans 5YR4/6, 7.5YR4/6; tooth enamel 10YR4/4, dentine 10YR7/4, and root 10YR8/3
(to west)	A	FT-R37	10YR5/3	granules 2.5YR5/8,7.5YR5/8,5Y7/4,5G5/1,5Y2/1; root traces 10YR5/4,10YR6/3
	Bk	FT-R38	10YR5/3	granules 7.5YR3/4,10YR7/6,2.5Y7/8,5Y6/4,5Y7/4,5Y5/2,5Y2/1; calcareous nodules 10YR6/3, sesquans 10YR4/6,5YR4/6
	Bk	FT-R39	10YR6/3	granules 7.5YR5/8,10YR6/6,10YR7/8,10YR6/3,2.5Y6/4,5GY5/1; manganese nodules 7.5YR3/2, with ferran 2.5YR3/6
	C	FT-R40	2.5Y6/2	granules 7.5YR4/6,10YR6/6,2.5Y7/4,N7/0,5B4/1,5Y2/1
type Chogo	A	FT-R2	10YR3/2	granules 10YR5/6,2.5Y6/4,5Y6/2,5Y2.5/1
clay	A	FT-R3	10YR4/2	granules 7.5YR5/8,10YR5/6,10YR5/4,2.5YR4/4,5G5/1,5Y2/1
	clast	FT-R4	5Y4/4	phonolite, phenocrysts 2.5Y7/4, sesquans 5YR5/6, weathering rind 5Y8/2
	clast	FT-R5	5Y7/2	nephelinite tuff, granules 7.5YR5/6,10YR7/6,5Y5/2,10YR2/1, sesquans 2.5Y5/4
	Bk	FT-R6	10YR5/2	granules 10YR6/8,10YR5/8,10YR5/4,5Y6/3,5G5/2,5G6/1,5GY4/1, N6
	clast	FT-R7	10YR5/4	phonolite, phenocrysts 5YR2.5/1, mottles 10YR3/2
	clast	FT-R8	5G5/2	nephelinite lava, veins 5YR6/2,10YR5/6,10YR8/6
	Bk	FT-R9	10YR4/2	granules 7.5YR4/6,5Y5/6,5Y5/2,5Y2.5/1; calcareous stringers 7.5YR7/2,7.5YR8/2
	C	FT-R10	10YR3/1	granules 10YR5/4,5Y6/3,5G5/1,10YR2/1
	C	FT-R11	10YR7/2	granules 10YR6/6,5G6/2,10YR2/1; sesquan above and below this bed 10YR6/6
Chogo clay	A	FT-R23	10YR4/2	granules 2.5YR4/6,2.5Y8/4,5Y7/6,5GY6/1,5B4/1; calcareous root traces 10YR8/2
eroded phase	nod.	FT-R24	10YR7/3	granules 2.5YR4/6,10YR4/4,5Y7/4,5Y6/3,5Y2/1; bone 10YR8/4, 10YR8/2
	A	FT-R25, 26,27	10YR5/3	granules 2.5YR4/8,5Y8/4,5G6/2,5Y2/1; tubules 10YR7/3; root traces 7.5YR5/6,10YR6/3
	A	FT-R28	10YR4/2	granules 2.5YR4/6,2.5Y8/8,2.5Y4/4,5B4/1; calcareous stringers 10YR6/3
	Bk	FT-R29	10YR6/2	granules 7.5YR5/6,2.5Y7/8,5Y6/4,5BG5/1,5BG4/1
	clast	FT-R30	5Y3/1	phonolite, phenocrysts 2.5Y7/6,2.5YR4/6; weathering rind 10YR8/4
	clast	FT-R30	10YR6/1	nephelinite lava, phenocrysts 5Y2/1,2.5Y8/2,7.5YR5/6; sesquans 10YR6/6
	Bk	FT-R31	10YR5/2	granules 7.5YR5/8,5Y8/3,5Y6/6,10YR2/1; calcareous nodules and stringers 10YR8/2, with sesquan 10YR8/3; bone 10YR8/2, 10YR7/4
	den?	FT-R32	10YR6/3	granules 7.5YR5/8,5G5/2,10YR2/1; bone 10YR8/2,10YR8/3
	C	FT-R33,34	10YR2/1	granules 5YR6/4,10YR5/8,5Y7/4,N6/0; Liesegang banding

APPENDIX 2. continued

Paleosol	Horizon	Specimen Number	Fresh Color	Minor Colors
				10YR6/3,10YR4/2; sesquans 2.5YR3/6,10YR6/8
sandst. 16.6 m (to east)	-	FT-R1	5Y6/2	granules 10YR5/4,5GY6/1,5GY5/1,5Y2.5/1; cement 5Y7/2; sesquans 7.5YR5/2,10YR5/6
sandst. 16.6 m (to west)	-	FT-R21	10YR3/2	granules 2.5YR4/6;2.5YR6/4,2.5Y7/6,5G6/2,5G4/1; cement 10YR8/2; sesquans 10YR7/6,10YR6/3
sandst. 16.7 m	-	not taken	5Y4/1	granules 5Y3/1,5Y8/3; sesquans 7.5YR4/6
sandst. 16.9 m	-	not taken	5Y5/1	granules 5YR4/4,5Y5/1,5G6/1,5Y2/1; silty interbeds 5Y6/1, with clayey flasers 5Y5/1, biotite 10YR2/2, clasts 5Y8/4
sandst. 17.0 m	-	not taken	2.5Y5/2	granules 5Y8/4
sandst. 17.1 m	-	not taken	2.5Y7/2	
brecc. 17.3 m	-	not taken	5Y7/2	granules 10YR4/4,5Y8/4,5G5/1,5Y2/1; larger clasts 7.5YR5/6, 10YR6/4,2.5Y7/4,2.5YR6/2,2.5Y6/4,5Y7/4,5GY6/1
Dhero series	A	not taken	5Y5/2	granules 7.5YR5/4,10YR5/4,5G6/1,5B5/1,5Y2/1
un. #29	C	not taken	10YR8/2	sesquan 10YR8/3
sandst. 18.0 m	-	not taken	10YR5/2	granules 7.5YR4/6,5Y7/3,5YR3/1,5Y2/1
sandst. 18.1 m	-	not taken	10YR6/3	granules 7.5YR4/6,10YR7/8,5Y7/3,5Y2/1; sesquans 7.5YR4/6
breccia at 19.0 m	-	not taken	5Y5/1	granules 5Y6/4,10YR8/2,5B4/1,5Y2/1; cobbles 5Y2.5/1,5Y4/1, 5GY5/1,10YR6/4; sesquans 7.5YR5/6
type Dhero	A	FT-R62	10YR5/2	granules 7.5YR5/6,5B4/1,7.5YR2/1
clay	A	FT-R63	10YR5/3	accretionary lapilli 5B4/1,5Y5/4
	cast	FT-R64	10YR4/2	tree stump, micrite 10YR8/2, sparry calcite 2.5Y6/4
	A	FT-R65	10YR5/2	interbeds 10YR6/3,10YR5/3
	C	FT-R66	10YR4/1	granules 5Y8/3,2.5Y7/8,5B4/1,5Y6/4,5Y2/1; nephelinite cobble 5BG4/1, with phenocrysts 5Y6/4, inner weathering rind 5Y5/3, with phenocrysts 5Y7/4 and sesquans 2.5YR5/6, outer weathering rind 10YR6/4
breccia at 20.1 m	-	FT-R61	10YR5/3	granules 7.5YR4/6,10YR5/8,5Y8/3,5B4/1,7.5YR2/1; calcite veins 7.5YR8/2, with sesquans 7.5YR5/8,2.5YR4/6
type Onuria clay	A	FT-75	2.5Y4/4	granules 10R5/6,2.5Y7/6,5Y8/6,5Y2/1; root traces 2.5Y6/6, with sesquans 7.5YR5/6
	A	FT-R76	2.5Y5/4	granules 7.5YR5/8,7.5YR5/6,2.5Y7/6,5G6/2; root traces 7.5YR5/8
	Bk	FT-R77	2.5Y6/6	granules 7.5YR6/6,10YR4/4,10YR3/4,5Y5/4,5B5/1,5Y2/1
	R	FT-R78	10YR5/2	granules 10YR6/6.10YR5/6,5Y2/1,5Y8/1; calcite veins 5Y8/1, with mangans 5Y2/1
sandst. 21.0 m	-	FT-R73, FT-R74	5Y5/1	granules 7.5YR5/6,2.5Y8/2,5Y7/6,5Y2/1; fossil grasses 2.5Y5/4, 10YR5/4, with clay fill 2.5Y8/4
sandst. 21.1 m	-	FT-R72	5Y6/2	granule interbeds 5Y7/2; granules 7.5YR7/6,5Y6/4,10YR4/2, 5Y5/1, 5Y2/1; fossil grasses 5Y6/3, with mangans 5Y4/2, clay fill 2.5Y8/4
Dhero clay lapillistone	A	FT-R70	2.5Y4/2	lapilli 2.5Y7/2, with concentric bands of 2.5Y4/2; root traces 2.5Y7/2, with mangans 5Y4/2; interbeds 10YR5/6
variant	C	FT-R71	2.5Y6/2	interbeds 2.5Y3/2; root traces 5Y7/4, with haloes 2.5Y6/6
sandst. 21.4 m	-	FT-R69	5Y6/2	interbeds 10YR5/6; basal sesquan 5YR4/2
sandst. 21.5 m	-	FT-R68	5Y5/2	
sandst. 22.0 m	-	FT-R67	5Y5/3	granules 7.5YR4/4,5Y7/4,5Y2/1,10YR5/4
sandst. 1.5 m	-	not taken	5Y7/2	weathers 2.5Y7/2
Sarang series	A	not taken	5YR5/3	root traces 5GY7/1; weathers 7.5YR5/4
un. #33	C	not taken	10YR6/6	weathers 10YR6/4
Sarang series	A	not taken	7.5YR4/4	root traces 5GY7/1; weathers 7.5YR6/4
un. #34	C	not taken	5Y7/1	weathers 7.5YR5/4
type Sarang clay	A	SI-353766	5YR5/4	root traces 5G7/1,5GY7/1,10R4/6,10YR7/6; ferric concretions 10YR5/3,10YR5/6; weathers 7.5YR6/4

APPENDIX 2. continued

Paleosol	Horizon	Specimen Number	Fresh Color	Minor Colors
	C	SI-353767	10YR5/3	mottles 10YR5/6; weathers 7.5YR6/4
	C	SI-353768	7.5YR4/4	mottles 10YR5/6; weathers 7.5YR6/4
	C	SI-353769	7.5YR4/4	root traces 5Y7/2; mottles 10YR5/6; ferric concretions 10YR5/3; weathers 7.5YR6/4
	C	SI-353770	10YR6/4	root traces 5G7/1,5GY7/1; burrows 7.5YR4/4; weathers 7.5YR6/4
Sarang series	A	SI-353764	5YR5/4	burrows 7.5YR4/4,5Y7/1; weathers 7.5YR5/4
un. #36	C	SI-353765	10YR6/4	root traces 5GY7/1; burrows 7.5YR4/4; weathers 7.5YR6/4
Sarang series	A	not taken	7.5YR4/4	weathers 7.5YR6/4
un. #37	C	not taken	7.5YR5/4	weathers 7.5YR6/4
Bhura series	A	not taken	7.5YR4/4	root traces 5GY7/1; weathers 7.5YR6/4
un. #38	C	not taken	7.5YR4/4	weathers 7.5YR6/4
Bhura series un. #39	A	not taken	mottled 7.5YR5/4-4/4	root traces 5GY7/1; weathers 7.5YR6/4
	C	not taken	7.5YR4/4	sandy interbeds 7.5YR5/4; weathers 7.5YR6/4
Sar. un. #40	A	not taken	7.5YR4/4	strata concordant mottle 5GY7/1; weathers 7.5YR6/4
Bhura series	A	not taken	7.5YR3/4	root traces 5Y7/1,5G6/1; weathers 7.5YR6/6
un. #41	C	not taken	7.5YR3/4	weathers 7.5YR6/6
Sar. un. #42	A	not taken	7.5YR4/4	root traces 5GY6/1,5GY5/1; weathers 7.5YR6/6
Bhura series	A	not taken	7.5YR3/4	root traces 5GY6/1,5GY4/1; weathers 10YR5/3
un. #43	C	not taken	7.5YR4/4	clay skins 10YR3/3; mangans 10YR3/1; weathers 7.5YR6/4
Khakistari	A	not taken	2.5Y5/4	root traces 5G6/1,5BG5/1; clay skins 5YR4/4; weathers 10YR6/4
un. #44	C	not taken	10YR3/3	weathers 10YR6/4
Khakistari	A	not taken	10YR4/2	clay skins 5YR4/4; weathers 10YR5/3
un. #45	C	not taken	7.5YR3/4	root traces 5GY4/1,5Y2.5/1; clay skins 5YR4/4; weathers 7.5YR5/4
Khak. un. #46	A	not taken	10YR4/3	root traces 5G6/1; clay skins 5YR4/4; weathers 10YR6/4
Kha.cl.thin s.	A	SI-353776	10YR5/4	root traces 5G6/1; clay skins 5YR4/4; weathers 10YR6/4
type Khakistari clay	A	SI-353772	10YR3/3	root traces 5G7/1; sandstone dike 2.5YR7/2; clay skins 5YR5/4; weathers 5YR5/4
	A	SI-353773	10YR5/3	root traces 5GY6/1,5G4/1,5Y2.5/1; clay skins 5YR4/4; claystone clasts 5YR4/6; weathers 7.5YR6/4
	Bg	SI-353774	2.5Y4/2	root traces 5YR2.5/2,5G6/1,5GY6/1,5Y2.5/1; mottles 5YR3/3, 10YR5/4; weathers 10YR6/4
	C	SI-353775	7.5YR5/4	root traces 5G6/1; interbeds 10YR5/4,5YR5/3; weathers 10YR6/4
Sonita clay ferric concretionary v.	A	SI-353786	5YR3/4	root traces 5GY6/1,5BG2.5/1; calcareous nodules 5YR7/4; weathers 5YR6/4
	C	SI-353771	5YR3/4	weathers 5YR5/4
Lal clay cumulic	A	SI-353785	5YR4/4	root traces 5GY5/1,5BG4/1; mottles 5YR3/4; relict beds 5YR3/4, 5YR4/4; weathers 5YR5/4
surface	Bt	not taken	5YR3/4	root traces 5GY6/1; weathers 5YR6/4
phase	Bk	not taken	5YR3/4	calcareous nodules 5YR7/4; weathers 5YR6/4
type Lal clay	A	SI-353778	5YR4/4	root traces 10YR5/4,10YR6/6; mottles 5Y7/1,5GY7/1; weathers 5YR5/4
	A	SI-353779	5YR3/4	root traces 10YR5/6,5GY7/1; weathers 5YR5/4
	Bt	SI-353780, SI-353781	5YR3/4	root traces 5Y7/1,5GY7/1,10YR5/4,10YR3/2; calcareous nodules 7.5YR6/4; weathers 5YR5/4
	Bk	SI-353782	5YR3/4	root traces 5GY6/1,10YR5/3,5GY7/1; weathers 5YR5/4
	2Bk	SI-353783	5YR3/4	root traces 5GY7/1; weathers 5YR5/4
	2Co	SI-353784	5YR3/4	root traces 5G6/1,5Y5/3,5YR3/2,7.5YR7/4; weathers 5YR5/4
Lal series unnamed	A	not taken	5YR3/4	root traces 5G7/1,5BG3/1; calcareous nodules 5YR7/3; weathers 5YR6/4
#52	Bt	not taken	5YR3/4	root traces 5G6/1; weathers 5YR6/4

APPENDIX 2. continued

Paleosol	Hor-izon	Specimen Number	Fresh Color	Minor Colors
	C	SI-353777	5YR3/4	root traces 10YR6/6; weathers 5YR6/4
Sonita series unnamed	A	not taken	5YR3/4	root traces 5GY6/1,10YR5/6; mangans 7.5YR3/2; weathers 5YR6/4
#53	C	not taken	5YR3/4	weathers 5YR6/4
Bhura series	A	not taken	7.5YR4/6	root traces 5G6/1,5GY7/1,5YR3/2; weathers 7.5YR6/6
un. #54	C	not taken	5YR3/4	weathers 5YR6/4
sandst. 54.0 m	-	not taken	10YR6/4	weathers 10YR6/3
Sar. un. #55	A	not taken	7.5YR6/6	root traces 5Y7/1,5Y4/1; weathers 10YR6/4
Bhura series	A	not taken	7.5YR6/6	root traces 5GY7/1; weathers 10YR7/4
un. #56	C	not taken	10YR6/4	weathers 10YR7/4
Bhura series	A	not taken	7.5YR3/4	root traces 10YR5/6; weathers 7.5YR6/6
un. #57	C	not taken	10YR6/4	weathers 10YR5/3
Bhura series	A	not taken	10YR5/6	root traces 7.4YR4/6,5GY7/1; weathers 10YR6/6
un. #58	C	not taken	7.5YR5/6	mottles 5Y7/1,5GY7/1; weathers 7.5YR6/6
Sonita series	A	not taken	5YR3/4	mottles 5GY6/1; weathers 5YR4/6
unnamed #59	C	not taken	7.5YR4/6	root traces 5GY6/1,5Y2.5/1; calcareous nodules 5Y7/1,7.5YR7/4; weathers 7.5YR6/6
Sonita series	A	not taken	5YR3/4	root traces 5G6/1; calcareous nodules 5YR7/3; weathers 5YR4/6
un. #60	C	not taken	5YR3/4	root traces 5YR3/1; weathers 5YR4/6
Lal series	A	not taken	5YR3/4	root traces 5GY7/1,5G7/1; weathers 5YR4/6
un. #61	Bt	not taken	5YR3/4	weathers 5YR4/6
Sonita series	A	not taken	5YR3/4	weathers 5YR4/6
un. #62	C	not taken	5YR3/4	root traces 5GY6/1,5GY7/1; weathers 5YR4/6
Lal series	A	not taken	5YR3/4	root traces 5GY6/1; weathers 5YR4/6
un. #63	Bt	not taken	5YR3/4	root traces 5GY5/1; weathers 5YR4/6
Son. un. #64	A	not taken	5YR3/4	root traces 5GY7/1,5GY4/1; weathers 5YR5/6
Son. un. #65	A	not taken	5YR3/4	root traces 5GY6/1,5GY7/1,5Y7/1,10YR5/6; weathers 5YR5/6
Son. un. #66	A	not taken	5YR3/4	root traces 5GY5/1; weathers 5YR5/6
Lal series unnamed	A	SI-353787	5YR3/4	root traces 5GY7/1,10YR5/6; charcoal 5BG2.5/1,5BG4/1; weathers 5YR5/6
#67	Bt	not taken	5YR3/4	weathers 5YR5/6
Lal series	A	not taken	5YR5/6	root traces 5Y7/1; weathers 7.5YR6/6
un. #68	Bt	not taken	5YR3/4	weathers 5YR5/6
Sar. un. #69	A	not taken	7.5YR5/4	root traces 5GY6/1,7.5YR3/4,5Y6/3; weathers 10YR7/4
Bhura series	A	not taken	7.5YR3/4	root traces 5GY6/1; weathers 7.5YR5/4
un. #70	C	not taken	7.5YR5/4	root traces 5Y6/3; weathers 10YR7/4
Bhura series	A	not taken	7.5YR3/4	root traces 5GY7/1; weathers 7.5YR5/4
un. #71	C	not taken	7.5YR4/4	root traces 5Y6/3; burrows 7.5YR3/4; weathers 10YR7/4
Bhura series	A	not taken	7.5YR3/4	root traces 5GY5/1; weathers 7.5YR5/4
un. #72	C	not taken	7.5YR4/4	root traces 5Y7/1; weathers 7.5YR5/4
Bhura series	A	not taken	7.5YR4/4	root traces 5G6/1; clasts 5YR3/4; weathers 7.5YR5/4
unnamed #73	C	not taken	7.5YR4/6	root traces 5GY6/1,10YR5/6; mottles 7.5YR4/4; weathers 7.5YR5/4
Sar. un. #74	A	not taken	7.5YR5/4	root traces 5GY7/1; clasts 5YR3/4; weathers 7.5YR6/4
Sar. un. #75	A	not taken	10YR5/6	root traces 5GY7/1,5Y5/1; burrows 7.5YR3/4; weathers 10YR6/4
Lal series unnamed	A	not taken	5YR3/4	root traces 5GY6/1,5GY7/1; calcareous nodules 5YR6/4; ferric concretions 5YR3/3; weathers 7.5YR5/6
#76	C	not taken	mottled 7.5Y-R5/4,10YR5/4	root traces 5GY6/1; clay skins 5YR3/4; weathers 10YR6/4
Sonita series un. #77	A	not taken	5YR3/4	root traces 5GY6/1; calcareous nodules 5YR6/3; weathers 7.5YR5/6
	C	not taken	mottled 7.5Y-	root traces 5GY7/1,5GY6/1; weathers 7.5YR6/4

APPENDIX 2. continued

Paleosol	Horizon	Specimen Number	Fresh Color	Minor Colors
			R5/4,10YR5/4	
Sarang series	A	not taken	10YR4/4	root traces 5GY7/1; clay skins 7.5YR3/4; weathers 10YR6/3
un. #78	C	not taken	7.5YR5/4	clay skins 7.5YR4/4,7/5YR4/6; weathers 10YR6/4
Lal series	A	not taken	5YR3/4	root traces 5GY6/1.5GY7/1; weathers 7.5YR5/6
unnamed #79	Bk	not taken	5YR3/4	root traces 5GY7/1,5GY6/1; calcareous nodules 5YR6/6; weathers 7.5YR5/6
	C	not taken	5YR3/4	root traces 5G6/1; weathers 7.5YR5/6
	C	not taken	7.5YR4/4	root traces 5GY6/1; weathers 7.5YR6/4
Sarang clay krotovina	A	SI-353794	7.5YR4/4	root traces 5GY7/1,5GY6/1; burrows 7.5YR3/4; weathers 7.5YR6/4
variant	C	not taken	7.5YR3/4	root traces 5GY7/1; clay skins 7.5YR4/4,7.5YR6/4; weathers 7.5YR6/4
Pila clay nodular	A	SI-353789	7.5YR3/4	root traces 5G6/1; calcareous nodules 7.5YR6/6,7.5YR5/8, 2.5YR5/8; clay skins 5YR3/4; weathers 10YR7/6
variant	Bt	SI-353790	7.5YR3/4	root traces 5G6/1,5GY4/1; burrows 7.5YR7/6; mangans 10R2.5/1; weathers 7.5YR7/6
	2Bk	SI-353791, SI-353792	7.5YR3/4	root traces 5GY7/1; calcareous nodules 7.5YR7/6, with rims 2.5YR4/8,5YR5/6; weathers 10YR7/6
	2C	SI-353793	10YR6/6	root traces 5GY7/1; calcareous nodules 7.5YR6/4; claystone clasts 7.5YR4/4; weathers 7.5YR6/4
Pila series	A	not taken	7.5YR3/4	root traces 5GY7/1,5G6/1; weathers 10YR7/4
unnamed #82	Bk	SI-353788	7.5YR4/4	root traces 5GY7/1; calcareous nodules 7.5YR8/2,7.5YR6/4, 7.5YR5/4, with rims 2.5YR3/6; weathers 10YR7/6
shale 35.0 m	-	not taken	5YR3/4	weathers 7.5YR5/4
Sar. un. #83	A	not taken	7.5YR3/4	root traces 5GY6/1,5Y7/1; weathers 7.5YR5/4
Bhura series	A	not taken	7.5YR3/4	root traces 5GY5/1; weathers 10YR5/4
un. #84	Bk	not taken	7.5YR3/4	root traces 5GY5/1; weathers 10YR5/4
Bhura series	A	not taken	5YR3/4	root traces 5GY6/1; dike 5Y7/1; weathers 10YR5/4
un. #85	Bk	not taken	mottled 7.5-YR3/2-3/4	weathers 10YR5/4
Bhura series	A	not taken	7.5YR5/4	root traces 5Y7/3,5GY6/1; mottles 5YR4/4; weathers 7.5YR5/4
un. #86	C	not taken	7.5YR4/4	clay skins 5YR3/4; weathers 10YR5/4
sandst. 38.0 m	-	not taken	7.5YR5/4	mottles 5GY6/1; clasts 5YR3/4; weathers 10YR6/4
Sar. un. #87	A	not taken	7.5YR4/4	root traces 5GY5/1; weathers 7.5YR6/6
sandst. 38.6 m	-	not taken	10YR5/4	mottles 5GY6/1; weathers 10YR6/3
Sarang series un. #88	A	not taken	mottled 10YR-6/6,7.5YR4/4	root traces 5GY7/1,5GY6/1; weathers 7.5YR5/4
sandst. 39.0 m	-	not taken	10YR6/3	burrows 7.5YR4/4; weathers 10YR5/2,10YR6/3
Sar. un. #89	A	not taken	7.5YR4/4	root traces 5GY7/1,5GY6/1; weathers 7.5YR6/4
sandst. 39.5 m	-	not taken	10YR5/6	interbeds 5Y6/1; weathers 10YR6/4
Sar. un. #90	A	not taken	5YR3/4	root traces 5GY7/1,5GY6/1; weathers 7/5YR6/4
sandst. 39.9 m	-	not taken	10YR5/4	weathers 10YR5/3
shale 40.0 m	-	not taken	7.5YR4/4	sandstone dike 5Y6/1; weathers 7.5YR6/4
sandst. 40.3 m	-	not taken	10YR5/2	root traces 5GY7/1; weathers 10YR6/4
Sar. un. #91	A	not taken	7.5YR4/4	root traces 5GY5/1; weathers 7.5YR6/4
	C	not taken	7.5YR4/4	root traces 5GY5/1; weatheres 7.5YR6/4
Sarang series	A	not taken	7.5YR4/4	root traces 5GY5/1; weathers 7.5YR6/4
un. #92	C	not taken	10YR4/4	interbeds 5GY6/1,5Y5/1; weathers 7.5YR6/4
Sarang series	A	not taken	7.5YR4/6	root traces 5G6/1; weathers 7.5YR6/4
un. #93	C	not taken	10YR5/4	mottles 5GY6/1; weathers 10YR6/3
Sarang series	A	not taken	7.5YR4/6	sandstone dikes 5GY5/1,5G5/1; weathers 10YR7/4
unnamed	C	not taken	interbeds	root traces 5G6/1,5Y5/1; clay skins 10YR7/4; weathers 10YR4/4

APPENDIX 2. continued

Paleosol	Horizon	Specimen Number	Fresh Color	Minor Colors
#94			10YR4/4-5/4	
Bhu. un. #95	A	SI-353823	7.5YR4/6	root traces 5G6/1,5G4/1,5Y7/1,5Y8/1; weathers 10YR7/4
Lal series	A	not taken	5YR4/4	root traces 5GY6/1,5GY7/1; weathers 7.5YR7/6
un. #96	Bt	not taken	7.5YR4/4	claystone clasts 5YR3/4; weathers 7.5YR7/6
Lal series	A	not taken	5YR4/4	root traces 5G5/1,5GY6/1,5Y7/1; weathers 7.5YR7/6
un. #97	Bt	not taken	5YR4/3	weathers 7.5YR7/6
Bhura clay ferric con-	A	SI-353822	7.5YR4/4	root traces 5Y7/1,5GY6/1; ferric concretions 2.5YR2.5/4, 2.5YR3/6; calcareous nodules 7.5YR7/6; weathers 10YR7/6
cretionary variant	Bk	not taken	7.5YR4/4	root traces 5GY7/1,5GY6/1; ferric concretions 2.5YR2.5/4, 2.5YR3/6; weathers 10YR7/6
type Pila clay	A	SI-353817, SI-353818	7.5YR5/4	root traces 5Y5/1,5GY5/1, 10YR6/6; sandstone dikes 5Y6/1; clay skins 2.5YR4/6; ferric concretions 7.5YR3/4,5YR3/3; calcareous nodules 10YR7/4; weathers 7.5YR5/4
	Bt	SI-353819	7.5YR4/4	root traces 5G6/1,5G7/1; ferric concretions 2.5YR3/6; calcareous nodules 7.5YR7/6; mangans 2.5YR2.5/4; weathers 10YR7/6
	Bk	SI-353820	10YR4/4	root traces 5G5/1; ferric concretions 2.5YR3/6,2.5YR2.5/4; calcareous nodules 10YR7/4; weathers 10YR7/6
	C	SI-353821	10YR4/4	root traces 5G5/1; mottles 5G7/1; weathers 10YR7/6
Pandu clay	A	SI-353803	2.5Y5/2	root traces 5Y7/1,10YR4/6; weathers 2.5Y6/2
loam	Bk	SI-353814	5Y7/3	root traces 2.5Y5/6; clay skins 7.5YR6/8; burrows 5YR4/4; weathers 2.5Y6/4,2.5Y7/2
	C	SI-353815	5Y6/3	claystone clasts 5YR3/3; weathers 2.5Y6/2
	C	SI-353816	10YR4/3	interbeds 10YR6/3; clasts 5YR3/4; weathers 2.5Y6/2
Pandu clay	A	SI-353800	2.5Y5/4	root traces 5Y6/1,10YR5/4; clay skins 5YR3/4; weathers 2.5Y6/4
thin sur-	A	SI-353801	2.5Y6/2	root traces 10YR6/6; burrows 10YR4/3; weathers 2.5Y6/4
face phase	Bk	SI-353802	5Y7/3	root traces 5Y7/1; sandstone dikes 7.5YR4/6; weathers 2.5Y7/4, 2.5Y7/2
type Pandu clay	A	SI-353796	10YR3/3	root traces 5GY6/1,5GY4/1; sandstone dikes 5YR3/4; weathers 10YR5/3
	A	SI-353797	2.5Y4/2	root traces 5Y7/1,5GY5/1; sandstone dikes 5YR4/4; weathers 10YR5/3
	Bk	SI-353798, SI-353799	2.5YR6/4	root traces 5Y5/1,5Y7/1,10YR6/6; burrows 5YR3/4; weathers 2.5Y7/4,5Y7/1
Sonita series	A	not taken	5YR3/4	root traces and sandstone dikes 5GY5/1; weathers 5YR5/4
un. #103	C	SI-353795	5YR3/3	interbeds 5G5/1; weathers 5YR5/4
Son. un. #104	A	not taken	5YR3/4	root traces 5GY5/1; weathers 5YR5/4
Lal series	A	not taken	5YR3/4	root traces 5GY4/1,5GY7/1,10YR6/6; weathers 5YR5/4
un. #105	C	not taken	5YR3/4	root traces 7.5YR4/4; burrows 5YR3/4; weathers 5YR5/4
Lal un. #106	A	not taken	5YR3/4	root traces 5G5/1,10YR5/4,5Y7/4; weathers 5YR5/4
Bhura series unnamed	A	not taken	7.5YR3/2	root traces 5GY4/1; iron-manganese nodules 5BG2.5/1; weathers 7.5YR5/4
#107	Bk	not taken	5YR4/3	root traces 2.5Y4/4; calcareous nodules 10YR5/4,7.5YR5/6; weathers 7.5YR5/4
Sonita series	A	not taken	5YR3/4	root traces 5GY5/1,10YR5/8; weathers 10YR7/4
unnamed #108	Bk	not taken	7.5YR3/2	calcareous nodules 5GY6/1,10YR6/6; mangans 5BG2.5/1; weathers 7.5YR5/4
	C	not taken	7.5YR5/4	burrows 5YR5/4; weathers 7.5YR6/4
Bhura series	A	not taken	7.5YR4/4	root traces 5GY5/1; mottles 7.5YR5/6; weathers 10YR7.4
unnamed	Bk	not taken	7.5YR4/4	root traces 5GY5/1; weathers 10YR7/4
#109	C	not taken	10YR6/4	lowest bed 5GY6/1; weathers 10YR7/2
Sarang clay loam	A	SI-353813	10YR4/4	root traces 5G6/1; interbeds 10YR6/4,10YR4/4; weathers 10YR5/3

APPENDIX 2. continued

Paleosol	Horizon	Specimen Number	Fresh Color	Minor Colors
	C	not taken	10YR4/4	mottles 5G6/1; weathers 10YR5/4
type Kala clay	A	SI-353807	10YR3/3	root traces 5G4/1,10YR5/4; mangans 5B2.5/1; weathers 10YR5/4
	Bg	SI-353808	7.5YR3/4	root traces 5GY5/1,5Y5/3; iron-manganese nodules 5BG3/1; clay skins 2.5YR4/4; interbed 5GY6/1; weathers 10YR5/4
	Bg	SI-353809, SI-353810	10YR3/2	root traces 5G5/1,10YR4/4,10YR2/2,10YR6/6,10YR5/6; iron-manganese nodules 5B2.5/1; clay skins 10YR4/3,2.5Y3/2; charcoal 5BG2.5/1; calcareous nodules 10YR6/6; weathers 10YR5/3
	Bk	SI-353811	10YR6/6	root traces 5GY7/1,7.5YR5/8; weathers 10YR5/4
	Bk	SI-353812	10YR6/6	calcareous nodules 7.5YR6/6; weathers 10YR5/4
Sarang clay	A	SI-353805	7.5YR4/4	root traces 5GY6/1; charcoal 5G2.5/1; weathers 7.5YR5/6
eroded phase	C	SI-353806	10YR4/3	clay skins 10YR6/4; interbeds 5GY6/1,5GY7/1,5YR4/2; weathers 7.5YR5/6
sandst. 57.5 m	-	SI-353804	2.5Y7/4	burrows 5YR4/4,5Y7/1; weathers 7.5YR5/6
sandst. 58.0 m	-	not taken	5Y7/2	weathers 5Y2.5/1,5Y7/1
sandst. 1.0 m	-	not taken	10YR6/4	clasts 5YR4/4,7.5YR5/6; weathers 10YR7/2
sandst. 1.9 m	-	not taken	7.5YR5/4	burrows 5YR4/4; weathers 5YR6/3
Sar. un. #113	A	not taken	5YR4/4	root traces 5GY6/1; weathers 5YR6/4
Son. un. #114	A	not taken	5YR3/3	root traces 5GY5/1,5GY7/1; weathers 5YR6/4
Sonita series	A	not taken	5GY3/4	root traces 5GY7/1; weathers 5YR5/4
un. #115	C	not taken	5YR3/3	root traces 5GY7/1,5GY6/1; weathers 5YR5/4
Sonita series unnamed	A	not taken	5YR3/3	root traces 5GY6/1,5GY5/1; mottles 5YR4/4; calcareous nodules 7.5YR4/4; weathers 5YR5/4
#116	C	not taken	7.5YR4/4	interbeds 5YR3/4; weathers 5YR5/4
siltst. 5.5 m	-	not taken	5YR5/3	interbeds 5YR4/4; weathers 5YR6/3
Sonita series	A	not taken	5YR3/4	root traces 5GY5/1,5G5/1; weathers 7.5YR6.6
unnamed	Bk	not taken	5YR4/4	root traces 5GY6/1; weathers 5YR6/4
#117	C	not taken	7.5YR6/4	claystone clasts 5YR3/4; weathers 5YR6/3
Sarang series	A	not taken	5YR4/4	root traces 5GY6/1; mottles 7.5YR4/4; weathers 7.5YR6/6
un. #118	C	not taken	7.5YR4/4	weathers 7.5YR6/6
Sarang series	A	not taken	5YR3/4	root traces 5GY6/1; weathers 7.5YR5/4
un. #119	C	not taken	7.5YR4/4	root traces 5Y7/1,5GY5/1; weathers 7.5YR5/4
sandst. 9.2 m	-	not taken	10YR6/3	claystone clasts 5YR3/4; weathers 10YR6/2
Bhura series un. #120	A	not taken	mottled 7.5YR4/4-4/6	root traces 5G6/1,5Y6/2; weathers 7.5YR5/4
	C	not taken	mottled 7.5YR4/4-4/6	root traces 5G6/1; weathers 10YR6/2
Sonita series	A	not taken	5YR4/4	root traces 5GY5/1; weathers 5YR5/6
un. #121	C	not taken	7.5YR5/4	root traces 5G6/1; burrows 5YR4/6; weathers 7.5YR6/4
Sonita series unnamed	A	not taken	mottled 5YR-4/4,5YR4/6	root traces 5GY6/1; weathers 10YR6/3
#122	C	not taken	7.4YR4/4	weathers 5YR5/6
sandst. 13.0 m	-	not taken	10YR6/3	burrows 5YR5/4; weathers 10YR6/2
sandst. 13.6 m	-	not taken	7.5YR5/4	root traces 5G6/1; weathers 7.5YR6/4
Sarang series unnamed	A	not taken	7.5YR5/6	root traces 5GY7/1,5GY5/1; interbeds 7.5YR7/4,10YR5/4, 5YR4/4; weathers 5YR6/6
#123	C	not taken	7.5YR5/4	interbeds 7.5YR5/4,7.5YR7/4; weathers 7.5YR6/6
	C	not taken	7.5YR4/6	root traces 5GY6/1; weathers 10YR6/6
sandst. 16.0 m	-	not taken	2.5Y6/4	weathers 10YR6/4
type Naranji clay	A	SI-353827	5YR3/4	root traces 5GY7/1,5GY4/1; charcoal 5B2.5/1; sandstone dikes 10YR5/4; calcareous nodules 7.5YR5/6; weathers 7.5YR7/6

APPENDIX 2. continued

Paleosol	Horizon	Specimen Number	Fresh Color	Minor Colors
	AB	SI-353828, SI-353829	7.5YR4/4	root traces 5GY6/1,10YR4/2,10YR6/4; charcoal 5BG4/1; claystone clasts 10YR5/4,10YR7/4; weathers 7.5YR6/4
	BA	SI-353830	5YR4/4	root traces 5GY5/1,5YR3/2; burrows 5GY6/1; weathers 7.5YR5/4
	BA	SI-353831	mottled 2.5YR 4/6,7.5YR4/4	root traces 5Y7/1,5GY6/1; weathers 5YR5/6,7,5YR6/6
	Bt	SI-353832, SI-353833	2.5YR4/6	root traces 5GY7/1,10YR6/6,5YR5/3; mottles 7.5YR4/4; weathers 5YR5/8
	BC	SI-353834, SI-353835, SI-353836	5YR4/4	root traces 5GY7/1,5GY6/1; burrows 5YR3/4; mottles 10YR5/4, 5YR3/4; weathers 10YR7/6
	C	SI-353837	7.5YR5/4	weathers 10YR6/3
Sarang silty clay massive surface phase	A	SI-353824	5YR3/4	root traces 10YR5/4; calcareous nodules 7.5YR5/6; weathers 7.5YR6/6
	C	SI-353825	7.5YR5/4	burrows 5YR3/4; weathers 7.5YR7/6
	C	SI-353826	5G6/1	weathers 7.5YR7/6
type Bhura clay	A	SI-353855	7.5YR4/4	root traces 5GY5/1,5Y5/1; charcoal 5Y2.5/1; calcareous nodules 7.5YR4/6; weathers 10YR5/3
	A	SI-353856	mottled 7.5YR3/4-3/2	root traces 5GY6/1,5GY5/1; clay skins 10YR5/6,7.5YR2/2; calcareous nodules 7.5YR4/6; weathers 10YR5/3
	Bk	SI-353857	7.5YR4/4	root traces 5Y7/1,5GY6/1; clay skins 5YR4/4; weathers 7.5YR5/4
	Bk	SI-353858	7.5YR4/4	root traces 5GY7/1,5GY6/1; strata concordant mottles 5GY6/1; calcareous nodules 7.5YR5/6; weathers 7.5YR5/4
	C	SI-353859	10YR4/4	burrows 5YR4/4; weathers 7.5YR5/4
Sarang silty clay rhizo. v.	A	SI-353854	7.5YR4/4	root traces 5GY6/1,5G5/1,10YR5/4; clay skins 5YR4/4; calcareous nodules 7.5YR5/6; mangans 7.5YR2/2; weathers 7.5YR6/4
type Sonita clay	A	SI-353850, SI-353823	5YR4/4	root traces 5Y7/1,5GY6/1,5GY5/1,5YR3/3; weathers 5YR5/4
	Bk	SI-353851	5YR4/3	root traces 5GY6/1,5Y6/1; calcareous nodules 7.5YR6/6; clay skins and claystone clasts 5YR4/4; weathers 7.5YR5/4
	C	SI-353852	7.5YR4/4	burrows 5Y5/1,5YR4/4; weathers 7.5YR5/4
	C	SI-353853	7.5YR4/4	burrows 5YR4/4; interbeds 5GY6/1; weathers 7.5YR6/4
Sarang clay cumulic surface ph.	A	SI-353848	5YR3/4	root traces 5GY7/1,5GY6/1,2.5Y6/4; claystone clasts 7.5YR3/4; weathers 5YR5/4
	C	SI-353849	7.5YR4/4	root traces 10YR5/8,10YR3/2; weathers 5YR5/4
Sarang clay rhizoconcretionary v.	A	SI-353846	5YR4/3	root traces 5GY6/1,5GY7/1; mottles 7.5YR4/6; clay skins 10YR4/3; calcareous nodules 5YR7/4; weathers 5YR5/4
	C	SI-353847	7.5YR4/4	interbeds 5YR4/4,5YR3/4,7.5YR4/6; weathers 5YR5/4
Sonita series unnamed #131	A	not taken	5YR3/4	root traces 5G5/1; weathers 5YR5/4
	Bk	not taken	5YR4/4	root traces 5Y7/1,5GY7/1,5GY6/1; weathers 5YR5/4
	C	SI-353845	7.5YR4/4	root traces 5GY6/1,5YR4/2; burrows 7.5YR3/4; mottles 7.5YR5/8; calcareous nodules 7.5YR4/2; weathers 7.5YR6/6
Sonita series unnamed #132	A	not taken	5YR3/3	root traces 5Y7/1,5GY6/1; weathers 5YR5/4
	Bk	not taken	5YR3/3	root traces 5G5/1,10YR5/4; weathers 5YR5/4
	C	not taken	5YR3/4	weathers 5YR3/4
Sarang series unnamed #133	A	not taken	mottled 5YR4/4, 5YR4/6	root traces 5GY4/1,5GY6/1; weathers 5YR5/4
	C	not taken	mottled 7.5YR5/3,2.5Y6/2	interbeds 7.5YR3/4; weathers 7.5YR6/6
sandst. 26.0 m	-	not taken	10YR6/4	clasts 5YR4/4; mottles 5Y5/1,5GY5/1; weathers 10YR7/4
Son. un. #134	A	not taken	5YR3/4	root traces 5GY6/1; weathers 7.5YR4/6
Son. un. #135	A	not taken	5YR3/4	root traces 5GY6/1; weathers 7.5YR4/6

APPENDIX 2. continued

Paleosol	Hor-izon	Specimen Number	Fresh Color	Minor Colors
Sarang series un. #136	A	not taken	mottled 7.5YR4/4-4/6	mottles 5GY6/1; weathers 7.5YR5/4
Sarang series un. #137	A	not taken	mottled 7.5YR4/4-4/6	mottles 5GY6/1; weathers 7.5YR5/4
Son. un. #138	A	not taken	5YR4/4	root traces 5GY4/1,5GY5/1; weathers 10YR5/4
sandst. 30.2 m	-	not taken	10YR5/4	mottles 5GY6/1; clasts 5YR4/4; weathers 10YR5/3
Naranji series	A	not taken	mottled 7.5YR4/4-4/6	weathers 5YR5/6
unnamed	Bt	not taken	2.5YR4/6	weathers 5YR5/6
#139	2Bt	not taken	5YR3/3	weathers 5YR5/6
	2Bt	not taken	5YR4/6	weathers 5YR5/6
	2BC	not taken	mottled 10YR-5/4,7.5YR5/6	mottles 5GY7/1.5GY6/1; weathers 10YR6/6
	2C	not taken	mottled 2.5Y6/2-6/6	weathers 10YR6/4
Sar. un. #140	A	not taken	5YR4/4	weathers 5YR6/4
Sarang clay	A	SI-353844	5YR4/4	root traces 5GY6/1; burrows 5YR3/4,5Y6/3,5Y7/1; clay skins 5YR3/4; weathers 5YR7/4,5YR6/4
Sonita silty	A	SI-353840	5YR4/4	root traces 5G6/1; weathers 5YR6/4
clay	A	SI-353841	5YR4/4	root traces 5GY6/1; sandstone dikes 5GY6/1,5GY5/1; weathers 5YR6/4,5YR7/4
	C	SI-353842	7.5YR4/4	burrows 5YR4/4; weathers 7.5YR6/4
	C	SI-353843	7.5YR5/4	root traces 5GY6/1,5GY6/3; burrows 5YR4/4, 7.5YR5/6; claystone clasts 5YR4/4; weathers 7.5YR6/4
silst. 36.0 m	-	SI-353839	7.5YR4/4	interbeds 5YR4/4,7.5YR4/4,7.5YR6/4; weathers 7.5YR6/4
Sarang series un. #143	A	not taken	mottled 5YR-6/4,7.5YR7/4	root traces 5GY7/1; weathers 10YR7/4
sandst. 37.8 m	-	not taken	7.5YR7/4	weathers 7.5YR6/4

Note: All colors were taken using a Munsell (1975) chart on rock samples within minutes of excavation, with exception of weathered colors of Pakistani paleosols that in some cases include a component of modern soil creep, given here to enable relocation in these colorful badlands outcrops. Rock specimens are housed in the Kenyan National Museum, Nairobi (-R-) and the Department of Palaeobiology, Smithsonian Institution, Washington, D.C. (SI-).

APPENDIX 3. Textures (volume percent) from point counting petrographic thin sections and calcareousness from reaction with dilute acid of Miocene paleosols in Pakistan and Kenya.

Paleosol	Horizon	Specimen Number	Calcar-eousness	Percent Clay	Percent Silt	Percent Sand	Percent Gravel	Texture
Mobaw bouldery	A	KO-R96	4	71.0	4.0	25.0	0	clay
clay	Bw	KO-R99	4	67.8	17.0	15.0	0.2	clay
	clast	KO-R100	5	64.6	7.8	27.4	0.2	clay
	C	KO-R101	4	75.0	6.8	18.2	0	clay
	C	KO-R102	4	63.2	13.6	23.2	0	clay
	2Cg	KO-R105	1	91.8	1.6	6.4	0.2	clay
	2Cg	KO-R108	1	94.4	1.0	4.6	0	clay
carbonatite tuff	-	KO-R109	5	28.0	36.6	35.4	0	clay loam
Kwar clay calcareous	A	KO-R90	4	51.8	22.8	21.6	3.8	clay
stringer variant	Bt	KO-R91	4	75.8	3.0	19.4	1.8	clay
	C	KO-R92	4	78.6	6.8	14.8	0	clay
	2C	KO-R94	4	79.4	4.4	16.2	0	clay
	2C	KO-R95	3	63.4	2.0	34.6	0	clay
type Kwar clay	A	KO-R87	4	57.2	29.0	13.8	0	clay
	Bw	KO-R88	4	46.8	40.8	12.4	0	silty clay
	C	KO-R89	4	48.4	33.2	14.6	3.8	clay
type Mobaw silty	A	KO-R84	4	36.6	45.8	17.6	0	silty clay loam
clay loam	Bt	KO-R85	4	52.8	30.8	16.4	0	clay
	C	KO-R86	5	39.2	46.4	14.4	0	silty clay loam
Kwar silty clay	A	KO-R80	4	53.2	41.4	3.0	2.4	silty clay
	A	KO-R81	4	66.2	15.0	17.8	1.0	clay
	C	KO-R82	4	66.2	13.6	17.2	3.0	clay
	C	KO-R83	4	50.8	23.4	23.2	2.6	clay
calcareous siltstone	-	KO-R79	5	31.4	52.6	16.0	0	silty clay loam
calcareous claystone	-	KO-R79	5	48.8	49.8	1.4	0	silty clay
Kiewo clay thin s.p.	A	SO-R148	2	84.8	3.2	12.0	0	clay
Kiewo clay single	A	SO-R144	3	58.2	18.6	23.2	0	clay
stringer variant	Ck	SO-R145	4	69.6	11.6	18.8	0	clay
	Ck	SO-R146	4	60.8	11.6	27.6	0	clay
	Ck	SO-R147	3	50.4	38.4	3.8	0	clay
type Kiewo clay	A	SO-R142	3	64.0	22.0	14.0	0	clay
	Ck	SO-R143	5	39.4	13.2	47.4	0	sandy clay
type Choka clay	A	SO-R134	3	49.8	29.2	21.0	0	clay
	A	SO-R135	4	58.2	28.0	13.8	0	clay
	BA	SO-R136	4	52.8	19.8	27.4	0	clay
	Bt	SO-R137	4	64.4	18.8	16.8	0	clay
	Bt	SO-R138	4	60.9	20.4	19.2	0	clay
	C	SO-R139	4	47.6	23.6	28.8	0	clay
	C	SO-R140	4	43.6	16.6	39.8	0	clay
	C	SO-R141	4	28.8	35.4	35.8	0	clay loam
type Tut clay	A	SO-R121	4	24.0	23.4	50.8	1.8	sandy clay loam
	A	SO-R122	3	44.4	34.2	21.4	0	clay
	Bt	SO-R123	3	49.4	25.0	25.6	0	clay
	nodule	SO-R124	4	25.6	41.0	33.4	0	clay loam
	Bt	SO-R125	3	60.4	24.0	15.6	0	clay
	nodule	SO-R126	4	27.0	41.4	31.6	0	clay loam
	Bt	SO-R127	3	59.6	19.4	21.0	0	clay
	Bk	SO-R128	3	48.8	28.2	23.0	0	clay
	dike	SO-R129	5	58.8	23.8	17.4	0	clay
	BC	SO-R130	4	45.4	21.2	33.4	0	clay

APPENDIX 3. continued

Paleosol	Horizon	Specimen Number	Calcar-eousness	Percent Clay	Percent Silt	Percent Sand	Percent Gravel	Texture
	C	SO-R132	4	55.0	32.6	12.4	0	clay
	C	SO-R133	4	64.2	26.4	9.4	0	clay
sandstone	-	SO-R120	3	51.2	16.4	32.4	0	clay
Buru clay thin sur. ph.	A	SO-R118	1	62.8	7.8	29.4	0	clay
type Buru clay	A	SO-R116	1	68.2	12.4	18.6	0	clay
	C	SO-R117	1	61.0	11.6	27.6	0	clay
sandstone	-	SO-R115	1	68.6	10.0	21.0	0	clay
type Lwanda clay	A	FT-R57	1	93.0	1.8	5.2	0	clay
	AB	FT-R58	2	43.8	45.6	10.6	0	silty clay
	Bt	FT-R59	3	51.8	36.4	11.8	0	clay
	C	FT-R60	1	55.6	22.2	22.2	0	clay
Rairo bouldery clay	A	FT-R54	2	84.6	11.2	4.2	0	clay
	C	FT-R56	2	64.8	12.2	15.8	7.2	clay
type Rairo clay	A	FT-R51	3	83.2	4.2	12.6	0	clay
	A	FT-R52	3	83.0	3.4	12.6	0	clay
	Ck	FT-R53	3	69.2	2.6	22.4	3.4	clay
type Rabuor clay	A	FT-R45	2	83.2	2.2	13.2	1.4	clay
	A	FT-R47	2	86.6	2.4	10.2	0.8	clay
	Bt	FT-R48	2	87.0	2.2	10.6	0.2	clay
	Ck	FT-R50	2	79.4	3.2	16.2	1.2	clay
Dhero clay loam	A	FT-R42	3	34.6	25.4	35.4	4.6	clay loam
	C	FT-R43	3	29.4	7.6	51.4	11.6	sandy clay loam
lapilli tuff	-	FT-R41	4	31.8	17.6	38.4	12.2	clay loam
Chogo clay ferrug-	A	FT-R12	4	56.2	26.2	11.6	6.0	clay
inized nodule	A	FT-R13	4	48.8	27.8	19.0	4.4	clay
variant (to east)	Bk	FT-R14	5	34.8	37.8	20.0	6.8	clay loam
	Bk	FT-R15	3	59.8	12.4	17.8	10.0	clay
	Bk	FT-R16	4	36.6	25.0	30.0	8.4	clay loam
	C	FT-R18	4	26.6	9.2	17.2	46.8	clay
Chogo clay ferrug-	A	FT-R35	3	69.2	10.6	14.4	5.8	clay
inized nodule	A	FT-R36	5	53.0	26.6	18.8	1.6	clay
variant (to west)	A	FT-R37	4	65.2	10.0	22.0	2.8	clay
	Bk	FT-R38	4	38.8	32.8	27.2	1.2	clay loam
	Bk	FT-R39	4	27.4	20.0	33.8	18.8	clay loam
	C	FT-R40	5	42.4	13.4	40.4	3.8	clay
type Chogo clay	A	FT-R2	2	72.8	9.8	12.6	4.8	clay
	A	FT-R3	3	65.6	12.2	17.0	5.2	clay
	Bk	FT-R6	4	33.0	19.2	29.4	18.4	clay
	Bk	FT-R9	4	42.2	13.6	10.6	33.0	clay
	C	FT-R10	4	36.4	18.2	26.2	19.2	clay
	C	FT-R11	4	37.4	15.8	37.0	9.6	clay
Chogo clay eroded	A	FT-R22	3	66.2	10.0	14.6	9.2	clay
phase	A	FT-R23	3	72.0	7.6	15.6	4.8	clay
	A	FT-R25	4	48.8	25.6	20.0	5.6	clay
	A	FT-R28	4	49.0	28.0	19.2	3.8	clay
	Bk	FT-R29	4	60.2	15.6	19.4	4.8	clay
	Bk	FT-R31	3	38.2	18.2	23.8	19.8	clay
	C	FT-R33	3	23.6	12.6	41.4	22.6	sandy clay loam
sandstone (to east)	-	FT-R1	4	24.4	9.0	44.2	22.4	sandy clay loam
sandstone (to west)	-	FT-R21	4	4.4	23.6	25.4	46.6	loam
type Dhero clay	A	FT-R62	4	51.0	22.8	25.2	1.0	clay
	A	FT-R63	5	48.4	25.2	26.0	0.4	clay

APPENDIX 3. continued

Paleosol	Horizon	Specimen Number	Calcar-eousness	Percent Clay	Percent Silt	Percent Sand	Percent Gravel	Texture
	C	FT-R65	5	42.2	23.0	34.6	0.2	clay
	C	FT-R66	4	63.0	8.8	26.0	1.8	clay
breccia matrix	-	FT-R61	4	48.8	12.2	38.2	0.8	clay
type Onuria clay	A	FT-R75	3	79.6	11.8	8.4	0.2	clay
	A	FT-R76	4	78.2	12.4	9.2	0	clay
	Bk	FT-R77	5	45.8	24.2	29.6	0.4	clay
	R	FT-R78	4	47.6	21.6	28.4	2.4	clay
sandstone	-	FT-R74	4	51.4	21.2	27.4	0	clay
Dhero clay lapilli-	A	FT-R70	4	52.0	23.0	24.8	0.2	clay
stone variant	C	FT-R71	5	45.0	36.4	18.0	0.6	clay
	C	FT-R72	4	26.6	25.4	48.0	0	sandy clay loam
siltstone	-	FT-R69	4	35.4	27.4	37.2	0	clay loam
sandstone	-	FT-R68	4	15.0	34.2	50.8	0	sandy loam
sandstone	-	FT-R67	3	19.6	3.8	21.0	55.6	sandy clay
type Sarang clay	A	SI-353766	4	57.0	37.8	5.2	0	clay
	C	SI-353767	4	53.3	33.8	13.0	0	clay
	C	SI-353768	4	72.4	27.6	0	0	clay
	C	SI-353769	4	67.2	31.4	1.4	0	clay
	C	SI-353770	4	57.6	36.2	6.2	0	clay
buff sandstone	-	SI-353765	4	31.0	19.4	49.6	0	sandy clay loam
Khakistari thin s. p.	A	SI-353776	4	91.6	5.4	3.0	0	clay
type Khakistari clay	A	SI-353772	3	81.0	15.4	3.6	0	clay
	A	SI-353773	3	85.8	5.8	8.4	0	clay
	Bg	SI-353774	4	99.2	0.8	0	0	clay
	C	SI-353775	4	95.6	4.4	0	0	clay
Sonita clay fer. con. v.	C	SI-353771	4	85.6	12.0	2.4	0	clay
Lal clay cumulic s. p.	A	SI-353785	4	70.8	27.4	1.8	0	clay
type Lal clay	A	SI-353778	3	79.4	19.6	1.0	0	clay
	A	SI-353779	4	80.0	15.8	3.0	0	clay
	Bt	SI-353780	4	85.4	14.6	0	0	clay
	Bt	SI-353781	4	84.8	15.0	0.2	0	clay
	Bk	SI-353782	4	75.0	24.8	0.2	0	clay
	2Bk	SI-353783	4	80.0	18.2	1.6	0	clay
	2Co	SI-353784	4	77.0	23.0	0	0	clay
claystone	-	SI-353777	4	81.8	17.8	0.4	0	clay
Sarang clay kroto. v.	A	SI-353794	4	45.8	23.6	30.6	0	clay
Pila clay nodular	A	SI-353790	3	66.8	22.2	11.0	0	clay
variant	Bk	SI-353791	3	55.4	27.6	17.0	0	clay
	2Bk	SI-353792	3	56.4	23.2	20.4	0	clay
	3C	SI-353793	4	42.0	37.2	20.8	0	clay
claystone	-	SI-353788	4	60.8	24.4	14.8	0	clay
Bhura clay fer. con. v.	A	SI-353822	3	63.0	22.8	14.8	0	clay
type Pila clay	A	SI-353817	3	67.0	22.0	10.8	0	clay
	A	SI-353818	3	71.2	20.0	8.8	0	clay
	Bt	SI-353819	3	64.6	24.8	10.6	0	clay
	Bk	SI-353820	3	66.6	21.0	12.4	0	clay
	C	SI-353821	3	69.6	19.2	11.2	0	clay
Pandu clay loam	A	SI-353803	4	44.4	29.6	26.0	0	clay
	Bk	SI-353814	5	65.4	21.4	13.2	0	clay
	C	SI-353815	5	55.0	29.0	16.0	0	clay
	C	SI-353816	3	59.4	30.8	9.8	0	clay

APPENDIX 3. continued

Paleosol	Horizon	Specimen Number	Calcar-eousness	Percent Clay	Percent Silt	Percent Sand	Percent Gravel	Texture
Pandu clay thin	A	SI-353800	3	51.0	27.6	21.4	0	clay
surface phase	Bk	SI-353802	4	61.0	30.2	8.8	0	clay
type Pandu clay	A	SI-353796	3	75.4	20.4	4.2	0	clay
	A	SI-353797	3	68.8	22.8	8.4	0	clay
	Bk	SI-353798	4	66.0	27.4	6.6	0	clay
	C	SI-353799	4	47.4	32.8	19.8	0	clay
claystone	-	SI-353795	4	93.2	5.2	1.6	0	clay
Sarang clay loam	A	SI-353813	5	31.0	40.2	28.8	0	clay loam
type Kala clay	A	SI-353807	3	84.8	11.4	3.8	0	clay
	Bg	SI-353808	3	87.0	10.4	2.6	0	clay
	Bg	SI-353810	4	54.0	31.4	14.6	0	clay
	Bk	SI-353811	4	44.6	32.4	23.0	0	clay
Sarang clay eroded	A	SI-353805	3	54.8	9.0	36.2	0	clay
phase	C	SI-353806	3	10.4	3.2	86.4	0	loamy sand
blue-gray sandstone	-	SI-353804	3	5.6	8.4	86.0	0	loamy sand
type Naranji clay	A	SI-353827	4	84.0	15.0	1.0	0	clay
	AB	SI-353828	3	51.6	35.4	13.0	0	clay
	AB	SI-353829	3	55.4	19.6	25.0	0	clay
	BA	SI-353830	3	51.8	13.4	34.6	0	clay
	BA	SI-353831	3	51.8	10.2	38.0	0	clay
	Bt	SI-353832	3	49.6	9.4	41.0	0	clay
	Bt	SI-353833	3	48.8	5.6	45.6	0	sandy clay
	Bt	SI-353834	3	51.8	9.4	38.8	0	clay
	BC	SI-353835	3	46.8	7.4	45.6	0	sandy clay
	BC	SI-353836	3	41.2	11.4	47.4	0	sandy clay
	C	SI-353837	3	28.0	6.8	65.2	0	sandy clay loam
Sarang silty clay	A	SI-353824	4	53.4	46.6	0	0	silty clay
massive surface	C	SI-353825	4	27.4	67.4	5.2	0	silty clay loam
phase	C	SI-353826	4	26.0	72.2	1.6	0	silty clay loam
type Bhura clay	A	SI-353856	3	69.4	23.6	7.0	0	clay
	Bk	SI-353857	4	55.8	39.8	4.4	0	clay
	Bk	SI-353858	4	46.6	51.4	2.0	0	silty clay
	C	SI-353859	5	39.0	43.0	18.0	0	silty clay loam
Sarang silty clay r. v.	A	SI-353854	5	50.2	49.0	0.8	0	silty clay
type Sonita clay	A	SI-353850	4	69.6	29.0	1.4	0	clay
	Bk	SI-353851	4	54.4	44.6	1.0	0	silty clay
	C	SI-353852	4	51.0	46.8	2.2	0	silty clay
	C	SI-353853	5	59.0	36.2	4.8	0	clay
Sarang clay cumulic	A	SI-353848	4	90.8	9.2	0	0	clay
surface phase	C	SI-353849	4	65.0	35.0	0	0	clay
Sarang clay rhizo-	A	SI-353846	3	91.6	8.4	0	0	clay
concretionary var.	C	SI-353847	4	65.6	34.4	0	0	clay
siltstone	-	SI-353845	4	28.4	66.8	4.8	0	silty clay loam
Sarang clay	A	SI-353844	3	87.8	12.2	0	0	clay
Sonita silty clay	A	SI-353840	4	58.8	41.2	0	0	silty clay
	C	SI-353841	4	60.4	39.6	0	0	clay
	C	SI-353842	5	28.0	72.0	0	0	silty clay loam
	C	SI-353843	5	27.6	72.4	0	0	silty clay loam
claystone	-	SI-353839	4	50.2	49.8	0	0	silty clay

Note: Relative scale of calcareousness (1-5) by reaction with 1.2M HCl is from Retallack (1988a, 1990a). Standard error of these 500-point counts is about 2 volume % (Murphy 1983). Counts made with a Swift automatic point counter by analysts listed for Appendix 4.

APPENDIX 4. Mineral composition (volume percent) by point counting of petrographic thin sections of Miocene paleosols in Pakistan and Kenya

Paleosol	Hoz-izon	Specimen Number	Clay & Micrite	Calcite	Feldspar	Pyroxene	Mica	Rock Fragment	Other	Opaque	Quartz
Mobaw	A	KO-R96	73.8	2.2	1.2	1.4	1.2	10.4	0.4	2.0	7.4
boul-	Bw	KO-R99	66.8	13.6	0	0	0	15.6	0	3.8	0.2
dery	clast	KO-R100	66.0	7.6	0	0	0	22.4	0	4.0	0
clay	C	KO-R101	72.8	7.4	0	0	1.4	16.2	0	0.6	1.6
	C	KO-R102	64.0	11.8	0	0	0	22.4	0	1.4	0.4
	2Cg	KO-R105	93.6	0.2	0	0	0	1.2	0	1.8	3.2
	2Cg	KO-R108	94.2	0.2	0	0.4	0	1.6	0.4	1.8	1.4
tuff	-	KO-R109	29.8	68.6	0	0	0	0	0.2	1.4	0
Kwar	A	KO-R90	49.8	21.4	1.2	2.4	0.2	18.4	0	1.4	5.2
clay	Bt	KO-R91	75.6	4.6	1.4	0.8	5.8	10.2	0	1.2	5.9
calc.	C	KO-R92	79.0	3.4	1.4	0.2	1.2	10.6	0	1.0	3.2
stringer	2C	KO-R94	79.2	3.6	1.0	0.2	2.4	5.6	1.0	1.4	5.6
variant	2C	KO-R95	65.8	2.0	2.6	2.6	0.4	10.8	2.0	2.2	11.6
type	A	KO-R87	58.6	28.4	0.2	0	0	2.8	9.4	0	0.6
Kwar	Bw	KO-R88	49.2	38.4	0.2	0	0	10.0	0	0.6	1.6
clay	C	KO-R89	45.4	34.6	0.8	0.8	0	15.4	0.6	0.2	2.0
type	A	KO-R84	37.4	44.4	0.6	0.6	2.8	12.9	0	1.0	0.8
Mobaw	Bt	KO-R85	53.8	30.8	0	0.2	0	14.2	0	0.8	0.2
scl.	C	KO-R86	39.6	46.2	0.4	0	0	12.8	0.2	0.2	0.6
Kwar	A	KO-R80	55.8	39.2	0	0	0	0.4	0	2.0	2.6
silty	A	K0-R81	66.4	13.4	0	0.8	0	16.8	0	0.6	2.0
clay	C	KO-R82	64.2	13.4	1.6	1.4	1.0	14.2	0	0.2	4.0
	C	KO-R83	48.4	19.6	0.6	0.8	0	27.4	0	0.6	2.8
siltst.	-	KO-R79	34.4	63.6	0.2	0	0	0.2	0	0.2	1.4
sandst.	-	KO-R79	49.4	50.0	0	0	0	0	0	0.2	0.4
Kie. ctsp.	A	SO-R148	87.0	0.8	1.0	1.8	0.4	2.2	0	1.0	5.8
Kiewo	A	SO-R144	60.2	8.2	9.6	1.8	1.8	8.8	0	4.0	5.6
clay	Ck	SO-R145	70.4	10.8	5.8	1.6	0.4	5.0	0	2.0	4.0
ssv.	Ck	SO-R146	61.2	26.6	1.0	2.0	1.6	3.0	0	0.8	3.8
	Ck	SO-R147	53.0	40.6	0.6	1.8	0	0	0	0.2	3.8
type Ki-	A	SO-R142	63.2	8.6	7.6	3.0	1.6	13.0	0	2.8	0.2
ewo c.	Ck	SO-R143	41.3	42.2	2.2	1.2	1.0	6.8	0	3.6	1.8
type	A	SO-R134	48.0	8.8	17.6	8.4	1.6	9.0	0	1.8	4.8
Choka	A	SO-R135	58.0	17.8	7.6	0.8	1.0	6.4	0	3.0	5.4
clay	BA	SO-R136	52.0	25.2	6.4	2.2	0.6	9.4	0	4.0	0.2
	Bt	SO-R137	66.4	9.6	9.8	2.0	0.4	7.6	0	3.8	0.4
	Bt	SO-R138	59.6	9.6	11.8	2.0	0.8	12.6	0	2.0	1.6
	C	SO-R139	49.0	16.2	14.4	2.0	1.8	14.4	0	1.2	1.0
	C	SO-R140	45.4	11.4	5.4	2.2	0.6	28.6	1.2	3.8	1.4
	C	SO-R141	29.6	45.6	3.6	0.8	0.2	17.2	0	2.6	0.4
type Tut	A	SO-R121	24.8	10.6	14.0	3.4	1.8	42.8	0	1.4	1.2
clay	A	SO-R122	42.8	37.8	1.2	9.0	0	4.4	0	1.2	3.2
	Bt	SO-R123	49.8	6.6	9.4	5.2	0.8	19.6	0.2	6.2	2.2
	nod.	SO-R124	24.0	30.8	8.2	3.0	0.6	30.6	0	2.8	0
	Bt	SO-R125	60.0	5.8	11.4	1.2	2.6	5.2	0	5.6	8.2
	nod.	SO-R126	28.4	29.8	5.2	2.6	0.6	30.6	0	2.8	0
	Bt	SO-R127	60.6	10.0	12.2	4.8	2.2	5.0	0.8	3.0	1.4
	Bk	SO-R128	48.6	12.2	9.4	4.6	1.0	17.6	0	3.8	2.8
	dike	SO-R129	59.4	10.6	16.4	4.4	0.6	7.0	0	0.8	0.8
	BC	SO-R130	45.0	27.6	7.6	1.8	0.8	13.2	0	2.6	1.4

APPENDIX 4. continued

Paleosol	Hor-izon	Specimen Number	Clay & Micrite	Calcite	Feldspar	Pyroxene	Mica	Rock Fragment	Other	Opaque	Quartz
	C	SO-R132	56.6	30.6	0.6	1.0	0	7.2	0	1.2	2.4
	C	SO-R133	64.4	15.0	4.8	1.8	0.4	6.4	0	5.0	1.6
sandst.	-	SO-R120	53.2	17.6	1.0	16.0	0	9.6	0.6	1.2	1.4
Bur. ctsp.	A	SO-R118	62.0	0	8.6	3.2	2.8	17.2	0	1.8	4.4
type Bu-	A	SO-R116	69.6	0	9.0	0.8	2.4	10.8	0	3.8	3.6
ru clay	C	SO-R117	62.0	0	5.2	4.4	5.6	8.4	0	9.6	4.8
sandst.	-	SO-R115	69.4	0	5.8	1.6	2.6	12.8	0	3.4	4.4
type	A	FT-R57	93.0	0	0	0	0	5.4	0	1.4	0.2
Lwanda	AB	FT-R58	41.0	0	32.0	2.6	0	0	18.6	5.4	0
clay	Bt	FT-R59	52.2	0	22.4	1.8	0	0	18.4	4.2	0
	C	FT-R60	56.8	0	13.0	2.4	0.4	0	27.0	0.4	0
Rairo bc.	A	FT-R54	86.0	0	2.8	0.6	0.4	7.0	0	3.2	0.6
	C	FT-R56	65.4	0	8.0	0.6	0.8	7.8	0	17.4	0
type	A	FT-R51	84.8	1.4	0	0.4	0.2	9.8	0	3.0	0.4
Rairo	A	FT-R52	81.0	0	0.2	2.0	2.0	8.4	0	4.4	2.0
clay	Ck	FT-R53	69.0	0	0.2	0.4	0.2	23.8	0	5.0	1.4
type	A	FT-R45	85.0	0.2	0	1.8	0.4	9.0	0	1.8	1.8
Rabuor	A	FT-R47	86.8	0	0.2	0.8	0.2	7.0	0	4.4	0.6
clay	Bt	FT-R48	88.2	0	0.4	0.6	0.6	7.6	0	1.4	1.2
	Ck	FT-R50	78.8	0.4	0.6	2.2	0.2	12.4	0.2	4.0	1.2
Dhero	A	FT-R42	34.8	25.0	0	1.6	0	36.2	0	1.4	0
cl.loam	C	FT-R43	30.8	5.8	0.2	5.6	0	53.8	0	2.6	1.2
tuff	-	FT-R41	31.4	16.6	0	2.0	0.2	48.4	0	1.4	0
Chogo	A	FT-R12	58.6	6.2	0	4.4	1.4	18.0	6.4	5.0	0
clay	A	FT-R13	48.8	14.0	0	6.8	1.4	20.0	2.2	6.8	0
ferrug.	Bk	FT-R14	32.0	16.8	0	9.6	1.0	33.2	3.2	4.2	0
nodule	Bk	FT-R15	60.4	1.2	0	7.2	2.0	19.0	3.0	7.2	0
variant	Bk	FT-R16	38.8	14.4	0	10.6	2.6	26.0	2.0	5.6	0
(east)	C	FT-R18	24.2	9.4	0	15.8	4.2	43.4	1.8	1.2	0
Chogo	A	FT-R35	69.6	0.8	4.8	3.8	0.2	13.2	0.8	6.8	0
clay	A	FT-R36	51.8	19.6	0	4.4	0.6	19.4	0	4.2	0
ferrug.	A	FT-R37	66.4	0.2	4.4	6.0	0.6	15.2	0	7.2	0
nodule	Bk	FT-R38	39.6	24.8	1.2	2.4	0	24.0	0	8.0	0
variant	Bk	FT-R39	27.2	21.8	0.4	4.2	0.2	38.6	0	7.6	0
(west)	C	FT-R40	41.8	15.8	2.8	11.4	2.8	18.2	0.8	6.4	0
type	A	FT-R2	75.6	1.0	0	4.6	0.2	12.6	3.0	3.0	0
Chogo	A	FT-R3	65.4	0	0	4.8	0.8	22.2	1.4	5.4	0
clay	Bk	FT-R6	34.4	18.6	0	25.6	0.4	11.8	2.6	6.6	0
	Bk	FT-R9	39.4	9.2	0	9.8	1.0	33.0	2.6	5.0	0
	C	FT-R10	33.8	11.6	0	8.2	2.8	37.6	2.6	3.4	0
	C	FT-R11	39.6	8.6	0	11.4	2.4	43.0	0.8	4.2	0
Chogo	A	FT-R22	66.2	0.6	3.2	9.6	1.0	12.0	0.8	6.6	0
clay	A	FT-R23	72.8	0	3.4	7.0	0.8	9.4	0.8	5.8	0
eroded	A	FT-R25	47.6	16.0	0.8	4.2	1.2	24.8	0	5.4	0
phase	A	FT-R28	52.2	15.6	1.6	4.0	1.0	20.8	0	4.8	0
	Bk	FT-R29	60.6	8.4	2.0	5.8	1.8	17.6	0	3.8	0
	Bk	FT-R31	39.0	13.6	1.6	4.2	1.0	37.6	0.2	2.8	0
	C	FT-R33	23.0	17.0	3.0	5.0	2.0	38.6	0	11.4	0
sandst. E	-	FT-R1	23.4	3.8	0	4.0	1.0	69.4	0.8	1.4	0
sandst. W	-	FT-R21	4.8	27.2	0	3.8	0	60.4	0	3.8	0
type	A	FT-R62	51.8	23.8	0	4.2	0	17.0	0.2	2.2	0.8
Dhero	A	FT-R63	47.8	31.0	0	3.4	0	15.2	0	2.4	0.2

APPENDIX 4. continued

Paleosol	Hor-izon	Specimen Number	Clay & Micrite	Calcite	Feldspar	Pyroxene	Mica	Rock Fragment	Other	Opaque	Quartz
clay	C	FT-R65	43.4	23.0	0	3.2	0	28.0	0	2.0	0.4
	C	FT-R66	62.6	9.6	0	9.8	0.8	9.0	0	8.0	0.2
breccia	-	FT-R61	49.8	11.8	0	10.0	0.4	23.0	0	2.4	2.6
type	A	FT-R75	77.8	0	7.8	4.0	0.2	5.2	0	4.4	0.6
Onuria	A	FT-R76	79.8	0	3.6	4.4	0	7.6	0	4.2	0.4
clay	Bk	FT-R77	44.6	21.2	0	3.8	0	26.6	0	2.6	1.0
	Bk	FT-R78	46.6	25.2	0	7.8	0	13.8	0	4.8	1.8
sandst.	-	FT-R74	53.4	11.6	0	9.4	3.6	16.8	0	5.2	0
Dhero	A	FT-R70	53.6	21.4	0	0.8	0.8	19.2	0	3.0	1.2
clv.	C	FT-R71	44.0	22.4	0	0.6	0.4	30.2	0	2.2	0.2
	C	FT-R72	25.2	23.0	0	6.4	1.2	38.4	0	2.8	3.0
siltst.	-	FT-R69	34.8	29.4	0	4.6	1.0	22.4	0	5.8	1.4
sandst.	-	FT-R68	14.4	37.2	0	2.6	1.6	40.6	0	2.2	1.4
sandst.	-	FT-R67	19.0	12.6	0.6	11.4	2.2	47.8	0	6.4	0
type	A	SI-353766	56.0	14.4	8.0	0	2.6	0.4	0	1.4	17.2
Sarang	C	SI-353767	54.0	9.8	9.0	0	4.6	1.6	0	2.2	18.8
clay	C	SI-353768	69.0	3.8	1.4	0	0.4	0	0	3.0	22.4
	C	SI-353769	64.8	3.0	5.8	0	2.2	0	0	2.4	21.8
	C	SI-353770	57.4	4.2	7.6	0	1.2	0	0	7.8	21.2
sandst.	-	SI-353765	30.4	18.4	8.6	0	6.8	3.0	0	5.8	27.0
Kh. ctsp.	A	SI-353776	93.6	1.8	0	0	0.4	2.0	0	0.4	1.8
type	A	SI-353772	81.0	0.6	1.2	0	0	0.4	0	1.8	15.0
Khaki-	A	SI-353773	85.4	8.0	0	0	0	5.0	0	0.4	1.2
stari	Bg	SI-353774	98.2	1.2	0	0	0	0	0	0.2	0
clay	C	SI-353775	98.0	0.8	0	0	0	0	0	0.8	0.4
Son. cfcv.	C	SI-353771	85.4	2.6	0.8	0	1.8	0	0	0.4	9.0
Lal ccsp.	A	SI-353785	71.8	0	16.6	0.8	0.6	1.6	0	1.2	7.4
type Lal	A	SI-353778	77.6	0	7.2	1.8	0.8	4.4	0	1.2	7.0
clay	A	SI-353779	79.0	0.4	6.8	0.6	0.4	1.2	0	0.8	10.8
	Bt	SI-353780	84.2	1.0	5.4	0.8	0.4	1.6	0	1.0	5.6
	Bt	SI-353781	84.0	0	6.2	1.0	0.4	1.0	0	2.0	5.4
	Bk	SI-353782	74.4	0.6	11.2	2.2	1.8	1.4	0	3.6	4.8
	2Bk	SI-353783	81.2	0	8.6	1.0	0.4	1.8	0	1.2	5.8
	2Co	SI-353784	76.2	0	14.4	1.4	1.2	1.0	0	1.8	4.0
clayst.	-	SI-353777	82.4	1.8	8.0	1.0	1.8	0.2	0	1.8	2.4
Sar. ckv.	A	SI-353794	47.4	6.2	6.0	0.4	2.2	10.6	0	2.4	24.8
Pila clay	A	SI-353790	66.4	0	5.0	0.2	1.0	8.0	0	0.8	18.6
nodular	Bk	SI-353791	54.0	9.6	7.2	0.2	0.8	7.4	0	0.6	20.2
variant	2Bk	SI-353792	55.4	3.8	13.0	0.8	1.4	6.6	0	2.6	16.4
	3C	SI-353793	43.2	16.0	9.6	0.6	0.8	7.2	0	3.0	19.6
clayst.	-	SI-353788	60.4	1.4	6.2	0.4	1.6	5.0	0	4.2	20.8
Bhu. cfcv.	A	SI-353822	67.6	7.4	5.0	0	0.2	1.0	0	1.6	17.2
type Pila	A	SI-353817	64.2	6.6	4.4	0	0.6	2.4	0	2.2	19.6
clay	A	SI-353818	74.2	2.8	3.6	0	0.4	1.4	0	1.6	16.0
	Bt	SI-353819	64.4	13.0	4.4	0	0.4	0	0	1.0	16.8
	Bk	SI-353820	64.8	3.4	4.4	0	0	0.2	0	0.6	26.6
	C	SI-353821	71.6	0.2	3.8	0	0.4	0.6	0	1.6	21.8
Pandu	A	SI-353803	45.2	8.8	5.0	0	1.4	0	0	10.2	29.4
clay	Bk	SI-353814	66.6	14.2	5.0	0.2	1.6	0.8	0	3.4	8.2
loam	C	SI-353815	52.8	11.4	1.8	0	1.2	0.4	0	4.4	28.0
	C	SI-353816	59.0	14.6	2.2	0	0.8	1.4	0	2.0	20.0

APPENDIX 4. continued

Paleosol	Horizon	Specimen Number	Clay & Micrite	Calcite	Feldspar	Pyroxene	Mica	Rock Fragment	Other	Opaque	Quartz
Pandu	A	SI-353800	49.2	4.2	9.2	0	3.2	3.8	0	2.8	27.6
ctsp.	Bk	SI-353802	61.6	11.2	7.8	0.8	2.2	2.8	0	4.2	9.4
type	A	SI-353796	75.4	0	3.2	0	0.6	0.8	0	1.2	18.8
Pandu	A	SI-353797	68.2	10.6	4.6	0	0.4	1.8	0	1.8	12.6
clay	Bk	SI-353798	65.4	1.6	6.2	0	1.8	0.4	0	0.6	24.0
	C	SI-353799	47.4	18.0	9.6	0	3.4	0.2	0	3.6	17.8
clayst.	-	SI-353795	91.6	2.2	0	0	0	1.2	0	0.2	4.8
Sar. cl.	A	SI-353813	29.0	24.6	0.6	0	3.6	0.6	6.0	0	28.6
type	A	SI-353807	85.6	0	0.4	0	0.4	2.0	0	0.4	11.2
Kala	Bg	SI-353808	87.0	0	0.2	0	0	4.0	0	0.8	8.0
clay	Bg	SI-353810	54.8	6.4	5.6	0	2.0	2.6	0	4.8	23.8
	Bk	SI-353811	47.2	10.0	9.4	0	2.0	3.0	0	2.4	24.4
Sarang	A	SI-353805	55.4	0	2.6	0	3.4	10.8	0.8	0.6	26.2
cep.	C	SI-353806	9.6	0	3.6	0	1.6	23.2	3.4	0	58.6
sandst.	-	SI-353804	6.6	4.0	13.6	0	1.0	11.8	1.2	4.0	61.6
type	A	SI-353827	86.2	8.4	1.6	0	0	0.2	0	0	3.6
Naranji	AB	SI-353828	52.4	9.6	8.4	0	0	4.4	0	1.4	23.6
clay	AB	SI-353829	58.2	3.0	3.2	0	0	6.6	0	1.0	28.0
	BA	SI-353830	52.2	4.0	6.0	0	0.4	11.4	0	0.8	25.2
	BA	SI-353831	50.2	0.8	2.4	0	0.4	11.8	0	1.0	33.4
	Bt	SI-353832	51.0	1.2	3.8	0	0.2	13.4	0	0.6	29.8
	Bt	SI-353833	49.6	0	1.6	0	0	16.6	0	1.8	30.4
	Bt	SI-353834	51.4	0	5.2	0	2.0	10.8	0	1.0	29.6
	BC	SI-353835	44.8	1.8	3.8	0	0.4	17.0	0	2.2	30.0
	BC	SI-353836	40.6	4.0	5.2	0	0.4	16.2	0	1.0	32.6
	C	SI-353837	26.6	1.8	2.2	0	0.8	25.8	0	3.0	39.8
Sarang	A	SI-353824	52.8	20.8	6.2	0	1.6	0	0	3.0	15.6
scmsp.	C	SI-353825	28.0	26.2	6.8	0	1.0	1.8	0	6.8	29.4
	C	SI-353826	25.8	24.8	10.6	0	2.2	0	0	7.0	29.6
type	A	SI-353856	69.0	1.4	2.2	0	1.0	3.0	0	1.8	21.6
Bhura	Bk	SI-353857	54.0	9.2	2.8	0	3.2	3.0	0	1.4	26.4
clay	Bk	SI-353858	47.0	16.8	5.8	0	0.8	0.2	0	3.0	26.4
	C	SI-353859	38.4	18.0	4.4	0	0	2.2	0	4.0	32.8
Sar. scrv.	A	SI-353854	48.4	19.0	1.8	0	2.0	0	0	2.0	26.8
type	A	SI-353850	70.8	8.0	5.8	0	1.4	0	0	0.6	13.4
Sonita	Bk	SI-353851	53.8	15.8	2.8	0	2.4	0	0	2.4	22.8
clay	C	SI-353852	49.2	9.2	3.4	0	1.4	0	0	1.0	35.8
	C	SI-353853	60.8	3.0	7.6	0.4	2.2	2.4	0	1.8	21.8
Sarang	A	SI-353848	91.6	1.4	1.4	0	0.8	0	0	1.0	5.2
ccsp.	C	SI-353849	66.2	6.8	3.4	0	3.0	1.0	0	2.0	17.6
Sarang	A	SI-353846	93.2	3.8	0	0	0.4	0	0	0.2	2.4
crv.	C	SI-353847	68.0	9.8	3.2	0	3.4	0	0	1.0	14.6
siltst.	-	SI-353845	28.8	15.6	11.2	0	12.6	0.6	0	4.0	27.2
Sar. c.	A	SI-353844	86.0	2.6	1.8	0	1.8	0	0	0.2	7.6
Sonita	A	SI-353840	59.4	10.4	0	0	2.0	0.2	0	1.4	26.6
silty	C	SI-353841	58.6	8.4	0.2	0	1.8	0	0	1.2	29.8
clay	C	SI-353842	30.0	30.2	4.4	0	3.2	0	0	4.8	27.4
	C	SI-353843	26.6	35.8	2.6	0	3.8	0	0	4.8	26.4
claystone	-	SI-353839	49.9	14.4	2.6	0	4.6	0	0	2.0	28.5

Note: Paleosol names, counts and error as for Appendices 1 and 3. Counts were by G. J. Retallack (SO-R115-145, FT-R1-40, SI-353778-353794), G.S. Smith (other -R-), R. Goodfellow (SI-353785-353794, and S. Cook (other SI-).

APPENDIX 5. Major element chemical analyses, loss on ignition (LOI), organic carbon (all weight percent), and bulk density (g/cc) of Miocene paleosols from Pakistan and Kenya

Paleosol	Hz	Speci-men No.	SiO_2	TiO_2	Al_2O_3	Fe_2O_3	FeO	MgO	CaO	Na_2O	K_2O	MnO	P_2O_5	LOI	Sum of Oxides	C org.	g/cc
error (σ)	-	all -R-	0.25	0.02	0.17	0.15	0.08	0.05	0.10	0.05	0.01	.005	.004	-	-	.009	0.04
Mobaw	A	KOR96	44.38	1.33	12.41	9.29	0.07	2.00	8.93	1.32	1.77	0.19	1.04	17.43	100.16	0.11	1.94
bould-	Bw	KOR99	20.12	0.70	5.89	6.77	0	1.23	34.36	0.34	0.57	0.63	2.03	26.93	99.67	0.11	1.98
ery clay	C	KOR100	1.28	0.70	0.24	0.94	0.39	0.27	54.60	0.19	0.06	0.18	1.48	39.92	99.52	0	2.36
	C	KOR102	15.47	0.66	4.10	5.06	0.92	1.03	35.16	0.36	0.35	0.48	1.86	40.00	105.45	0.32	2.12
tuff	-	KOR110	0.48	0.84	0.11	0	0.31	0.14	54.52	0.26	0.14	0.18	0.83	47.00	105.12	0.06	2.68
Kwar	A	KOR90	27.22	0.76	7.26	7.23	0.15	1.49	26.58	0.76	1.07	0.28	1.09	25.72	99.61	-	2.06
clay	Bt	KOR91	27.74	1.26	7.62	7.67	0.28	1.84	23.07	1.00	1.30	0.28	0.99	27.40	100.45	0.22	2.06
calcar.	C	KOR92	30.34	1.06	8.32	8.12	0.30	2.11	20.95	0.88	1.75	0.26	1.30	32.00	107.39	0.17	2.05
string.	2C	KOR94	35.40	1.05	10.40	8.20	0	1.95	17.54	1.18	1.61	0.37	1.01	20.79	99.50	0.11	2.02
variant	2C	KOR95	41.28	1.21	12.15	8.64	0	1.94	11.26	1.34	1.53	0.68	0.98	18.62	99.63	0.11	1.92
type	A	KOR87	9.32	0.40	2.66	5.60	0.18	0.74	38.76	0.38	0.42	0.51	1.64	40.00	100.61	0.34	2.14
Kwar	Bw	KOR88	11.25	0.47	3.11	5.81	0.30	0.86	37.74	0.43	0.51	0.37	1.64	37.40	99.89	0.50	2.17
clay	C	KOR89	14.39	0.55	3.94	6.24	0	1.07	35.78	0.51	0.64	0.58	1.52	35.00	100.22	0.22	2.13
Kie. cssv.	A	SOR144	43.21	1.41	10.55	9.28	0.87	3.14	8.09	0.88	1.68	0.21	0.58	25.00	103.90	0.39	2.13
type Ki-	A	SOR142	39.26	1.41	10.38	7.30	1.48	2.97	9.17	0.82	1.45	0.36	0.67	25.50	100.77	-	1.50
ewo cl.	Ck	SOR143	18.85	0.68	5.07	3.55	0.63	1.47	35.78	0.51	0.67	0.31	0.26	32.40	100.18	0.38	1.96
type	A	SOR134	37.70	1.20	8.94	8.53	0.77	2.69	13.99	0.92	1.73	0.20	0.81	26.00	103.48	0.50	1.95
Choka	A	SOR135	28.28	1.09	7.74	7.64	0	2.17	23.24	0.80	1.58	0.78	1.04	26.00	100.36	0.17	1.95
clay	BA	SOR136	24.10	0.84	5.66	6.05	0.43	1.86	29.58	0.72	1.25	0.48	0.92	28.50	100.39	0.17	2.06
	Bt	SOR137	28.56	0.91	7.09	5.90	0.75	2.15	25.21	0.72	1.10	0.27	0.59	27.00	100.25	0.17	1.90
	Bt	SOR138	34.34	1.25	7.76	7.16	0.75	2.49	19.71	0.83	1.30	0.25	0.63	27.00	103.47	0.11	1.85
	C	SOR139	30.46	0.91	7.70	6.18	0.92	2.28	22.50	0.85	1.09	0.29	0.55	26.50	100.23	0.11	2.02
	C	SOR140	28.83	1.09	8.05	6.13	0.82	2.40	23.51	0.73	1.04	0.28	0.70	28.50	102.08	0.17	1.95
	C	SOR141	18.24	0.70	4.43	3.99	0.60	1.56	35.76	0.50	0.71	0.48	0.26	33.00	100.23	0.22	2.11
type Tut	A	SOR121	29.71	1.05	6.99	6.60	0.32	2.20	22.87	0.61	1.16	0.26	0.88	27.00	99.65	0.06	2.00
clay	A	SOR122	43.39	1.75	11.02	11.06	0.77	3.58	7.83	1.27	2.69	0.20	1.52	20.00	105.08	0.11	1.89
	Bt	SOR123	41.16	1.33	10.95	11.78	0.32	3.55	9.27	1.13	2.62	0.42	2.04	20.00	104.57	0.11	2.05
	Bt	SOR125	42.99	1.36	11.34	12.06	0.70	3.49	8.41	1.17	2.64	0.24	1.83	14.00	100.23	0.17	1.87
	Bt	SOR127	32.64	1.09	8.62	9.83	0.92	3.13	16.73	0.99	2.02	0.24	2.13	25.00	103.34	0.11	2.00
	Bk	SOR128	39.37	1.57	10.48	9.57	1.05	2.94	11.58	1.15	2.42	0.19	1.55	19.30	106.37	0.56	2.02
	BC	SOR130	33.42	1.00	8.76	7.75	0.80	2.45	18.17	0.98	1.98	0.29	1.50	22.00	99.10	0.11	1.96
	C	SOR132	32.78	1.09	8.45	8.20	0.55	2.40	19.45	0.96	1.94	0.45	1.17	24.00	100.84	0.17	1.94
	C	SOR133	21.20	0.81	5.63	5.57	0.25	1.76	30.11	0.67	1.19	0.77	0.95	30.70	99.61	0.34	2.09
type	A	FTR45	41.63	3.46	14.93	13.31	0.60	2.79	2.77	0.47	3.03	0.41	0.52	18.50	102.421	0.17	1.85
Rabuor	A	FTR47	42.27	4.14	15.40	13.10	0.77	2.74	2.60	0.53	3.21	0.30	0.35	20.00	05.41	0.17	1.81
clay	Bt	FTR48	44.03	2.93	16.17	11.72	0.72	2.46	2.35	0.77	3.84	0.29	0.35	15.20	100.83	0.22	1.69
	Ck	FTR50	39.68	3.21	14.85	11.56	1.80	2.85	4.00	0.51	2.53	0.22	0.79	25.60	107.60	0.65	1.85
Chogo	A	FTR12	41.16	3.612	13.21	13.61	1.93	3.89	9.06	0.47	2.92	0.23	1.33	8.54	99.51	0.06	1.90
clay	A	FTR13	30.38	.54	9.61	10.28	1.67	2.91	18.82	0.51	2.13	0.21	0.97	20.07	99.56	0.06	1.85
ferrug.	Bk	FTR14	24.23	2.22	7.64	8.32	1.28	2.39	28.10	0.29	1.74	0.21	0.78	23.29	100.49	0	1.97
nodule	Bk	FTR15	43.41	3.58	13.90	14.16	2.02	4.04	6.57	0.53	3.03	0.27	1.31	7.44	100.26	0.06	1.81
variant	Bk	FTR16	38.22	3.86	10.78	13.27	2.50	4.66	13.08	0.65	2.31	0.32	1.71	8.42	99.78	0.06	1.89
	C	FTR18	29.03	1.87	8.06	8.58	1.83	3.14	24.89	0.68	1.77	0.29	1.33	18.93	100.40	0.06	2.17
type	A	FTR2	41.44	3.88	13.43	14.78	2.02	3.98	7.05	0.48	3.38	0.20	1.85	6.92	99.41	0.17	1.79
Chogo	A	FTR3	42.81	3.75	14.03	14.46	2.19	4.14	5.64	0.64	3.54	0.20	1.54	6.41	99.35	0.11	1.86
clay	Bk	FTR6	32.72	2.32	9.96	8.67	1.51	3.36	21.10	0.63	2.23	0.20	1.31	16.38	100.39	0.06	1.91
	Bk	FTR9	34.22	2.51	10.41	9.41	1.82	3.86	18.46	0.52	1.67	0.20	1.25	14.96	99.29	0.06	1.87
	C	FTR10	31.62	2.50	9.52	9.25	1.65	3.66	21.34	0.43	1.64	0.33	1.19	16.85	99.98	0.06	1.95
	C	FTR11	31.96	2.32	9.89	9.11	1.44	3.36	21.10	0.40	1.91	0.26	1.02	17.08	99.85	0	1.95

APPENDIX 5. continued

Paleosol	Hz	Speci-men No.	SiO_2	TiO_2	Al_2O_3	Fe_2O_3	FeO	MgO	CaO	Na_2O	K_2O	MnO	P_2O_5	LOI	Sum of Oxides	C org.	g/cc
sandst.	-	FTR1	31.60	1.39	10.16	6.25	0.94	2.62	23.94	0.32	1.37	0.33	0.78	19.88	100.03	0.11	1.91
type	A	FTR75	40.89	4.25	13.72	14.63	0.77	3.56	3.50	0.48	4.56	0.15	0.82	16.00	103.33	0.17	1.82
Onuria	A	FTR76	42.19	3.88	13.51	14.01	0.97	3.52	3.50	0.49	4.63	0.17	0.87	15.00	102.74	0.11	1.81
clay	Bk	FTR77	26.38	2.36	8.81	7.70	0.67	2.41	23.65	0.42	2.95	0.12	0.90	24.00	100.37	0.11	1.94
	R	FTR78	33.73	2.81	10.09	11.11	1.23	4.39	15.56	0.54	1.79	0.57	0.87	17.20	99.89	0.11	2.03
type	A	SI353772	54.35	0.72	18.08	7.29	0.02	3.40	4.58	0.87	3.06	0.08	0.20	9.14	101.79	0.16	2.07
Khaki-	A	SI353773	50.41	0.66	17.47	7.07	0.02	3.24	7.22	1.02	3.13	0.08	0.20	11.08	101.60	0.20	2.37
stari	Bg	SI353774	47.12	0.64	16.54	6.77	0.16	3.58	9.57	1.25	3.10	0.18	0.22	12.54	101.67	0.21	2.32
clay	C	SI353775	48.72	0.63	17.71	6.60	0.05	3.44	7.11	1.24	3.31	0.15	0.20	10.94	100.10	0.35	2.23
type Lal	A	SI353778	50.57	0.71	18.10	7.69	0.07	3.51	6.12	0.61	2.72	0.07	0.17	10.76	101.10	0.18	2.21
clay	A	SI353779	51.30	0.73	18.83	7.70	0.01	3.69	6.18	0.61	2.80	0.08	0.18	10.70	102.80	0.18	2.66
	Bt	SI353780	51.28	0.75	18.33	7.68	0.02	3.55	5.75	0.61	2.74	0.07	0.18	10.96	101.92	0.21	2.41
	Bt	SI353781	49.71	0.74	18.15	7.53	0.01	3.53	6.10	0.60	2.64	0.08	0.19	10.75	100.03	0.29	2.31
	Bk	SI353782	51.47	0.01	17.95	7.47	0.01	3.53	6.88	0.69	2.71	0.003	0.05	10.98	101.75	0.13	2.63
	2B	SI353783	50.64	0.72	18.02	7.31	0.06	3.51	6.48	0.63	2.68	0.08	0.19	10.99	101.31	0.23	2.21
	2C	SI353784	51.77	0.68	18.09	7.29	0.08	3.52	7.05	0.70	2.53	0.09	0.18	10.94	102.92	0.18	2.41
type Pila	A	SI353817	60.68	0.56	14.56	5.21	.004	2.69	4.77	1.33	2.28	0.08	0.17	7.82	99.85	0.10	2.35
clay	A	SI353818	60.81	0.56	14.20	5.49	0.05	2.22	5.86	0.97	1.61	0.12	0.11	9.01	101.01	0.16	2.33
	Bt	SI353819	57.72	0.53	13.97	5.17	0.01	2.11	7.40	0.93	1.90	0.12	0.11	9.96	99.93	0.11	2.37
	Bk	SI353820	63.16	.004	14.91	5.97	0.02	2.14	3.67	0.81	1.84	.002	0.05	7.81	100.21	0.11	2.38
	C	SI353821	65.61	0.63	15.40	6.07	0.03	2.17	1.49	0.84	1.98	0.05	0.10	6.32	100.69	0.11	2.25
type	A	SI353796	59.58	0.68	18.35	6.86	0.10	3.22	2.12	1.04	3.08	0.06	0.19	6.63	101.91	0.11	2.30
Pandu	A	SI353797	62.28	0.66	17.42	6.65	0.04	2.87	2.48	1.14	3.12	0.05	0.19	6.74	101.671	0.14	2.36
clay	Bk	SI353798	61.62	0.56	15.95	6.03	0.02	2.62	3.69	1.35	2.75	0.06	0.19	6.95	01.79	0.16	2.30
	C	SI353799	61.86	0.53	13.86	4.78	0.02	2.65	5.20	1.35	2.20	0.06	0.19	7.73	100.43	0.14	2.40
Sar. cl.	A	SI535813	61.22	0.42	10.14	4.11	0.02	1.89	9.32	1.11	1.41	0.09	0.13	10.37	100.23	0.06	2.34
type	A	SI353807	60.10	0.68	17.14	6.53	0.01	2.30	1.67	0.81	2.26	0.05	0.15	6.75	98.45	0.09	2.50
Kala	Bg	SI353808	56.14	0.68	18.35	7.00	0.03	3.27	3.25	0.66	2.29	0.05	0.15	8.51	100.38	0.07	2.28
clay	Bg	SI353810	65.31	0.54	12.39	4.67	0.01	2.20	4.16	1.16	1.77	0.05	0.15	6.75	99.16	0.16	2.34
	Bk	SI353811	51.02	0.43	9.26	3.67	0.04	1.92	15.89	0.81	1.02	0.15	0.13	15.75	100.09	0.12	2.31
type	A	SI353827	55.71	0.62	15.75	6.53	0.08	2.98	6.20	0.75	2.16	0.07	0.16	9.40	100.41	0.03	2.26
Naranji	AB	SI353828	64.61	0.59	14.23	5.72	0.02	2.31	2.95	0.88	2.22	0.06	0.18	6.61	100.38	0.09	2.32
clay	AB	SI353829	70.06	0.61	13.61	5.20	0.03	1.97	1.54	0.67	2.09	0.05	0.14	5.03	100.55	0.19	2.31
	BA	SI353830	71.89	0.60	12.04	4.78	.005	1.53	1.28	0.69	2.08	0.05	0.13	4.61	99.68	0.23	2.27
	BA	SI353831	75.35	0.55	10.89	4.12	0.02	0.95	1.57	0.50	1.65	0.04	0.09	4.44	100.17	0.15	2.20
	Bt	SI353832	76.05	0.55	11.00	4.14	.001	0.96	0.91	0.47	1.66	0.04	0.09	4.09	99.96	0.08	2.12
	Bt	SI353833	75.31	0.56	11.59	4.15	0.02	0.93	1.02	0.43	1.64	0.04	0.09	4.69	100.47	0.08	2.25
	Bt	SI353834	73.05	0.55	11.90	4.30	0.01	0.99	1.80	0.55	1.80	0.04	0.09	5.11	100.19	0.10	2.13
	BC	SI353835	75.46	0.54	10.77	3.69	0.01	1.01	1.61	0.81	1.77	0.05	0.08	4.33	100.13	0.12	2.14
	BC	SI353836	74.06	0.57	11.89	4.44	0.03	1.26	1.56	0.53	1.66	0.04	1.10	5.13	101.27	0.04	2.19
	C	SI353837	77.05	0.49	10.10	3.83	0.01	0.98	2.22	1.27	1.72	0.06	0.08	3.97	101.78	0.07	1.97
type	A	SI353856	62.66	0.60	15.37	6.09	0.01	2.65	2.28	0.90	2.29	0.05	0.25	6.01	99.16	0.16	2.20
Bhura	Bk	SI353857	61.40	0.57	15.50	6.26	0.01	2.75	4.59	0.84	2.34	0.06	0.18	8.03	102.53	0.10	2.07
clay	Bk	SI353858	56.91	0.58	14.66	5.92	0.06	2.60	5.97	0.84	2.00	0.07	0.17	9.23	99.01	0.16	2.21
	C	SI353859	64.96	0.50	11.67	4.39	0	1.82	6.43	1.28	1.75	0.09	0.14	7.87	100.98	0.06	2.26
Sar. scrv.	A	SI353854	52.43	0.55	14.98	6.29	0.01	2.78	9.98	0.70	2.16	0.12	0.16	12.25	102.41	0.11	2.24
type	A	SI353850	56.48	0.65	17.53	7.15	0.02	3.40	3.92	0.70	2.53	0.06	0.18	8.54	101.16	0.14	2.04
Sonita	Bk	SI353851	55.74	0.42	15.90	5.99	0.44	2.90	3.69	0.68	2.20	0.06	0.18	10.25	99.31	0.08	2.49
clay	C	SI353852	56.04	0.57	14.15	5.69	0.01	2.59	6.34	0.80	1.90	0.07	0.16	9.18	97.50	0.07	2.35
	C	SI353853	54.63	0.56	13.49	5.39	0.05	2.45	9.31	0.81	1.73	0.12	0.16	11.25	99.95	0.23	2.32
Sar. ccv.	A	SI353848	52.91	0.76	20.06	8.45	0.03	4.16	2.26	0.50	3.05	0.05	0.20	8.05	100.48	0.04	2.28

APPENDIX 5. continued

Paleosol	Hz	Specimen No.	SiO_2	TiO_2	Al_2O_3	Fe_2O_3	FeO	MgO	CaO	Na_2O	K_2O	MnO	P_2O_5	LOI	Sum of Oxides	C org.	g/cc
	C	SI353849	53.57	0.72	18.61	7.90	0	3.78	4.43	0.62	2.76	0.07	0.18	9.11	101.75	0.04	2.23
Sar. crv.	A	SI353846	49.57	0.62	16.42	6.83	0.04	3.23	8.45	0.62	2.37	0.12	0.18	11.88	100.43	0.07	2.32
	C	SI353847	51.67	0.59	16.79	7.01	0.06	3.23	8.13	0.63	2.35	0.12	0.17	11.28	102.03	0.03	2.33
error (σ)	–	all SI–	2.96	.004	0.14	0.25	.001	0.32	0.15	.002	0.03	.002	.001	–	–	0.04	0.04

Note: For Appendices 5, 6, and 7, dashes (-) signify analyses not attempted and zeroes (0) are values beyond detection. Analyses were from inductively coupled plasma-fusion spectroscopy at the University of Washington, Seattle, by Colin Cool (numbers prefixed SI-) and by Anthony Irving (numbers FT-R1-18), and from atomic absorption at the University of Oregon, Eugene by Christine McBirney. Organic carbon values were determined from the Walkely-Black titration in the Soil Testing Laboratory, Oregon State University, Corvallis, by Donald Horneck. Bulk density was calculated by weighing paraffin-coated clods in and out of water at the University of Oregon, Eugene, by Gregory J. Retallack. Errors were estimated from counting statistics for plasma-fusion, from 10 replicate analyses of standard rock W2 for atomic absorption, from 50 replicates of a standard soil for Walkley-Black titration and from 10 replicates of rock FT-R14 for bulk density.

APPENDIX 6. Trace element analyses (ppm) of Miocene paleosols from Pakistan and Kenya

Paleosol	Hoz	Specimen	Au	Ba	Cd	Co	Cr	Cu	La	Li	Nb	Ni	Pb	Rb	Sc	Sr	V	Y	Zn	Zr
error (σ)	-	all -R-	-	12	-	1	2	2	-	1	-	1	-	2	-	4	-	-	3	-
Mobaw	A	KO-R96	-	737	-	26	67	49	-	40	-	25	-	124	-	699	-	-	370	-
bouldery	Bw	KO-R99	-	3094	-	15	35	35	-	34	-	11	-	37	-	2358	-	-	394	-
clay	C	KO-R100	-	865	-	8	9	8	-	10	-	0	-	5	-	2557	-	-	134	-
	C	KO-R102	-	1363	-	17	28	31	-	39	-	16	-	25	-	2706	-	-	396	-
tuff	-	KO-R110	-	1212	-	11	5	6	-	14	-	2	-	2	-	2246	-	-	57	-
Kwar clay	A	KO-R90	-	1027	-	15	50	28	-	35	-	20	-	66	-	2330	-	-	355	-
calcar.	Bt	KO-R91	-	963	-	23	56	41	-	38	-	20	-	83	-	2014	-	-	373	-
stringer	C	KO-R92	-	849	-	21	64	44	-	54	-	22	-	116	-	1773	-	-	411	-
variant	2C	KO-R94	-	1578	-	22	65	61	-	41	-	23	-	126	-	1285	-	-	355	-
	2C	KO-R95	-	3078	-	22	61	111	-	28	-	25	-	92	-	717	-	-	358	-
type Kwar	A	KO-R87	-	1441	-	17	19	18	-	38	-	9	-	28	-	3017	-	-	381	-
clay	Bw	KO-R88	-	1258	-	19	22	18	-	39	-	11	-	38	-	2999	-	-	390	-
	C	KO-R89	-	1692	-	20	28	36	-	44	-	40	-	40	-	2912	-	-	426	-
Kie. cssv.	A	SO-R144	-	620	-	32	195	116	-	49	-	83	-	82	-	310	-	-	161	-
type Kie-	A	SO-R142	-	1023	-	36	173	118	-	41	-	75	-	70	-	279	-	-	152	-
wo clay	Ck	SO-R143	-	713	-	27	97	70	-	23	-	43	-	31	-	220	-	-	84	-
type	A	SO-R134	-	1536	-	24	163	76	-	43	-	70	-	78	-	687	-	-	166	-
Choka	A	SO-R135	-	563	-	27	97	95	-	30	-	60	-	70	-	587	-	-	145	-
clay	BA	SO-R136	-	1717	-	23	79	61	-	24	-	36	-	48	-	749	-	-	110	-
	Bt	SO-R137	-	1061	-	23	117	65	-	33	-	40	-	53	-	572	-	-	114	-
	Bt	SO-R138	-	557	-	24	124	70	-	36	-	53	-	59	-	514	-	-	123	-
	C	SO-R139	-	386	-	23	136	73	-	33	-	48	-	46	-	574	-	-	106	-
	C	SO-R140	-	548	-	24	154	83	-	33	-	53	-	50	-	663	-	-	112	-
	C	SO-R141	-	513	-	25	116	59	-	21	-	33	-	26	-	387	-	-	69	-
type Tut	A	SO-R121	-	690	-	24	164	68	-	38	-	63	-	52	-	658	-	-	131	-
clay	A	SO-R122	-	487	-	27	165	109	-	54	-	55	-	118	-	619	-	-	219	-
	Bt	SO-R123	-	918	-	40	110	117	-	49	-	78	-	108	-	611	-	-	239	-
	Bt	SO-R125	-	1175	-	33	116	93	-	48	-	55	-	109	-	545	-	-	232	-
	Bt	SO-R127	-	820	-	23	76	74	-	39	-	33	-	80	-	920	-	-	214	-
	Bk	SO-R128	-	840	-	23	109	73	-	39	-	43	-	94	-	510	-	-	182	-
	BC	SO-R130	-	662	-	23	96	62	-	34	-	43	-	78	-	592	-	-	150	-
	C	SO-R132	-	624	-	27	92	75	-	33	-	45	-	76	-	570	-	-	159	-
	C	SO-R133	-	981	-	23	75	78	-	24	-	45	-	48	-	716	-	-	118	-
type	A	FT-R45	-	858	-	41	114	124	-	34	-	55	-	129	-	224	-	-	186	-
Rabuor	A	FT-R47	-	669	-	39	99	114	-	38	-	65	-	141	-	200	-	-	196	-
clay	Bt	FT-R48	-	775	-	34	80	106	-	41	-	50	-	131	-	221	-	-	167	-
	Ck	FT-R50	-	480	-	32	78	130	-	40	-	43	-	117	-	259	-	-	170	-
Chogo	A	FT-R12	-	506	-	41	175	127	-	22	195	52	-	-	19	436	263	44	180	359
clay fer-	A	FT-R13	-	464	-	57	128	92	-	30	104	61	-	-	19	464	223	37	134	289
ruginized	Bk	FT-R14	-	365	-	50	137	88	-	38	180	69	-	-	15	398	182	35	104	239
nodule	Bk	FT-R15	-	606	-	41	153	101	-	19	149	50	-	-	18	510	241	35	176	358
variant	Bk	FT-R16	-	670	-	84	185	109	-	44	350	94	-	-	29	623	305	42	163	321
	C	FT-R18	-	501	-	55	46	46	-	28	107	46	-	-	8	551	181	40	123	287
type	A	FT-R2	-	513	-	62	155	129	-	25	116	51	-	-	22	448	311	55	182	383
Chogo	A	FT-R3	-	511	-	39	136	98	-	18	122	45	-	-	19	434	264	38	173	373
clay	Bk	FT-R6	-	562	-	24	40	55	-	11	59	22	-	-	9	575	161	27	106	208
	Bk	FT-R9	-	537	-	44	75	108	-	35	119	61	-	-	18	589	215	34	127	264
	C	FT-R10	-	529	-	58	92	97	-	46	118	82	-	-	21	588	231	38	486	256
	C	FT-R11	-	497	-	62	101	116	-	55	140	95	-	-	23	507	222	39	128	312
sandstone	-	FT-R1	-	2030	-	30	5	17	-	10	140	10	-	-	2	490	103	22	103	335
type	A	FT-R75	-	448	-	20	87	78	-	25	-	30	-	194	-	269	-	-	174	-
Onuria	A	FT-R76	-	437	-	33	88	67	-	25	-	30	-	175	-	284	-	-	161	-

APPENDIX 6. continued

Paleosol	Hoz	Specimen	Au	Ba	Cd	Co	Cr	Cu	La	Li	Nb	Ni	Pb	Rb	Sc	Sr	V	Y	Zn	Zr
clay	Bk	FT-R77	-	324	-	23	60	60	-	18	-	16	-	110	-	305	-	-	103	-
	R	FT-R78	-	1123	-	34	55	114	-	23	-	30	-	51	-	523	-	-	128	-
type Kha-	A	SI-353772	3	323	1	22	131	53	34	59	14	96	0	-	20	185	167	22	99	111
kistari	A	SI-353773	1	351	1	19	127	52	33	64	13	70	0	-	31	184	150	21	95	87
clay	Bg	SI-353774	1	401	0	20	82	59	34	68	14	53	0	-	17	207	123	24	111	87
	C	SI-353775	0	412	1	21	90	55	31	70	12	52	0	-	18	189	126	21	107	86
type Lal	A	SI-353778	2	254	1	20	150	48	34	55	13	113	0	-	19	175	148	21	91	122
clay	A	SI-353779	0	270	1	22	156	50	35	57	13	115	0	-	20	177	149	23	94	112
	Bt	SI-353780	0	278	1	22	161	52	35	61	15	124	0	-	20	179	151	22	97	102
	Bt	SI-353781	1	273	1	22	161	52	36	59	14	121	0	-	20	184	152	25	97	110
	Bk	SI-353782	-	280	-	26	170	-	-	-	17	100	-	-	17	185	150	25	120	150
	2B	SI-353783	0	265	1	21	154	49	35	60	14	115	0	-	20	179	152	23	94	120
	2C	SI-353784	0	266	1	21	146	51	33	55	13	108	0	-	20	175	143	22	90	85
type Pila	A	SI-353817	1	322	0	21	85	41	29	42	12	59	0	-	15	183	105	18	71	50
clay	A	SI-353818	0	262	0	33	98	41	33	43	12	70	0	-	15	162	111	23	72	62
	Bt	SI-353819	1	244	0	17	94	40	30	43	10	68	0	-	15	144	102	23	67	62
	Bk	SI-353820	2	270	2	32	120	45	31	48	13	80	0	-	14	145	120	25	95	110
	C	SI-353821	7	226	1	22	104	41	29	50	11	73	0	-	17	141	128	17	77	83
type	A	SI-353796	1	306	1	25	128	50	32	54	14	94	0	-	19	162	144	20	93	94
Pandu	A	SI-353797	1	334	0	22	110	49	33	49	13	78	0	-	18	175	137	21	89	85
clay	Bk	SI-353798	0	393	0	22	88	47	33	45	12	62	0	-	16	183	119	19	76	59
	C	SI-353799	0	333	0	36	72	40	31	34	12	48	0	-	14	197	117	18	58	49
Sarang cl.	A	SI-353813	0	212	0	17	66	31	22	26	8	47	0	-	10	165	76	16	59	40
type Kala	A	SI-353807	3	261	1	22	139	51	33	53	14	100	0	-	20	154	145	19	89	88
clay	Bg	SI-353808	1	259	1	25	140	50	37	54	15	104	0	-	20	166	142	24	95	98
	Bg	SI-353810	0	219	0	38	110	40	29	35	12	69	0	-	14	147	112	19	66	65
	Bk	SI-353811	0	181	0	25	70	33	26	26	6	48	0	-	8	203	77	23	52	45
type	A	SI-353827	2	233	0	29	121	44	30	44	12	83	0	-	18	179	134	22	82	90
Naranji	AB	SI-353828	3	252	0	34	96	39	28	40	12	71	0	-	16	154	117	21	73	75
clay	AB	SI-353829	3	311	0	35	95	35	30	40	13	66	0	-	15	135	113	19	73	62
	BA	SI-353830	4	326	1	64	82	33	30	37	13	54	0	-	14	123	105	18	74	59
	BA	SI-353831	5	227	1	87	65	26	28	34	12	40	0	-	12	103	98	15	47	59
	Bt	SI-353832	5	234	1	92	77	27	28	35	13	42	0	-	12	103	101	15	49	64
	Bt	SI-353833	3	227	1	49	71	28	27	38	11	42	0	-	12	101	101	16	48	61
	Bt	SI-353834	3	257	1	70	67	28	27	34	11	42	0	-	12	107	105	14	47	60
	BC	SI-353835	4	341	1	90	71	25	29	27	11	38	0	-	11	117	95	19	37	49
	BC	SI-353836	4	220	2	73	83	31	31	35	14	56	0	-	13	105	92	20	48	69
	C	SI-353837	3	228	0	79	52	20	29	23	10	28	0	-	10	154	106	20	36	48
type	A	SI-353856	2	268	1	20	111	53	32	40	14	79	0	-	18	161	135	22	84	70
Bhura	Bk	SI-353857	1	257	0	20	111	49	30	44	12	80	0	-	18	165	126	20	77	70
clay	Bk	SI-353858	4	245	0	20	106	47	31	43	12	72	0	-	17	169	122	22	69	87
	C	SI-353859	1	236	1	30	74	34	27	27	10	46	0	-	12	163	89	18	47	57
Sar. scrv.	A	SI-353854	1	227	2	17	102	47	31	40	10	69	0	-	15	169	122	23	68	86
type	A	SI-353850	2	257	2	23	128	51	32	48	13	92	0	-	19	175	144	22	90	101
Sonita	Bk	SI-353851	3	261	2	20	125	49	32	46	12	86	0	-	18	168	132	22	80	97
clay	C	SI-353852	1	230	2	21	101	46	31	39	12	76	0	-	25	160	111	21	73	74
	C	SI-353853	0	220	1	16	92	45	30	36	10	61	0	-	13	168	104	21	62	72
Sar. ccv.	A	SI-353848	4	258	4	24	165	56	36	59	17	131	0	-	23	166	145	22	100	109
	C	SI-353849	2	256	3	23	145	50	34	53	13	106	0	-	20	168	147	21	90	93
Sar. crv.	A	SI-353846	1	253	2	21	132	53	32	48	11	92	0	-	18	199	129	22	87	70
	C	SI-353847	3	238	1	19	121	52	30	45	12	85	0	-	18	192	123	21	80	77
error (σ)	-	all SI-	1	9	1	.3	2	2	1	1	1	2	-	-	.4	5	4	1	1	3

APPENDIX 7. Molecular weathering ratios of Miocene paleosols in Pakistan and Kenya

Paleosol	Hor-izon	Specimen Number	Na_2O / K_2O	$CaO+MgO$ / Al_2O_3	Al_2O_3 / SiO_2	Al_2O_3 / $CaO+MgO+Na_2O+K_2O$	Ba / Sr	FeO / Fe_2O_3
Mobaw bouldery clay	A	KO-R96	1.13	1.71	0.16	0.49	0.67	0.017
	Bw	KO-R99	0.91	11.13	0.17	0.09	0.83	.0003
	C	KO-R100	4.81	416.46	0.11	0.002	0.22	0.922
	C	KO-R102	1.56	16.23	0.16	0.06	0.32	0.404
carbonatite tuff	-	KO-R110	2.82	904.31	0.14	0.001	0.34	684.85
Kwar clay calcareous stringer variant	A	KO-R90	1.08	7.17	0.16	0.13	0.28	0.049
	Bt	KO-R91	1.17	6.12	0.16	0.15	0.31	0.081
	C	KO-R92	0.76	5.22	0.16	0.18	0.31	0.082
	2C	KO-R94	1.11	3.54	0.17	0.26	0.78	0.0002
	2C	KO-R95	1.33	2.09	0.17	0.41	2.74	0.0002
type Kwar clay	A	KO-R87	1.38	27.20	0.17	0.04	0.30	0.071
	Bw	KO-R88	1.28	22.91	0.16	0.04	0.27	0.115
	C	KO-R89	1.21	17.20	0.16	0.06	0.37	0.0004
Kiewo clay ssv.	A	SO-R144	0.80	2.15	0.14	0.41	1.28	0.208
type Kiewo clay	A	SO-R142	0.86	2.29	0.16	0.38	2.34	0.451
	Ck	SO-R143	1.16	13.56	0.16	0.07	2.07	0.394
type Choka clay	A	SO-R134	0.81	3.61	0.14	0.25	1.43	0.201
	A	SO-R135	0.77	6.17	0.16	0.15	0.61	.0003
	BA	SO-R136	0.88	10.33	0.14	0.09	1.46	0.158
	Bt	SO-R137	0.99	7.23	0.15	0.13	1.18	0.283
	Bt	SO-R138	0.97	5.43	0.13	0.17	0.69	0.233
	C	SO-R139	1.19	6.06	0.15	0.16	0.43	0.331
	C	SO-R140	1.07	6.06	0.16	0.16	0.53	0.297
	C	SO-R141	1.07	15.57	0.14	0.06	0.85	0.334
type Tut clay	A	SO-R121	0.80	6.74	0.14	0.14	0.67	0.107
	A	SO-R122	0.72	2.11	0.15	0.39	0.50	0.155
	Bt	SO-R123	0.66	2.36	0.16	0.36	0.96	0.060
	Bt	SO-R125	0.67	2.13	0.16	0.39	1.38	0.129
	Bt	SO-R127	0.74	4.45	0.16	0.20	0.57	0.208
	Bk	SO-R128	0.72	2.71	0.16	0.32	1.05	0.244
	BC	SO-R130	0.75	4.48	0.15	0.20	0.71	0.229
	C	SO-R132	0.75	4.90	0.15	0.19	0.70	0.149
	C	SO-R133	0.86	10.51	0.16	0.09	0.87	0.100
type Rabuor clay	A	FT-R45	0.24	0.81	0.21	0.92	2.44	0.100
	A	FT-R47	0.25	0.76	0.21	0.96	2.13	0.130
	Bt	FT-R48	0.30	0.65	0.22	1.02	2.24	0.137
	Ck	FT-R50	0.31	0.98	0.22	0.82	1.18	0.346
Chogo clay ferruginized nodule variant	A	FT-R12	0.24	1.99	0.19	0.44	0.74	0.315
	A	FT-R13	0.36	4.33	0.19	0.21	0.64	0.361
	Bk	FT-R14	0.25	7.48	0.19	0.13	0.59	0.341
	Bk	FT-R15	0.27	1.59	0.19	0.53	0.76	0.317
	Bk	FT-R16	0.43	3.30	0.17	0.28	0.69	0.419
	C	FT-R18	0.58	6.60	0.16	0.14	0.58	0.474
type Chogo clay	A	FT-R2	0.22	1.70	0.19	0.49	0.73	0.303
	A	FT-R3	0.27	1.48	0.19	0.55	0.75	0.337
	Bk	FT-R6	0.43	4.70	0.17	0.20	0.62	0.387
	Bk	FT-R9	0.47	4.27	0.17	0.22	0.58	0.430
	C	FT-R10	0.52	5.08	0.18	0.19	0.49	0.432
	C	FT-R11	0.32	4.77	0.18	0.20	0.62	0.351
sandstone	-	FT-R1	0.36	4.94	0.19	0.19	2.64	0.334

APPENDIX 7. continued

Paleosol	Hor-izon	Specimen Number	Na_2O / K_2O	$CaO+MgO$ / Al_2O_3	Al_2O_3 / SiO_2	Al_2O_3 / $CaO+MgO+Na_2O+K_2O$	Ba / Sr	FeO / Fe_2O_3
type Onuria clay	A	FT-R75	0.16	1.12	0.20	0.65	1.06	0.117
	A	FT-R76	0.16	1.13	0.19	0.64	0.98	0.154
	Bk	FT-R77	0.22	5.57	0.20	0.17	0.68	0.193
	R	FT-R78	0.46	3.90	0.18	0.24	1.37	0.246
type Khakistari clay	A	SI-353772	0.43	0.94	0.20	0.83	1.12	0.006
	A	SI-353773	0.50	1.22	0.20	0.66	1.22	0.006
	Bg	SI-353774	0.61	1.60	0.21	0.52	1.23	0.052
	C	SI-353775	0.57	1.22	0.21	0.65	1.39	0.017
type Lal clay	A	SI-353778	0.34	1.11	0.21	0.76	0.93	0.020
	A	SI-353779	0.33	1.09	0.22	0.77	0.97	0.003
	Bt	SI-353780	0.34	1.06	0.21	0.78	0.99	0.006
	Bt	SI-353781	0.35	1.10	0.22	0.76	0.95	0.003
	Bk	SI-353782	0.39	1.19	0.21	0.70	0.97	0.003
	2Bk	SI-353783	0.36	1.14	0.21	0.73	0.94	0.018
	2Co	SI-353784	0.42	1.20	0.21	0.71	0.97	0.024
type Pila clay	A	SI-353817	0.89	1.09	0.14	0.71	1.12	0.002
	A	SI-353818	0.92	1.15	0.14	0.72	1.04	0.020
	Bt	SI-353819	0.74	1.35	0.14	0.62	1.08	0.004
	Bk	SI-353820	0.67	0.81	0.14	0.97	1.19	0.008
	C	SI-353821	0.64	0.53	0.14	1.31	1.02	0.011
type Pandu clay	A	SI-353796	0.51	0.65	0.18	1.08	1.19	0.032
	A	SI-353797	0.57	0.68	0.16	1.02	1.22	0.013
	Bk	SI-353798	0.75	0.84	0.15	0.86	1.37	0.007
	C	SI-353799	0.93	1.16	0.13	0.67	1.08	0.009
Sarang clay loam	A	SI-353813	1.19	2.14	0.10	0.40	0.82	0.011
type Kala clay	A	SI-353807	0.54	0.52	0.17	1.36	1.08	0.003
	Bg	SI-353808	0.44	0.77	0.19	1.03	0.99	0.010
	Bg	SI-353810	1.00	1.06	0.11	0.73	0.95	0.005
	Bk	SI-353811	1.20	3.64	0.11	0.26	0.57	0.024
type Naranji clay	A	SI-353827	0.53	1.19	0.17	0.70	0.83	0.027
	AB	SI-353828	0.60	0.79	0.13	0.95	1.04	0.008
	AB	SI-353829	0.49	0.59	0.11	1.18	1.47	0.013
	BA	SI-353830	0.50	0.51	0.10	1.26	1.69	0.002
	BA	SI-353831	0.46	0.48	0.09	1.38	1.41	0.011
	Bt	SI-353832	0.43	0.37	0.09	1.65	1.44	.0005
	Bt	SI-353833	0.40	0.36	0.09	1.73	1.43	0.011
	Bt	SI-353834	0.46	0.49	0.10	1.38	1.54	0.005
	BC	SI-353835	0.70	0.51	0.08	1.23	1.85	0.006
	BC	SI-353836	0.49	0.51	0.09	1.37	1.34	0.015
	C	SI-353837	1.12	0.64	0.08	0.97	0.94	0.058
type Bhura clay	A	SI-353856	0.60	0.71	0.14	1.03	1.06	0.004
	Bk	SI-353857	0.55	0.99	0.15	0.81	0.99	0.004
	Bk	SI-353858	0.64	1.19	0.15	0.70	0.92	0.023
	C	SI-353859	1.11	1.40	0.11	0.58	0.93	.0004
Sarang scrv.	A	SI-353854	0.49	1.68	0.17	0.52	0.86	0.004
type Sonita clay	A	SI-353850	0.42	0.90	0.18	0.89	0.93	0.006
	Bk	SI-353851	0.47	0.88	0.17	0.91	1.08	0.163
	C	SI-353852	0.64	1.28	0.15	0.66	0.92	0.004
	C	SI-353853	0.71	1.71	0.15	1.95	0.83	0.021

APPENDIX 7. continued

Paleosol	Horizon	Specimen Number	Na_2O / K_2O	$CaO+MgO$ / Al_2O_3	Al_2O_3 / SiO_2	Al_2O_3 / $CaO+MgO+Na_2O+K_2O$	Ba / Sr	FeO / Fe_2O_3
Sarang clay cv.	A	SI–353848	0.25	0.73	0.22	1.07	1.00	.0007
	C	SI–353849	0.34	0.95	0.20	0.86	0.97	.0003
Sarang clay rv.	A	SI–353846	0.40	1.43	0.20	0.61	0.81	0.013
	C	SI–353847	0.41	1.37	0.19	0.63	0.79	0.019

Note: Molecular weathering ratios were calculated by converting weight percent values (from Appendices 5 and 6) to moles using molecular weights (Retallack, 1990a).

APPENDIX 8. Early Miocene (20Ma) fossils from Songhor and Koru, Kenya, and paleosols in which they have been found

ANGIOSPERMAE (flowering plants)
Celtis sp. cf. *C. mildbraedii* (hackberry, Ulmaceae): Kwar (1)

PLANTAE-TRACE FOSSILS
fine root traces: Buru (X), Choka (X), Kiewo (X), Kwar (X), Mobaw (X), Tut (X)
large root traces: Buru (X), Choka (X), Kiewo (X), Kwar (X), Mobaw (X), Tut (X)
large drab-haloed root traces: Choka (X), Kiewo (X), Tut (X)
twig casts

GASTROPODA (snails)
Maizania lugubroides (turbinate, wide umbilicus, Maizaniidae)
Tropidophora (Ligatella) miocenica (turbinate, 5-6 radial ridges, Pomatiasidae): Choka (1)
Conulinus (ovoid, conical, Enidae)
Cerastua miocenica (elongate, ovate, Enidae): Kwar (1)
Cecilioides sp. (small, high spired, Ferussaciidae): Kwar (1)
Subulininae indet. (elongate, very high spired, Subulinidae): Choka (1), Kiewo (4), Kwar (75), Mobaw (165), Tut (2)
Krapfiella angusta (high spired, obtuse summit, Subulinidae): Kwar (19), Mobaw (18)
Pseudoglessula sp. (very high spired, ribbed, Subulinidae)
Curvella sp. (ovate with sinuous lip, Subulinidae): Kwar (6)
Opeas sp. (high spired, bluntly rounded apex, Subulinidae): Kwar (1), Mobaw (1)
Pseudopeas sp. (high spired, ribbed, striate protoconch, Subulinidae): Mobaw (9)
Achatina leakeyi (large, high spired, Achatinidae): Choka (3), Kiewo (16), Kwar (50), Mobaw (62), Tut (3)
Burtoa nilotica (high spired, growth folds and radial lines, Achatinidae): Kwar (182), Mobaw (58)
? Sitala sp. (minute, conical, Ariophantidae)
Urocyclinae sp. indet. (slugs, cap shell, Urocyclidae): Kwar (4)
Trochozonites sp. (trochiform, Urocyclidae): Kwar (8), Mobaw (2)
Trochonanina sp. (conical with keeled whorls, Urocyclidae): Kwar (17), Mobaw (3)
Trochonanina (? Zingis) sp. (low spired, globular, Urocyclidae): Kwar (3), Mobaw (7)
Thapsia sp. (very low spired to discoidal, Urocyclidae): Kwar (14), Mobaw (9)
Chlamydarion sp. (low spired, rapidly expanding, Helicarionidae): Kwar (1), Mobaw (3)
Halolimnohelix sp. (globular, Bradybaenidae): Kwar (12), Mobaw (1)
Tayloria sp. (low spired, subangular around umbilicus, Streptaxidae): Choka (6), Kiewo (5), Kwar (22)
Gonaxis sp. (large irregularly ovoid, Streptaxidae): Kiewo (1), Kwar (5), Mobaw (16)
Gonaxis protocavalii (irregularly ovoid, Streptaxidae): Choka (1), Kwar (62), Mobaw (48)
Marconia sp. (subcylindrical to ovoid, Streptaxidae): Kwar (12), Mobaw (20)
Edentulina sp. (ovate, high spired, Streptaxidae)
Gonospira sp. (subcylindical to pupiform, untoothed aperture, Streptaxidae): Kwar (42), Mobaw (21)
Streptostele (very high spired, Streptaxidae)
Ptychotrema sp. (high spired, spindle shaped, toothed aperture, Streptaxidae): Mobaw (3)
Gulella (Primigullela) miocenica (pupiform, toothed aperture, Streptaxidae): Choka (5), Kwar (103), Mobaw (38)
Gulella sp. (small, pupiform, toothed aperture, Streptaxidae): Kwar (17), Mobaw (64)

MYRIAPODA (millipedes and centipedes)
Diplopoda indet. (large roundback millipede): Kwar (X), Mobaw (X)

INVERTEBRATA-TRACE FOSSILS
small spherical cast (gastropod egg): Kwar (1)
bullet-shaped, lined chamber (large insect pupal cell): Choka (2)
Skolithus (6-11 mm diameter burrow of insect or spider): Choka (X), Kiewo (X), Tut (X)
small ovoid cast (cocoon of beetle?): (Kwar 3), Mobaw (1)
Pallichnus sp. entire ovoid (pupal cell of beetle?): Choka (~10)
Pallichnus sp. broken ovoid (dung beetle nest?): Choka (~50)
Planolites sp. (3-10 mm galleries of ant or termite): Kwar (X), Mobaw (X)
Termitichnus sp. (calie of fungus-growing ant or termite): Kwar (3)
Celliforma sp. (brood cell of wasp or bee): Choka (X), Kwar (X), Mobaw (X), Tut (X)

CHELONIA (turtles): Buru (X), Choka (1), Kiewo (5), Tut (2)

LACERTILIA (lizards)
Chameleonidae indet. (chameleon): Kwar (27), Mobaw (12)
Varanidae indet. (monitor lizard): Kwar (10), Mobaw (2)
Scincidae indet. (skink)
Gekkonidae indet. (gecko)
Agamidae indet. (agama lizard)
Lacertidae indet. (true lizard)

APPENDIX 8 continued

Amphisbaenia indet. (worm lizard)

SERPENTES (snakes): Kwar (3), Tut (3)
Boidae indet. (pythons and boas): Choka (1), Kiewo (7), Kwar (19), Tut (2)
Elapidae indet. (cobras): Kwar (82)
Colubridae indet. (rat, king and water snakes)
Viperidae indet. (vipers)

AVES (birds): Choka (1), Kiewo (4), Tut (3)
Falconiformes indet. (eagles, hawks, falcons)
Cuculiformes indet. (cuckoos, coucals)
Columbiformes indet. (pigeons): Kwar (X)
Passeriformes indet. (songbirds): Kwar (X)
Musophagidae indet. (turacos): Kwar (X)
Bucerotidae indet. (hornbills): Kwar (X)
Phasianidae indet. (quails, guinea fowl): Kwar (X)
Phoenicopteridae indet. (flamingos)

INSECTIVORA (insect eaters): Tut (3)
Miorhynchocyon clarki (giant elephant shrew): Kwar (4), Tut (1)
Miorhynchocyon meswae (giant elephant shrew)
Miorhynchocyon rusingensis (giant elephant shrew)
Myohyrax oswaldi (elephant shrew)
Macroscelidae indet. sp. a (elephant shrew)
Macroscelidae indet. sp. b (elephant shrew)
Galerix africanus (hedgehog): Kwar (1), Tut (1)
Amphechinus rusingensis (hedgehog): Kwar (8)
Amphechinus sp. (hedgehog)
Gymnurechinus leakeyi (hedgehog): Kwar (2), Mobaw (1)
Gymnurechinus camptolophos (hedgehog): Kiewo (3), Tut (2)
Protenrec tricuspis (tenrec): Kwar (2), Tut (1)
Parageogale aletris (tenrec)
Erythrozootes chamerpes (tenrec): Kwar (1), Mobaw (X)
Ndamathaia kubwa (giant tenrec)
Prochrysochloris miocaenicus (golden mole): Kwar (8), Mobaw (X), Tut (2)

CHIROPTERA (bats): Choka (2)
Propotto leakeyi (flying fox or fruit bat)
cf. *Propotto* (flying fox or fruit bat)
Pteropodae indet. sp. a (flying fox or fruit bat): Kwar (1)
Pteropodae indet. sp. b (flying fox or fruit bat)
Taphozous incognita (tomb bat, insectivorous): Kwar (1)
Emballonuridae indet. (tomb bat, insectivorous): Kwar (1)
Nycterididae indet. (slit-faced bat, insectivorous)
Hipposideros sp. (leaf-nosed bat, insectivorous)
Hipposideridae indet. (leaf-nosed bat, insectivorous)
Chamtwaria pickfordi (insectivorous bat)
Vespertilionidae indet. sp. a (insectivorous bat)
Vespertilionidae indet. sp. b (insectivorous bat)

PRIMATES (primates)
Komba minor (bush baby): Kiewo (3), Tut (1)
Komba robustus (bush baby): Kiewo (1), Tut (3)
Progalago songhorensis (bush baby)
Progalago dorae (bush baby): Kiewo (1)
Mioeuoticus sp. (bush baby)
Dendropithecus macinnesi (gibbon-like monkey-ape, 9 kg): Kiewo (1), Tut (1)
Rangwapithecus gordoni (monkey-ape, 15 kg): Kiewo (4), Tut (1)
Nyanzapithecus vancouveringorum (monkey-ape, 9 kg): Kiewo (1)
Micropithecus clarki (very small monkey-ape, 3.5 kg): Kwar (X), Mobaw (X)
Kalepithecus songhorensis (small monkey-ape, 5 kg): Kwar (X), Mobaw (X)
Limnopithecus evansi (small monkey-ape, 4 kg)
Limnopithecus legetet (small monkey-ape, 5 kg): Kiewo (8), Kwar (X), Mobaw (X), Tut (5)
Proconsul africanus (monkey-ape, 18 kg): Kiewo (3), Kwar (X), Mobaw (X)
Proconsul major (monkey-ape, 50 kg): Kiewo (4), Kwar (X), Mobaw (X), Tut (2)

RODENTIA (rodents)
Andrewsimys parvus (theridomyid cane rat)
Phiomys andrewsi (theridomyid cane rat): Kiewo (1)
Paraphiomys pigotti (theridomyid cane rat): Kiewo (3), Tut (1)
Paraphiomys hopwoodi (theridomyid cane rat): Kiewo (1), Tut (5)
Epiphiomys coryndoni (theridomyid cane rat)
Diamantomys luederitzii (theridomyid cane rat): Choka (2), Kiewo (45), Tut (38)
Kenyamys mariae (theridomyid cane rat)
Simonimys genovefae (theridomyid cane rat): Kiewo (2), Tut (5)
Myophiomys arambourgi (theridomyid cane rat): Tut (1)
Elmerimys woodi (theridomyid cane rat): Kiewo (1), Tut (8)
Bathyergoides neotertiarius (mole rat): Kiewo (3)
Proheliophobius leakeyi (mole rat)
Paranomalurus bishopi (scaly-tailed flying squirrel): Kiewo (3), Tut (6)
Paranomalurus soniae (scaly-tailed flying squirrel): Kiewo (2)
Paranomalurus walkeri (scaly-tailed flying squirrel): Kiewo (1), Tut (4)
Zenkerella wintoni (scaly-tailed squirrel)
Megapedetes pentadactylus (spring hare): Kiewo (2)
Afrocricetodon songhorensis (cricetid mouse): Tut (1)
Protarsomys macinnesi (cricetid mouse)
Notocricetodon petteri (cricetid mouse): Kiewo (2), Tut (5)
Vulcanisciurus africanus (squirrel)

APPENDIX 8 continued

CREODONTA (archaic meat eaters)
Kelba quadeemae (long-snouted, proviverrine creodont)
Teratodon spekei (proviverrine creodont): Mobaw (X)
Teratodon enigmae (proviverrine creodont)
Anasinopa leakeyi (large, civetlike, proviverrine creodont)
Hyainalouros sp. (giant hyaenodont creodont)
Hyaenodon andrewsi (large hyaenodont creodont, 4-7 kg): Kiewo (1)
Hyaenodon matthewi (hyaenodont creodont, 2-5 kg)
Hyaenodon pilgrimi (small hyaenodont creodont, 1-2 kg): Tut (1)

CARNIVORA (meat eaters)
Cynelos euryodon (amphicyonine bear dog)
Kenyalutra songhorensis (otter, 15-25 kg)
Kichechia zamanae (mongoose, 2-5 kg): Kwar (X), Mobaw (X)
Legetetia nandii (mongoose, 1-2 kg)
Mioprionodon pickfordi (mongoose, 0.5-1 kg)
Stenoplesictis muhoroni (mongoose, 1-2 kg)
Afrosmilus turkanae (sabre-tooth cat)

PROBOSCIDEA (elephants)
Prodeinotherium hobleyi (four tusk, deinothere)
cf. *Archaeobelodon* sp. (gomphothere): Kwar (3)
Eozygodon morotoensis (two tusk, mastodon)
Elephantoidea indet. (bunodont form)

TUBULIDENTATA (aardvarks)
Orycteropus minutus (small aardvark): Kiewo (1)

HYRACOIDEA (hyraxes)
Megalohyrax championi (large, long-snouted hyrax)

CHALICOTHERIIDAE (extinct "horse-bears")
Chalicotherium rusingense (large) Kwar (1)

RHINOCEROTIDAE (rhinos)
Rhinocerotidae indet.

ANTHRACOTHERIIDAE (extinct "river-hogs")
Masritherium sp. (large, long faced)
Anthracotheriidae indet.

SUIDAE (pigs)
Nguruwe kijivium (small, short faced): Tut (2)

TRAGULIDAE (chevrotains)
Dorcatherium songhorense (chevrotain): Kiewo (3), Kwar (58), Mobaw (18), Tut (2)

GELOCIDAE (early deerlike ungulates)
Gelocus whitworthi
Walangania africanus (hornless?): Choka (1), Kiewo (3), Kwar (236), Mobaw (19), Tut (2)

VERTEBRATA-TRACE FOSSILS
bulbous burrow (of rodent?): Choka (X), Tut (X)
ellipsoidal shell (7-9 mm) of gecko: Kwar (X)

Note: Sources of species names are Butler (1984), Harrison (1981, 1986), Hirsch & Harris (1989), Kelley (1986), L.L. Jacobs, Anyonge, & Barry (1987), Lavocat (1973), Martin (1981), Meyer (1978), Patterson (1978), Pickford (1984, 1986a, 1986b), Pickford & Andrews (1981), Schmidt-Kittler (1987), Tassy (1986), A. C. Walker (1978), and Wenz & Zilch (1960). Sources of numbers of specimens are Pickford (1984a, 1986a) whose Legetet locality 10 is a Mobaw paleosol and whose locality 14 is a Kwar paleosol; and Pickford and Andrews (1981) whose beds (paleosol series) are 10 (Buru), 7,8 (Choka), 5,6 (Kiewo) and 9 (Tut).

APPENDIX 9. Middle Miocene (14Ma) fossils from Fort Ternan, Kenya, and paleosols in which they have been found

ALGAE (algae)
Botryococcus (palynomorph colonies): type Chogo

PTERIDOPHYTA (ferns and club mosses, monolete spores): type Chogo (2)

CONIFERAE (conifers)
Juniperus (Cupressaceae, pollen): type Chogo (4)
Podocarpus (Podocarpaceae, pollen): type Chogo (9)

ANGIOSPERMAE (flowering plants)
? *Annonospermum* sp. (Annonaceae, custard apple family, seed)
Amaranthaceae or Chenopodiaceae (cockscomb and pokeweed families, pollen): type Chogo (7)
cf. *Sterculia* (Sterculiaceae, cocoa family, leaf): Chogo (1)
Celtis rusingensis (hackberry, Ulmaceae)
Celtis sp. (Ulmaceae, pollen): type Chogo (3)
Urticaceae indet. (stinging nettle family, pollen): type Chogo (9)
Cucurbitaceae indet. (gourd family, nodose exocarp)
cf. *Acacia*, stipular spines (Leguminosae): Chogo (1)
Papilionaceae (pea family, pods)
Combretaceae (bush willow, pollen): type Chogo (1)
Icacinocarya sp. (icacina vine, Icacinaceae, endocarp): Chogo (1)
Euphorbiotheca pulchra (Euphorbiaceae, calyx)
Acalypha sp. (Euphorbiaceae, pollen): type Chogo (3)
cf. *Croton* (Euphorbiaceae, pollen): type Chogo (1)
Berchemia pseudodiscolor (bird plum, Rhamnaceae, exocarp)
Zizyphus rusingensis (blinkbaar, Rhamnaceae, exocarp)
Cnestis rusingensis (itch pod, Connaraceae, endocarp)
cf. *Lannea* sp. (Anacardiaceae, pollen): type Chogo (1)
Leakeyia vesiculosa (Apocynaceae, oleander family, endocarp)
Olea sp. cf. *O. africana* (Oleaceae, olive family, pollen): type Chogo (2)
Plantago sp. cf. *P. coronopus* (Plantaginaceae, plantain family, pollen): type Chogo (1)
Anthospermum sp. (Rubiaceae, coffee family, pollen): type Chogo (1)
Compositae or Tubiflorae (daisy and eyebright families, pollen): type Chogo (2)
Artemisia sp. (wormwood, pollen): Chogo (X)
Potamogeton sp. cf. *P. pectinatum* (Potamogetonaceae, pondweed family, pollen): type Chogo (1)
Gramineae indet. (grass pollen): type Chogo (154)
Panicoideae sp. indet. a (Gramineae): Onuria (X)
Panicoideae sp. indet. b (Gramineae): Onuria (X)
Panicoideae sp. indet. c (Gramineae): Onuria (X)
Chloridoideae sp. indet. a (Gramineae): Onuria (X)
Chloridoideae sp. indet. b (Gramineae): Onuria (X)
Cyperaceae (sedge family, pollen): type Chogo (76)
Typha sp. (cattail, Typhaceae, pollen): type Chogo (2)

PLANTAE - TRACE FOSSILS
fine root traces (less than 2 mm diameter): Chogo (X), Chogo eroded (X), Chogo ferruginized nodule (X), Chogo type (X), Dhero (X), Lwanda (X), Onuria (X) Rabuor (X), Rairo (X)
large root traces (2-10 mm diameter): Chogo (X), Chogo eroded (X), Chogo ferruginized nodule (X), Dhero (X), Lwanda (X), Rabuor (X), Rairo (X)
grass or sedge casts: Chogo eroded (X), Dhero (X)
twig casts (3-5 mm): Chogo eroded (X), Dhero (X)
log casts: Dhero (3)
stump casts: Chogo eroded (2), Dhero (2)

GASTROPODA (snails)
Subulininae indet. (elongate, very high spired, Subulinidae): Chogo (100)
Curvella sp. (ovate with sinuous lip, Subulinidae): Chogo (5)
Burtoa nilotica (high spired, intersecting growth ridges and radial lines, Achatinidae): Chogo (14)
Helicarion sp. (low spired, very rapidly expanding whorls, Helicarionidae): Chogo (8)
Trochonanina sp. (conical, with keeled whorls, Urocyclidae): Chogo (5)
Thapsia sp. (very low spired to discoidal, Urocyclidae): Chogo (4)
Halolimnohelix sp. (low spired, globular, Bradybaenidae): Chogo (7)
Tayloria sp. (low spired, angular around umbilicus, Streptaxidae): Chogo (3)
Gulella sp. (small, pupiform, Streptaxidae): Chogo (6)

MYRIAPODA (millipedes and centipedes): Chogo (2)
Myriapoda indet. (round back millipede): Chogo (96)

ARACHNIDA (spiders and mites)
Amblypygi indet. (whip scorpion): Chogo (1)

INSECTA (insects)
Tenebrionidae indet. (darkling beetle): Chogo (1)

INVERTEBRATA-TRACE FOSSILS
Skolithus (spider or insect burrow): Chogo eroded (X)
ovoid casts (cocoons): Chogo (6), Chogo eroded (2)
Celliforma (wasp or bee larval cell): Chogo (1)

LACERTILIA (lizards): Chogo eroded (2)
Chameleo intermedius (chameleon): Chogo eroded (1), Dhero (1)

OPHIDEA (snakes): Chogo eroded (5)

APPENDIX 9. continued

AVES (birds): Chogo eroded (7)
Struthio sp. (ostrich bone and egg shell): Chogo (3), Chogo eroded (1)
Accipitridae indet. (vulture): Chogo (1)

INSECTIVORA (insect eaters): Chogo (3)
Pronasilio ternanensis (elephant shrew)
Miorhynchocyon rusingensis (elephant shrew): Chogo (3), Chogo eroded (1)
Myohyrax oswaldi (elephant shrew)
Amphechinus sp. (hedgehog): Chogo eroded (1)

CHIROPTERA (bats): Chogo (1)

PRIMATES (primates): Chogo (7), Chogo eroded (1)
Lorisinae (slow loris): Chogo (1)
cf. *Oreopithecus* sp. (monkey-ape): Chogo (1)
Micropithecus sp. (small monkey-ape, 5 kg): Chogo (2)
Proconsul sp. cf. *P. africanus* (monkey-ape, 18 kg): Chogo (2), Chogo eroded (1)
Kenyapithecus wickeri (monkey-ape, small canines, 30 kg): Chogo (24), Chogo eroded (1)

RODENTIA (rodents): Chogo (71), Chogo eroded (14)
Sciuridae indet. (squirrel): Chogo (1)
Paranomalurus sp. (flying squirrel): Chogo (6)
Myocricetodon irhoudi (cricetid mouse)
Myocricetodon parvus (cricetid mouse)
Myocricetodon sp.: Chogo (1), Chogo eroded (1)
cf. *Afrocricetodon* (cricetid mouse): Chogo (11)
Leakeymys ternani (cricetid gerbil): Chogo (8), Chogo eroded (16)
? *Diamantomys* sp. (theridomyid cane rat): Chogo (1), Chogo eroded (1)
Paraphiomys pigotti (theridomyid cane rat): Chogo (6), Chogo eroded (1)
Megapedetes sp. (spring hare): Chogo (1), Chogo eroded (1)
Bathyergidae indet. (mole rat): Chogo (2), Chogo eroded (1)

CREODONTA (archaic meat eaters): Chogo (19), Chogo eroded (7)
Megistotherium osteothalastes (huge wolflike): Chogo (5)
Dissopsalis pyroclasticus (proviverrine): Chogo (1)

CARNIVORA (meat eaters)
Amphicyonidae indet. (large bear dog): Chogo (3), Chogo eroded (1)
Amphicyonidae indet. (small bear dog): Chogo (2)
Percrocuta sp. (hyaena): Chogo (5), Chogo eroded (5)
Kanuites sp. (mongoose): Chogo (20)

TUBULIDENTATA (aardvarks): Chogo (3), Chogo eroded (1)
Myorycteropus chemeldoi: Chogo (2)

PROBOSCIDEA (elephants): Chogo (37), Chogo eroded (3)
Choerolophodon ngorora (four tusk, gomphothere)
Protanancus macinnesi (shovel tusk, gomphothere): Chogo (62), Chogo eroded (13)
Deinotheriidae indet. (hoe tusk, deinothere): Chogo (1)

RHINOCEROTIDAE (rhinos): Chogo eroded (4)
Paradiceros mukirii (two horns, narrow lip): Chogo (97), Chogo eroded (18)

SUIDAE (pigs): Chogo (1), Chogo eroded (1)
Lopholistriodon kidogosana (small pig): Chogo (7)
Listriodon akatikubas (large pig): Chogo (1)
cf. *Taucanamo* (peccary, Tayassuidae)

HIPPOPOTAMIDAE (hippos)
Kenyapotamus ternani (small, 225 kg, piglike)

TRAGULIDAE (chevrotains)
Dorcatherium chappuisi (large): Chogo (28)
Dorcatherium pigotti (small): Chogo (1)

GELOCIDAE (archaic deerlike ungulates)
Walangania africanus (hornless?): Chogo (5)

PALAEOMERYCIDAE (archaic deerlike ruminants): Chogo (73), Chogo eroded (11)
Propalaeoryx sp. (horned): Chogo (2)
Climacoceras gentryi (branched horn): Chogo (4), Chogo eroded (2)

GIRAFFIDAE (giraffes): Chogo (39), Chogo eroded (10)
Samotherium africanum (erect horn): Chogo (11)
Palaeotragus primaevus (hornless?, okapilike): Chogo (167), Chogo eroded (9)

BOVIDAE (antelopes): Chogo (1,539), Chogo eroded (152), Chogo ferruginized nodule (1,104), Chogo type (235)
? *Eotragus* sp. (early boselaphine?): Chogo (1)
Kipsigicerus labidotus (vane-horn, boselaphine): Chogo (444), Chogo eroded (33), Chogo ferruginized nodule (228), Chogo type (91)
Gazella sp. (gazelle, antilopine): Chogo (21), Chogo ferruginized nodule (38), Chogo type (10)
Oioceros tanyceras (bow-horn, caprine): Chogo (428), Chogo eroded (36), Chogo ferruginized nodule (202), Chogo type (83)
Capratragoides potwaricus (erect horn, caprine): Chogo (76), Chogo eroded (18), Chogo ferruginized nodule (47), Chogo type (7)

APPENDIX 9. continued

VERTEBRATA - TRACE FOSSILS	rounded coprolites (antelope?)
large burrow (den of *Kanuites*?): Chogo eroded (1)	elongate coprolites (hyaena?)

Note: Sources of species names are Andrews and Walker (1976), Barry (1988), Bonnefille (1984), Butler (1984), Dugas (1989), Gentry (1970), W.R. Hamilton (1978), Harris (1978), Harrison (1986), Hillenius (1978), Kelley (1986), Lavocat (1978), Patterson (1978), Pickford (1983, 1984a, 1985, 1986a, 1986b), Shipman (1977, 1986a), Tassy (1986), and Thomas (1984). Sources of numbers of specimens are Shipman's (1977) "Leakey Assemblage" for Chogo and "1974 Excavation" for Chogo clay eroded phase; Pickford (1985) for Chogo snails, Pickford (1984a) for Rabuor, and Shipman *et al.* (1981) for Dhero. Supposed crocodile teeth and crab claws were reassigned to primates and insect fragments, respectively, by Pickford (1985). A single specimen of the ape, *Rangwapithecus* sp. cf. *R. gordoni*, supposedly from Fort Ternan, is more likely from Songhor (Pickford, 1985).

APPENDIX 10. Late Miocene (8.5Ma at lower Level and 7.8Ma at upper level) fossils from the Dhok Pathan Formation near Khaur, Pakistan, and paleosols in which they have been found

PLANTAE - TRACE FOSSILS
fine root traces: lower level - Bhura (X), Kala (X), Khakistari (X), Lal (X), Pandu (X), Pila (X), Sarang (X), Sonita (X); upper level - Bhura (X), Naranji (X), Sarang (X), Sonita (X)
calcareous rhizoconcretions: lower level - Pandu (X)
slender, copiously branched, drab-haloed root traces: lower level - Pila (X); upper level - Bhura (X), Naranji (X), Sarang (X), Sonita (X)
large weakly branched drab haloed root traces: lower level - Bhura (X), Khakistari (X), Lal (X), Pandu (X), Sarang (X), Sonita (X)

INVERTEBRATA - TRACE FOSSILS
Edaphichnium sp. (burrow of earthworm?): lower level - Pandu (X), Sarang (X); upper level - Sarang (X)
Macanopsis sp. (burrow of freshwater crab? or spider?): lower level - Pandu (X), Sarang (X); upper level - Naranji (X), Sarang (X)
Skolithus sp. (burrow of insect or spider): lower level - Pila (X), Sarang (X), Sonita (X); upper level - Sarang (X), Sonita (X)
Coprinsphaera sp. (pupal cell of dung beetle?): upper level - Bhura (X)
Termitichnus sp. (fungus garden termite): lower level - Pandu (X), Sarang (X), Sonita (X); upper level - Naranji (X), Sarang (X)
slender branching galleries and pellets (mound-building termites): lower level - Lal (X), Pila (X); upper level - Sonita (X)

GASTROPODA (snails)
Pila prisca (apple snail, mainly opercula, Ampullariidae): lower level - Bhura (2), Khakistari (1), Lal (3), Pandu (1)

PELECYPODA (clams)
Indonaia hasnotensis (unionid)

PISCES (fish)
Cypriniformes indet. (minnows)
Siluriformes indet. (catfish)
Perciformes, Channidae indet. (murrel)

AMPHIBIA (amphibians)
Anura indet. (frog)

LACERTILIA
Varanus sp. (monitor lizard)

OPHIDIA (snakes)
Python sp. (python)
Acrochordus sp. (wart snake)
Colubridae indet. (rat, king and water snakes)

CROCODYLIA (crocodiles)
Crocodylus palaeindicus (mugger): lower level - Kala (5)
Gavialus gangeticus (gavial)

CHELONIA (turtles, tortoises): lower level - Bhura (X), Pandu (X); upper level - Sonita (X)
Testudo sp. (land tortoise)
Emydidae indet. (pond and box turtles)
? *Chitra* sp. (river softshell turtle)
Geoemyda sp. (spined turtle)
Trionyx sp. (softshell turtle)
Lissemys spp. (soft terrapin)

AVES (birds)
Leptophilos siwalikensis (stork)
Lophura sp. (pheasant)
Porphyrio sp. (rail)

INSECTIVORA (insect eaters)
Erinaceidae indet. (hedgehog)
Soricidae indet. (shrew)
Tupaiidae indet. (tree shrew)

PRIMATES (primates)
Nycticeboides simpsoni (slow loris)
Gigantopithecus giganteus (giant ape, 166 kg)
Sivapithecus sivalensis (orang-utanlike ape, 58 kg): lower level - Sarang (19)
Sivapithecus punjabicus (orang-utanlike ape, 40 kg): lower level - Sarang (9)

RODENTIA (rodents)
Sciuridae spp. (squirrel)
Democricetodon sp. (cricetid mouse)
Kanisamys sivalensis (rhizomyid or bamboo rat): upper level - Sonita (1)
Brachyrhizomys sp. (rhizomyid or bamboo rat)
Rhizomyoides sp. (rhizomyid or bamboo rat)
Progonomys debruijni (murid mouse)
Progonomys sp. (murid mouse)
Karnimata darwini (murid mouse)
Parapodemus sp. (murid mouse)
Gliridae indet. (dormouse)

CARNIVORA (meat eaters)
Arctamphicyon lydekkeri (bear dog, 150-250 kg)
Eomellivora sp. (ratel or honey badger, 10-20 kg)
Sivaonyx bathygnathus (otter, 5-15 kg)
cf. *Ischyrictis* (weasel, 1-2 kg)
Mustelidae indet. (10-20 kg)
Viverra chinjiensis (civet, 5-9 kg): lower level - Sarang (1)
Viverridae indet. (large civet, 15-20 kg)
Herpestes sp. (mongoose, 0.5-2 kg)

APPENDIX 10. continued

Progenetta sp. (genet, 20-30 kg)
Percrocuta grandis (hyaena, 65-85 kg)
Palhyaena sivalensis (hyaena, 15-25 kg): lower level - Sarang (1)
Hyaenidae indet. (hyaena): lower level - Sonita (1)
Paramachaerodus sp. (sabre-tooth cat, 180-220 kg)
Felis cf. *F. sivalensis* (big cat, 60-80 kg): lower level - Kala (1)

TUBULIDENTATA (aardvarks)
Orycteropus browni (large, 50-70 kg)

PROBOSCIDEA (elephants): lower level - Kala (1), Sarang (>4); upper level - Sonita (1)
Choerolophodon corrugatus (four tusk, gomphothere, >2,000 kg): lower level - Kala (1), Khakistari (1), Sonita (2)
cf. *Platybelodon* sp. (shovel tusk, gomphothere, >2,000 kg)
Stegolophodon sp. (two tusk, stegodont, >2,000 kg): lower level - Sarang (10)
Deinotherium indicum (hoe tusk, deinothere, >2,000 kg)

EQUIDAE (horses): lower level - Kala (1), Sonita (1)
"*Cormohipparion*" *(Sivalhippus)* spp. (large, three-toed horse): lower level - Bhura (1), Kala (3), Pandu (4); upper level - Sonita (4)
"*Cormohipparion*" *(Sivalhippus) perimense* (large three-toed horse)
"*Cormohipparion*" *(Sivalhippus) theobaldi* (large three-toed horse)
? *Hipparion antilopinum* (small, three-toed horse, 175-300 kg)

CHALICOTHERIIDAE (extinct "horse-bears")
Chalicotherium salinum (large, 210-250 kg)

RHINOCEROTIDAE (rhinos): lower level - Kala (1), Sarang (18), Sonita (2)
Gandaitherium browni (one horned, >1,000 kg)
Aceratherium sp. cf. *A. simorrense* (hornless, > 1,000 kg)
Brachypotherium perimense (hornless, > 2,000 kg)
Chilotherium intermedium (hornless, > 1,000 kg): upper level - Sonita (1)

SUIDAE (pigs)
Tetraconodon magnus (large, bone or nut crusher?, 180-240 kg)
Propotamochoerus hysudricus (bush pig, 40-60 kg): lower level - Sonita (2)
Hippopotamodon sivalensis (bush pig, 180-240 kg)
Schizochoerus gandakasensis (pig?, 25-35 kg)

ANTHRACOTHERIIDAE (extinct "river hogs"): lower level - Sarang (4)
"*Anthracotherium*" *punjabiense* (60-80 kg)
Hemimeryx pusillus (180-220 kg)
Anthracotheriidae sp. (60-80 kg)

TRAGULIDAE (chevrotains): lower level - Kala (1), Sarang (2)
Dorcabune nagrii (25-30 kg)
Dorcatherium majus (25-30 kg): upper level - Naranji (1)
Dorcatherium sp. cf. *D. nagrii* (3-6 kg)
Dorcatherium sp. indet. (8-10 kg)

GIRAFFIDAE (giraffes): lower level - Sonita (1); upper level - Naranji (1), Sonita (1)
Brahmatherium megacephalum (horned giraffe, 700-1100 kg) : lower level - Khakistari (1)
Giraffidae sp. indet.

BOVIDAE (antelopes)
Miotragocerus punjabicus (erect horned, boselaphine, 40-50 kg): lower level - Sonita (1)
Elachistoceras khauristanensis (four-horned, boselaphine, chowsinghalike, 3-5 kg): lower level - Bhura (1), Sarang (4)
Selenoportax vexillarius (crescent-horn, boselaphine, 60-90 kg): lower level - Lal (?1)
Selenoportax lydekkeri (large crescent-horn, boselaphine): upper level - Naranji (1), Sarang (2), Sonita (2)
Protragelaphus sp. cf. *P. skouzesii* (helical-horn, boselaphine)
Protoryx sp. (backward-arch horn, hippotragine)
Gazella sp. (gazelle, antelopine, 8-15 kg): lower level - Kala (3), Sonita (4)

Note: Sources of species names include Badgley (1982, 1986), Barry, Lindsay, & Jacobs (1982), Gromova (1962), MacPhee & Jacobs (1986), Pilbeam *et al.* (1979), Prashad (1924), Roe (1987,1988), Thomas (1984), and Vokes (1935); specimen numbers are from collection by author (G.J.R.), excavations at localities 260 (for Sarang Series), 262-2 (Kala), 262-3 & 4 (Sonita) of Badgley (1982), and surface collections at localities 410,260,444 (for Sarang), 443 (Naranji) and 477,544 (Sonita) of Barry, Lindsay, & Jacobs (1982). Apes are divided into species using simple size criterion of Badgley *et al.* (1984) and Fleagle (1988). Horses were referred to "*Cormohipparion*" *(Sivalhippus)* following Bernor & Hussain (1985).

REFERENCES

Aandahl, A. R. (1982). *Soils of the Great Plains.* University of Nebraska, Lincoln, 282 pp.

Agarwal, R. R. and Mukherji, S. K. (1951). Gangetic alluvium of India: pedochemical characters of the genetic soil types of Gorakhpur district in the United Provinces. *Soil Science* **72**, 21-32.

Ahmad, M., Ryan, J., and Paeth, R. C. (1977). Soil development as a function of time in the Punjab river plains of Pakistan. *Proceedings of the Soil Science Society of America* **41**, 1162-1166.

Allard, P., Dajlevic, D., and Delarue, C. (1989). Origin of carbon dioxide emanation from the 1979 Dieng eruption, Indonesia: implications for the origin of the 1986 Nyos catastrophe. *Journal of Volcanology and Geothermal Research* **39**, 195-206.

Allchin, B. and Allchin, R. (1982). *The rise of civilization in India and Pakistan.* Cambridge, New York, 379 pp.

Allen, J. R. L. (1965). Fining upwards cycles in alluvial successions. *Geological Journal* **4**, 229-246.

Allen, J. R. L. (1986a). Pedogenic carbonates in the Old Red Sandstone facies (Late Silurian-Early Carboniferous) of the Anglo-Welsh area, southern Britain. In *Paleosols: their recognition and interpretation* (ed. V. P. Wright), pp. 58-86. Blackwells, Oxford, England.

Allen, J. R. L. (1986b). Time scales of color change in late Flandrian intertidal muddy sediments of the Severn Estuary. *Proceedings of the Geologists Association* **97**, 23-28.

Anderson, G. D. (1963). Some weakly developed soils of the eastern Serengeti Plains, Tanzania. *African Soils* **8**, 339-347.

Anderson, G. D. and Herlocker, D. J. (1973). Soil factors affecting the distribution of the vegetation types and their utilization by wild animals in Ngorongoro Crater, Tanzania. *Journal of Ecology* **61**, 627-651.

Anderson, G. D. and Talbot, L. W. (1965). Soil factors affecting the distribution of grassland types and their utilization by wild animals on the Serengeti Plains, Tanzania. *Journal of Ecology* **53**, 33-56.

Andrews, P. (1971). *Ramapithecus wickeri* mandible from Fort Ternan, Kenya. *Nature* **231**, 192-194.

Andrews, P. (1981a). A short history of Miocene field palaeontology in western Kenya. *Journal of Human Evolution* **10**, 3-9.

Andrews, P. (1981b). Species diversity and diet in monkeys and apes during the Miocene. In *Aspects of human evolution* (ed. C.B. Stringer), pp. 25-61. Taylor & Francis, London.

Andrews, P. (1982). Ecological polarity in primate evolution. *Zoological Journal of the Linnaean Society of London* **74**, 233-244.

Andrews, P. (1983). The natural history of *Sivapithecus*. In *New interpretations of ape and human ancestry* (eds. R. L. Ciochon and R. S. Corruccini), pp. 441-463. Plenum, New York.

Andrews, P. (1986). Fossil evidence on human origins and dispersal. *Cold Spring Harbor Symposia on Quantitative Biology* **51**, 419-428.

Andrews, P. (1989). Lead review: palaeoecology of Laetoli. *Journal of Human Evolution* **18**, 173-181.

Andrews, P. (1990a). Paleoecology of the Miocene fauna from Paşalar, Turkey. *Journal of Human Evolution* **19**, 569-582.

Andrews, P. (1990b). Lining up the ancestors. *Nature* **345**, 664-665.

Andrews, P. and Cronin, J. (1982). The relationships of *Sivapithecus* and *Ramapithecus* and the evolution of the orang-utan. *Nature* **297**, 541-546.

Andrews, P. and Evans, E. N. (1979). The environment of *Ramapithecus* in Africa. *Paleobiology* **5**, 22-30.

Andrews, P., Groves, C. P. and Horne, J. F. M. (1975). Ecology of the lower Tana River flood plain (Kenya). *Journal of the East African Natural History Society and National Museum* **30(151)**, 1-31

Andrews, P., Lord, J. M. and Evans, E. M. N. (1979). Patterns of ecological diversity in fossil and modern mammalian faunas. *Biological Journal of the Linnaean Society of London* **11**, 177-205.

Andrews, P. and Martin, L. (1987). The phyletic position of the Ad Dabtiyah hominoid. In *Miocene geology and palaeontology of Ad Dabtiyah, Saudi Arabia* (ed. P. J. Whybrow). *Bulletin of the British Museum of Natural History* **41**, 383-393.

Andrews, P., Meyer, G. E., Pilbeam, D. R., Van Couvering, J. A. and Van Couvering, J. A. H. (1981). The Miocene fossil beds of Maboko Island, Kenya: geology, age, taphonomy, and palaeontology. *Journal of Human Evolution* **10**, 35-48.

Andrews, P. and Van Couvering, J. A. H. (1975). Palaeoenvironments in the East African Miocene. In *Approaches to primate paleobiology* (ed. F. S. Szalay), pp. 62-103. S. Karger, New York.

Andrews, P. and Walker, A. (1976). The primate and other fauna from Fort Ternan, Kenya. In

Human origins: Louis Leakey and the East African evidence (eds. G. Isaac and R. McCown), pp. 279-304. Benjamin, Menlo Park, California.

Ardrey, R. A. (1976). *The hunting hypothesis*. Athenaeum, New York, 242 p.

Arkley, R. J. (1963). Calculation of carbonate and water movement in soil from climatic data. *Soil Science* **96**, 239-248.

Arrow, G. J. (1931). *Fauna of British India. Coleoptera Lamellicornia. III. Coprinae*. Taylor & Francis, London, 498 pp.

Aubert, H. and Pinta, M. (1977). *Trace elements in soils*. (Trans. L. Zuckermann and P. Segalen). Elsevier, Amsterdam, 395 pp.

Awasthi, N. (1982). Tertiary plant megafossils from the Himalaya - a review. *The Palaeobotanist* **30**, 254-267.

Axelrod, D. I. (1975). Evolution and biogeography of Madrean-Tethyan sclerophyll vegetation. *Annals of the Missouri Botanical Garden* **62**, 280-284.

Axelrod, D. I. (1986). Analysis of some palaeogeographic and palaeoecologic problems of palaeobotany. *The Palaeobotanist* **35**, 115-129.

Axelrod, D. I. (1989). Age and origin of chaparral. *Science Series of the Museum of Los Angeles County* **34**, 1-19.

Axelrod, D. I. and Raven, P. H. (1978). Late Cretaceous and Tertiary vegetation history. In *Biogeography and ecology in southern Africa* (eds. M. J. A. Werger and A. C. Van Bruggen), pp. 77-130. Junk, The Hague.

Badgley, C. E. (1982). Community reconstruction of a Siwalik mammalian assemblage. Unpublished Ph.D. thesis, Yale University, New Haven, Connecticut, 364 pp.

Badgley, C. (1986). Taphonomy of mammalian fossil remains from Siwalik rocks of Pakistan. *Paleobiology* **12**, 119-142.

Badgley, C. and Behrensmeyer, A. K. (1980). Paleoecology of Middle Siwalik sediments and faunas, northern Pakistan. *Palaeogeography, Palaeoclimatology, Palaeoecology* **30**, 133-155.

Badgley, C., Kelley, J., Pilbeam, D., and Ward, S. (1984). The paleobiology of South Asian Miocene hominoids. In *The people of South Asia* (ed. J. R. Lukacs), pp. 3-28. Plenum, New York.

Badgley, C., Qi, G., Chen, W., and Han, D. (1988). Paleoecology of a Miocene, tropical, upland fauna: Lufeng, China. *National Geographic Research* **4**, 178-195.

Badgley, C. and Tauxe, L. (1990). Paleomagnetic stratigraphy and time in sediments: studies in alluvial Siwalik rocks of Pakistan. *Journal of Geology* **98**, 457-472.

Badgley, C., Tauxe, L., and Bookstein, F. L. (1986). Estimating the error of age interpretation in sedimentary rocks. *Nature* **319**, 139-141.

Baker, B. H. (1986). Tectonics and volcanism of the southern Kenya Rift Valley and its influence on rift sedimentation. In *Sedimentation in the African Rifts* (eds. L. E. Frostick, R. W. Renaut, I. Reid and J. J. Tiercelin). *Special publication of the Geological Society of London* **25**, 45-57.

Baker, B. H. (1987). Outline of the petrology of the Kenya Rift alkaline province. In *Alkaline igneous rocks* (eds. J. G. Fitton and B. G. J. Upton). *Geological Society of London Special Publication* **3**, 293-311.

Baker, B. H., Williams, L. A. J., Miller, J. A. and Fitch, F. J. (1971). Sequence and chronology of the Kenya Rift volcanics. *Tectonophysics* **11**, 191-215.

Baker, V. R. (1986). Fluvial landforms. In *Geomorphology from space* (eds. N. M. Short and R. W. Blair), pp.255-316. NASA, Washington DC.

Bakker, R. T. (1983). The deer flees, the wolf pursues: incongruencies in predator-prey coevolution. In *Coevolution* (eds. D. J. Futuyma and M. Slatkin), pp. 350-382. Sinauer, Sunderland, Massachusetts.

Baldwin, B. and Butler, C. O. (1985). Compaction curves. *Bulletin of the American Association of Petroleum Geologists* **69**, 622-626.

Bancroft, H. (1932a). A fossil cyatheoid stem from Mount Elgon, East Africa. *New Phytologist* **31**, 241-253.

Bancroft, H. (1932b). Some dicotyledonous woods from the Miocene (?) beds of East Africa. *Annals of Botany* **46**, 745-767.

Bancroft, H. (1933). A contribution to the geological history of the Dipterocarpaceae. *Forhandlungen Geologiska Foreningens i Stockholm* **55**, 59-100.

Barker, D. S. (1989). Field relations of carbonatites. In *Carbonatites, genesis and evolution* (ed. K. Bell), pp. 38-69. Allen & Unwin, London.

Barker, D. S., and Nixon, P. H. (1983). Carbonatite lava and "welded" air-fall tuff (natural Portland cement), Fort Portal field, western Uganda. *Eos* **64**, 896.

Barndt, J., Johnson, N. M., Johnson, G. D., Opdyke, N. D., Lindsay, E. H., Pilbeam, D., and Tahirkheli, R. A. H. (1978). The magnetic polarity stratigraphy and age of the Siwalik Group, near Dhok Pathan village, Potwar Plateau, Pakistan. *Earth and Planetary Science Records* **41**, 355-384.

Barnosky, C. W. (1984). Late Miocene vegetational and climatic variations inferred from a pollen record in northwest Wyoming. *Science* **223**, 49-51.

Barry, J. C. (1988). *Dissopsalis*, a middle and late Miocene proviverrine creodont (Mammalia) from

Pakistan and Kenya. *Journal of Vertebrate Paleontology* **8**, 25-45.

Barry, J. C., Behrensmeyer, A. K., and Monaghan, M. (1980). A geologic and biostratigraphic framework for Miocene sediments near Khaur village, northern Pakistan. *Postilla* **183**, 19 pp.

Barry, J. C. and Cheema, I. U. (1984). Notes on a small fossil collection from near Gali Jagir on the Potwar Plateau of Pakistan. In *Contributions to the geology of Siwaliks of Pakistan* (eds. S. M. I. Shah and D. Pilbeam). *Memoirs of the Geological Survey of Pakistan* **11**, 15-18.

Barry, J. C., Jacobs, L. L., and Kelley, J. (1986). An early middle Miocene catarrhine from Pakistan, with comments on the dispersal of catarrhines into Eurasia. *Journal of Human Evolution* **15**, 501-508.

Barry, J. C., Johnson, N. M., Raza, S. M. and Jacobs, L. L. (1985). Neogene mammalian faunal change in southern Asia: correlations with climate, tectonic and eustatic events. *Geology* **13**, 637-640.

Barry, J. C., Lindsay, E. H., and Jacobs, L. L. (1982). A biostratigraphic zonation of the middle and upper Siwaliks of the Potwar Plateau of northern Pakistan. *Palaeogeography, Palaeoclimatology, Palaeoecology* **37**, 95-130.

Bathurst, R. G. C. (1975). *Carbonate sediments and their diagenesis.* 2nd ed., Elsevier, Amsterdam, 628 pp.

Beauchamp, J., Lemoigne, Y., and Petrescu, J. (1973). Les paléoflores tertiares de Debré Libanos (Éthiopie). *Annales de Societe Geologique du Nord, Lille* **93**, 17-32.

Beck, R. A. and Burbank, D. W. (1990). Continental-scale diversion of rivers: a control of alluvial stratigraphy. *Abstracts of the Geological Society of America* **22**, A238.

Begun, D. R., Moyá-Sola, S., and Kohler, M. (1990). New Miocene hominoid specimens from Can Llobateres (Vallès Penedès, Spain), and their geological and paleoecological context. *Journal of Human Evolution* **19**, 255-268.

Behrensmeyer, A. K. (1987). Miocene fluvial facies and vertebrate taphonomy in northern Pakistan. In *Recent developments in fluvial sedimentology* (eds. F. G. Ethridge, R. M. Flores and M. D. Harvey). *Special Publication of the Society of Economic Paleontologists and Mineralogists, Tulsa* **39**, 169-176.

Behrensmeyer, A. K. (1988). Vertebrate preservation in fluvial channels. *Palaeogeography, Palaeoclimatology, Palaeoecology* **63**, 183-199.

Behrensmeyer, A. K. and Raza, S. M. (1984). A procedure for documenting fossil localities in Siwalik deposits of Pakistan. In *Contributions to the geology of Siwaliks of Pakistan* (eds. S. M. I. Shah and D. Pilbeam). *Memoir of the Geological Survey of Pakistan* **11**, 65-74.

Behrensmeyer, A. K. and Tauxe, L. (1982). Isochronous fluvial systems in Miocene deposits of northern Pakistan. *Sedimentology* **29**, 331-352.

Behrensmeyer, A. K., Western, D., and Dechant-Boaz, D. E. (1979). New perspectives in vertebrate paleoecology from a Recent bone assemblage. *Paleobiology* **5**, 12-21.

Bell, R. H. V. (1982). The effect of soil nutrient availability on community structure in African ecosystems. In *Ecology of tropical savannas* (eds. B. J. Huntley and B. J. Walker), pp. 193-216. Springer, Berlin.

Bellon, H. and Pouclet, A. (1980). Datations K-Ar de quelques lavas du Rift-Ouest de l'Afrique Centrale: implications sur l'evolution magmatique et structurale. *Geologische Rundschau* **69**, 49-62.

Berggren, W. A., Kent, D. V., and Van Couvering, J. A. (1985). The Neogene, Part 2. Neogene geochronology and chronostratigraphy. In *The chronology of the geological record* (ed. N. J. Snelling). *Memoirs of the Geological Society of London* **10**, 211-251.

Bernor, R. L. (1983). Geochronology and zoogeographic relationships of Miocene Hominoidea. In *New interpretations of ape and human ancestry* (eds. R. L. Ciochon and R. S. Corruccini), pp. 21-64. Plenum, New York.

Bernor, R. L., Flynn, L. J., Harrison, T., Hussain, S. T., and Kelley, J. (1988). *Dionysopithecus* from southern Pakistan and the biochronology and biogeography of early Eurasian catarrhines. *Journal of Human Evolution* **17**, 339-359.

Bernor, R. L. and Hussain, S. T. (1985). An assessment of the systematic, phylogenetic and biogeographic relationships of Siwalik hipparionine horses. *Journal of Vertebrate Paleontology* **5**, 32-87.

Bestland, E. A. (1990a). Sedimentology and paleopedology of Miocene alluvial deposits at the Paşalar hominoid site, western Turkey. *Journal of Human Evolution* **19**, 363-377.

Bestland, E. A. (1990b). Miocene volcaniclastic deposits and paleosols of Rusinga Island, Kenya. Unpublished Ph.D. thesis, Department of Geological Sciences, University of Oregon, Eugene, 107 pp.

Bethke, C. M. and Altaner, S. P. (1986). Layer-by-layer mechanism of smectite illitization and application to a new rate law. *Clays and Clay Minerals* **34**, 136-142.

Bhimaya, C. P. and Kaul, R. N. (1965). Root systems of four desert tree species. *Annals of the Arid Zone* **4**, 185-194.

Binford, L. R. (1981). *Bones: ancient men and modern myths*. Academic, New York, 320 pp.

Binge, F. W. (1962). Geology of the Kericho area. *Report of the Geological Survey of Kenya* **50**, 67 pp.

Birkeland, P. W. (1984). *Soils and geomorphology*. Oxford, New York, 372 pp.

Bishop, W. W. (1968). The evolution of fossil environments in East Africa. *Transactions of the Leicester Literary and Philosophical Society* **62**, 27-44.

Bishop, W. W. (1978). Geological framework of the Kilombe Acheulian archaeological site, Kenya. In *Geological background to fossil man* (ed. W. W. Bishop), pp. 329-336. Scottish Academic, Edinburgh.

Bishop, W. W., Miller, J. A. and Fitch, F. J. (1969). New potassium-argon age determinations relevant to the Miocene fossil mammal sequence in East Africa. *American Journal of Science* **267**, 669-699.

Bishop, W. W. and Trendall, A. F. (1966). Erosion-surfaces, tectonics and volcanic activity in Uganda. *Quarterly Journal of the Geological Society of London* **122**, 385-420.

Bishop, W. W. and Whyte, F. (1962). Tertiary mammalian faunas and sediments in Karamoja and Kavirondo, East Africa. *Nature* **196**, 1283-1286.

Bloomfield, K. (1973). Economic aspects of Uganda carbonatite complexes. *Overseas Geology and Mineral Resources* **41**, 139-167.

Bogdan, A. V. (1958). Some edaphic vegetational types at Kiboko, Kenya. *Journal of Ecology* **46**, 115-126.

Bonnefille, R. (1984). Cenozoic vegetation and environments of early hominoids in East Africa. In *The evolution of the East Asian environment. Vol. II. Palaeobotany, palaeozoology and palaeoanthropology* (ed. R. O. Whyte), pp. 579-612. Centre of Asian Studies, University of Hong Kong.

Bosák, P., Ford, D. C., Glazek, J., and Horáček, I. (1989). *Paleokarst: a systematic and regional survey*. Elsevier, Amsterdam, 725 pp.

Botha, G. A. (1985). Significance of parent rock type in the development and distribution of soils in the Kruger National Park, south of the Oliphants River. In *Palaeoecology of Africa* (ed. H. J. Deacon), v. 17, pp. 95-108. A. A. Balkema, Rotterdam.

Bowen, G. D. (1983). Microbial determinants of plant nutrient uptake. In *Soils: an Australian perspective* (ed. CSIRO), pp. 693-710. Academic, Melbourne.

Bown, T. M. (1982). Ichnofossils and rhizoliths of the nearshore Jebel Qatrani Formation (Oligocene), Fayum Province, Egypt. *Palaeogeography, Palaeoclimatology, Palaeoecology* **40**, 255-309.

Bown, T. and Beard, K. C. (1990). Systematic lateral variation in the distribution of fossil mammals in alluvial paleosols, lower Eocene, Willwood Formation, Wyoming. In *Dawn of the Age of Mammals in the northern part of the Rocky Mountain Interior, North America* (eds. T. M. Bown and K. D. Rose), pp. 135-151. *Special Paper of the Geological Society of America* **243**.

Bown, T. M. and Kraus, M. J. (1981). Vertebrate fossil-bearing paleosol units (Willwood Formation, northwest Wyoming, U.S.A.): implications for taphonomy, biostratigraphy and assemblage analysis. *Palaeogeography, Palaeoclimatology, Palaeoecology* **34**, 31-56.

Bown, T. M. and Kraus, M. J. (1983). Ichnofossils of the alluvial Willwood Formation (Lower Eocene), Bighorn Basin, northwest Wyoming, U.S.A. *Palaeogeography, Palaeoclimatology, Palaeoecology* **43**, 95-128.

Bown, T. M., Kraus, M. J., Wing, S. L. ,Fleagle, J. G., Tiffney, B. H., Simons, E. L., and Vondra, C. F. (1982). The Fayum primate forest revisited. *Journal of Human Evolution* **11**, 603-632.

Bown, T. M. and Ratcliffe, B. C. (1988). The origin of *Chubutolithes* Ihering, ichnofossils from the Eocene and Oligocene of Chubut Province, Argentina. *Journal of Paleontology* **62**, 163-167.

Brain, C. K., Churcher, C. S, Clark, J. D., Grine, F. E., Shipman, P., Susman, R. L., Turner, A., and Watson, V. (1988). New evidence of early hominoids, their culture and environment from the Swartkrans Cave, South Africa. *South African Journal of Science* **84**, 828-836.

Braithwaite, C. J. R. (1984). Depositional history of the late Pleistocene limestones of the Kenya coast. *Journal of the Geological Society of London* **141**, 685-699.

Braithwaite, C. J. R. (1989). Displacive calcite and grain breakage in sandstones. *Journal of Sedimentary Petrology* **59**, 258-286.

Brammer, H. (1971). Coatings in seasonally flooded soils. *Geoderma* **6**, 5-16.

Brandis, D. (1972). *Illustrations of the forest flora of northwest and central India* (commenced by J. L. Stewart, drawn by W. Fitch). Reprint B. Singh & M. P. Singh, Dehra Dun, India, 5 pp.

Brewer, R. (1976). *Fabric and mineral analysis of soils*. 2nd ed., Krieger, New York, 482 pp.

Brewer, R. and Sleeman, J. R. (1969). The arrangement of constituents in Quaternary soils. *Soil Science* **107**, 435-441.

Brewer, R., Sleeman, J. R., and Foster, R. C. (1983).

The fabric of Australian soils. In *Soils: an Australian perspective* (ed. CSIRO), pp. 439-476. Academic, Melbourne.

Brimhall, G. H., Lewis, C. J., Ague, J. J., Dietrich, W. E., Hampel, J., Teague, T., and Rix, P. (1988). Metal enrichment in bauxites by deposition of chemically mature aeolian dust. *Nature* **333**, 819-824.

Brock, P. W. and MacDonald, R. (1969). Geological environment of the Bukwa mammal fossil locality, eastern Uganda. *Nature* **223**, 593-596.

Brook, G. A., Folkoff, M. E., and Box, E. O. (1983). A world model of soil carbon dioxide. *Earth Surface Processes and Landforms* **8**, 79-88.

Brown, F., Harris, R., Leakey, R., and Walker, A. (1985). Early *Homo erectus* skeleton from west Lake Turkana, Kenya. *Nature* **316**, 788-792.

Buehrer, T. F., Martin, W. P. and Parks, R. Q. (1939). The oxidation-reduction potential of alkaline calcareous soils in relation to puddling and organic matter decomposition. *Journal of the American Society of Agronomy* **31**, 903-914.

Buol, S. W., Hole, F. D. and McCracken, R. D. (1980). *Soil genesis and classification* 2nd ed., Iowa State University Press, Ames, 406 pp.

Burbank, D. W. (1983). The chronology of inter-montane-basin development in the northwestern Himalaya and the evolution of the Northwest Syntaxis. *Earth and Planetary Science Letters* **64**, 77-92.

Burbank, D. W. (1988). Stratigraphic keys to the timing of thrusting in terrestrial foreland basins. In *New perspectives in basin analysis* (eds. K. L. Kleinspehn and C. Paola), pp. 331-351. Springer, New York.

Burbank, D. W., Beck, R. A., Raynolds, R. G. H., Hobbs, R., and Tahirkheli, R. A. K. (1988). Thrusting and gravel progradation in foreland basins: a test of post-thrusting gravel dispersal. *Geology* **16**, 1143-1146.

Burbank, D. W. and Raynolds, R. G. H. (1984). Sequential late Cenozoic structural disruption of the northern Himalayan foredeep. *Nature* **311**, 114-119.

Burbank, D. W., Raynolds, R. G. H. and Johnson, G. D. (1986). Late Cenozoic tectonics in the northwestern Himalayan foredeep. II. Eastern limb of the Northwest Syntaxis and regional synthesis. In *Foreland basins* (eds. P. A. Allen and P. Homewood). *Special Publication of the International Association of Sedimentologists* **8**, 293-306.

Burkill, H. M. (1985). *The useful plants of west tropical Africa. Vol. 1. Families A-D.* Royal Botanical Gardens, Kew, 960 pp.

Burnham, L. (1978). Survey of social insects in the fossil record. *Psyche* **48**, 85-133.

Butler, P. M. (1984). Macroscelidea, Insectivora and Chiroptera from the Miocene of East Africa. *Palaeovertebrata* **14**, 117-198.

Button, A. and Tyler, N. (1981). The character and significance of Precambrian paleoweathering and erosion surfaces in southern Africa. In *Economic geology, seventy fifth anniversary volume 1905-1980* (ed. B. J. Skinner), pp. 686-709. Economic Geology, El Paso, Texas.

Butzer, K. W. (1977). Environment, culture and human evolution. *American Scientist* **65**, 572-584.

Cahen, L. and Snelling, N. J. (1984). *The geochronology and evolution of Africa*. Clarendon Press, Oxford, England, 512 pp.

Cas, R. A. F. and Wright, J. V. (1987). *Volcanic successions, modern and ancient*. Allen & Unwin, London, 528 pp.

Champion, H. G. and Seth, S. K. (1968). *A revised survey of the forest types of India*. Government of India, Delhi, 404 pp.

Chaney, R. W. (1933). A Tertiary flora from Uganda. *Journal of Geology* **41**, 702-709.

Chaney, R. W. and Sanborn, E. I. (1933). The Goshen flora of central Oregon. *Publication of the Carnegie Institution of Washington* **439**, 237 pp.

Chaudhri, R. S. (1975). Sedimentology and genesis of the Cenozoic sediments of northwestern Himalayas (India). *Geologische Rundschau* **64**, 958-977.

Chaudhri, R. S. (1984). Heavy mineral zones in the Cenozoic sequence of the Punjab and Kumaon Himalaya. In *Current trends in geology. Vol. 5. Sedimentary geology of the Himalaya* (ed. R. A. K. Srivastava), pp. 1-14. Today's and Tomorrow's, New Delhi, India.

Chesters, K. I. M. (1957). The Miocene flora of Rusinga Island, Lake Victoria, Kenya. *Palaeontographica* **B101**, 30-71.

Choquette, P. W. and James, N. P. (1987) Introduction. In *Paleokarst* (eds. N. P. James and P. W. Choquette), pp. 1-21. Springer, New York.

Churcher, C. S. (1970). Two upper Miocene giraffids from Fort Ternan, Kenya, East Africa: *Palaeotragus primaevus* n. sp. and *Samotherium africanum* n. sp. In *Fossil vertebrates of Africa* (eds. L. S. B. Leakey and R. J. G. Savage), v. 2, pp. 1-106. Academic, London.

Churchill, W. L. S. (1942). A new experience-victory. In *Winston S. Churchill: his complete speeches, 1897-1963. Volume VI. 1935-1942* (ed. R. R. James), pp.6692-6695. Chelsea House & R. R. Bowker, New York (1969).

Clauer, N. and Hoffert, M. (1985). Sr isotopic constraints for the sedimentation rate of the deep sea red clays in the southern Pacific Ocean. In *The chronology of the geological record* (ed. N. J. Snelling). *Memoir of the Geological Society of London* **10**, 290-296.

Clayton, W. D. (1970). Gramineae. Part 1. In *Flora of tropical East Africa* (eds. E. Milne-Redhead and R. M. Polhill), 176 pp. Crown Agents for Oversea Governments and Administrations, London.

Clayton, W. D., Phillips, S. M., and Renvoize, S. A. (1974). Gramineae. Part 2. In *Flora of tropical East Africa* (eds. E. Milne-Redhead and R. M. Polhill), pp. 177-450. Crown Agents for Oversea Governments and Administrations, London.

Clayton, W. D. and Renvoize, S. A. (1982). Gramineae. Part 3. In *Flora of tropical East Africa* (ed. R. M. Polhill), pp. 451-898. A. A. Balkema, Rotterdam.

Cohen, A. S. (1982). Paleoenvironments of root casts from the Koobi Fora Formation, Kenya. *Journal of Sedimentary Petrology* **52**, 401-414.

Cole, M. M. (1986). *The savannas*. Academic, London, 438 pp.

Cole, S. (1975). *Leakey's luck - the life of Louis Seymour Bazett Leakey*. Harcourt, Brace & Jovanovich, London, 448 pp.

Collins, D. A. and McGrew, W. C. (1988). Habitats of three groups of chimpanzees (*Pan troglodytes*) in western Tanzania compared. *Journal of Human Evolution* **17**, 553-574.

Collins, J. F. and Larney, F. (1983). Micromorphological changes with advancing pedogenesis in some Irish alluvial soils. In *Soil micromorphology. Vol. 1. Techniques and applications* (eds. P. Bullock and C. P. Murphy), pp. 297-301. A. B. Academic, Berkhamsted, England.

Collinson, M. E. (1983). Revision of East African Miocene floras: a preliminary report. *Newsletter of the International Association for Angiosperm Paleobotany* **8(1)**, 4-10.

Cope, M. J. and Chaloner, W. G. (1985). Wildfire: an interaction of biological and physical processes. In *Geological factors and the evolution of plants* (ed. B. H. Tiffney), pp. 257-277. Yale University, New Haven, Connecticut.

Cotter, G. de P. (1933). The geology of the Attock district west of longitude 72° 45' E. *Memoirs of the Geological Survey of India* **55**, 63-161.

Courty, M. A. and Féderoff, N. (1985). Micromorphology of recent and buried soils in a semiarid region of northwestern India. *Geoderma* **35**, 287-332.

Covert, H. H., and Kay, R. F. (1981). Dental microwear and diet: implications for determining feeding behaviors of extinct primates, with a comment on the dietary pattern of *Sivapithecus*. *American Journal of Physical Anthropology* **55**, 331-336.

Dale, I. R. and Greenway, P. J. (1961). *Kenya trees and shrubs*. Buchanans Kenya Estates, Nairobi, 654 pp.

Dalrymple, G. B. (1979). Critical tables for conversion of K-Ar ages from old to new constants. *Geology* **7**, 558-560.

Dart, R. A. (1957). The osteodontokeratic culture of *Australopithecus prometheus*. *Memoir of the Transvaal Museum* **10**, 105 pp.

Darwin, C. (1872). *The descent of man, and selection in relation to sex*. 2 vols. John Murray, London, 423 and 475 pp.

Davies, R. G. and Crawford, A. R. (1971). Petrography and age of the rocks of the Bulland Hill, Kirana Hills, Sargodha district, Pakistan. *Geological Magazine* **108**, 235-246.

Davis, R.A. (1983) *Depositional systems*. Prentice-Hall, Englewood Cliffs, New Jersey, 669 pp.

Dawson, J. B. (1962). The geology of Oldoinyo Lengai. *Bulletin volcanologique* **24**, 349-387.

Dawson, J. B. (1964). Carbonate tuff cones in northern Tanganyika. *Geological Magazine* **101**, 129-137.

Dawson, J. B., Bowden, P., and Clark, G. C. (1968). Activity of the carbonatite volcano Oldoinyo Lengai, 1966. *Geologische Rundschau* **57**, 865-879.

Dawson, J. B., Garson, M. S., and Roberts, B. (1987). Altered former alkali carbonatite lava from Oldoinyo Lengai, Tanzania: inferences for calcite carbonatite lavas. *Geology* **15**, 765-768.

Deans, T. and Roberts, B. (1984). Carbonatite tuffs and lava clasts of the Tinderet foothills, western Kenya: a case study of calcified natrocarbonatites. *Journal of the Geological Society of London* **141**, 563-580.

de Bonis, L., Bouvrain, G., Geraads, D., and Koufos, G. (1990). New hominid skull material from the Late Miocene of Macedonia in northern Greece. *Nature* **345**, 712-714.

Degens, E. T., von Herzen, R. P., Wong, H.-K., Deuser, W., and Jannasch, H. W. (1973). Lake Kivu: structure, chemistry and biology of an East African rift lake. *Geologische Rundschau* **62**, 245-277.

Deshpande, B. D. and Sharma, K. K. (1975). Root architecture and sand-binding capacity of plants growing in sand dunes and sandy plains. *Geobios* **2**, 129-132.

Deshpande, S. B., Fehrenbacher, J. B., and Beavers, A. H. (1971). Mollisols of the tarai region of Uttar Pradesh, northern India. 1. Morphology and mineralogy. *Geoderma* **6**, 179-193.

Deshpande, S. B., Fehrenbacher, J. B., and Ray, S. W. (1971). Mollisols of the tarai region of Uttar Pradesh, northern India. 2. Genesis and classification. *Geoderma* **6**, 195-201.

de Terra, H. and de Chardin, P. T. (1936). Observations on the upper Siwalik Formation and late Pleistocene deposits in India. *Proceedings of the American Philosophical Society* **76**, 791-822.

de Terra, H. and Paterson, T. T. (1939). Studies on the Ice Age in India and associated human cultures. *Publication of the Carnegie Institution of Washington* **493**, 354 pp.

Dewey, J. F., Shackleton, R. M., Chengfa, C., and Yiyin, S. (1988). The tectonic evolution of the Tibetan Plateau. *Philosophical Transactions of the Royal Society of London* **A327**, 379-413.

de Villiers, J. M. (1965). The genesis of some Natal soils. II, Estcourt, Avalon, Bellevue, and Rensburgspruit Series. *South African Journal of Agricultural Sciences* **8**, 507-524.

de Wit, H. A. (1978). *Soils and grassland types of the Serengeti Plain (Tanzania).* Center for Agricultural Publishing and Documentation, Wageningen, Netherlands, 300 pp.

Dokuchaev, V. V. (1883). Russian chernozem (Russkii chernozem). In *Collected writings (Sochineya)*, v. 3. (Trans. N. Kaner) Israel Program for Scientific Translation, Jerusalem (1967).

Downie, C. and Wilkinson, P. (1962). The explosion craters of Basotu, Tanganyika Territory. *Bulletin Volcanologique* **24**, 391-420.

Drake, R. L., Van Couvering, J. A., Pickford, M., Curtis, G. H. and Harris, J. A. (1988). New chronology for the early Miocene mammalian faunas of Kisingiri, western Kenya. *Journal of the Geological Society of London* **145**, 479-491.

Duchafour, P. (1982). *Pedology: pedogenesis and classification*. (Trans. T. R. Paton) Allen & Unwin, London, 448 pp.

Duffin, M. E., Lee, M., Klein, G. de V., and Hay, R. L. (1989). Potassic diagenesis of Cambrian sandstones and Precambrian granitic basement in UPH3 deep hole, upper Mississippi Valley, U.S.A. *Journal of Sedimentary Petrology* **59**, 848-861.

Dugas, D. P. (1989). Middle Miocene fossil grasses and associated paleosol from Fort Ternan, Kenya: geology, taphonomy and taxonomy. Unpublished M.Sc. thesis, Interdisciplinary Program, University of Oregon, Eugene, 118 pp.

Eberl, D. D., Srodon, J., Kralik, M., Taylor, B. E., and Peterman, Z. E. (1990). Ostwald ripening of clays and metamorphic minerals. *Science* **248**, 474-477.

Eggeling, W. (1951). *The indigenous trees of the Uganda Protectorate.* 2nd ed., Government, Entebbe, Uganda, 491 pp.

Ellis, B. S. (1950). A guide to some Rhodesian soils. II. A note on mopane soils. *Rhodesian Agricultural Journal* **47**, 49-61.

Ellis, R. P. (1987). A review of comparative leaf blade anatomy in the systematics of Poaceae: the past twenty five years. In *Grass systematics and evolution* (eds T. R. Soderstrom, K. W. Hilu, C. S. Campbell and M. E. Barkworth), pp. 3-10. Smithsonian Institution, Washington, D.C.

Emerson, A. E. (1952). Geographical origins and dispersion of termite genera. *Fieldiana Zoology* **37**, 465-521.

Estes, R. (1962). A fossil gerrhosaur from the Miocene of Kenya. *Breviora* **158**, 1-9.

Evans, E. M. N., Van Couvering, J. A. H., and Andrews, P. (1981). Palaeoecology of Miocene sites in western Kenya. *Journal of Human Evolution* **10**, 99-116.

Evernden, J. F. and Curtis, G. H. (1965). The potassium-argon dating of Late Cenozoic rocks in East Africa and Italy. *Current Anthropology* **6**, 343-384.

Falconer, H. (1868). *Palaeontological memoirs and notes. Volume 1. Fauna antiqua sivalensis*. Edited by C, Murchison. Robert Hardwicke, London, 590 pp.

Farah, A., Mirza, M. A., Ahmad, M. A., and Butt, M. H. (1977). Gravity field of the buried shield in the Punjab plains, Pakistan. *Bulletin of the Geological Society of America* **88**, 1147-1155.

Fatmi, A. N. (1974). Lithostratigraphic units of the Kohat-Potwar Province, Indus Basin, Pakistan. *Memoir of the Geological Survey of Pakistan* **10**, 80 pp.

Fawley, A. P. and James, T. C. (1955). A pyrochlore (columbium) carbonatite, southern Tanganyika. *Economic Geology* **50**, 571-585.

Feakes, C. R. and Retallack, G. J. (1988). Recognition and characterization of fossil soils developed on alluvium: a Late Ordovician example. In *Paleosols and weathering through geologic time* (eds. J. Reinhardt and W. R. Sigleo). *Special Paper of the Geological Society of America* **216**, 35-48.

Fisher, D. C. (1984). Taphonomic analysis of late Pleistocene mastodon occurrences: evidence of butchery by North American Paleo-Indians. *Paleobiology* **10**, 338-357.

Fisher, G. C. and Yam, O.-L. (1984). Iron mobilization by heathland plant extracts. *Geoderma* **32**, 339-345.

Fisher, R. V. and Schmincke, H.-U. (1984). *Pyroclastic rocks*. Springer, New York, 472 pp.

Fitch, F. J., Hooker, P. J., and Miller, J. A. (1978).

Geochronological problems and radioisotopic dating in the Gregory Rift Valley of East Africa. In *Geological background to fossil man* (ed. W. W. Bishop), pp. 441-461. Scottish Academic, Edinburgh.

Fleagle, J. G. (1988). *Primate adaptation and evolution.* Academic Press, San Diego, 486 pp.

Fleagle, J. G. and Kay, R. F. (1983). New interpretations of the phyletic position of Oligocene hominoids. In *New interpretations of ape and human ancestry* (eds. R. L. Ciochon and R. S. Corruccini), pp. 181-210. Plenum Press, New York.

Fletcher, T. B. (1914). *Some South Indian insects.* Government Printer, Madras, 440 pp.

Flynn, L. J., Pilbeam, D., Jacobs, L. L., Barry, J. C., Behrensmeyer, A. K., and Kappelman, J. W. (1990). The Siwaliks of Pakistan: time and faunas in a Miocene terrestrial setting. *Journal of Geology* **98**, 589-604.

Foley, R. (1987). *Another unique species.* Longman, London, 313 pp.

Foley, R. A. and Lee, P. C. (1989). Finite social space, evolutionary pathways and reconstructing hominid behavior. *Science* **243**, 901-906.

Folk, R. L. (1965). Some aspects of recrystallization in ancient limestones. In *Dolomitization and limestone diagenesis* (eds. L. C. Pray and R. C. Murray). *Special Paper Society of Economic Paleontologists and Mineralogists Tulsa* **13**, 14-48.

Folk, R. L., Roberts, H. H., and Moore, C. H. (1973). Black phytokarst from Hell, Cayman Islands, British West Indies. *Bulletin of the Geological Society of America* **84**, 2351-2360.

Food and Agriculture Organization (1974). *Soil map of the world. Vol. I. Legend.* UNESCO, Paris, 59 pp.

Food and Agriculture Organization (1977a). *Soil map of the world. Vol. VI. Africa.* UNESCO, Paris, 299 pp.

Food and Agriculture Organization (1977b). *Soil map of the world. Vol. VII. South Asia.* UNESCO, Paris, 117 pp.

Food and Agriculture Organization (1979). *Soil map of the world. Vol. IX. Southeast Asia.* UNESCO, Paris, 149 pp.

Forman, R. T. and Godron, M. (1986). *Landscape ecology.* John Wiley, New York, 619 pp.

Frakes, L. A. (1979). *Climates through geological time.* Elsevier, Amsterdam, 310 pp.

Frenguelli, J. (1938a). Nidi fossili di scarabeidi e vespidi. *Bolletino della Societa Geologica Italiana* **57**, 77-96.

Frenguelli, J. (1938b). Bolas de escarabeides y nidos de véspidos fósiles. *Physis* **12**, 348-352.

Frenguelli, J. (1939). Nidos fósiles de insectos en el Terciario del Neuquén y Río Negro. *Notas del Museo de La Plata* **4**, *Paleontologia* **18**, 379-402.

Frey, M. (1987). Very low grade metamorphism of clastic sedimentary rocks. In *Low temperature metamorphism* (ed. M. Frey), pp. 9-58. Blackie, Glasgow, Scotland.

Freytet, P. and Plaziat, J.-C. (1982). *Continental carbonate sedimentation and pedogenesis-Late Cretaceous and Early Tertiary of southern France.* E. Schweitzerbart'sche, Stuttgart, West Germany, 213 pp.

Frick, C. (1986). The Phalarborwa syenite intrusions along the west-central boundary of the Kruger National Park. *Koedoe* **29**, 45-58.

Friedman, G. M. (1958). Determination of sieve size distribution from thin section data for sedimentary petrological studies. *Journal of Geology* **66**, 394-416.

Gabunia, L. K. and Chochieva, K. I. (1982). Coevolution of the *Hipparion* fauna and vegetation in the Paratethys Region. *Evolutionary Theory* **6**, 1-13.

Gentry, A. W. (1970). The Bovidae (Mammalia) of the Fort Ternan fossil fauna. In *Fossil vertebrates of Africa* (eds. L. S. B. Leakey and R. J. G. Savage), v. 2, pp. 243-324. Academic, London.

Gentry, A. W. (1987). Ruminants from the Miocene of Saudi Arabia. In *Miocene geology and palaeontology of Ad Dabtiyah, Saudi Arabia* (ed. P. J. Whybrow). *Bulletin of the British Museum of Natural History Geology* **41**, 433-439.

Gertenbach, W. P. D. (1983). Landscapes of the Kruger National Park. *Koedoe* **26**, 9-121.

Gile, L. H., Hawley, J. W., and Grossman, J. B. (1980). Soils and geomorphology in the Basin and Range area of southern New Mexico-guidebook to the Desert Project. *Memoir of the New Mexico Bureau of Mines and Mineral Resources* **39**, 222 pp.

Gile, L. H., Peterson, F. F., and Grossman, R. B. (1966). Morphological and genetic sequences of carbonate acccumulation in desert soils. *Soil Science* **101**, 347-360.

Gilg, E. (1960). Dipterocarpaceae. In *Die natürliche Pflanzenfamilien* (ed. A. Engler), v. 21, pp. 237-269. Duncker & Humblot, Berlin.

Gill, W. D. (1952). The stratigraphy of the Siwalik Series in the northern Potwar Plateau, Punjab, Pakistan. *Quarterly Journal of the Geological Society of London* **107**, 375-394.

Glinka, K. D. (1931). *Treatise on soil science (Pochvovedeniye)* 4th ed. (Trans. A. Gourevich). Israel Program for Scientific Translation, Jerusalem, 674 pp. (1963).

Glover, P. E. (1950). The root systems of some British Somaliland plants. I. *East African*

Agricultural Journal **16**, 98-112.

Glover, P. E. (1951a). The root systems of some British Somaliland plants. II. *East African Agricultural Journal* **16**, 154-173.

Glover, P. E. (1951b). The root systems of some British Somaliland plants. III. *East African Agricultural Journal* **16**, 205-217.

Glover, P. E. (1952). The root systems of some British Somaliland plants. IV. *East African Agricultural Journal* **17**, 38-50.

Goldich, S. S. (1938). A study in rock weathering. *Journal of Geology* **46**, 17-58.

Gorman, T. P. (1972). *A glossary in English, Kiswahili, Kikuyu and Dholuo*. Cassell, London, 111 pp.

Gowlett, J. A. J., Harris, J. W. K., Walton, D., and Wood, B. A. (1981). Early archeological sites, hominid remains and traces of fire from Chesowanja, Kenya. *Nature* **294**, 125-129.

Gregory, J. W. (1921). *The rift valleys and geology of East Africa.* Seeley Service, London, 479 pp.

Gromova, V. I. (1962). *Osnovy paleontologii. Volume 13 Mlekopitayuschie (Fundamentals of paleontology. Volume 13. Mammals).* Academia Nauk, Moscow, 421 pp.

Gupta, R. N. (1961). Clay minerals in soils of the lower Gangetic Basin of Uttar Pradesh. *Journal of the Indian Society for Soil Science* **9**, 141-150.

Gupta, R. N. (1968). Clay mineralogy of the Indian Gangetic alluvium of Uttar Pradesh. *Journal of the Indian Society for Soil Science* **16**, 115-127.

Gupta, R. N., Agarwal, R. R., and Mehrotra, C. L.(1957). Genesis and pedochemical characteristics of *Dhankar*: gray hydromorphic soils in lower Gangetic plains of Uttar Pradesh. *Journal of the Indian Society for Soil Science* **5**, 5-12.

Halffter, G. and Edmonds, W. D. (1982). *The nesting behavior of dung beetles (Scarabaeinae).* Instituto de Ecologia, Mexico, 176 pp.

Hamilton, A. C. (1968). Some plant fossils from Bukwa. *Uganda Journal* **32**, 157-164.

Hamilton, A. C. (1982). *Environmental history of East Africa.* Academic, New York, 328 pp.

Hamilton, W. R. (1978). Fossil giraffes from the Miocene of Africa and a revision of the phylogeny of the Giraffoidea. *Philosophical Transactions of the Royal Society of London* **B283**, 165-229.

Häntzschel, W. (1975). Trace fossils and problematica. In *Treatise of invertebrate paleontology. Part W. Miscellanea, Supplement 1* (eds. R. C. Moore and C. Teichert), 269 pp. Geological Society of America and University of Kansas Press, Boulder and Lawrence.

Haq, B. U., Hardenbol, J. and Vail, P. R. (1987). Chronology of fluctuating sea levels since the Triassic. *Science* **245**, 1156-1167.

Harden, J.W. (1982). A quantitative index of soil development from field descriptions: examples from a chronosequence in central California. *Geoderma* **28**, 1-28.

Harms, H. (1960). Meliaceae, Akaniaceae. In *Die natürliche Pflanzenfamilien* (ed. A. Harms), V. 19bI, 183 pp. Duncker & Humblot, Berlin.

Harris, J.M. (1978). Deinotherioidea and Barytherioidea. In *Evolution of African mammals* (eds. V. J. Maglio and H. B. S. Cooke), pp. 315-332. Harvard University, Cambridge, Massachusetts.

Harris, T. M. (1981). Burnt ferns in the English Wealden. *Proceedings of the Geologists Association* **92**, 47-58.

Harrison, T. (1981). New finds of small fossil apes from the Miocene locality at Koru in Kenya. *Journal of Human Evolution* **10**, 129-137.

Harrison, T. (1986). New fossil anthropoids from the Miocene of East Africa and their bearing on the Oreopithecidae. *American Journal of Physical Anthropology* **71**, 265-284.

Harrison, T. (1989). New species of *Micropithecus* from the Middle Miocene of Kenya. *Journal of Human Evolution* **18**, 537-557.

Harrop, J. F. (1960). The soils of the Western Province of Uganda. *Memoirs of the Research Division of the Department of Agriculture, Uganda Protectorate, Series 1, Soils* **6**, 27 pp.

Hawkins, R. E. (ed.) (1986). *Encyclopedia of Indian natural history.* Oxford University, Delhi, 620 pp.

Hay, R. L. (1976). *Geology of Olduvai Gorge.* University of California, Berkeley, 203 pp.

Hay, R. L. (1978). Melilitite-carbonatite tuffs in the Laetolil beds of Tanzania. *Contributions to Mineralogy and petrology* **67**, 357-367.

Hay, R. L. (1983). Natrocarbonatite tephra of Kerimasi volcano, Tanzania. *Geology* **11**, 599-602.

Hay, R. L. (1986). Role of tephra in the preservation of fossils in Cenozoic deposits of East Africa. In *Sedimentation in the African rifts* (eds. L. E. Frostick, R. W. Renaut, I. Reid and J. J. Tiercelin). *Special Publication of the Geological Society of London* **25**, 339-344.

Hay, R. L. (1989). Holocene carbonatite-nephelinite tephra deposits of Oldoinyo-Lengai, Tanzania. *Journal of Volcanology and Geothermal Research* **37**, 77-91.

Hay, R. L. and O'Neil, J. R. (1983). Natrocarbonatite tuffs in the Laetoli beds of Tanzania and the Kaiserstuhl in Germany. *Contributions to Mineralogy and Petrology* **82**, 403-406.

Hay, R. L. and Reeder, R. J. (1978). Calcretes of

Olduvai Gorge and the Ndolanya Beds of northern Tanzania. *Sedimentology* **25**, 649-673.

Hays, J. D., Imbrie, J., and Shackleton, N. J. (1976). Variations in the Earth's orbit: pacemaker of the Ice Ages. *Science* **194**, 1121-1132.

Heller, J. (1984). Deserts as refugia for relict land snails, In *Worldwide snails: biogeographical studies on nonmarine mollusca* (eds. A. Solem and A. C. Van Bruggen), pp. 107-123. Brill, Leiden, Netherlands

Hemphill, W. R. and Kidwai, A. H. (1973). Stratigraphy of the Bauru and Dera Ismail Khan areas, Pakistan. *Professional Paper of the U.S. Geological Survey* **716B**, 36 pp.

Herrera, C. M. (1985). Grass/grazer radiations: an interpretation of silica-body diversity. *Oikos* **45**, 446-447.

Hewes, G. W. (1961). Food transportation and the origins of bipedalism. *American Anthropologist* **63**, 687-710.

Hewes, G. W. (1964). Hominid bipedalism: independent evidence for food carrying theory. *Science* **146**, 416-418.

Hill, A. and Ward, S. (1988). Origin of the Hominidae: the record of African large hominoid evolution between 14 My and 4 My. *Yearbook of Physical Anthropology* **31**, 49-83.

Hillenius, D. (1978). Notes on chameleons. IV. A new chameleon from the Miocene of Fort Ternan, Kenya (Chameleonidae, Reptilia). *Beaufortia* **28(343)**, 9-15.

Hirsch, K. and Harris, J. (1989). Fossil eggs from the lower Miocene Legetet Formation of Koru, Kenya: snail or lizard ? *Historical Biology* **3**, 61-78.

Holland, H. D. (1984). *The chemical evolution of the atmosphere and oceans*. Princeton University, Princeton, New Jersey, 582 pp.

Holmes, D. A. and Western, S. (1969). Soil texture patterns in the alluvium of the Lower Indus plains. *Journal of Soil Science* **20**, 23-37.

Hooijer, D. A. (1968). A rhinoceros from the late Miocene of Fort Ternan, Kenya. *Zoologische Mededelingen Rijksmuseum van Naturlijke Historie te Leiden* **43**, 77-92.

Hunt, J. M. (1979). *Petroleum geochemistry and geology*. Freeman, San Francisco, 617 pp.

Hurelbrink, R. L. and Fehrenbacher, J. B. (1970). Soils and stratigraphy of the Gola River fan of Uttar Pradesh, India. *Proceedings of the Soil Science Society of America* **34**, 911-916.

Hussain, S. T., Munthe, J., Shah, S. M. I., West, R. M. I., and Lukacs, J. R. (1979). Neogene stratigraphy and fossil vertebrates of the Daud Khel area, Mianwali district, Pakistan. *Memoirs of the Geological Survey of Pakistan* **13**, 27 pp.

Hutchinson, J., Dalziel, J. M. and Keay, R. W. J. (1958). Connaraceae. In *Flora of west tropical Africa* (eds. W. B. Turrill and E. Milne-Redhead), pp. 297-828. Crown Agents for Oversea Governments and Administrations, London.

Iljinskaya, I. A. (1988). Contributions to the characterization and origin of the Turgai flora of the U.S.S.R. *Tertiary Research* **9**, 169-180.

Irvine, F. R. (1961). *Woody plants of Ghana*. Oxford University, London, 868 pp.

Isaac, G. L. (1981). Emergence of human behavior patterns. *Philosophical Transactions of the Royal Society of London* **B292**, 177-188.

Ishida, H., Pickford, M., Nakaya, H., and Nakano, Y. (1984). Fossil anthropoids from Nachola and Samburu Hills, Samburu district, Kenya. *Supplementary Issue, African Study Monographs, Kyoto University* **2**, 73-85.

Jackson, M. B. (ed.) (1986). *New root formation in plants and cuttings*. Martinus Nijhoff, Dordrecht, Netherlands, 265 pp.

Jackson, M. L., Tyler, S. A., Willis, A. L., Bourbeau, G. A., and Pennington, R. P. (1948). Weathering sequence of clay size minerals in soils and sediments. I. Fundamental generalizations. *Journal of Physical and Colloidal Chemistry* **52**, 1237-1261.

Jacobs, B. F. and Kabuye, C. H. S. (1987). A middle Miocene (12.2 m.y. old) forest in the East African Rift Valley, Kenya. *Journal of Human Evolution* **16**, 147-155.

Jacobs, L. L., Anyonge, W., and Barry, J. C. (1987). A giant tenrecid from the Miocene of Kenya. *Journal of Mammalogy* **68**, 10-16.

Jager, T. J. (1982). *Soils of the Serengeti woodlands, Tanzania.* Centre for Agricultural Publishing and Documentation, Wageningen, Netherlands, 239 pp.

Jaillard, B. (1987). *Les structures rhizomorphes calcaires: modèle de réorganization de minéraux du sol par les racines.* Institut National de la Recherche Agronomique, Laboratoire de Science du Sol, Montpellier, France, 221 pp.

Jenik, J. (1978). Roots and root systems in tropical trees: morphologic and ecologic aspects. In *Tropical trees as living systems* (eds. P. B. Tomlinson and M. H. Zimmermann), pp. 323-349. Cambridge University Press, Cambridge, England.

Jennings, D. J. (1971). Geology of the Molo area. *Report of the Geological Survey of Kenya* **86**, 39 pp.

Jennings, J. N. (1985). *Karst geomorphology.* Blackwell, Oxford, England, 293 pp.

Jenny, H. (1941). *Factors of soil formation*. McGraw-

Hill, New York, 281 pp.

Johanessen, C. L., Davenport, W. A., and McWilliams, S. (1971). The vegetation of the Willamette Valley. *Annals of the Association of American Geographers* **61**, 282-302.

Johnson, G. D. (1977). Paleopedology of *Ramapithecus*-bearing sediments, north India. *Geologische Rundschau* **66**, 192-216.

Johnson, G. D., Opdyke, N. D., Tandon, S. K., and Nanda, A. C. (1983). The magnetic polarity stratigraphy of the Siwalik Group at Haritalyangar (India) and a new last appearance datum for *Ramapithecus* and *Sivapithecus* in Asia. *Palaeogeography, Palaeoclimatology, Palaeoecology* **44**, 223-249.

Johnson, G. D., Raynolds, R. G., and Burbank, D. W. (1986). Late Cenozoic tectonics and sedimentation in the north-western Himalayan foredeep. I. Thrust ramping and associated deformation in the Potwar region. In *Foreland Basins* (eds. P. A. Allen and P. Homewood). *Special Publication of the International Association of Sedimentologists* **8**, 273-291.

Johnson, G. D., Rey, P. H., Ardrey, R. H., Visser, C. F., Opdyke, N. D., and Tahirkheli, R. A. K. (1981). Paleoenvironments of the Siwalik Group, Pakistan and India. In *Hominid sites: their geologic settings* (eds. G. Rapp and C. F. Vondra), pp. 197-254. Westview, Boulder, Colorado.

Johnson, G. D., Zeitler, P., Naeser, C. W., Johnson, N. M., Summers, D. M., Frost, C. D., Opdyke, N. D., and Tahirkheli, R. A. K. (1982). The occurrence and fission-track ages of late Neogene and Quaternary volcanic sediments, Siwalik Group, northern Pakistan. *Palaeogeography, Palaeoclimatology, Palaeoecology* **37**, 63-93.

Johnson, N. M., Sheikh, K. A., Dawson-Saunders, E., and McRae, L. E. (1988). The use of magnetic-reversal time lines in stratigraphic analysis: a case study in measuring variability in sedimentation rates. In *New perspectives in basin analysis* (eds. K. L. Kleinspehn and C. Paola), pp. 189-200. Springer, New York.

Johnson, N. M., Stix, J., Tauxe, L., Cerveny, P. F., and Tahirkheli, R. A. K. (1985). Palaeomagnetic chronology, fluvial processes and tectonic implications of the Siwalik deposits near Chinji village, Pakistan. *Journal of Geology* **93**, 27-40.

Johnson, R. W. (1969). Volcanic geology of Mount Suswa, Kenya. *Philosophical Transactions of the Royal Society of London* **A265**, 385-412.

Johnston, M. C. (1972). Rhamnaceae. In *Flora of tropical East Africa* (eds. E. Milne-Redhead and R.M. Polhill), 40 pp. Crown Agents for Oversea Governments and Administrations, London.

Jolly, C. (1970). The seed-eaters, a new model of hominid differentiation based on a baboon analogy. *Man* **5**, 5-26.

Jones, B. and Pemberton, S. G. (1987). The role of fungi in the diagenetic alteration of spar calcite. *Canadian Journal of Earth Sciences* **24**, 903-914.

Jones, W. B. and Lippard, S. J. (1979). New age determinations and the geology of the Kenya Rift-Kavirondo Rift junction, west Kenya. *Journal of the Geological Society of London* **136**, 693-704.

Joshi, N. V. and Kelkar, B. V. (1952). The role of earthworms in soil fertility. *Indian Journal of Agricultural Science* **22**, 189-196.

Justin-Visentin, E., Nicoletti, M., Tolomeo, L., and Zanettin, B. (1974). Miocene and Pliocene volcanic rocks of the Addis Abbaba-Debra Berhan region (Ethiopia): geopetrographic and radiometric study. *Bulletin Volcanologique* **38**, 237-253.

Kabata-Pendias, A. and Pendias, H. (1984). *Trace elements in soils*. CRC, Boca Raton, Florida, 315 pp.

Kanwar, J. S. (1959). Two dominant clay minerals in Punjab soils. *Journal of the Indian Society for Soil Science* **7**, 249-254.

Kappelman, J. (1991). The paleoenvironment of *Kenyapithecus* at Fort Ternan. *Journal of Human Evolution* **20**, 95-129.

Kedves, M. (1971). Présence de types sporomorphes importants dans les sédiments préquaternaires Egyptiens. *Acta Botanica Academia Sciencia Hungarica* **17**, 371-378.

Keller, J. (1981). Carbonatitic volcanism in the Kaiserstuhl alkalic complex: evidence for highly fluid carbonatitic melts at the Earth's surface. *Journal of Volcanological and Geothermal Research* **9**, 423-431.

Kelley, J. (1986). Species recognition and dimorphism in *Proconsul* and *Rangwapithecus*. *Journal of Human Evolution* **15**, 461-495.

Kennett, J. P. (1982). *Marine geology*. Prentice-Hall, Engelwood Cliffs, 813 pp.

Kent, P. E. (1944). The Miocene beds of Kavirondo, Kenya. *Quarterly Journal of the Geological Society of London* **100**, 85-118.

Kerfoot, O. (1963). The root systems of tropical forest trees. *Commonwealth Forestry Review* **42**, 19-25.

King, B. C., Le Bas, M. J., and Sutherland, D. S. (1972). The history of the alkaline volcanoes and intrusive complexes of eastern Uganda and western Kenya. *Journal of the Geological Society of London* **128**, 173-205.

Kingdon, J. (1971). *East African mammals. Vol. 1.*

Introduction, primates, hyraxes, pangolins, protoungulates and sirenians. Academic, London, 446 pp.

Kingdon, J. (1974a). *East African mammals. Vol. IIA. Insectivores and bats.* University of Chicago, Chicago, 341 pp.

Kingdon, J. (1974b). *East African mammals. Vol. IIB.Hares and rabbits.* University of Chicago, Chicago, 704 pp.

Kingdon, J. (1977). *East African mammals. Vol. IIIA. Carnivores.* University of Chicago, Chicago, 476 pp.

Kingdon, J. (1979). *East African mammals. Vol. IIIB. Large mammals.* University of Chicago, Chicago, 436 pp.

Kling, G. W., Tuttle, M. L., and Evans, W. C. (1989). The evolution of thermal structure and water chemistry in Lake Nyos, Cameroon. *Journal of Volcanology and Geothermal Research* **39**, 151-165.

Kokwaro, J. O. (1986). Anacardiaceae. In *Flora of tropical East Africa* (ed. R. M. Polhill), 55 pp. A.A. Balkema, Rotterdam.

Kong, Z.-C. and Du, N.-Q. (1981). Preliminary study on the vegetation of the Qinghai-Xizang Plateau during Neogene and Quaternary periods. In *Geological and ecological studies of Qinghai-Xizang Plateau* (ed. Liu, D.-S.), v. 1, pp. 239-246. Science Press and Gordon & Breach, Beijing and New York.

Kooistra, M. J. (1982). *Micromorphological analysis and characterization of 70 Benchmark soils of India.* Centre for Agricultural Publishing and Documentation, Wageningen, Netherlands, 778 pp.

Kortlandt, A. (1983). Facts and fallacies concerning Miocene ape habitats. In *New interpretations of ape and human ancestry* (eds. R. L. Ciochon and R. S. Corruccini), pp. 465-514. Plenum, New York.

Kovar-Eder, J. B. (1988). Three dimensional distribution maps for fossil plants: examples from middle to upper Miocene leaf-floras of central Europe. *Tertiary Research* **9**, 213-236.

Krafft, M. and Keller, J. (1989). Temperature measurements in carbonatite lava lakes and flows from Oldoinyo Lengai, Tanzania. *Science* **245**, 168-170.

Kraus, M. J. (1988). Nodular remains of early Tertiary forests, Bighorn Basin, Wyoming. *Journal of Sedimentary Petrology* **58**, 888-893.

Krishna, P. C. and Perumal, S. (1948). Structure in black cotton soils of the Nizamsager Project area, Hyderabad State, India. *Soil Science* **66**, 29-38.

Krynine, P. D. (1937). Petrography and genesis of the Siwalik Series. *American Journal of Science* **34**, 422-446.

Kubiena, W. C. (1970). *Micromorphological features of soil geography.* Rutgers University, New Brunswick, New Jersey, 254 pp.

Kukla, G. J. (1977). Pleistocene land-sea correlations. I. Europe. *Earth Science Reviews* **13**, 307-374.

La Breque, J. L., Kent, D. V., and Cande, S. C. (1977). Revised magnetic polarity time-scale for Late Cretaceous and Cenozoic time. *Geology* **5**, 330-335.

Lakhanpal, R. N. (1966). Some middle Tertiary plant remains from south Kivu, Congo. *Annales de la Musée de l'Afrique Central Belgique, Tervuren, Sciences Geologiques* **52**, 21-30.

Lakhanpal, R. N. (1970). Tertiary floras of India and their bearing on the historical geology of the region. *Taxon* **19**, 675-694.

Lakhanpal, R. N. and Dayal, R. (1966). Lower Siwalik plants from near Jawalamukhi, Punjab. *Current Science* **35**, 209-211.

Lakhanpal, R. N. and Guleria, J. S. (1986). Fossil leaves of *Dipterocarpus* from the Lower Siwalik beds near Jawalamukhi, Himachal Pradesh. *The Palaeobotanist* **35**, 258-262.

Lakhanpal, R. N. and Prakash, U. (1970). Cenozoic plants from Congo. I. Woods from the Miocene of Lake Albert. *Annales de la Musée Royal de l'Afrique Central Belgique, Tervuren, Sciences Geologiques* **64**, 1-20.

La Lumiére, L. P. (1981). Evolution of human bipedalism: a hypothesis about where it happened. *Philosophical Transactions of the Royal Society of London* **B272**, 103-107.

Lambert, D. (1987). *The field guide to early man.* Facts on File, New York, 255 pp.

Lambert, M. (1981). *Prehistoric life encyclopedia.* Rand McNally, Chicago, 138 pp.

Lavocat, R. (1973). Les rongeurs du Miocene d'Afrique orientale. 1. Miocene inferieur. *Mémoires et Travaux de l'Institut Montpellier de l'École Pratique des Haute Études* **1**, 284 pp.

Lavocat, R. (1978). Rodentia and Lagomorpha. In *The evolution of African mammals* (eds. V. J. Maglio and H. B. S. Cooke), pp. 69-89. Harvard University, Cambridge, Massachusetts.

Leakey, L. S. B. (1952). Lower Miocene invertebrates from Kenya. *Nature* **169**, 624-625.

Leakey, L. S. B. (1962). A new lower Pliocene fossil from Kenya. *Annals and Magazine of Natural History* **4**, 689-696.

Leakey, L. S. B. (1968). Bone smashing by late Miocene Hominidae. *Nature* **218**, 528-530.

Leakey, M. D. and Harris, J. M. (eds.) (1987). *Laetoli, a Pliocene site in northern Tanzania.* Clarendon, Oxford, England, 561 pp.

Leakey, R. E., Leakey, M. G., and Behrensmeyer, A.

K. (1978). The hominid catalogue. In *Koobi Fora Research Project. Volume 1. The fossil hominids and an introduction to their context, 1968-1974* (eds. M. G. Leakey and R. E. Leakey), pp. 82-182. Clarendon, Oxford, England.

Leary, R. L. (1981). Early Pennsylvanian geology and paleobotany of the Rock Island County, Illinois, area. *Reports of Investigations of the Illinois State Museum* **37**, 88 pp.

Le Bas, M. J. (1977). *Carbonatite-nephelinite volcanism: an African case history.* John Wiley, London, 347 pp.

Le Bas, M. J. and Dixon, F. (1965). A new carbonatite in the Legetet Hills, Kenya. *Nature* **207**, 68.

Lee, K. E. and Wood, T. G. (1971). *Termites and soils*. Academic, London, 251 pp.

Lemoigne, Y. and Beauchamp, J. (1972). Paléoflores tertiaires de la région Welkite (Éthiopie, Province du Shoa). *Bulletin de la Societé Geologique de la France* **14**, 336-346.

Lemoigne, Y., Beauchamp, J., and Samuel, E. (1974). Étude paléobotanique des dépôts volcaniques d'âge tertiaires, bordures est et ouest du système des rifts éthiopiens. *Geobios* **7**, 267-288.

Lewin, R. (1987). *Bones of contention: controversies in the search for human origins*. Simon & Schuster, New York, 348 pp.

Lewis, G. E. (1937). A new Siwalik correlation. *American Journal of Science* **33**, 192-204.

Leys, C. A. (1983). Volcanic and sedimentary processes during formation of the Saefell tuff-ring, Iceland. *Transactions of the Royal Society of Edinburgh, Earth Sciences* **74**, 15-22.

Lind, E. M. and Morrison, M. E. (1974). *East African vegetation.* Longman, London, 257 pp.

Lippard, S. J. (1973a). The petrology of phonolites from the Kenya Rift. *Lithos* **6**, 217-234.

Lippard, S. J. (1973b). Plateau phonolite lava flows. *Geological Magazine* **110**, 543-549.

Livingstone, F. B. (1962). Reconstructing man's Pliocene pongid ancestor. *American Anthropologist* **64**, 301-395.

Lockwood, J. P. and Lipman, P. W. (1980). Recovery of datable charcoal beneath young lavas: lessons from Hawaii. *Bulletin Volcanologique* **43**, 609-615.

Lovejoy, C. O. (1981). The origin of man. *Science* **211**, 341-350.

Machette, M. N. (1985). Calcic soils of the southwestern United States. In *Soils and Quaternary geology of the southwestern United States* (ed. D. L. Weide). *Special paper of the Geological Society of America* **203**, 1-21.

MacPhee, R. D. E. and Jacobs, L. L. (1986). *Nycticeboides simpsoni* and the morphology, adaptations and relationships of Miocene Siwalik Lorisinae. In *Vertebrates, phylogeny and philosophy: a tribute to George Gaylord Simpson* (eds. K. M. Flanagan and J. A. Lillegraven). *Special Paper, Contributions to Geology of the University of Wyoming* **3**, 131-161.

MacVicar, C. N., De Villiers, J. M., Loxton, R. F., Verster, E., Lambrecht, J. J. N., Merryweather, F. R., Le Roux, J., Van Rooyen, T. H., and Harmse, H. J. Von M. (1977). *Soil classification: a binomial system for South Africa*. Department of Agricultural and Technical Services, Pretoria, South Africa, 150 pp.

Mahaney, W. C. (1989). Quaternary geology of Mount Kenya. In *Quaternary and environmental research on East African mountains* (ed. W. C. Mahaney), pp. 121-140. A. A. Balkema, Rotterdam.

Mahaney, W. C. and Boyer, M. G. (1989). Microflora distributions in paleosols: a method for calculating the validity of radiocarbon dated surfaces. In *Quaternary and environmental research on East African mountains* (ed. W. C. Mahaney), pp.343-352. A. A. Balkema, Rotterdam.

Mahaney, W. C. and Spence, J. R. (1989). Late Holocene sand dune stratigraphy and its paleoclimatic-ecologic significance on the northeast flanks of Mount Kenya. In *Quaternary and environmental research on East African mountains* (ed. W. C. Mahaney), pp.217-229. A. A. Balkema, Rotterdam.

Mai, D. H. (1981). Entwicklung und klimatische Differenzierung der Laubwaldflora Mitteleuropas und Tertiär. *Flora, Jena* **171**, 535-582.

Makinouchi, T., Koyaguchi, T., Matsuda, T., Mitsushio, H., and Ishida, S. (1984). Geology of the Nachola area and the Samburu Hills, west of Baragoi, northern Kenya. *Supplementary Issue, African Study Monographs, Kyoto University* **2**, 15-44.

Mani, M. S. (1974a). Physical features. In *Ecology and biogeography in India* (ed. M. S. Mani), pp.11-59. Junk, The Hague.

Mani, M. S. (1974b). Limiting factors. In *Ecology and biogeography in India* (ed. M. S. Mani), pp.135-158. Junk, The Hague.

Mani, M. S. (1974c). Biogeographical evolution in India. In *Ecology and biogeography in India* (ed. M. S. Mani), pp. 698-724. Junk, The Hague.

Mani, M. S. (1974d). Biogeography of Indo-Gangetic Plain. In *Ecology and biogeography in India* (ed. M.S. Mani), pp. 689-698. Junk, The Hague.

Marbut, C. F. (1935). *Atlas of American agriculture. Part III. Soils of the United States*. U.S. Government Printing Office, Washington DC, 98pp.

Mariano, A. N. (1989). Nature of economic mineralization in carbonatites and related rocks. In *Carbonatites, genesis and evolution* (ed. K. Bell), pp.149-176. Allen & Unwin, London.

Mariano, A. N. and Roeder, P. L. (1983). Kerimasi: a neglected carbonatite volcano. *Journal of Geology* **91**, 449-455.

Martin, L. (1981). New specimens of *Proconsul* from Koru, Kenya. *Journal of Human Evolution* **10**, 139-150.

Mathur, A. K. (1978). Some fossil leaves from the Siwalik Group. *Geophytology* **8**, 98-102.

Mathur, Y. K. (1984). Cenozoic palynofossils, vegetation, ecology and climate of north and northwestern subHimalayan region, India. In *The evolution of the East Asian environment. Vol. II. Palaeobotany, palaeozoology and palaeoanthropology* (ed. R.O. Whyte), pp. 504-551. Centre of Asian Studies, University of Hong Kong.

Mbuvi, J. P. and Njeru, E. B. (1977). Soil resources of the Mau-Narok area, Narok district: a preliminary investigation. *Site Evaluation of the Kenya Soil Survey* **29**, 35 pp.

McDougall, I., Morton, W. H., and Williams, M. A. J. (1975). Age and rates of denudation of Trap Series basalts at Blue Nile Gorge, Ethiopia. *Nature* **254**, 207-209.

McDougall, J. W. (1989). Tectonically-induced diversion of the Indus River west of the Salt Range, Pakistan. *Palaeogeography, Palaeoclimatology, Palaeoecology* **71**, 301-307.

McFadden, L. D. (1988). Climatic influences on rates and processes of soil development in Quaternary deposits of southern California. In *Paleosols and weathering through geologic time: principles and applications* (eds. J. Reinhardt and W. R. Sigleo). *Special Paper of the Geological Society of America* **216**, 153-177.

McFarlane, M.J. (1976). *Laterite and landscape*. Academic, New York, 151 pp.

McGowran, B. (1979). Comment on Early Tertiary tectonism and lateritization. *Geological Magazine* **116**, 227-230.

McNaughton, S. J. and Tarrants, J. L. (1983). Grass leaf silicification: natural selection for an inducible defense against herbivores. *Proceedings of the National Academy of Sciences, U.S.A.* **80**, 790-791.

Mermut, A. R., Arshad, M. A., and St. Arnaud, R. J. (1984). Micropedological study of termite mounds of three species of *Macrotermes* in Kenya. *Journal of the Soil Science Society of America* **48**, 613-620.

Metcalfe, C. R. (1960). *Anatomy of the monocotyledons. Volume 1. Gramineae*. Clarendon, Oxford, England, 731 pp.

Meyer, G. E. (1978). Hyracoidea. In *Evolution of African mammals* (eds. V. J. Maglio and H. B. S. Cooke), pp. 284-314. Harvard University, Cambridge, Massachusetts.

Milton, J. (1627). On the morning of Christ's nativity. In *The complete poetical works of John Milton* (ed. H. F. Fletcher), pp. 47-55. Houghton-Mifflin, Boston (1941).

Miyamoto, M. M., Slightom, J. L., and Goodman, M. (1987). Phylogenetic relations of humans and African apes from DNA sequences in the $\psi\eta$-globin region. *Science* **238**, 369-373.

Mizota, C. and Chapelle, J. (1988). Characterization of some Andepts and andic soils in Rwanda, central Africa. *Geoderma* **41**, 193-209.

Mizota, C., Kawasaki, I., and Wakatsuki, T. (1988). Clay mineralogy and chemistry of seven pedons formed in volcanic ash, Tanzania. *Geoderma* **43**, 131-141.

Mohr, E. C. J. and Van Baren, F. A. (1954). *Tropical soils*. Interscience, New York, 498 pp.

Morgan, E. (1972). *The descent of woman*. Stein & Day, New York, 258 pp.

Moore, D. G., Curray, J. R., Raitt, R. W., and Emmel, F. J. (1974). Stratigraphic-seismic section correlations and implications to Bengal Fan history. In *Initial Reports of the Deep Sea Drilling Project* (ed. A. C. Pimm), v. 22, pp. 403-412. U.S. Government Printing Office, Washington DC.

Morris, S. F. (1979). A new fossil terrestrial isopod with implications for the East African Miocene landform. *Bulletin of the British Museum of Natural History, Geology* **32**, 71-75.

Morton, J. P. (1985). Rb-Sr evidence for punctuated illite/smectite diagenesis in the Oligocene Frio Formation, Texas Gulf Coast. *Bulletin of the Geological Society of America* **96**, 114-122.

Morton, W. H., Mitchell, J. G., Rex, D. C, and Mohr, P. (1979). Riftward younging of volcanic units of the Addis Abbaba region, Ethiopian Rift Valley. *Nature* **280**, 284-288.

Muchena, F. N. and Sombroek, W. F. (1981). The Oxisols (Ferralsols) of Kenya: profile descriptions and analytical data. *Miscellaneous Soil Paper, Kenya Soil Survey* **M23**, 70 pp.

Muhs, D. R., Crittenden, R. C., Rosholt, J. N., Bush, C. A., and Stewart, K. C. (1987). Genesis of marine terrace soils, Barbados, West Indies: evidence from mineralogy and geochemistry. *Earth Surface Processes and Landforms* **12**, 605-618.

Muller, J. (1981). Fossil pollen record of extant angiosperms, *Botanical Review* **47**, 1-142.

Müller, M.J. (1982). *Selected climatic data for a*

global set of standard stations for vegetation science. Junk, Hague, 306 pp.

Munsell Color (1975). *Munsell soil color charts.* Munsell, Baltimore, 24 pp.

Munthe, J., Dongo, B., Hutchison, J. H., Kean, W. F., Munthe, K., and West, R. M. (1983). New discoveries from the Miocene of Nepal include a hominoid. *Nature* **303**, 331-333.

Murphy, C. P. (1983). Point-counting pores and illuvial clay in thin section. *Geoderma* **31**, 133-150.

Murphy, C. P. and Kemp, R. A. (1984). The overestimation of clay and underestimation of pores in soil thin sections. *Journal of Soil Science* **35**, 481-495.

Murthy, R. S., Hirekurer, L. R., Deshpande, S. B., and Veneka Rao, B. V. (eds.) (1982). *Benchmark soils of India: morphology, characteristics and classification for resource management.* National Bureau of Soil Survey and Land Use Planning (ICAR), Nagpur, India, 374 pp.

Mutakyahwa, M. K. D. (1987). Fluvio-deltaic environment and *in situ* pedogenesis of middle Miocene kaolinitic sandstones of the Pugu coastal area of Tanzania. *Journal of African Earth Science* **6**, 229-242.

Nagatoshi, K. (1987). Miocene hominoid environments of Europe and Turkey. *Palaeogeography, Paleoclimatology, Palaeoecology* **61**, 145-154.

Naqvi, S. M. and Rogers, J. J. W. (1987). *Precambrian geology of India.* Oxford University, New York, 223 pp.

Neilson, A. (1921). Appendix IV. Igneous rocks from British East Africa. In *The rift valleys and geology of East Africa* (by J. W. Gregory), pp. 388-407. Seeley & Service, London.

Neville, A. C. (1975). *Biology of the arthropod cuticle.* Springer, New York, 448 pp.

Newton, R. B. (1914). On some non-marine molluscan remains from the Victoria Nyanza region associated with Miocene vertebrates. *Quarterly Journal of the Geological Society of London* **70**, 187-198.

Nixon, P. H. and Hornung, G. (1973). The carbonatite lavas and tuffs near Fort Portal, western Uganda. *Overseas Geology and Mineral Resources* **41**, 168-179.

Northcote, K. H. (1974). *A factual key for the recognition of Australian soils.* Rellim, Adelaide, 123 pp.

Northcote, K. H. and Skene, J. K. M. (1972). Australian soils with saline and sodic properties. *Soil Publications of the Commonwealth Scientific and Industrial Research Organization, Australia* **27**, 62 pp.

Nyamweru, C. (1988). Activity of Oldoinyo Lengai volcano, Tanzania, 1983-1987. *Journal of African Earth Sciences* **7**, 603-610.

Oldham, C. F. (1893). The Sarasvati and the Lost River of the Indian Desert. *Journal of the Royal Asiatic Society of Great Britain and Ireland* **25**, 49-76.

Olsen, P. E. (1986). A 40-million-year lake record of early Mesozoic orbital climatic forcing. *Science* **234**, 842-848.

Olson, S. L. and Rasmussen, D. T. (1986). Paleoenvironment of the earliest hominoids: new evidence from the Oligocene avifauna of Egypt. *Science* **233**, 1202-1204.

Osborn, H. F. (1936-1942). *Proboscidea* (2 vol.). American Museum of Natural History, New York, 1630 pp.

Owen-Smith, N. (1987). Pleistocene extinctions: the pivotal role of megaherbivores. *Paleobiology* **13**, 351-362.

Oxnard, C. E. (1987). *Fossils, teeth, and sex: a new perspective on human evolution.* University of Washington, Seattle, 296 pp.

Palmer, E. and Pitman, N. (1972). *Trees of southern Africa* (3 vol.). A. A. Balkema, Cape Town, 223 pp.

Palmer, J. A., Phillips, G. N., and McCarthy, T. S. (1989). Paleosols and their relevance to Precambrian atmospheric composition. *Journal of Geology* **97**, 77-92.

Palmer, P. G. and Gerbeth-Jones, S. (1986). A scanning electron microscope survey of the epidermis of East African grasses. IV. *Smithsonian Contributions to Botany* **62**, 120 pp.

Palmer, P. G. and Gerbeth-Jones, S. (1988). A scanning electron microscope survey of the epidermis of East African grasses. V and West African supplement. *Smithsonian Contributions to Botany* **67**, 157 pp.

Palmer, P. G., Gerbeth-Jones, S., and Hutchinson, S. (1985). A scanning electron microscope survey of East African grasses. III. *Smithsonian Contributions to Botany* **55**, 136 pp.

Palmer, P. G. and Tucker, H. E. (1981). A scanning electron microscope survey of the epidermis of East African grasses. I. *Smithsonian Contributions to Botany* **49**, 84 pp.

Palmer, P.G. and Tucker, H.E. (1983). A scanning electron microscope survey of East African grasses. II. *Smithsonian Contributions to Botany* **53**, 72 pp.

Parkash, B., Sharma, R. P., and Roy, A. K. (1980). The Siwalik Group (molasse): sediments shed by collision of continental plates. *Sedimentary Geology* **25**, 127-150.

Parrish, J. T., Ziegler, A. M., and Scotese, C. R. (1982). Rainfall patterns and the distribution of coals and evaporites in the Mesozoic and Cenozoic. *Palaeogeography, Palaeoclimatology, Palaeoecology* **40**, 67-101.

Pascoe, E. H. (1920a). Petroleum in the Punjab and Northwest Frontier Provinces. *Memoir of the Geological Survey of India* **40**, 491 pp.

Pascoe, E. H. (1920b). The early history of the Indus, Brahmaputra and Ganges. *Quarterly Journal of the Geological Society of London* **75**, 138-155.

Paton, T. R. (1974) Origin and terminology for gilgai in Australia. *Geoderma* **11**, 221-242.

Paton, T. R. and Williams, M. A. J. (1972). The concept of laterite. *Annals of the Association of American Geographers* **62**, 42-56.

Patterson, B. (1978). Pholidota and Tubulidentata. In *Evolution of African mammals* (eds. V. J. Maglio and H. B. S. Cooke), pp. 269-278. Harvard University, Cambridge, Massachusetts.

Paulian, R. (1976). Three fossil dung beetles (Coleoptera: Scarabaeidae) from the Kenya Miocene. *Journal of the East African Natural History Society and National Museum* **31(158)**, 1-4.

Pavich, M. J. and Obermeier, S. F. (1985). Saprolite formation beneath Coastal Plain sediments near Washington, D.C. *Bulletin of the Geological Society of America* **96**, 886-900.

Phadthare, N. R. (1989). Palaeoecologic significance of some fungi from the Miocene of Tanakpur (U.P.), India. *Review of Palaeobotany and Palynology* **59**, 127-131.

Pickford, M. (1977). Pre-human fossils from Pakistan. *New Scientist* **75**, 578-590.

Pickford, M. (1981). Preliminary Miocene mammalian biostratigraphy for western Kenya. *Journal of Human Evolution* **10**, 73-97.

Pickford, M. (1982). The tectonics, volcanics and sediments of the Nyanza Rift Valley, Kenya. *Supplementband, Zeitschrift für Geomorphologie* **42**, 1-33.

Pickford, M. (1983). Sequence and environments of the lower and middle Miocene hominoids of western Kenya. In *New interpretations of ape and human ancestry* (eds. R. L. Ciochon and R. S. Corruccini), pp. 421-439. Plenum, New York.

Pickford, M. (1984). *Kenya palaeontology gazeteer. Vol. 1. Western Kenya*. National Museums of Kenya, Nairobi, 282 pp.

Pickford, M. (1985). A new look at *Kenyapithecus* based on recent discoveries in western Kenya. *Journal of Human Evolution* **14**, 113-143.

Pickford, M. (1986a). Cenozoic paleontological sites in western Kenya. *Münchner Geowissenschaftliche Abhandlungen, Reihe A, Geologie und Paläontologie* **8**, 151 pp.

Pickford, M. (1986b). Sedimentation and fossil preservation in the Nyanza Rift System, Kenya. In *Sedimentation in the African Rifts* (eds. L. E. Frostick, R. W. Renaut, I. Reid, and J. J. Tiercelin). *Special Publication of the Geological Society of London* **25**, 345-362.

Pickford, M. (1986c). Did *Kenyapithecus* utilize stones? *Folia Primatologia* **47**, 1-7.

Pickford, M. (1986d). The geochronology of Miocene higher primate faunas of East Africa. In *Primate evolution* (eds. J. G. Else and P. C. Lee), pp. 19-33. Cambridge University, Cambridge, England.

Pickford, M. (1987). Fort Ternan (Kenya) palaeoecology. *Journal of Human Evolution* **16**, 305-309.

Pickford, M. and Andrews, P. (1981). The Tinderet Miocene sequence in Kenya. *Journal of Human Evolution* **10**, 11-33.

Pickford, M., Ishida, H., Nakano, Y., and Nakaya, H. (1984). Fossiliferous localities of the Nachola-Samburu Hills area, northern Kenya. *Supplementary Issue, African Study Monographs, Kyoto University* **2**, 45-56.

Pilbeam, D. (1982). New hominoid skull material from the Miocene of Pakistan. *Nature* **295**, 232-234.

Pilbeam, D. (1984). The descent of hominoids and hominids. *Scientific American* **250(3)**, 84-96.

Pilbeam, D. R., Barry, J., Meyer, G. E., Shah, S. M. I., Pickford, M. H. L., Bishop, W. W., Thomas, H., and Jacobs, L. L. (1977a). Geology and palaeontology of Neogene strata of Pakistan. *Nature* **270**, 684-689.

Pilbeam, D. R., Behrensmeyer, A. K., Barry, J. C., and Shah, S. M. I. (1979). Miocene sediments and faunas of Pakistan. *Postilla* **179**, 45 pp.

Pilbeam, D. R., Meyer, G. E., Badgley, C., Rose, M. D., Pickford, M. H. L., Behrensmeyer, A. K., and Shah, S. M. I. (1977b). New hominoid primates from the Siwaliks of Pakistan and their bearing on hominoid evolution. *Nature* **270**, 689-695.

Pilbeam, D. R., Rose, M. D., Badgley, C., and Lipschutz, B. (1980). Miocene hominoids from Pakistan. *Postilla* **181**, 94 pp.

Pilgrim, G.E. (1910). Preliminary notes on a revised classification of the Tertiary freshwater deposits of India. *Records of the Geological Survey of India* **40**, 185-205.

Pilgrim, G. E. (1913). The correlation of the Siwaliks with mammal horizons of Europe. *Records of the Geological Survey of India* **43**, 264-326.

Pilgrim, G. E. (1919). Suggestions concerning the history of drainage of northern India. *Journal of the Asiatic Society of Bengal* **15**, 81-89.

Pilgrim, G. E. (1934). Correlation of the ossiferous sections of the upper Cenozoic of India.

American Museum Novitates **704**, 5 pp.

Platts, J. T. (1960). *A dictionary of Urdu, Classical Hindi and English*. Oxford University, Oxford, England, 1259 pp.

Pohl, J., Stöffler, D., Gall, H., and Ernston, K. (1977). The Ries Impact Crater. In *Impact and explosion cratering* (eds. D. J. Roddy, R. O. Roddy, and R. B. Merrill), pp.342-404. Pergamon Press, New York.

Polhill, R. M. (1966). Ulmaceae. In *Flora of tropical East Africa* (eds. C. E. Hubbard and E. Milne-Redhead), 14 pp. Crown Agents for Oversea Governments and Administrations, London.

Potter, P. E. and Pettijohn, E. J. (1963). *Paleocurrents and basin analysis*. Academic, New York, 274 pp.

Powell, C. McA. (1979). A speculative tectonic history of Pakistan and surroundings: some constraints from the Indian Ocean. In *Geodynamics of Pakistan* (eds. A. Farah and K. A. de Jong), pp. 5-24. Geological Survey of Pakistan, Quetta.

Prakash, I. (1974). The ecology of vertebrates of the Indian desert. In *Ecology and biogeography in India* (ed. M.S. Mani), pp. 369-420. Junk, The Hague.

Prakash, U. (1975). Fossil woods from the Lower Siwalik beds of Himachal Pradesh. *The Palaeobotanist* **22**, 278-392.

Prakash, U. (1978). Fossil woods from the Lower Siwalik Beds of Uttar Pradesh, India. *The Palaeobotanist* **25**, 378-392.

Prakash, U. (1979). Some more fossil woods from the Lower Siwalik beds of Himachal Pradesh, India. *Proceedings of Himalayan Geology* **8**, 61-81.

Prakash, U. (1981). Further occurrence of fossil woods from the Lower Siwalik beds of Uttar Pradesh, India. *The Palaeobotanist* **28-29**, 374-388.

Prakash, U. and Prasad, M. (1984). Wood of *Bauhinia* from the Siwalik Beds of Uttar Pradesh, India. *The Palaeobotanist* **32**, 140-145.

Prasad, K. N. (1971). Ecology of the fossil Hominoidea of the Siwaliks of India. *Nature* **323**, 413-414.

Prasad, K. N. (1982). Was *Ramapithecus* a tool user? *Journal of Human Evolution* **11**, 101-104.

Prasad, M. and Prakash, U. (1984). Leaf impressions from the Lower Siwalik beds of Koilabas, Nepal. *Special Publication, Proceedings of the Indian Geophytology Conference, Lucknow (1983)*, 246-256.

Prashad, B. (1924). On a fossil ampullariid from Poonch, Kashmir. *Records of the Geological Survey of India* **56**, 210-212.

Prinn, R. G. and Fegley, B. (1987). The atmospheres of Venus, Earth and Mars: a critical comparison. *Annual Review of Earth and Planetary Sciences* **15**, 171-212.

Pulfrey, W. (1953). A Kenya alnöite and associated skarns. *Journal of the East African Natural History Society and National Museum* **22**, 23-34.

Pundeer, G. S., Sidhu, P. S., and Hall, G. F. (1978). Mineralogy of soils developed on two geomorphic surfaces of Sutlej alluvium in central Punjab, N.W. India. *Journal of the Indian Society for Soil Science* **26**, 151-159.

Puri, G. S. (1950). Soil pH and forest communities in the sal (*Shorea robusta*) forests of the Dehra Dun Valley, U. P., India. *The Indian Forester* **76**, 292-309.

Quade, J., Cerling, T. E., and Bowman, J. R. (1989). Development of Asian monsoon revealed by marked ecological shift during late Miocene in northern Pakistan. *Nature* **342**, 163-166.

Rahmatullah, Dixon, J. B., and Golden, D. C. (1990). Manganese-containing nodules in two calcareous rice soils of Pakistan. In *Soil micromorphology: a basic and applied science* (ed. L. A. Douglas), pp. 387-394. Elsevier, Amsterdam.

Rao, Y. P. (1981). The climate of the Indian subcontinent. In *Climates of southern and western Asia* (eds. K. Takahashi and H. Arakawa), pp. 67-182. In *Survey of Climatology* (ed. H. Landsberg), v. 9. Elsevier, New York.

Ratcliffe, B. C. and Fagerstrom, J. A. (1980). Invertebrate lebensspuren of Holocene floodplains: their morphology, origin and paleoecological significance. *Journal of Paleontology* **54**, 614-630.

Ravey, M. (1978). Bipedalism: an early warning system for Miocene hominoids. *Science* **199**, 372.

Raynolds, R. G. H. (1982). Did the ancestral Indus River flow into the Ganges drainage? *Geological Bulletin, University of Peshawar* **14**, 141-150.

Raynolds, R. G. H., Johnson, G. D., Johnson, N. M., and Opdyke, N. D. (1980). The Siwalik molasse, a sedimentary record of orogeny. In *Proceedings of the International Committee on Geodynamics, Group 6 Meeting* (eds. R. A. K. Tahirkheli, J. M. Qasim, and M. Majid). *Special Issue, Geology Bulletin, University of Peshawar* **13**, 47-50.

Raza, S. M. and Meyer, G. E. (1984). Early Miocene geology and paleontology of the Bugti Hills, Pakistan. In *Contributions to the geology of Siwaliks of Pakistan* (eds. S. M. I. Shah and D. Pilbeam). *Memoirs of the Geological Survey of Pakistan* **11**, 43-63.

Razzaq, A. and Herbillon, A. J. (1979). Clay mineralogical trends in alluvium-derived soils of

Pakistan. *Pedologie* **29**, 5-23.

Rea, D. K., Leinen, M., and Janacek, T. R. (1985). Geologic approach to the long term history of atmospheric circulation. *Science* **227**, 721-725.

Reiff, W. (1977). The Steinheim Basin - an impact structure. In *Impact and explosion craters* (eds. D. J. Roddy, R. O. Pepin, and R. B. Merrill), pp. 309-320. Pergamon Press, New York.

Reilly, T. A., Raja, P. K. S., Mussett, A. E., and Brock, A. (1976). The palaeomagnetism of late Cenozoic volcanic rocks from Kenya and Tanzania. *Geophysical Journal of the Royal Astronomical Society* **45**, 483-494.

Reimer, T. O. (1983). Accretionary lapilli in volcanic ash falls: physical factors governing their formation. In *Coated grains* (ed. T. M. Peryt), pp. 56-68. Springer, Berlin.

Rendell, H. M., Hailwood, E. A. and Dennell, R. W. (1987). Magnetic polarity stratigraphy of Upper Siwalik Subgroup, Soan Valley, Pakistan: implications for early human occuppance of Asia. *Earth and Planetary Science Letters* **85**, 488-496.

Retallack, G. J. (1976). Triassic palaeosols in the upper Narrabeen Group of New South Wales. Part 1. Features of the palaeosols. *Journal of the Geological Society of Australia* **23**, 383-399.

Retallack, G. J. (1981). Preliminary observations on fossil soils in the Clarno Formation (Eocene to early Oligocene), near Clarno, Oregon. *Oregon Geology* **43**, 147-150.

Retallack, G. J. (1982). Paleopedological perspectives on the development of grasslands during the Tertiary. In *Proceedings of the 3rd North American Paleontological Convention* (eds. B. Mamet and M. J. Copeland), v. 2, pp. 417-421. Business and Economic Services, Toronto.

Retallack, G. J. (1983). Late Eocene and Oligocene paleosols from Badlands National Park, South Dakota. *Special Paper of the Geological Society of America* **193**, 82 pp.

Retallack, G. J. (1984a). Completeness of the rock and fossil record: some estimates using fossil soils. *Paleobiology* **10**, 59-78.

Retallack, G. J. (1984b). Trace fossils of burrowing beetles and bees in an Oligocene paleosol, Badlands National Park, South Dakota. *Journal of Paleontology* **58**, 571-592.

Retallack, G. J. (1985). Fossil soils as grounds for interpreting the advent of large plants and animals on land. *Philosophical Transactions of the Royal Society of London* **B309**, 105-142.

Retallack, G. J. (1986a). Fossil soils as grounds for interpreting long term controls on ancient rivers. *Journal of Sedimentary Petrology* **56**, 1-18.

Retallack, G. J. (1986b). The fossil record of soils. In *Paleosols: their recognition and interpretation* (ed. V. P. Wright), pp.1-57. Blackwells, Oxford, England.

Retallack, G. J. (1986c). Reappraisal of a 2200-Ma-old paleosol from near Waterval Onder, South Africa. *Precambrian Research* **32**, 195-232.

Retallack, G. J. (1988a). Field recognition of paleosols. In *Paleosols and weathering through geologic time: principles and applications* (eds J. Reinhardt and W. R. Sigleo). *Special Paper of the Geological Society of America* **216**, 1-20.

Retallack, G. J. (1988b). Down to earth approaches to vertebrate paleontology. *Palaios* **3**, 335-344.

Retallack, G. J. (1989). Paleosols and their relevance to Precambrian atmospheric composition: discussion 2. *Journal of Geology* **97**, 763-764.

Retallack, G. J. (1990a). *Soils of the past*. Unwin-Hyman, London, 520 pp.

Retallack, G. J. (1990b). The work of dung beetles and its fossil record. In *Evolutionary paleobiology of behavior* (by A. J. Boucot), pp. 214-226. Elsevier, Amsterdam.

Retallack, G. J. (1991). Untangling the effects of burial alteration and ancient soil formation. *Reviews of Earth and Planetary Sciences* **19**, 183-206.

Retallack, G. J. and Dilcher, D. L. (1988). Reconstructions of selected seed ferns. *Annals of the Missouri Botanical Garden* **75**, 1010-1057.

Retallack, G. J., Dugas, D. P., and Bestland, E. A. (1990). Fossil soils and grasses of a middle Miocene East African grassland. *Science* **247**, 1325-1328.

Retallack, G. J., Leahy, G. D., and Spoon, M. D. (1987). Evidence from paleosols for ecosystem changes across the Cretaceous/Tertiary boundary in eastern Montana. *Geology* **15**, 1090-1093.

Richard, J. J. (1942). Oldoinyo Lengai: the 1940-1941 eruption, volcanological observations in East Africa. *Journal of the East Africa and Uganda Natural History Society* **16**, 89-108.

Richards, B. N. (1987). *The microbiology of terrestrial ecosystems*. Longman, Harlow, England, 399 pp.

Ritchie, J.M. (1987). Trace fossils of burrowing Hymenoptera from Laetoli. In *Laetoli: a Pliocene site in northern Tanzania* (eds. M. D. Leakey and J. M. Harris), pp. 433-450. Clarendon, Oxford.

Robert, M. and Berthelin, J. (1986). Role of biological and biochemical factors in weathering. In *Interactions of soil minerals with natural organics and microbes* (eds. P. M. Huang and M. Schnitzer), pp. 453-495. *Special Publication of the Soil Science Society of America* **17**.

Robinson, D. and Wright, V. P. (1987). Ordered illite-smectite and kaolinite-smectite: pedogenic minerals in a Lower Carboniferous paleosol sequence, South Wales? *Clay Minerals* **22**, 109-118.

Robinson, J. M. (1989). Phanerozoic O_2 variation, fire, and terrestrial ecology. *Palaeogeography, Palaeoclimatology, Palaeoecology* **75**, 223-240.

Roche, H. and Tiercelin, J.-J. (1977). Découverte d'une industrie lithique ancienne *in situ* dans la Formation d'Hadar, Afar central, Éthiopie. *Comptes Rendus Hebdomaires de l'Academie des Sciences, Paris* **D284**, 1871-1874.

Rodolfo, K. S. (1989). Origin and early evolution of lahar channel at Mabinit, Mayon Volcano, Philippines. *Bulletin of the Geological Society of America* **101**, 414-426.

Roe, L. J. (1987). The Miocene fishes of the Siwalik Group, northern Pakistan. *Supplement, Abstracts, Journal of Vertebrate Paleontology* **3(7)**, 24A.

Roe, L. J. (1988). Neogene freshwater fishes of Pakistan. *Supplement, Abstracts, Journal of Vertebrate Paleontology* **8(3)**, 24A.

Rohr, D. M., Boucot, A. J., Miller, J., and Abbott, M. (1986). Oldest termite nest from the Upper Cretaceous of West Texas. *Geology* **14**, 87-88.

Romme, W. H. and Despain, D. G. (1989). The Yellowstone fires. *Scientific American* **261(5)**, 37-46.

Rose, M. D. (1986). Further hominoid postcranial specimens from the Late Miocene Nagri Formation of Pakistan. *Journal of Human Evolution* **15**, 333-367.

Ruhe, R. V. (1969). *Quaternary landscapes in Iowa*. Iowa State University Press, Ames, 255 pp.

Russell, R. S. (1977). *Plant root systems: their function and interaction with the soil*. McGraw-Hill, London, 298 pp.

Saggerson, E. P. (1952). Geology of the Kisumu district. *Reports of the Geological Survey of Kenya* **21**, 1-57.

Saggerson, E. P. (1965). Post-Jurassic erosion surfaces in eastern Kenya and their deformation in relation to structure. *Quarterly Journal of the Geological Society of London* **121**, 51-72.

Sah, S. C. D. (1967). Palynology of an upper Neogene profile from Rusizi Valley, Burundi. *Annales de la Musée Royal de l'Afrique Central, Tervuren, Sciences Geologiques* **57**, 1-173.

Sahni, A. and Mitra, H. C. (1980). Neogene palaeobiogeography of the Indian subcontinent, with special reference to fossil vertebrates. *Palaeogeography, Palaeoclimatology, Palaeoecology* **31**, 39-62.

Sahni, B. (1964). Revisions of Indian fossil plants. Part III. Monocotyledons. *Monograph of the Birbal Sahni Institute for Palaeobotany* 1: 89 pp.

Sakagami, S. F. and Michener, C. D. (1962). *The nest architecture of the sweat bees (Halictinae): a comparative study of behavior.* University of Kansas, Lawrence, 135 pp.

Salard-Cheboldaeff, M. (1979). Sur la palynoflore Maestrichtienne et Tertiaire du bassin sédimentaire littoral du Cameroun. *Pollen et Spores* **20**, 215-260.

Sanders, L. D. (1965). Geology of the contact between the Nyanza Shield and the Mozambique Belt in western Kenya. *Bulletin of the Geological Survey of Kenya* **7**, 45 pp.

Sands, W. A. (1987). Ichnocoenoses of probable termite origin from Laetoli. In *Laetoli, a Pliocene site in northern Tanzania* (eds. M. D. Leakey and J. M. Harris), pp. 409-433. Clarendon, Oxford, England.

Sanford, R. L. (1987). Apogeotropic roots in an Amazon rain forest. *Science* **235**, 1062-1064.

Sarich, V. M. and Wilson, A. C. (1967). Immunological time scale for human evolution. *Science* **158**, 1200-1203.

Sauer, W. (1955). *Coprinsphaera ecuadoriensis*, un fósil singular del Pleistóceno. *Boletin del Instituto de Ciencias Naturales de la Universidad Central, Quito* **1(2)**, 123-129.

Savage, D. E. and Russell, D. E. (1983). *Mammalian paleofaunas of the world.* Addison-Wesley, Reading, Pennsylvania, 432 pp.

Savage, R. J. G. and Long, M. R. (1986). *Mammal evolution*. Facts on File, New York, 259 pp.

Schau, M. K. and Henderson, J. B. (1983). Archaean weathering at three localities on the Canadian Shield. *Precambrian Research* **2**, 189-202.

Schmidt, V. and McDonald, D. A. (1979). The role of secondary porosity in the course of sandstone diagenesis. In *Aspects of diagenesis* (eds. P. A. Scholle and P. R. Schluger). *Special Publication of the Society of Economic Paleontologists and Mineralogists, Tulsa* **26**, 175-207.

Schmidt-Kittler, N. (1987). The Carnivora (Fissipedia) from the lower Miocene of East Africa. *Palaeontographica* **A197**, 1-37.

Schumm, S. A. (1977). *The fluvial system*. Wiley-Interscience, New York, 338 pp.

Schumm, S. A. (1981). Evolution and response of the fluvial system, sedimentologic implications. In *Recent and ancient non-marine depositional environments: models for exploration* (eds. F. C. Ethridge and R. M. Flores). *Special Publication of the Society of Economic Paleontologists and Mineralogists, Tulsa* **31**, 19-29.

Schwartz, J. H. (1984). The evolutionary relationships of man and orang-utans. *Nature* **308**, 501-505.

Schwartz, J. H. (1987). *The red ape: orangutans and human origins.* Houghton-Mifflin, Boston, 337 pp.

Scotese, C. R., Bambach, R. K., Barton, C., Van Der Voo, R., and Ziegler, A. M. (1979). Paleozoic base maps. *Journal of Geology* **81**, 217-277.

Scotese, C. R., Gahagan, L. M., and Larson, R. L. (1989). Plate tectonic reconstructions of the Cretaceous and Cenozoic ocean basins. In *Mesozoic and Cenozoic plate reconstructions* (eds. C. R. Scotese and W. W. Sager). *Tectonophysics* **155**, 27-48.

Scott, W. B. (1913). *A history of land mammals in the western hemisphere.* Macmillan, New York, 693 pp.

Seeber, L., Quittmeyer, R., and Armbruster, J. (1980). Seismotectonics of Pakistan: a review of results from network data and implications for the central Himalaya. In *Proceedings of the International Committee on Geodynamics, Group 6 Meeting* (eds. R. A. K. Tahirkheli, J. M. Qasim, and M. Majid). *Special Issue, Geology Bulletin, University of Peshawar* **13**, 151-168.

Sehgal, J. L. and Stoops, G. (1972). Pedogenic calcic accumulation in arid and semi-arid regions of the Indo-Gangetic alluvial plain of the erstwhile Punjab (India): their morphology and origin. *Geoderma* **8**, 59-72.

Sehgal, J. L. and Sys, C. (1970). The soils of Punjab (India). II. Application of the "7th approximation" to the classification of the soils of Punjab: some problems, considerations and criteria. *Pedologie* **20**, 244-267.

Sehgal, J. L., Sys, C., and Bhumbla, D. R. (1968). A climatic soil sequence from the Thar Desert to the Himalayan mountains in Punjab (India). *Pedologie* **18**, 351-373.

Seilacher, A. (1964). Biogenic sedimentary structures. In *Approaches to paleoecology* (eds. J. Imbrie and N. Newell), pp. 296-316. Wiley, New York.

Semeniuk, V. and Meagher, T. D. (1981). Calcrete in Quaternary coastal dunes in southwestern Australia: a capillary-rise phenomenon associated with plant roots. *Journal of Sedimentary Petrology* **51**, 47-68.

Sen, D. N. (1980). Root systems and root ecology. In *Environment and root behavior.* (ed. D. N. Sen), pp. 63-92. Geobios International, Jodhpur.

Sen-Sarma, P. K. (1974). Ecology and biogeography of the termites of India. In *Ecology and biogeography in India* (ed. M. S. Mani), pp. 421-472. Junk, The Hague.

Senut, B. (1988). Taxonomie et fonction chez les Hominoidea Miocènes Africains: example de l'articulation du crude. *Annales de Paléontologie* **74**, 128-154.

Shackleton, R. M. (1951). A contribution to the geology of the Kavirondo Rift Valley. *Quarterly Journal of the Geological Society of London* **106**, 343-392.

Shah, S. M. I. (1977). Stratigraphy of Pakistan. *Memoir of the Geological Survey of Pakistan* **12**, 138 pp.

Shankaranaryana, H. S. and Hirekirur, L. R. (1972). Characterization of some soils of the north Indian plains. *Journal of the Indian Society for Soil Science* **20**, 157-167.

Sheikh, K. A. and Shah, S. M. I. (1984). Paleocurrent directions of the Chinji Formation. In *Contributions to the geology of Siwaliks of Pakistan* (eds. S. M. I. Shah and D. Pilbeam). *Memoir of the Geological Survey of Pakistan* **11**, 75-77.

Shinn, E. A. and Lidz, B. H. (1988). Blackened limestone pebbles: fire at subaerial unconformities. In *Paleokarst* (eds. N. P. James and P. W. Choquette), pp. 117-130. Springer, New York.

Shipman, P. L. (1977). Paleoecology, taphonomic history and population dynamics of the vertebrate assemblage from the middle Miocene of Fort Ternan, Kenya. Unpublished Ph.D. thesis, New York University, New York, 410 pp.

Shipman, P. (1981). *Life history of a fossil.* Harvard University, Cambridge, Massachusetts, 222 pp.

Shipman, P. (1986a). Paleoecology of Fort Ternan reconsidered. *Journal of Human Evolution* **15**, 193-204.

Shipman, P. (1986b). Scavenging or hunting in early hominids: theoretical framework and tests. *American Anthropologist* **88**, 27-43.

Shipman, P., Walker, A., Van Couvering, J. A., Hooker, P. J., and Miller, J. A. (1981). The Fort Ternan hominoid site, Kenya: geology, age, taphonomy and paleoecology. *Journal of Human Evolution* **10**, 49-72.

Shroder, J. F. (1989). Hazards of the Himalaya. *American Scientist* **77**, 564-573.

Sidhu, P. S. (1977). Aeolian additions to the soils of northwest India. *Pedologie* **27**, 323-336.

Sidhu, P. S., Sehgal, J. L., and Randhawa, N. S. (1977a). Elemental distribution and association in some alluvium-derived soils of the Indo-Gangetic plain of Punjab, India. *Pedologie* **27**, 225-235.

Sidhu, P. S., Sehgal, J. L., Sinha, J. M. K., and Randhawa, N. S. (1977b). Composition and mineralogy of iron-manganese concretions from some soils of the Indo-Gangetic plains in northwest India. *Geoderma* **18**, 241-249.

Sigurdsson, H., Devine, J. D., Tchova, F. M., Presser, T. S., Pringle, M. K. W., and Evans, W. C. (1987). Origin of the lethal gas outburst from

Lake Monoun, Cameroun. *Journal of Volcanology and Geothermal Research* **31**, 1-16.

Simonson, R. W. (1941). Studies of buried soils formed from till in Iowa. *Proceedings of the Soil Science Society of America* **6**, 373-381.

Simonson, R. W. (1976). A multiple-process model of soil genesis. In *Quaternary soils* (ed. W. C. Mahaney), pp. 1-25. Geoabstracts, Norwich, England.

Singer, M. J. and Nkedi-Kizza, P. (1980). Properties and history of an exhumed Tertiary Oxisol in California. *Journal of the Soil Science Society of America* **44**, 587-590.

Singh, D. and Lal, G. (1946). "Kankar" composition as an index of the nature of soil profile. *Indian Journal of Agricultural Science* **16**, 328-342.

Singh, H. P. (1982). Tertiary palynology of the Himalaya: a review. *The Palaeobotanist* **30**, 268-278.

Sinha, S. D., Sahay, S., and Prasad, B. (1964). Mineralogical composition of sand fractions under different soil series in the district of Purnea, Bihar State. *Bulletin of the National Insitute of Sciences of India, New Dehli* **26**, 157-165.

Skaife, S. H. (1953) *African insect life*. Longmans, Green & Co., London, 387 pp.

Smith, G. S. (1988). Paleoenvironmental reconstruction of Eocene fossil soils from the Clarno Formation in eastern Oregon. Unpublished M.Sc. thesis, Department of Geological Sciences, University of Oregon, Eugene, 167 pp.

Smith, W. C. (1931). A classification of some rhyolites, trachytes and phonolite from part of Kenya colony, with a note on some associated basaltic rocks. *Quarterly Journal of the Geological Society of London* **87**, 212-258.

Soil Survey Staff (1951). Soil survey manual. *Handbook of the U.S. Department of Agriculture* **18**, 503 pp.

Soil Survey Staff (1962). *Supplement to U.S.D.A. Handbook 18, Soil Survey Manual* (replacing pp. 173-188). U.S. Government Printing Office, Washington, D.C.

Soil Survey Staff (1975). Soil taxonomy, a basic system of soil classification for making and interpreting soil surveys. *Handbook of the U.S. Department of Agriculture* **436**, 754 pp.

Solounias, N. and Dawson-Saunders, B. (1988). Dietary adaptations and paleoecology of the late Miocene ruminants from Pikermi and Samos in Greece. *Palaeogeography, Palaeoclimatology, Paleoecology* **40**, 67-101.

Sombroek, W. G., Braun, H. M. H. and Van Der Pouw, B. J. A. (1982). *Exploratory soil map and agro-climatic zone map of Kenya, 1980: scale 1:100,000*. Kenya Soil Survey, Nairobi, 56 pp.

Song, Z., Li, H., Zheng, Y., and Liu, G. (1984). The Miocene floristic regions of East Asia. In *The evolution of the East Asian environment. Vol. II. Palaeobotany, palaeozoology and palaeoanthropology* (ed. R. O. Whyte), pp. 448-460. Centre of Asian Studies, University of Hong Kong.

Song, Z.-C. and Liu, G.W. (1981). Tertiary palynological assemblages from Xizang, with reference to their paleogeographical significance. In *Geological and ecological studies of the Qinghai-Xizang Plateau* (ed. D. S. Liu), v. 1, pp. 207-214. Science Press and Gordon & Breach, Beijing and New York.

Srivastava, T. N. (1976). *Flora Gorakhpurensis*. Today's and Tomorrow's, New Delhi, India, 209 pp.

Stace, H. C. T., Hubble, G. D., Brewer, R., Northcote, K. H., Sleeman, J. R., Mulcahy, M. J., and Hallsworth, E. G. (1968). *A handbook of Australian soils.* Rellim, Adelaide, 435 pp.

Stach, E., Mackowsky, M.-T., Teichmüller, M., and Teichmüller, R. (1975). *Stach's textbook of coal petrology*. (Trans. D. G. Murchison, G. H. Taylor and F. Zierkie). Gebrüder Borntrãger, Berlin, 423 pp.

Stanley, S. M. (1986). *Earth and life through time*. Freeman, San Francisco, 690 pp.

Stebbins, G. L. 1981. Coevolution of grasses and grazers. *Annals of the Missouri Botanical Garden* **68**, 75-86.

Steininger, F. F. and Rögl, F. (1984). Paleogeography and palinspastic reconstruction of the Neogene of the Mediterranean and Paratethys. In *The geological evolution of the eastern Mediterranean* (eds. J. E. Dixon and A. H. F. Robertson), pp. 659-668. Blackwells, Oxford, England.

Stevenson, F. J. (1969). Pedohumus: accumulation and diagenesis during the Quaternary. *Soil Science* **107**, 470-479.

Stevenson, F. J. (1986). *Cycles of soil: carbon, nitrogen, phosphorus, sulfur and micronutrients.* Wiley, New York, 380 pp.

Stoops, G. (1983). Micromorphology of oxic horizons. In *Soil micromorphology. Vol. 1. Techniques and applications* (eds. P. Bullock and C. P. Murphy), pp. 419-440. A. B. Academic, Berkhamsted, England.

Susman, R. L., Stern, J. T., and Jungers, W. L. (1984). Arboreality and bipedality in the Hadar hominids. *Folia primatologia* **43**, 113-156.

Sweeting, M. M. (1973). *Karst landforms*. Columbia University, New York, 362 pp.

Szalay, F. S. (1975). Hunting-scavenging protohominids: a model for hominid origins. *Man* **10**, 420-429.

Talbot, W. A. (1976). *Forest flora of the Bombay Presidency and Sind* (2 vol.). Today's and Tomorrow's, New Dehli, 508 & 574 pp.

Tandon, S. K., Kumar, R., and Singh, P. (1985). Syntectonic controls of palaeoflow reversals and variability: sediment vector sequences in the late orogenic fluvial Siwalik basin, Punjab sub-Himalaya, India. *Sedimentary Geology* **41**, 97-112.

Tandon, S. K. and Narayan, D. (1981). Calcrete conglomerate, case-hardened conglomerate and cornstone - a comparative account of pedogenic and nonpedogenic carbonates from the continental Siwalik Group, Punjab, India. *Sedimentology* **28**, 353-367.

Tanwar, G. S. and Sen, D. N. (1980). Root patterns in plants of the Indian desert. In *Environment and root behavior* (ed. D. N. Sen), pp. 63-92. Geobios International, Jodphur, India.

Tassy, P. (1983). Les Élephantoidea Miocènes du Plateau du Potwar, Groupe de Siwalik, Pakistan. IIe Parte. Choerolophodontes et Gomphothères. *Annales de Paléontologie* **69**, 235-297.

Tassy, P. (1986). Nouveaux Élephantoidea (Mammalia) dans le Miocène du Kenya. *Cahiers de Paléontologie, Editions du C.N.R.S.,Paris*, 135 pp.

Tassy, P. and Pickford, M. (1983). Un nouveau Mastodonte zygolophodonte (Proboscidea, Mammalia), dans le Miocène inférieur d'Afrique orientale: systematique et paleoenvironment. *Geobios* **16**, 53-77.

Tatersall, I. (1969a). Ecology of north Indian *Ramapithecus*. *Nature* **221**, 451-452.

Tatersall, I. (1969b). More on the ecology of north Indian *Ramapithecus*. *Nature* **224**, 821-822.

Tauxe, L. and Badgley, C. (1984). Transition stratigraphy and the problem of remanence lock-in times in the Siwalik red beds. *Geophysical Research Letters* **11**, 611-613.

Tauxe, L. and Badgley, C. (1988). Stratigraphy and remanence acquisition of a palaeomagnetic reversal in alluvial Siwalik rocks of Pakistan. *Sedimentology* **35**, 697-715.

Tauxe, L., Kent, D. V., and Opdyke, N. D. (1980). Magnetic components contributing to the NRM of Middle Siwalik red beds. *Earth and Planetary Science Letters* **47**, 279-284.

Tauxe, L. and Opdyke, N. D. (1982). A time framework based on magnetostratigraphy for the Siwalik sediments of the Khaur area, northern Pakistan. *Palaeogeography, Palaeoclimatology, Palaeoecology* **37**, 43-61.

Taylor, C. R. and Rowntree, V. J. (1973). Running on two legs or four: which consumes more energy? *Science* **179**, 186-187.

Taylor, J. M. (1950). Pore-space reduction in sandstone. *Bulletin of the American Association of Petroleum Geologists* **34**, 701-706.

Tazieff, H. (1989). Mechanisms of the Nyos carbon dioxide disaster and of so-called phreatic steam eruptions. *Journal of Volcanology and Geothermal Research* **39**, 109-116.

Teaford, M. F. and Walker, A. C. (1984). Quantitative differences in dental microwear between primate species with different diets and a comment on the presumed diet of *Sivapithecus*. *American Journal of Physical Anthropology* **64**, 191-200.

Tessier, F. (1959). La laterite du Cap Manuel à Dakar et ses termitières fosiles. *Comptes Rendus de la Academie des Sciences Paris* **248**, 3320-3322.

Thackray, G. D. (1989). Paleoenvironmental analysis of paleosols and associated fossils in Miocene volcaniclastic deposits, Rusinga Island, western Kenya. Unpublished M.Sc. thesis, Department of Geological Sciences, University of Oregon, Eugene, 129 pp.

Thenius, E. (1972). Giraffes, phylogeny. In *Animal life encyclopedia. Volume IV. Mammals* (ed. B. Grzimek), pp. 246-247. Van Nostrand-Reinhold, New York.

Thomas, H. (1984). Les Bovidae (Artiodactyla, Mammalia) du Miocène du sous-continent Indien, de la Peninsule Arabique et de l'Afrique: biostratigraphie, biogeographie, et ecologie. *Palaeogeography, Palaeoclimatology, Palaeoecology* **45**, 251-299.

Thomas, H. (1985). The early and middle Miocene land connection of the Afro-Arabian plate and Asia: a major event for hominoid dispersal. In *Ancestors, the hard evidence* (ed. E. Delson), pp. 42-50. Alan R. Liss, New York.

Thomasson, J. R. (1985). Miocene fossil grasses: possible adaptation in reproductive bracts (lemma and palea). *Annals of the Missouri Botanical Garden* **72**, 843-851.

Thompson, J. B. (1972). Oxides and sulfides in regional metamorphism of pelitic schists. In *Proceedings of the 24th International Geological Congress, Montreal, Section 10, Geochemistry* (ed. J. E. Gill), pp. 27-35. Harpell's, Gardenvale, Ontario.

Thorp, J., Woodruff, G. A., Miller, F. T., Bellis, E., Mehlich, A., Robertson, W., and Pinkerton, A. (1960). *Soil survey of the Songhor area, Kenya*. Government, Nairobi, 191 pp.

Tissot, B. P. and Welte, D. H. (1978). *Petroleum formation and occurrence*. Springer, New York, 538 pp.

Tobias, P. V. (1981). The emergence of man in Africa and beyond. *Philosophical Transactions of the Royal Society of London* **B292**, 43-56.

Tolan, T. L., Beeson, M. H., and Vogt, B. F. (1984). Exploring the Neogene history of the Columbia River: discussion and geological trip guide to the Columbia River Gorge. 1. Discussion. *Oregon Geology* **46**, 87-96.

Tomar, K. P. (1987). Chemistry of pedogenesis in Indo-Gangetic alluvial plains. *Journal of Soil Science* **38**, 405-414.

Touber, L., Van Der Pouw, B. J. A., and Van Engelen, V. W. P. (1982). Soils and vegetation of the Amboseli-Kibwezi area. *Map of the Kenya Soil Survey* **R6**.

Townsend, F. C. and Reed, L. W. (1971). Effects of amorphous constituents on some mineralogical and chemical properties of a Panamanian latosol. *Clays and Clay Minerals* **19**, 303-310.

Trapnell, C. G., Brunt, M. A., Birch, W. R., and Trump, E. C. (1969). *Vegetation of Kenya. Map sheet 1. 1:250,000.* Edward Stanford, London.

Tree Society of South Africa (1974). *Trees and shrubs of the Witwatersrand.* Witwatersrand University, Johannesburg, 309 pp.

Tripathi, C. and Singh, G. (1987). Gondwana and associated rocks of the Himalaya and their significance. In *Gondwana six: stratigraphy, sedimentology and paleontology* (ed. G.D. McKenzie), pp. 195-205. *Monograph of the American Geophysical Union* **41.**

Tripathi, P.P. and Tiwari, V.D. (1983). Occurrence of *Terminalia* in the Lower Siwalik beds near Koilabas, Nepal. *Current Science* **52**, 167.

Trivedi, B. S. and Ahuja, M. (1979a). *Parinarioxylon splendidum* sp. nov. from Kalagarh. *Current Science* **47**, 638-639.

Trivedi, B. S. and Ahuja, M. (1979b). *Pentacme-oxylon ornatum* gen. et sp. nov. from the Siwalik of Kalagarh. *Current Science* **48**, 646-647.

Tuttle, R. H. (1981). Evolution of hominid bipedalism and prehensile capabilities. *Philosophical Transactions of the Royal Society of London* **B292**, 89-94.

Twidale, C. R. (1982). The evolution of bornhardts. *American Scientist* **70**, 268-276.

Vail, J. R. (1983), Pan-African crustal accretion in north-east Africa. *Journal of African Earth Sciences* **1**, 285-294.

Van Bruggen, A. C. (1978). Land molluscs. In *Biogeography and ecology of southern Africa* (eds. M. J. A. Werger and A. C. Van Bruggen), pp. 877-923. Junk, The Hague.

Van Couvering, J. A. (1972). Radiometric calibration of the European Neogene. In *Calibration of hominoid evolution* (eds. W. W. Bishop, J. A. Miller, and S. Cole), pp. 247-271. Scottish Academic, Edinburgh.

Van Couvering, J. A. H. (1980). Community evolution in East Africa during the Late Cenozoic. In *Fossils in the making* (eds. A. K. Behrensmeyer and A. P. Hill), pp. 272-298. University of Chicago, Chicago.

Van Couvering, J. A. H. (1982). Fossil cichlid fish of Africa. *Special Paper in Palaeontology, Palaeontological Association, London* **29**, 103 pp.

Van Donselaar-ten Bokkel Hiunink, W. A. (1966). *Structure, root systems and periodicity of savanna plants and vegetation in northern Surinam.* North Holland, Amsterdam, 162 pp.

Van Zinderen Bakker, E. M. (1976). The evolution of late Quaternary paleoclimates of southern Africa. *Palaeoecology of Africa* **9**, 160-202.

Vasishat, R. N. (1985). *Antecedents of early man in northwestern India: palaeontological and palaeoecological studies.* Inter-India, New Delhi, India, 230 pp.

Venter, F. J. (1986). Soil patterns associated with the major geological units of the Kruger National Park. *Koedoe* **29**, 125-138.

Verdcourt, B. (1963). The Miocene non-marine mollusca of Rusinga Island, Lake Victoria, and other localities in Kenya. *Palaeontographica* **A121**, 1-37.

Vishnu-Mittre (1984). Floristic changes in the Himalaya (southern slopes) and Siwaliks from the mid-Tertiary to Recent times. In *The evolution of the East Asian environment. Vol. II. Palaeobotany, palaeozoology and palaeoanthropology* (ed. R. O. Whyte), pp. 485-503. Centre of Asian Studies, University of Hong Kong.

Visser, C. F. and Johnson, G. D. (1978). Tectonic control of Late Miocene molasse sedimentation in a portion of the Jhelum reentrant, Pakistan. *Geologische Rundschau* **67**, 15-37.

Vogl, R. J. (1974). Effects of fire on grasslands. In *Fire and ecosystems* (eds. T. T. Kozlowski and C. E. Ahlgren), pp. 139-194. Academic, New York.

Vokes, H. E. (1935). Unionidae of the Siwalik Series. *Memoir of the Connecticut Academy of Arts and Sciences* **9**, 37-48.

Voorhies, M. R. (1975). Vertebrate burrows. In *The study of trace fossils* (ed. R. W. Frey), pp. 269-294. Springer, New York.

Voorhoeve, A. G. (1965). *Liberian high forest trees.* Centre for Agricultural Publication and Documentation, Wageningen, 416 pp.

Wadia, D. N. (1928). The geology of Poonch State (Kashmir) and adjacent portions of the Punjab. *Memoir of the Geological Survey of India* **51**, 185-370.

Wadia, D. (1975). *Geology of India.* 4th ed. Tata

McGraw-Hill, New Delhi, India, 508 pp.
Walker, A. C. (1978). Prosimian primates. In *Evolution of African mammals* (eds. V. J. Maglio and H. B. S. Cooke), pp. 90-99. Harvard University, Cambridge, Massachusetts.
Walker, A. C. and Andrews, P. W. (1973). Reconstruction of dental arcades of *Ramapithecus wickeri*. *Nature* **244**, 313-314.
Walker, A. C., Falk, D., Smith, R., and Pickford, M. (1983). The skull of *Proconsul africanus*. *Nature* **305**, 525-527.
Walker, A. C. and Pickford, M. (1983). New postcranial fossils of *Proconsul africanus* and *Proconsul nyanzae*. In *New interpretations of ape and human ancestry* (eds. R. L. Ciochon and R. S. Corruccini), pp. 325-351. Plenum, New York.
Walker, A. C. and Teaford, M. (1988). The Kaswanga primate site: an early Miocene hominoid site on Rusinga Island, Kenya. *Journal of Human Evolution* **17**, 539-544.
Walker, T. R. (1967). Formation of red beds in modern and ancient deserts. *Bulletin of the Geological Society of America* **78**, 353-368.
Walker, T. R., Ribbe, P. H., and Honea, R. M. (1967). Geochemistry of hornblende alteration in Pliocene red beds, Baja California, Mexico. *Bulletin of the Geological Society of America* **78**, 1055-1060.
Walter, H. (1973). *Vegetation of the Earth.* (Trans. J. Wieser). Springer, New York, 237 pp.
Ward, S. C. and Pilbeam, D. R. (1983). Maxillofacial morphology of Miocene hominoids from Africa and Indo-Pakistan. In *New interpretations of ape and human ancestry* (eds. R. L. Ciochon and R. S. Corruccini), pp. 211-238. Plenum, New York.
Washburn, S. L. (1967). Behavior and origin of man. *Proceedings of the Royal Anthropological Institute* **3**, 21-27.
Watts, N. L. (1978). Displacive calcite: evidence from recent and ancient calcretes. *Geology* **6**, 699-703.
Watts, N. L. (1979). Displacive calcite: evidence from recent and ancient calcretes: reply. *Geology* **7**, 421-422.
Wayland, E. J. (1928). Summary of fieldwork carried out in 1927. *Annual Report of the Geological Survey of Uganda* **for 1927**, 1-40.
Weaver, C. E. (1989). *Clays, muds and shales.* Elsevier, Amsterdam, 819 pp.
Weilemaker, W. G. and Wakatsuki, T. (1984). Properties, weathering and classification of some soils formed in peralkaline volcanic ash in Kenya. *Geoderma* **32**, 21-44.
Wenz, W. (1938). Gastropoda. Teil 1. Allgemeiner Teil und Prosobranchia. In *Handbuch der Paläozoologie* (ed. O. Schindewolf), v. 6, 1639 pp. Gebrüder Borntrāger, Berlin.
Wenz, W. and Zilch, A. (1960). Gastropoda. Teil 2. Euthynera. In *Handbuch der Paläozoologie* (ed. A. Zilch), 834 pp. Gebrüder Borntrāger, Berlin.
Weser, O. E. (1974). Sedimentological aspects of strata encountered on Leg 23 in the northern Arabian Sea. In *Initial Reports of the Deep Sea Drilling Project* (eds. P. R. Supko and O. E. Weser), v. 23, pp. 503-519. U.S. Government Printing Office, Washington, D.C.
West, L. T., Drees, L. R., Wilding, L. P., and Rabenhorst, M. C. (1988). Differentiation of pedogenic and lithogenic carbonate forms in Texas. *Geoderma* **43**, 271-287.
West, R. M. (1984). Siwalik faunas from Nepal: paleoecologic and paleoclimatic implications. In *The evolution of the East Asian environment. Vol. II. Palaeobotany, palaeozoology and palaeoanthropology* (ed. R. O. Whyte), pp. 725-744. Centre of Asian Studies, University of Hong Kong.
Wheeler, P. (1984). The evolution of bipedality and loss of functional body hair in hominids. *Journal of Human Evolution* **13**, 91-98.
Wheeler, P. (1985). The loss of functional body hair in man: the influence of thermal environment, body form, and bipedality. *Journal of Human Evolution* **14**, 23-28.
White, F. (1983). *The vegetation of Africa: a descriptive memoir to accompany the UNESCO/AETFAT/UNSO vegetation map of Africa.* UNESCO, Paris, 356 pp.
Whitworth, T. (1953). A contribution to the geology of Rusinga Island, Kenya. *Quarterly Journal of the Geological Society of London* **109**, 75-92.
Whitworth, T. (1958). Miocene ruminants of East Africa. *Fossil mammals of Africa, British Museum of Natural History* **15**, 50 pp.
Whybrow, P. J., Collinson, M. E., Daams, R., Gentry, A. W., and McClure, H. A. (1982). Geology, fauna (Bovidae, Rodentia) and flora from the early Miocene of Saudi Arabia. *Tertiary Research* **4**, 105-120
Whybrow, P. J. and McClure, H. H. (1981). Fossil mangrove roots and paleoenvironments of the Miocene of the eastern Arabian Peninsula. *Palaeogeography, Palaeoclimatology, Palaeoecology* **32**, 213-235.
Wieder, M. and Yaalon, D. H. (1982). Micromorphological fabrics and developmental stages of carbonate nodular forms related to soil characteristics. *Geoderma* **28**, 203-220.
Williams, G. E. (1986). Precambrian permafrost horizons as indicators of palaeoclimate. *Precambrian Research* **32**, 233-242.
Wilson, E. O. and Taylor, R. W. (1964). A fossil ant

colony: new evidence of social antiquity. *Psyche* **71**, 93-103.

Wolfe, J. A. (1981). Paleoclimatic significance of the Oligocene and Neogene floras of the northwestern United States. In *Paleobotany, paleoecology and evolution* (ed. K. J. Niklas), v. 2, pp. 79-101. Praeger, New York.

Wolfe, J. A. (1985). Distribution of major vegetational types during the Tertiary. In *The carbon cycle and atmospheric CO_2: natural variations Archaean to present* (eds. E. T. Sundquist and W. S. Broecker). *Geophysical Monograph, American Geophysical Union* **32**, 357-375.

Wolpoff, M. H. (1983). *Ramapithecus* and human origins: an anthropologist's perspective of changing interpretations. In *New interpretations of ape and human ancestry* (eds. R. L. Ciochon and R. S. Corruccini), pp. 651-676. Plenum, New York.

Woodburne, M. O. (ed.) (1987). *Cenozoic mammals of North America.* University of California, Berkeley, 336 pp.

Woolley, A. R. (1989). The spatial and temporal distribution of carbonatites. In *Carbonatites, genesis and evolution* (ed. K. Bell), pp. 15-37. Allen & Unwin, London.

Wrangham, R. W. (1980). Bipedal locomotion as a feeding adaptation in gelada baboons and its implications for human evolution. *Journal of Human Evolution* **9**, 329-332.

Wright, J. B. and Rix, P. (1967). Evidence for trough faulting in eastern central Kenya. *Overseas Geology and Mineral Resources* **10**, 30-41.

Wright, V. P. (1982). The recognition and interpretation of paleokarsts: two examples from the Lower Carboniferous of South Wales. *Journal of Sedimentary Petrology* **52**, 85-94.

Wright, V. P. (1984). The significance of needle-fibre calcite in a Lower Carboniferous palaeosol. *Geological Journal* **19**, 23-32.

Wyllie, P. J. and Tuttle, O. F. (1960). The system $CaO-CO_2-H_2O$ and the origin of carbonatites. *Journal of Petrology* **1**, 1-46.

Xu, R. (1981). Vegetational changes in the past and uplift of the Qinghai-Xizang Plateau. In *Geological and ecological studies of the Qinghai-Xizang Plateau* (ed. D.-S. Liu), v. 1, pp. 139-144. Science Press and Gordon & Breach, Beijing and New York.

Yanovsky, E., Nelson, E. K., and Kingsbury, R. M. (1932). Berries rich in calcium. *Science* **75**, 565-566.

Yeats, R. S. and Lawrence, R. D. (1984). Tectonics of the Himalayan thrust belt in northern Pakistan. In *Marine geology and oceanography of the Arabian Sea and coastal Pakistan* (eds. B. U. Haq and J. D. Milliman), pp. 177-198. Van Nostrand-Reinhold, New York.

Yemane, K., Bonnefille, R., and Faure, H. (1985). Palaeoclimatic and tectonic implications of Neogene microflora from the northwestern Ethiopian highlands. *Nature* **318**, 653-656.

INDEX

AA (atomic absorption chemical analysis) 84, 282-289
aardvark 137, 292, 294, 297
Aberdares, hills 235
Acacia 117, 120, 124, 135, 136, 151, 175, 240, 293
arabica 201
catechu 175
mellifera 128
nigrescens 104
tortilis 120, 124, 136, 238
Acacia-Balanites wooded grassland 150
Acacia-Commiphora wooded grassland 103
Acalypha 98, 136, 293
Acanthus pubecscens 98
Accipitridae 294
Accretionary lapilli 23, 65, 70, 79, 80, 101, 108, 111, 126, 127
Aceratherium simorrense 297
Achatina leakeyi 94, 99, 110, 290
Achatinidae 94, 95, 290, 293
Acheulian industry 256
acidic soil 13, 257
Acrisol 104
Acrochordus 296
Acrostichum 104
Adansonia 117
Adina cordifolia 210
aerobic microbial decay 43
aestivation 15, 182
Africa 156, 157, 226, 228, 249, 254, 255, 256
Afrocricetodon 138, 294
songhorensis 291
Afromontane vegetation 55, 59, 72, 88, 104, 110, 240
forest 136, 146, 150, 231
grassland 136, 143
Afrosmilus turkanae 292
agama lizard 290
Agamidae 290
agglomerate 4, 62, 63, 67-69, 70, 71, 73-76, 79, 80, 81, 83, 90, 100, 106, 107, 110-113, 125, 230
agglomeroplasmic microfabric 96, 101, 107, 108, 126, 141, 173, 183, 188, 198, 203, 207, 211-213
agriculture 256
Albizzia 89, 99
lebbek 210
procera 210
zygia 98
Alfic Ustochrept 190, 215
Alfisol 68, 90, 194
algae 135, 293
alkali oxides 125, 204, 208
alkaline earths/alumina ratio 109, 209
lake 252
pH 87, 96, 102, 109, 110, 116, 119, 127, 142, 175, 180, 185, 189, 194, 199, 204, 209, 214, 231
volcanic center 8
alluvial fan 168, 191, 194, 196, 200-202, 210, 216, 217, 218, 222-224, 241, 242, 244, 246, 247
Soil 45, 98, 112, 149, 175, 219
terrace 149
wooded grassland 220
alluvium 35, 41, 75, 120, 133, 149, 153, 158, 177, 181, 183, 185, 191, 199, 202, 205, 206, 217, 218, 222
alnöite 65
Alstonia congoensis 93
alteration of paleosols after burial 43-52, 85, 91, 96, 102, 108, 112, 115, 119, 122, 127, 132, 141, 174, 179, 185, 189, 193, 199, 203, 208, 214
alumina 90, 106, 125, 187, 196, 206, 217, 282-284
alumina/bases ratio 123, 132, 142, 209, 287-289
alumina/silica ratio 194, 287-289
alvikite 20, 65, 70, 85, 91, 101, 106-108, 126, 130, 131, 140, 141
Amaranthaceae 135, 293
Amblypygi 129, 293
Americas 259
Amphechinus 291, 294
rusingensis 94, 291
Amphibia 296
Amphicyonidae 137, 292, 294
Amphisbaenia 291
Ampullariidae 182, 195, 296
Anacardiaceae 136, 232, 293
Anachalcos 106
anaerobic microbial decay 44
analcite 46, 64, 68, 83, 141, 145
Anasinopa leakeyi 292
Andept 88, 98, 127, 137
andesite 78, 79
andic soils 133
Andosol 127
Andrewsimys parvus 291
Andropogon greenwayi 120, 124, 136
Angiospermae 232, 238, 244, 290, 293
Angola 34
angular blocky peds 35, 91, 108, 118, 119, 121, 183, 184, 198, 212, 213
animal life of Miocene time 94, 89, 99, 105, 110, 113, 117, 120, 124, 128, 137, 144, 175, 181, 191, 186, 195, 201, 205, 210, 216, 232-241, 245-249, 256-260
Aningeria pseudoracemosa 93
Annonaceae 105, 234, 240, 293
Annonospermum 135, 148, 293
anoxic 13
ant 290
Antarctic ice cap 237, 249
soils 9
antecedent stream 170, 183, 222, 224
antelope 54, 55, 134, 137, 138, 149, 152, 162, 176, 177, 191, 205, 216, 221, 240, 241, 246, 247, 249, 255, 294, 295, 297
Anthospermum 136, 293
anthracothere 176, 191, 221, 234, 249
Anthracotheriidae 292, 297
Anthracotherium punjabiense 297

Antiaris toxicaria 93
antilopine 294, 297
Antrocaryon 104, 110, 231
micraster 232
Anura 296
apatite 65
ape 54, 89, 94, 99, 110, 137, 138, 149, 154, 162, 221, 225-227, 241, 246, 247, 249, 256, 259, 295, 296
fossil rarity 258
habitats 255
ape-monkey 234, 235
Apocynaceae 240, 293
apple snail 182, 187, 221, 296
apron facies, volcanic 71, 80
Aquept 209
Aquic Eutrochrept 209
Haplustoll 200, 219
Hapludoll 200
Arabia 253
Arachnida 293
Aravalli Craton 155
Group 155
Archaeobelodon 94, 292
Arctamphicyon lydekkeri 296
Arctic Ocean 237
arena facies, volcanic 65, 71
Arenosol 104, 215
Argentina 182
argillan 11, 26, 85, 90, 91, 119, 130, 173, 178, 179, 183, 184, 188, 192, 193, 198, 203, 212
argillic horizon 87, 90, 107, 116, 118, 194, 196, 202, 215, 217, 242
Argiudoll 72
Argiustoll 123, 125, 147, 150
argon ($^{39}Ar/^{40}Ar$) radiometric date 73, 75, 82
Aridisol 204
Ariophantidae 290
Artemisia 293
articulated skeleton 16, 99, 110, 137, 235
Arundinoid grass 137, 143, 252
Arusha city 57
ash flow 82
Asia 154, 156, 235, 240, 246, 249
astragalus, bone 176, 181
Atlantic Ocean 237
attritional assemblage
bone 17, 81, 99, 137, 154
snails 15
augite 68, 74, 76, 79, 83
Australia 181, 190, 195, 201, 205, 209, 210, 215, 216, 259
australopithecine 8, 151, 256, 260
Australopithecus afarensis 151
authigenic feldspar 51
Aves 291, 294, 296
axis, bone 177

Ba (barium) 183, 285-286
baboon 260
babul 201
badlands 43, 181, 162, 172, 202, 207
Badlands National Park 241
Balanites aegyptica 136
ball-and-pillow structure 100, 102
Balugoloa 175, 181, 195
bamboo rat 191, 221, 296
bambusoid grass 137, 143 252
Banda Daud Shah, village 162
Banded Gneiss Group 155
Bangladesh 9, 153, 155, 185, 186, 205, 209, 246, 253
Baphia 104, 231
Baraget Phonolite Member 62, 64, 73, 75, 76, 83, 113, 115, 118, 121, 125, 130, 238
Valley 72, 73
barium 71, 79, 90, 285-286
barium/strontium ratio 71, 79, 90, 132, 142, 209, 287-289
bark, fossil impression 14, 78, 126, 128, 139
Barringtonia acutangula 216
Barsek soil catena 124, 142
Bartabwa village 151
basalt 59, 88, 124, 134, 146, 160, 185, 200, 209, 237
basanite 57, 62, 72, 83
lava 68
Member 62, 68
plug 83
base level 171
poor clays 39
rich clays 40, 96
saturation 109, 112, 116, 119, 123, 127, 142, 175, 180, 185, 189, 194, 199, 204, 209, 214
Basiaram Series soils 200
Basotu village 57, 60, 67
bat 94, 99, 138, 291, 294
Bathyergidae 294
Bathyergoides neotertiarius 99, 110, 149, 233, 291
Bauhinia 104, 231
bauxite 19, 40, 177
Bay of Bengal 205, 241
beach sand 228, 260
bear dog 137, 292, 294, 296
bee 80, 89, 95, 110, 139, 250, 290
beetle 16, 89, 95, 129, 139, 149, 179, 180, 205, 216, 290
Berchemia 105, 240
balugoloensis 195
discolor 238
pseudodiscolor 135, 148, 293
Berlinia 104, 231
Bersama 110, 231
bhabar 170, 171, 186, 200, 201, 216, 218
Bhalwal Series soils 183, 196, 206, 217
Bhilamur village 169
Bhura Series paleosols 11, 13, 22, 32, 35, 39, 177, 179, 180, 181, 186, 187, 189-192, 194, 202, 211, 215, 217-219, 220, 222-225, 242, 243, 246-248, 267-271
clay ferric concretionary variant 10, 179, 198, 261, 270, 276, 280
clay, type paleosol 44, 177, 178-180, 182, 183, 187, 189, 191, 261, 272, 277, 281, 283, 286, 288
big-game hunting 3
Bihar province 215, 242
bimasepic microfabric 113, 183, 184, 207
biomechanics 227
biostratigraphy 159, 160
biotite 18, 23, 37, 38, 41, 59, 64-66, 70, 72, 75-77, 79, 82, 91, 100, 101, 107, 108, 111, 114, 119, 130, 131, 140,

213, 217, 229
bioturbation 19, 250
biozone 160
bird 69, 94, 99, 110, 149, 252, 291, 294, 296
plum 293
Black Earth 123, 134
Sea 235
Tuffs and Agglomerates 66, 78, 79, 90
blackening of limestone 33
bladed sparry calcite 51
Blighia unijugata 98
blinkbaar 293
blocky peds 35, 70, 193
blue-gray paleochannel sandstone 40, 169, 173, 177, 183, 187, 196, 202, 206, 207, 210, 211, 217, 242
boa 291
Bogoro Scarp 94, 104
Boidae 291
Bolbites 182
bone 16, 53, 60, 76, 79, 81, 85, 91, 94, 100, 105, 117, 118, 120, 122, 124, 128, 129, 130, 138, 151, 162, 186, 195, 201, 205, 210, 227, 228, 231, 246, 257
histology 227
preservation 120, 144, 195, 201, 234
Bonga village 34
Bonheim bonheim profile 124, 200, 219
glengazi profile 123, 147
bordered pit 13
boselaphine 137, 294, 297
Boselaphus tragocamelus 240
Botryococcus 135, 136, 293
Botswana 103, 145
Bovidae 294, 297
box turtle 296
Brachypotherium perimense 297
Brachyrhizomys 296
Brachystegia 104, 231
Bradybaenidae 290, 293, 290
Brahmaputra Delta 185
River 185, 186, 205, 241
Brahmatherium megacephalum 186, 221, 246, 297
braid plain of river 223, 242, 244
brain expansion of human ancestors 256
Braunerde 132, 137
Braunlehm 123
bright clay 38
broadening of x-ray diffractometer (XRD) peaks 40
Bronze Age 3
brood cell 16, 182, 249, 290
Brooks' residential compound 59, 65, 67
Brown Clay soil 119, 147, 181, 219
clay of deep ocean 19
Earth 123, 134, 200
Forest Soil 75
browser 138
Bucerotidae 291
Budongo forest 110, 232
buff paleochannel sandstone 40, 170, 177, 183, 187, 196, 202, 206, 210, 211, 216, 217
Bugti Hills 156
Bukoban System 55, 59
Bukwa village 14, 99, 110, 231, 252
bulges in chemical abundance 27
bulk density 84, 172, 282-284
bulldozer herbivore 240, 253
Bundhelkhand Complex 155
burial 27, 43
compaction 35, 47-48, 96, 102, 108, 109, 112, 115, 119, 122, 132, 142, 174, 179, 189, 193, 199, 203, 208, 214, 234, 254
depth 47, 48, 52
gleization 43, 92, 96, 108, 112, 113, 142, 174, 180, 185, 189, 193, 199, 203, 204, 208, 214
reddening 12, 45, 108, 112, 115, 122, 199
burrow 9, 11, 15, 19, 21, 24, 43, 46, 48, 91, 92, 96, 99, 101-103, 106, 107, 111, 112, 120, 121, 131, 137, 139, 149, 167, 172, 173, 175-177, 179, 181, 182, 184, 185, 188, 190-192, 195, 198, 201, 203-206, 207, 210-213, 216, 218, 221 235, 250, 290, 292, 294, 296
Burseraceae 128
Burtoa nilotica 94, 290, 293
Burttdavya nyasica 93
Buru Series paleosols 13, 19, 39, 45, 70-72, 96, 99, 100, 103, 107, 110-113, 147, 148, 230, 231, 233, 234, 263, 264
clay thin surface phase 111, 112, 261, 264, 275, 279
clay, type paleosol 111, 261, 264, 275, 279
Hill 60, 63
Hill Carbonatite 63
Burundi 34, 146
bush baby 99, 110, 232, 291
bush pig 297
bush willow 136, 293
bushland 55, 135, 190
butterfly 106
buttressed tree trunk 93, 128

C_3 plants 225
C_4 plants 225
Caesalpinoidea 128
Calamus tenuis 186
calcan 22, 119, 130, 179, 184
calcaneum, bone 176
calcareous nodule 11, 20, 39, 47, 71, 108, 110, 118, 122, 124, 125, 167, 173, 181, 183, 184, 187, 189, 190, 192-195, 198, 199, 202-204, 207-209, 211, 215, 222, 242, 250, 251, 254
Calcareous Red Earth 147, 194
Red Soil 92
rhizoconcretion 9, 10, 11, 21, 22, 44, 48, 178, 180, 184, 188, 190, 192, 203, 204, 211, 212-216, 173, 235, 242, 250, 251, 296
Sand 204
Calcareousness, scale of 65, 166, 274-277
Calcaric Fluvisol 98, 112, 147, 175, 219
Gleysol 185, 186, 204, 209, 219
Phaeozem 200, 219
Regosol 147, 159
calciasepic microfabric 85, 91, 100-102, 107, 108, 130, 141, 173, 179, 183, 188, 192, 198, 203, 207, 211-213
Calcic Aeric Halaquept 200Calcic Luvisol 194
Cambisol 88, 92, 134, 147, 180, 190, 219
horizon 90, 107, 142-143, 187, 180, 190, 196, 202, 242
Kastanozem 133, 134, 142, 147

Luvisol 102, 109, 147, 219
Calcified Tuff Member 14, 63, 70, 71, 106, 111
calcimorphic soils 136
Calciorthid 4
calcite 64, 66, 70, 73, 78, 79, 80, 91, 96, 101, 114, 115, 121, 126, 127, 129, 140, 141, 144, 173, 229, 231
cement 46, 129, 140
crystal sheet 121, 122, 126, 130, 141
crystal tube 85, 87, 91, 96, 101, 102, 108, 119, 126, 131, 141, 179, 184, 188, 198, 203, 207, 211, 213
fibers 71
globule 19, 71, 101, 107, 108, 111
spar 9, 20, 21, 23, 48, 51, 66, 78, 96, 101, 102, 107, 118, 119, 126, 127, 130, 131, 141, 173, 176, 179, 180, 184, 188, 189, 198, 199, 211, 214, 229, 278-281
calcium carbonate 16
phosphate 16
Calciustoll 142, 147, 150
caldera 57, 125
caliche (see calcareous nodule) 20
calie of termite 37, 290
California 103, 107, 109, 143, 181, 196, 217
calyx 135, 293
Cambic Arenosol 104
Cambisol 119, 123, 132, 134, 147, 180, 190, 209, 219, 249
Cambrian 59, 69
Cameroon 81, 117
Canada 34
Canary Islands 259
cancrinite 64
candelabra cactus 128
cane rat 99, 105, 110, 138, 149, 233, 249, 291, 294
canine, tooth 138
canyon 162
Capratragoides potwaricus 137, 138, 294
caprine 137, 294
carbon dioxide, poison clouds 81
carbon dioxide/oxygen ratio in atmosphere 13
carbon isotopes 225, 247
carbonaceous root trace 11, 12
carbonatite 4, 12, 14, 16, 19, 21-24, 26, 27, 33, 34-36, 39, 41, 51, 57, 71, 72, 79, 100, 125, 133, 144, 146 234, 251
cinder cones 145, 146, 229
clasts 85, 88, 90, 91, 106, 107, 108
colluvium 85
intrusion 51, 60, 63
lapilli 59, 65, 66, 67, 70, 85
lava 33
tuff 53, 63, 65, 67, 71, 85, 88, 90, 91, 100-102, 105-107, 111, 118, 140, 149, 228, 229, 231, 234
welded tuff 231, 251
carbonatite-nephelinite sandstone 125
tuff 47, 48, 91, 99, 105, 107, 109, 120, 127-129, 140, 144, 145, 149, 150, 231, 238
volcano 53, 55, 57, 68, 70, 72, 228, 229, 235, 257
carbonatite-melilitite accretionary lapilli 79
tuff 71, 79, 92, 94, 144, 145, 149, 151, 228, 229, 257
Carboniferous 51, 156
Careya arborea 201
carnivore 81, 89, 94, 110, 137, 138, 176, 241, 292, 294, 296
accumulation of bone 17, 176, 210
carrying infants 259
Caspian Sea 235
Cassia 128, 231
cat 210, 221, 246, 292, 297
caterpillar 16, 106
catfish 296
Catharsius 106
cation exchange capacity 87, 92, 102, 109, 112, 116, 119, 123, 127, 142, 175, 180, 185, 189, 194, 199, 204, 209, 214
cattail 136, 293
cavernous subsoil weathering 251
Cavratia 105
Cecilioides 290
cell wall outline 13, 49, 143
Celliforma 89, 95, 110, 139, 290, 293
Celtis 15, 93, 94, 99, 136, 148, 231, 240, 290, 293
africana 98
durandii 93
mildbraedii 93, 148, 231, 232, 290
rusingensis 93, 135, 148, 293
cement 102, 109, 115, 119, 122, 127, 142, 144, 180, 185, 194, 199, 208
Cenozoic 34
centipede 290, 293
centripetal drainage pit 33
Cerastua miocenica 290
cercopithecid 241
Chadobets Plateau 34
chalicothere 94, 162, 234, 247
Chalicotheriidae 292, 297
Chalicotherium 230
rusingense 94, 233, 292
salinum 297
Chameleo intermedius 17, 129, 149, 293
chameleon 17, 78, 89, 94, 129, 149, 290, 293
Chameleonidae 290
Chamtwara Member 63, 79
Stream 63
Chamtwaria pickfordi 291
Chandigarh, town 158, 194
Chang Mai city 4
channel behavior 224
Channidae 296
Char Gali Formation 162
charcoal 13, 33, 206-212, 214
Chelonia 290, 290, 296
Chemeron Formation 151
Chemical analysis 19, 84, 284
error 282, 284, 285, 286
analysis methods 19, 84, 284
analysts 284
reduction of iron 44, 52
depth functions 27-32
Chenab River 158
Chenopodiaceae 135, 293
Chernozem 123, 133, 134, 142, 147, 249
chert 178, 188, 206
chevrotain 89, 94, 99, 110, 137, 149, 176, 210, 216, 221, 232-234, 241, 247, 292, 294, 297

Chihuahua Desert 23
Chilotherium 248
  intermedium 191, 221, 247, 297
chimpanzee 8, 151, 256, 259
China 7, 182, 256, 259
Chinji Formation 48, 22, 158, 165, 167-171, 206
  village 158, 159, 162, 165, 166, 168, 169, 171
Chironitis 106
Chiroptera 291, 294
Chitra 296
Chlamydarion 290
chloridoid grass 14, 143, 148, 240, 252, 293
Chloris gayana 128
chlorite 40, 177, 190, 194
Chlorophora excelsa 93
Chocolate Soil 116, 123, 134, 147
Choerolophodon 243, 249
  corrugatus 187, 221, 245, 246, 297
  ngorora 137, 138, 240, 294
Chogo Series paleosols 11-13, 15-18, 24, 27, 30, 35, 36, 37, 39, 41 46, 49, 52, 77, 78, 117, 120, 125, 128, 129, 130, 132-135, 139, 140, 142, 144, 147, 148, 206, 235, 238, 240
  clay eroded phase 17, 79, 131-133, 135-139, 148, 237, 239, 240, 261, 265, 275, 279
  clay ferruginized nodule variant 18, 23, 27, 38, 131-133, 136, 138, 140, 148, 237, 239, 261, 264, 265, 275, 279, 282, 285, 287
  clay, type paleosol 18, 23, 35, 36, 46, 50, 77, 106, 107, 117, 130, 131, 134-136, 138, 140, 148 237, 239, 261, 265, 275, 279, 282, 285, 287
Choka Series paleosols 11, 13, 18, 19, 24, 27, 29, 39, 45, 52, 70, 71, 91, 93, 96, 98, 100, 103, 110, 117, 100-104, 106, 107, 108-111, 230, 234
  clay, type paleosol 16, 97, 142, 147, 148, 229, 235, 261, 263, 274, 282, 285, 287
chowsingha 176, 181, 221, 246, 297
chroma, Munsell 52, 189, 262-273
Chromic Luvisol 215
chromium 211, 285, 286
chronosequence 21, 22, 25, 206
chronostratigraphy 159, 160
chrysomelid beetle 90, 95
Chrysophyllum 93
chute of river bar 202, 206, 210
cinder cone 53, 57, 70, 71, 88, 90, 95, 110, 111, 145, 146, 149, 228, 229, 231
circumgranular cracks 23, 24
Cissampelos 105
Cissus 105
civet 221, 246, 292, 296
clam 296
classification of paleosols 4, 84, 88, 92, 97, 102, 109, 112, 116, 119, 127, 132, 142, 175, 180, 185, 189, 194, 200, 204, 209, 214
clastic dike 35, 48, 103, 107, 108, 114-118, 179, 181, 183, 185, 188-190, 192, 193, 199, 200, 203, 207, 211, 212, 215, 229
Clausena 98
clay bulge 19
  minerals 39
  skin 11, 35-37, 70, 71, 106, 108, 112, 113, 119, 121, 122, 126, 130, 173, 178, 179, 183, 184, 188, 192, 193, 198, 203, 207, 211
clayey soil matrix 20
  surface horizons 17
claystone breccia 35, 36, 37, 114, 115
  pseudosand 37
Cliff Agglomerate Member 63, 74
Climacoceras gentryi 137, 239, 294
climate 224, 249, 254
climatic range, grassland 253, 255
clinobimasepic microfabric 39, 207
club moss 293
Cnestis 105, 135, 240, 293
  ferruginea 238
  rusingensis 148
Co (cobalt) 183, 285, 286
coal cleat 49
coalification 13, 49
coastal forest 93
  sands 146
  terrace 121
cobra 94, 291
cockscomb 136, 293
cocoa 293
cocoon 89, 95, 106, 139, 149, 290, 293
coevolution 253, 255
coffee 293
Cola 110, 231
colluvium 12, 41, 75, 88, 90, 107, 118, 120, 121, 124, 125, 127, 129, 139, 140, 149, 150, 153, 185, 238
colonizing forest 88, 89, 93, 148, 231, 233, 235, 246, 247, 259
  woodland 220
Colophospermum mopane 103, 145
Colubridae 291, 296
Columbia River Basalt 237
Columbiformes 291
columnar jointing 77
  peds 35, 180, 199, 209
Combretaceae 128, 136, 244, 293
Combretum 117, 128
Commiphora 128
compaction 47, 96, 102, 108, 109, 112, 115, 119, 122, 132, 142, 174, 179, 189, 193, 199, 203, 208, 214, 234, 254
  curves 48
Compositae 136, 293
compressional origin of rift valleys 55
concretion 22, 26, 44, 119, 120, 123-125, 141, 179, 173, 198
Congo Basin 55
conifers 293
conjugate shear 35
Connaraceae 238, 293
continent-to-continent collision 55, 59
Conulinus 290
cool temperate climate 24, 249, 252
cooperative hunting 260
Coprinsphaera 182, 296
Copris 106
coprolite 295
Coprophanaeus 182

coral 121
Cordia 105, 234
corestone 46, 59, 76, 82, 113-115, 118
Cormohipparion 176, 182, 191, 221, 243, 246, 249, 297
  perimense 205, 210, 246, 297
  theobaldi 205, 210, 245, 246, 297
coucal 291
Cr (chromium) 183, 285, 286
crab 129, 139, 205, 295, 296
cracking and veining 34
Crater lake 67, 81
creodont 110, 137, 234, 241, 247, 292, 294
Cretaceous 34, 104, 156, 176, 250
crevasse splay 202, 206
cricetid mouse 138, 291, 294, 296
crocodile 17, 138, 162, 210, 221, 246, 295, 296
Crocodylia 296
Crocodylus 243
  palaeindicus 210, 221, 246, 296
cross bedding 78, 79, 96, 107, 129, 131, 140, 164, 202, 206, 213
  set 113, 130, 140
Croton 89, 136, 293
cryptokarst 33, 251
crystic microfabric 96, 114, 126, 130, 131, 140, 173, 188, 198, 207
Cu-Ni radiation 40
Cu-Kα radiation 41
cuckoo 291
Cuculiformes 291
Cucurbitaceae 135, 148, 240, 293
cultivation 146, 222
cumulic horizon 191, 196, 214
Cupressaceae 293
cursoriality 241, 253
Curvella 290, 293
custard apple 293
cutan 34
cuticle, plant fossil 14, 49, 143, 144
Cyathea 104, 231
Cymbopogon nardus 201
Cynelos euryodon 292
Cynodon dactylon 120, 124
Cynometra 94
  alexanderi 93, 110
Cyperaceae 293
Cypriniformes 296
Czechoslovakia 224

dacite 78, 79
Dactyloctenium 128
daisy 136, 293
Dalbergia 104, 231
  sissoo 175, 195, 244, 246
dambo 137, 143, 200, 210, 234, 252
Damodar River 209
darkling beetle 129, 293
decalcification 180, 185, 186, 189, 194, 200, 209, 218, 257
decay, aerobic 13
Deccan Traps 156
deciduous forest 190, 195, 222, 225, 229, 231, 246, 259
  woodland 128, 145, 186, 128, 225
Deckenkarren 33
degassing structures 77, 81
Delhi, city 180
  Group 155
Deinotheriidae 294, 297
Deinotherium 249
  indicum 297
Democricetodon 296
den 17, 101, 294
Dendropithecus 150, 259
  macinnesi 54, 99, 110, 233-235, 291
density 84, 172, 282-284
derivation of paleosol names 85, 91, 95, 100, 107, 111, 113, 118, 121, 126, 130, 140, 172, 177, 183, 187, 192, 196, 202
desalinization 88
descriptions of paleosols 83-145, 172-217
desert 25, 107, 152
  grassland 143, 152
  scrub 98, 253
  shrubland 89, 93, 98, 117, 120, 124, 128, 150, 151
  soil 19, 79, 150, 190
  Southwest, United States 21
desertification 222
desiccation cracks 35
development of soil 145, 166, 217, 218, 223
dhankhar 170, 171, 186, 187, 218
Dhero Series paleosols 11, 19, 23, 39, 49, 76-78, 80, 81, 99, 128, 129, 130, 139, 147, 148, 236-238, 266
  clay lapillistone variant 125, 127, 129, 140, 141, 261, 266, 276, 280
  clay loam 121, 122, 127, 151, 261, 264, 275, 279
  clay, type paleosol 14, 126, 128, 129, 139, 261, 266, 275, 276, 279, 280
Dhok Pathan Formation 10-13, 16, 20, 21, 31, 32, 35, 37, 39, 47, 158, 159, 162, 166, 169-172, 174, 175, 178, 179, 181-184, 188-193, 195, 197-199, 202, 203, 206, 207, 208, 210-212, 214, 216-218, 222, 243, 248, 296, 297
  Pathan village 159, 170
  Maiki, village 162
Dholuo language 83, 85, 91, 100, 107, 111, 113, 118, 121, 126, 130, 140
Dhulian, village 160
Dhurnal, village 169
diagenesis of paleosols 43-52, 85, 91, 96, 102, 108, 112, 115, 119, 122, 127, 132, 141, 174, 179, 185, 189, 193, 199, 203, 208, 214
diagnosis of paleosols 85, 91, 95, 100, 107, 111, 113, 118, 121, 125, 130, 140, 147, 172, 177, 183, 187, 196, 202, 206, 211, 219
Diamantomys 233, 294
  luederitzii 99, 105, 110, 149, 233, 291
Dichotomius 182
Dichrostachys 104, 231
  cinerea 104
dicot 13, 14, 143, 148, 205, 240
Digitaria macroblephora 124, 128
dike, clastic 35, 48, 103, 107, 108, 114-118, 179, 181, 183, 185, 188-190, 192, 193, 199, 200, 203, 207, 211, 212, 215, 229
  volcanic 64

Dionysopithecus shuangouensis 249
Diospyros tomentosa 190
Diplopoda 290
Dipterocarpaceae 242, 244, 245, 259
Dipterocarpus 104, 231
 turbinatus 195, 244
discharge 224
displacive microfabric 20, 21, 23-25, 34, 46, 71, 96, 97, 98, 100, 102, 112, 128, 229
dissolution of bone 137
Dissopsalis pyroclasticus 137, 294
doline 33, 251
Dombeya mukole 89, 98
domed columnar peds 35, 180, 199, 209
dominance display 259
Dorcabune 249
 nagrii 297
Dorcatherium 297, 230, 232, 233, 248, 249
 chappuisi 137, 294
 majus 216, 221, 297
 nagrii 297
 pigotti 294
 songhorense 89, 94, 99, 110, 149, 292
dormouse 296
drab haloed root trace 10-12, 43, 44, 70, 71, 96, 98, 101, 112, 121, 167, 168, 173, 174, 177, 181, 183-185, 188-192, 194, 198, 203-205, 207, 208, 212, 213, 290, 296
drainage 145, 146, 217, 218, 224
dry deciduous forest 190, 195, 222, 225, 229,231, 246, 259
 deciduous woodland 128, 145, 186, 128, 225
 forest 98, 98, 110, 117, 148, 152, 215, 216, 218, 220, 222, 234, 235, 242, 257, 259
 grassland 25
 gully 138
 monsoon forest 253
 riverain forest 216
 woodland 93, 103, 137, 190, 195, 222, 234
Dudhwa National Park 222
Dundee dundee profile 98, 112, 147, 175, 181, 219
dung beetle 15, 16, 53, 80, 105, 106, 152, 182, 221, 234, 249, 250, 290, 296

eagle 291
early successional vegetation 112, 128, 225, 238, 242
earthworm 10, 11, 176, 221, 205, 296
East African Rift Valleys 55, 228, 235, 255
ecological opportunities for human ancestors 258
 succession 145, 146, 175, 190, 218, 231
ecotone 247, 254
Edaphichnium 10, 176, 177, 205, 296
Edentulina 290
egg shell 138, 290, 292, 294
Egypt 94
Eh 13, 26, 109, 132, 180, 185
Elachistoceras 243, 246
 khauristanensis 176, 182, 221, 245, 246, 297
Elapidae 291
elephant 94, 137, 138, 162, 186, 221, 235, 240, 241, 246, 249, 253, 255, 292, 294, 297
 grass 98
 shrew 94, 138, 291, 294
Elmerimys woodi 291
Emballonuridae 291
Emblica officionalis 201
emperor moth 106
Emuruilem Member 151
Emydidae 296
endocarp 93, 135, 293
endolithic algae 251
Enidae 290
Entandrophragma 104, 110, 231
 utile 232
Entisol 119, 150
Eocene 20, 40, 117, 120, 128, 156, 162, 165, 166, 170, 175-177, 217, 241, 251
eolian dust 18, 19, 172, 196
Eomellivora 296
Eotragus 249, 294
Eozygodon morotoensis 292
epidermal cells 143
Epiphiomys coryndoni 291
epiphyte 195, 222, 225, 247
episodic sedimentation 154
Equidae 297
erect walking human ancestors 260
Erinaceidae 296
error of chemical analyses (2σ) 84, 282-286
 of point count data 84, 277
Erythrozootes chamerpes 291
Ethiopia 99, 104, 231
Euchrozem 194, 219
Euphorbia candelabra 128
Euphorbiaceae 110, 136, 240, 293
Euphorbiotheca pulchra 135, 148, 293
Eurasia 152, 228, 235, 240, 241
Europe 156, 235, 252, 259
European Alps 55, 59, 228
Eustachys paspaloides 124
Eutrandept 146
Eutric Cambisol 134, 209
 Fluvisol 209
 Planosol 134
Eutrochrept 209, 219
Eutropept 147, 150
eutrophic brown soil 123
evaporite 190, 229
evolutionary divergence of humans and apes 256
exocarp 135, 293
exoskeleton 16
expansion of grasslands 252
exuviae, millipede 16, 94
eyebright 136, 293

Falconiformes 291
family in a home base 3
fault scarp 69
faunal overturn 241, 253
 zones 159
fecal pellet 10, 19, 37, 176, 195, 201, 250
feldspar 25, 36, 37, 64, 68, 70, 75, 76, 81, 83, 96, 100, 108, 114, 115, 150, 166, 150, 173, 176-179, 183, 188, 192, 193, 198, 206, 207, 211-213, 242
feldspathization 52, 51

Felis 210, 221, 246, 297
sivalensis 297
femur, bone 176
fenestral microfabric 129
fenite 21, 51, 65, 70, 71, 85, 90, 91, 100, 101, 108
Ferralsol 59, 123
ferran 122, 126, 198, 207
ferri-argillan 35
Ferric Acrisol 104
calcic intergrowth 22
concretion 21, 35, 141-144, 173, 181, 183-185, 187, 190, 192-194, 196, 198-200, 202-204, 207, 209, 213, 219, 242, 250, 251
hydroxide 45, 87, 91, 96, 102, 108, 112, 115, 119, 122 132, 153, 168, 174, 185, 189, 193, 214, 223, 250
mottle 39, 44, 173, 183, 185
nodule 91
oxide 250
rhizoconcretion 184
Ferrous iron/ferric iron ratio 27, 46, 87, 100, 102, 103, 109, 123, 125, 177, 180, 185, 191, 194, 199, 202, 204, 209, 132, 142, 144
ferruginization 36, 134, 141, 180
Ferussaciidae 290
Ficus precunea 195
fine clay (<0.5-μm) 19, 40
fining-upwards sequence 9, 27, 78, 140, 126, 193, 206
fire 6, 30, 222, 256, 296
fish 17, 162
fission track radiometric dating 169, 172
fissiped 241
Fissistigma senii 195
flamingo 291
flaser bed 207
flies 90, 95
flood 106, 196
basalt 236
floodplain 110, 153, 167, 181, 191, 194-196, 206, 215, 221, 224, 227
flower 99, 228
Fluvaquent 4
Fluvent 98, 112, 145, 147, 175, 189, 219
Fluventic Ustochrept 180, 219
fluvial accumulation of bones 17
geomorphology 227
Fluvisol 147, 175, 209, 219
flying fox 291
squirrel 99, 110, 138, 149, 232, 240, 291, 294
food storage chambers of harvester termites 94
footprint 80
Tuff 79, 100
footslope 12, 149, 150, 210
foraminifera 20, 162
forest 3, 13, 19, 54, 59, 68, 89, 93, 106, 109, 110, 112, 117, 135-137, 144, 148, 168, 171, 177, 181, 190, 195, 204, 205, 210, 218, 220, 222, 225, 229, 230, 232, 234, 238, 240, 245-247, 254, 256, 258, 260
adapted mammals 234, 257
Fort Ternan Member 62, 73, 76, 80-83, 113, 115, 118, 121, 122, 126, 130, 140, 236-237
National Monument 11, 12, 14, 16-18, 23, 30, 34, 35, 38, 39, 46, 48-50, 72-75, 77, 79, 81, 106, 113, 114, 117, 118, 120-122, 126-129, 130, 131-133, 140, 141, 146, 148, 150, 151 226-228, 238-241, 252-254, 293-295
railway siding 6, 57, 62, 70-72, 74, 82, 116, 235
Fort Portal village 92, 104, 123
fossil preservation 13-17, 228, 256-258
France 7, 95, 259
frog 296
frugivore 260
fruit 14, 53, 55, 135, 136, 228, 231, 240
bat 94, 291
functional morphology 225, 227, 260
fungally infested rootlets 10
fungus 23, 225
garden, of termite 15, 37, 94, 173, 176, 201, 215, 216, 221, 246, 247, 250, 296

Galerix africanus 291
galleries, of termites 15, 16, 37, 89, 176, 195, 198, 201, 206, 246, 250, 296
gallery forest 182, 222
woodland 240
Ganda Kas 184, 203, 204
Gandaitherium browni 297
Gandak River 222
Ganges-Brahmaputra River 158
Ganges alluvium 185
River 153, 156, 158, 165, 169, 170, 185, 186, 195, 205, 209, 217, 222, 223, 241, 242
Gangetic Plains 171
Garden of Eden 3
Gastropoda 60, 290, 293, 296
gathering of shellfish 258
gavial 17, 296
Gavialus gangeticus 296
gaylussite 80
Gazella 137, 138, 191, 210, 221, 246,294, 297
gazelle 191, 210, 221, 246, 294, 297
gecko 95, 290, 292
Gekkonidae 290
Gelocidae 241, 292, 294
Gelocus whitworthi 292
genet 297
Geoemyda 296
geographic range of grassland 253
geological maps 54, 61, 66, 69, 73, 163
geology Kenyan paleosols 54-83
Pakistani paleosols 154-172
geothermal gradient 49
geotrupine 105
gerbil 138, 294
Ghaggar River 158, 177, 202
Ghaghara River 222
gibbon 259, 291
gibbsite 250
Gigantopithecus giganteus 247, 296
giraffe 137, 138, 152, 186, 191, 216, 221, 240, 246, 249, 294, 297
Giraffidae 294, 297
Giraffokeryx 249
Girtasho profile 123
glaciation 172, 224, 250

glass, volcanic 64
gleization 95, 102, 109, 129, 177, 185, 194, 200, 201, 210
Glenrosa ponda profile 116, 147
gleyan 9
Gleyic Cambisol 209, 219
Gleysol 185, 186, 204, 209, 219
Gliridae 296
glycolation 40, 41
gnawing of bone 16
gneiss 59, 70, 133, 150, 170, 215
goat 205
goethite 45, 250
Gola River 200, 222
gold 27, 211
Golden Age 3
  mole 94, 291
gomphothere 94, 186, 187, 191, 210, 234, 240, 246, 249, 297
Gonaxis 99, 105, 290
  protocavalii 94, 290
Gondwana sequence 156
  supercontinent 59, 156, 228, 241
Gonospira 290
Gorakhpur town 222, 242
gorilla 8, 256, 259
gourd 293
graben 57
graded bed 9, 27, 206
grain size scales 84, 274-277
Gramineae 117, 135, 138, 293
granite 175, 215
granular peds 34, 35, 85, 91, 114, 119, 121, 123, 130, 131, 141, 173, 178, 179, 188, 198, 201, 203, 207, 208, 226
Granularia 205
graphic mean grain size 84
graphitized 49
grass 14, 34, 49, 54, 79, 80, 99, 117, 120, 124, 128, 134-136, 140-144, 148, 152, 181, 201, 205, 210, 216, 227, 228, 238, 240, 247, 255, 293
grasshopper 234, 255
grassland 93, 103, 117, 128, 136, 143, 144, 151, 201, 210, 216, 225, 227, 234, 238, 247, 252, 253-255, 256, 258-260
  climatic range 254, 255
  geographic range 253, 254
  mosaic vegetation 260
  soil 35, 134, 181, 195, 201, 249
  woodland ecotone 254
grassy forest 181
  vegetation 106, 153, 209, 232
  woodland 13, 89, 117, 120, 124, 128, 135-138, 143, 144, 148, 150, 151, 152, 159, 181, 186, 190, 195, 201, 205, 216, 238, 239, 254, 257
Gray Clay 219
  silt lithofacies 184, 204, 206
  Tuffs 77, 90
grazer 138, 144
grazing pressure 253
Great Plains, North America 23, 109, 134, 143, 181
green phyllosilicate 64
Gregory Rift 55-57, 76, 118, 150, 235, 238
gregoryite 47, 80, 229
Grewia 105, 234
  bicolor 104
  hexamita 104
  mollis 232
Grey Tuff 72
  Sandstone Member 57, 63, 70-72, 90, 100, 110-112
  Brown Calcareous Soil 204
  Clay 185, 186, 204
grike 33, 251
ground surge 77
groundwater 9, 38
  gley 9, 38
  laterite 59
growth ring 187
guinea fowl 94, 291
Guineo-Congolian rain forest 104, 105, 110, 146, 150, 231, 259
  vegetation 93, 110
Gujranwala Series soils 26
Gulella 293
  miocenica 290
Gurdaspur, town 194
Gwasi Hills 228
Gymnurechinus camptolophos 99, 291
  leakeyi 89, 291

habitat fragmentation for human ancestors 260
hackberry 15, 93, 136, 231, 290, 293
Halaquept 200
half graben 57
halloysite 109, 134
Halolimnohelix 290, 293
Hangram Series soils 26
Haplaquept 185, 186, 204, 219
Haplic Kastanozem 134
  Phaeozem 200
Hapludoll 18, 146, 200
Haplustalf 4, 102, 109, 145, 147, 190, 194, 215, 219
Haplustoll 4, 18, 133, 147, 150, 200, 219
hard setting soil 14, 127, 231
hardpan 24, 96, 109, 140, 142-144, 147, 229, 231
Haritalyangar village 170, 225, 247
harvester termites 94
hawk 291
  moth 106
heat stress 258
heavy metal toxicity 209
hedgehog 89, 94, 99, 291, 294, 196
Helicarion 293
Helicarionidae 290, 293
Heliocopris bucephalus 182, 249
hematite 45, 153, 185, 250
Hemimeryx pusillus 297
Herpestes 296
heterolithic cross-bedding 100, 113
high crowned teeth 253, 255
  temperature minerals 27, 36
highly birefringent clay 35, 38
Himachal Pradesh province 175, 181, 195, 201, 205, 210, 244, 246
Himalaya 40, 47, 153, 156-158, 160, 164, 165, 168, 169, 170, 171, 177, 183, 186, 187, 191, 196, 206, 210, 211,

215-218, 222, 225, 241, 247, 249, 255
Hindu Kush mountains 157, 160, 164, 168, 169, 225, 241, 247
Hipparion 171, 176, 191, 246, 249
antilopinum 297
s.l., zone of 171
hipparionine horse 169, 241
Hippopotamidae 294
Hippopotamodon sivalensis 297
hippopotamus 67, 294
Hipposideridae 291
Hipposideros 291
hippotragine 297
Hiwegi Hill 112
Formation 48, 112, 234
Hodotermes 94
Holocene 80
Homa Lime Company 59
Homo erectus 256
honey badger 296
Hooghly River floodplain 185
hornbill 291
hornblende 158, 169
horncore 210
horse 152, 169, 221, 241, 246, 297
horst 69, 74, 76, 81
hue, Munsell 52, 65, 145, 146, 166, 175, 179, 180, 189, 217, 218
human ancestor 152, 227, 256, 260
desertification 153
disturbance 252
origins 3, 151, 256-260
humerus, bone 191
Humic Cambisol 116
Nitosol 116
humid climate 25, 33, 34, 256, 257
humification 134, 145, 146, 217, 218
hummocky topography 144
hunter-gatherer society 259
hyaena 137, 176, 221, 246, 294, 297
Hyaenidae 191, 297
Hyaenodon andrewsi 233, 292
matthewi 292
pilgrimi 110, 292
Hyainalouros 292
hyaloclastic grain 79
hyalopilitic microfabric 75, 81
hydrocarbons 38
hydroclastic eruption 81
hydrodynamic group, bone 137
hydrolysis 25, 38
hydromorphic gley 187
hydrothermal alteration 9, 20, 34
clays 40
injection 36
hyperconcentrated outflow 79, 81
hypsodonty 54, 241
Hyracoidea 292
hyrax 234, 292

icacina vine 293
Icacinaceae 105, 293
Icacinocarya 135, 148, 293
ice wedges 250
Iceland-Faeroe ridge 238
ichnogenus 139, 176, 182, 191, 205, 216
ICP (inductively coupled plasma chemical analysis) 84, 172
ijolite 73, 141
illite 40, 41, 49, 177, 180, 187, 190, 194, 242
crystallinity 50
illitization 49, 52, 180, 185, 189, 194, 199, 204, 208, 214
immigration 240, 249
Imperata cylindrica 201
Inceptisol 88, 116, 121, 150, 180
India 7, 14, 19, 21, 22, 103, 109, 134, 153, 155, 166, 170, 171, 175-177, 180-182, 185-187, 190, 194, 200-202, 204, 205, 209, 210, 216, 218, 224-226, 240-242, 244, 247, 249, 255, 256
Indian Ocean 235, 242
Indo-Gangetic alluvium 19, 40, 155, 175, 190, 194, 215
Plains 20, 21, 43, 45, 103, 109, 134, 153, 158, 175, 180, 104, 204, 210, 217, 218, 242, 244. 251, 253, 255
Indo-Pakistan 27
Indonaia hasnotensis 296
Indus River 134, 137, 153, 158, 187, 206, 209, 216, 217, 223, 241, 242
Inhoek coniston profile 119, 147
drydale profile 134, 147
insect 16, 17, 53, 94, 99, 105, 106, 110, 113, 129, 139, 149, 176, 191, 201, 221, 228, 290, 293, 296
insectivore 94, 99, 110, 138, 291, 294, 296
inselberg 69, 150
insepic microfabric 39, 85, 91, 107, 108, 126, 173, 179, 183, 184, 188, 192, 198, 203, 207, 211, 213
intergranular voids 46, 51
intergrowth, calcareous nodules and ferric concretions 22
intertextic microfabric 85, 91, 107, 112, 118, 119, 122, 126, 173, 179, 188, 192, 198, 207, 211, 213
intrastratal alteration 9, 36, 38
inundulic microfabric 126, 141, 192, 193
Iran 235
iron-manganese concretion 20, 22, 44, 119, 120, 198, 123-125
layer 151
nodule 26, 88, 101, 107, 118, 191, 206-209, 210, 213, 251
mottle 91, 108, 114, 119, 120
stringer 26, 33
Iron Age 3
ironwood 93, 110
Isaack's Ranch surface 196
Ischyrictis 296
Isoberlinia 104, 231
isopod 16
isostriotubule 85
isotic microfabric 173
Israel 23
itch pod 293
Itwa Series soils 204

Jalalpur, village 169
Jammu-Kashmir province 201

Jawalamukhi, village 175, 181, 195, 201, 205, 244
Jhatla, village 165
Jhelum city 174, 184, 189, 193, 199, 203, 225
River 158
juniper 135
Juniperus 135, 136, 293
Jurassic 34, 123, 156, 224

K/Ar radiometric date 60, 65, 68, 70, 76, 82, 118
Kaiserstuhl volcano 66
Kakamega forest 232
Kala Series paleosols 11, 13, 17, 18, 26, 27, 31, 35, 39, 44, 49, 182, 187, 191, 195, 201, 202, 205, 206, 210, 211, 214, 218-220, 223, 224, 225, 234, 242-246, 252, 257
Chitta Hills 160
clay, type paleosol 201, 207-211, 261, 271, 277, 281, 283, 286, 288
Kalagarh village 181, 190, 201, 205, 216, 245
Kalepithecus songhorensis 54, 89, 94, 291
Kallar Kahar village 171
Kamlial Formation 20, 22, 156, 162, 165, 167, 168, 206
village 22
Kanagarh Series soils 4, 26, 185
village 186
Kanisamys sivalensis 191, 221, 296
Kano Plains 59
Kanuites 17, 137, 237, 239, 294
kaolinite 39, 40, 109, 134, 177, 194, 229, 242, 249, 250
Kapchure sandy loam and sandy clay soils 59
Kapkut volcano 57, 80
Kapsibor village 76, 80, 82, 122
Kapurtay Agglomerate 63, 67-69, 70, 71, 73-76, 79, 80, 81, 83, 90, 100, 106, 107, 110-113, 125, 230
village 75
Karakorum mountains 157, 241
Karnimata darwini 296
Karrenrohren 33
karst 33, 251
Kas Dovac 171
Kastanozem 123, 133, 134, 142, 147
Kaswanga site 235
Katspruit killarney profile 204, 219
Kaulial Kas 47, 160, 162, 164, 169-172, 176, 178, 179, 183, 186, 189, 192, 197, 202, 207, 208, 211, 213, 223
village 10-13, 16, 20, 21, 31, 32, 35, 37, 39, 44, 47, 49, 50, 153, 159, 160, 162, 165, 166, 170, 172, 174, 178, 183, 184, 188, 189, 192, 193, 197, 202, 207, 208, 211-214, 258
Kavirondian System 55, 59
Kavirondo Rift 55
Kelba quadeemae 292
Kenya, Miocene paleoenvironment 145-152, 228-241
Miocene stratigraphy 54-83
studied localities 6-8, 54-83
tectonic development 54-58
Kenyalutra songhorensis 292
Kenyamys mariae 291
Kenyapithecus 8, 125, 150, 151, 256, 259, 260
Kenyapithecus wickeri 5, 55, 124, 138, 139, 149, 239, 240, 294
Kenyapotamus ternani 294
Kericho Phonolite 62, 64, 73, 75, 76, 81-83, 113, 115, 116, 118
Plateau 117, 235
town 62, 82
Kerimasi volcano 57, 66, 67
khair 175
Khair e Murat, village 171
Khakistari Series paleosols 11, 13, 15, 27, 31, 35, 39, 40, 44, 52, 142, 182-187, 195, 199, 201, 202, 204-206, 209, 211, 217, 219, 220, 223-225, 243, 244, 246, 247, 257, 267
clay thin surface phase paleosol 184, 261, 267, 276, 280
clay, type paleosol 44, 50, 183-185, 187, 261, 267, 276, 280, 283, 285, 288
Khaur Anticline 162, 165
Rest House 20
village 22, 48, 154, 158, 160, 164, 167-169, 171, 184, 187, 204, 206, 207, 242, 246, 247, 253, 254, 296, 297
Khaya antotheca 93
nyasica 93
Kiahera Formation 106
Kichechia zamanae 89, 94, 292
Kiewo Series paleosols 11, 12, 13, 14, 23, 27, 29, 45, 49, 51, 70, 91, 99, 100, 102, 103, 105-107, 110-113, 128, 130, 147, 148, 230, 231, 233, 234, 258, 262
clay single stringer variant 25, 96, 97, 261, 263, 274, 278, 282, 285, 287
clay thin surface phase 96, 97, 261, 263, 274, 278
clay, type paleosol 25, 96, 97, 261, 263, 274, 278, 282, 285, 287
killer apes 258
Kilombe village 118
volcano 57
king snake 291, 296
Kipchorion Gorge 75, 76, 82
Kipingai village 70
Kipsegi farm 62
Nephelinite 62, 64, 82, 83
Kipsesin 70
Kipsigicerus 138, 237
labidotus 137, 138, 149, 239, 240, 294
Kirana Hills 155, 156
Kisii Group 59
Hills 76
town 76
Kisingiri volcano 48, 57, 80, 234
Kisozi town 146
Kisumu town 59, 76, 82
Kitale town 228
Kluftkarren 33
Kohat Formation 162
town 162, 169, 171
Koilabas village 216
Komba 232
minor 99, 110, 291
robustus 99, 110, 291
kopje 150
Koru Formation 60, 63, 65, 68
village 11, 12, 14, 16, 17, 22, 26-28, 33, 48-50, 53, 57, 59, 60, 62, 63, 65, 67-71, 79, 83, 85, 89, 91,

93-95, 108, 139, 145, 146, 148, 228, 231, 234, 235, 240, 254, 258
Kosi River 170, 222
Krapfiella angusta 290
krotovina 12
Kruger National Park 124, 134, 142, 175, 181, 185, 194, 200, 209, 215
Kubler index 50
Kuldana Formation 162
Kulu Beds 53
Kwar Series paleosols 11, 13, 15, 23, 27, 28, 39, 41, 45, 46, 89, 91, 93-95, 98, 100, 102, 103, 107, 110, 139, 147, 148, 151, 230, 231, 233, 234, 258
 clay calcareous stringer variant 87, 91, 92, 94, 261, 262, 274, 278, 282, 285, 287
 clay, type paleosol 50, 85, 87, 91, 92, 94, 261, 262, 274, 278, 282, 285, 287
 silty clay 14, 85, 87, 91-93, 108, 261, 262, 274, 278
kyanite 158, 169
Kyllinga nervosa 144

Lacertidae 290
Lacertilia 290, 293, 296
lacustrine 133
Laetoli Beds 66, 70, 71, 79, 98
 village 23, 67, 71
Lagenaria 105
Lagerstroemia 181, 195, 205
 parviflora 205
 speciosa 181, 244, 246
lagoon 259
lahar 14, 17, 78, 80-82, 126, 128, 129, 141, 142, 144, 149, 150, 238
Lahore city 177, 183, 202, 206, 217
Lake Albert 94, 104, 231
 deposits 53, 224
 Kivu 81
 Mobutu 94, 104, 231
 Monoun 81
 Nyos 81
 Victoria 53, 55, 110, 234
Lal Series paleosols 11, 13, 18, 22, 27, 31, 35, 39, 40, 45, 49, 52, 180, 181, 186-190, 192-195, 199, 202, 206, 211, 215, 218-220, 223-225, 242, 243, 246, 247, 267-270
 clay cumulic surface phase 193, 261, 267, 276, 280
 clay, type paleosol 20, 21, 36, 37, 50, 174, 187, 192-194, 196, 261, 267, 276, 280, 283, 286, 288
Lamai profile 116
lamellar caps to calcic horizon 25
laminae 81, 140
land bridge 235
 clearance by humans 256
landscape ecology 145
Lannea 105, 136, 234, 293
 stuhlmannii 104, 232
lanthanum 71, 79
lapilli 47, 59, 129
 tuff 60, 62, 73, 82
Laroa swynnertonii 93
larval cell 15, 89, 95, 110, 147, 293
Las Cruces city 107, 196, 206, 217
Lashaine volcano 67
lateritic soil 19, 55, 89, 105, 231, 249, 255
laurel forest 259
leaf 13, 14, 16, 60, 67, 99, 104, 134, 136, 141, 181, 186, 205, 225, 293
 litter 16
leaf-nosed bat 291
Leakeyia vesiculosa 135, 148, 293
Leakeymys ternani 138, 239, 294
Legetet Carbonatite 33, 63, 65, 66, 67, 70, 71, 85, 90, 91, 106, 230
 Hill 27, 59, 62, 65, 67, 68, 83
Legetetia nandii 292
Leguminosae 238, 244, 293
Lemuta Hill 142
lentil peds 186
lepidopteran cocoon 149
Leptophilos siwalikensis 296
levee 100, 110, 113, 149, 175, 177, 181, 187, 196, 202, 206, 221, 222, 227, 244
Liatongus 106
Ligatella 105
lignite 49
lime 27, 90, 106
 quarry 27, 33
limestone 20, 34, 53, 145, 166, 173, 196, 198, 204
 clasts 173, 177, 178, 179, 183, 188, 203, 206, 207, 211, 212, 213, 217, 242
Limnopithecus 150, 230, 259
 evansi 291
 legetet 54, 94, 99, 110, 149, 233, 234, 291
Limpopo River 145
Ling River 165
linguoid ripple marks 207
Lissemys 296
Listriodon akatikubas 239, 294
lithic contact 142
 Ustropept 88, 147
Lithoclast 88
Lithorelict 114
Lithosol 88, 127, 128, 147
Lithostratigraphy 54-83, 54-172
Litra Formation 158
lizard 53, 94, 129, 149, 152, 228, 290, 293
locule 135
loess 107, 252, 224
log casts 14, 78, 82, 128, 293
Londiani town 82, 83
 volcano 55, 57
Lopholistriodon kidogosana 137, 294
Lophura 296
loris 138, 240, 296, 296
Lorisinae 138, 240, 294, 296
Losuguta type phonolite 62, 75, 82
lowland alluvial wooded grassland 210, 244, 247
low-nutrient soil 12
Lufengpithecus 256
Lumbwa Phonolite 82
 Phonolitic Nephelinite 82
 village 62, 82
 Volcanics 62
Luvic Kastanozem 123, 134, 147

Luvisol 147, 194, 215, 219
Lwanda Series paleosols 11, 12, 41, 46, 76, 81, 82, 113, 117, 124, 128, 147, 148, 175, 235, 236, 238
clay, type paleosol 50, 113-117, 119, 261, 264, 275, 279

Maboko Island 150, 240, 241
Mabokopithecus 150
Macanopsis 176, 177, 205, 296
Macroscelidae 291
Macrotermes 94, 176
Madagascar 154, 156
Maesopis emini 89, 98, 113
mafic intrusion 145
minerals 38
volcanic clasts 217
magnesia 125, 187, 196, 206, 217
magnetite 65, 153
magnetization components 45
magnetostratigraphy 118, 153, 160-162, 164, 165, 168, 169, 171, 172
Mahupur Series soils 180
Maizania lugubroides 290
Maizaniidae 290
maize crib, fossil locality 85, 89
Majiwa village 14
Malani Volcanics 155
Mali 120
mallee scrub 201
Mallotus 195, 205
philippensis 205
Malvales 110
mammal 16, 59, 60, 69, 94, 99, 117, 135-137, 149, 151, 152-154, 159, 160, 165, 168, 169, 172, 176, 205, 225-227, 232, 233, 239-242, 245, 253, 258, 260
mandible 151, 162, 176, 191, 210, 235
mangal paleosol 259
mangan 11, 26, 85, 91, 108, 114, 119, 121, 141, 178, 179, 188, 191, 203, 204, 207, 212
mangrove 185, 205
manipulation by human ancestors 3, 258
Maranthus 104, 231
goetzeniana 93
Marconia 290
Margaritaria discoides 98
marine sediments 34
Markhamia 99
platycalyx 98
marl 63
marsh 136, 151
martial origin of erect stance 3, 258
masepic microfabric 108
Masritherium 292
mass death assemblage of bone 81, 137
mastodon 187, 292
Matongo village 34
Mau Escarpment 76, 137
maxilla, bone 176
Mbalageti Ash 125
profile 123
Mbeya village 34
meadow 205, 210
mechanical analysis, grain size 223
Mediterranean area 156
climate 251
Ethiopian mammal fauna 222, 247
sclerophyll woodland 254
Sea 228, 235, 249
Meerut village 185
Megalohyrax championi 233, 292
Megapedetes 233, 294
pentadactylus 99, 233, 291
Megistotherium osteothalastes 137, 294
melanephelinite 62, 71, 72, 74, 77, 80, 83, 130
dike 68
lava 68, 73
Member 62, 68, 72
melanite 73, 79
Meliaceae 232
melilite 66, 71, 79
melilitite 23
Menengai volcano 57
Menispermaceae 105
Merced River 196, 217
meristem 255
mesophytic vegetation 14, 225
Mesozoic 217, 228, 241
Messina town 145
Meswa Bridge 17
River 60, 63
meta-isotubules 101, 107, 121, 188, 213
Metacatharsius 106
metagranotubule 101, 107, 108, 110, 111, 112, 173, 178, 179, 183, 188, 192, 198, 203, 207, 211, 212
metamorphic illite 40, 51
rock fragments 20
metamorphism 27, 43
metapodial, bone 176, 191, 210
meteorite impact 236
Mfangano Island 53, 94, 104, 106, 110, 129, 231
mica 51, 173, 183, 188, 192, 193, 198, 203, 207
micrite 10, 20, 21, 23, 37, 71, 91, 100, 101, 119, 121, 126, 130, 141, 173, 192, 198, 203, 212, 213
microcline 49, 59, 67, 71, 72, 85, 90, 91, 96, 100-102, 107, 108, 111-113, 184
microfabric of clayey matrix 38-39
microfoyaite 60
microped 36, 37, 183, 190, 250, 251
Micropithecus 138, 150, 240, 294
clarki 54, 89, 94, 234, 291
microwear of teeth 134, 152, 240, 241, 253, 260
Middle East 241
middle gray silts tongue 197, 202, 203
middle lamella between wood cells 13
Middle Paleolithic 172
mid-successional forest 93
mid-length grassland 143
mid-Tertiary erosion surface 59
Milankovitch cycle 224, 242, 251
Milkwood graythorne profile 142, 147
millipede 16, 69, 89, 94, 139, 149, 228, 290, 293
litter splitting 16
round-backed 16
mineral composition, paleosols 278-281

volcanic rocks 64
minnow 296
Miocene animal life 5, 94, 89, 99, 105, 110, 113, 117, 120, 124, 128, 137, 144, 175, 181, 191, 186, 195, 201, 205, 210, 216, 232-241, 245-249, 256-260
stratigraphy for Kenya 54-83
stratigraphy for Pakistan 154-172
timescale for Kenya 58
timescale for Pakistan 161
vegetation 2, 13, 54, 55, 92, 88, 98, 103, 109, 112, 117, 120, 124, 128, 134, 154, 143, 146, 149-152, 154, 175, 181, 186, 190, 195, 201, 204, 209, 215, 221-223, 225, 226, 229-232, 234, 236-238, 242-247,252-256, 258-260
Mioeuoticus 291
miombo woodland 104, 231
Mioprionodon pickfordi 292
Miorhynchocyon clarki 94, 233, 291
meswae 291
rusingensis 138, 291, 294
Miotragocerus punjabicus 191, 221, 246, 297
Mispah kalkbank profile 127, 147
muden profile 88, 147
Mississippi Delta 186
mite 293
mixed-layer montmorillonite-vermiculite 41
Mnara Hill 68
Mobaw Series paleosols 11, 13, 23, 28, 39, 44, 45, 85, 88-90, 92, 93, 95, 98, 100, 103, 107, 110, 113, 139, 147, 148, 230, 231, 233, 234
bouldery clay paleosol 26, 85, 86, 87, 90, 261, 262, 274, 278, 282, 285, 287
silty clay loam, type paleosol 22, 85, 87, 90, 91, 261, 262, 274, 278
Modesto surface 196
mofette 81
moist Gangetic deciduous forest 222, 244
monsoon forest 242, 253
sal savanna 201
tarai sal forest 186
Mojave Desert 23, 103, 109, 134, 143, 181
molar tooth 151, 182, 186, 205
mole rat 99, 110, 149, 233, 257, 291, 294
molecular weathering ratio 84
mollic epipedon 123, 132, 142, 226, 252
soil 200
Solonetz 134
Mollisol 132, 142, 209, 214
mollusc 185
mongoose 17, 81, 89, 137, 139, 292, 294, 296
den 133
monitor lizard 89, 94, 290, 296
monkey 151, 241, 260
monkey-ape 240, 291, 294
monsoon 22,23, 103, 120, 175, 181, 195, 210, 235, 251, 255
forest 226, 253
montmorillonite-vermiculite 115
mopane 104
Morrocco 123, 134
mosaic of vegetation 54, 128, 135, 137, 148, 150-152, 225, 238, 240, 241, 258, 260
mosepic microfabric 101, 108, 179, 188, 198, 203, 207, 211, 212
moth 90, 95, 106
mottle 34, 85, 101, 108, 114, 121, 125, 141, 173, 179, 183-185, 187, 188, 192, 198, 207-209, 213, 219
mound building termites 94
mountain vegetation 55, 59, 72, 88, 104, 110, 240
forest 136, 146, 150, 231
grassland 136, 143
top 260
Mozambique Belt 55, 59, 69
Mt Blackett 82
Elgon volcano 57, 104, 228, 231
Kenya volcano 90, 107, 116, 118, 121
Londiani volcano 118
Suswa volcano 74
Mteitei Valley 72, 74
mudcrack 9, 198
mudflow 128, 129, 144
mugger 296
Muhoroni Agglomerate 59, 60, 63, 67
village 6, 59, 60, 63
mukkara structure 35, 103, 186
Mumek Gorge 62
Munsell color 83, 262-273
chroma 42, 52
hue 45, 52, 171
value 171
murid mouse 296
Murree Cantonment 162
Formation 156, 162, 164, 165, 166, 167, 168
murrel 296
muscovite 37, 49, 59, 177, 179, 206, 207, 212, 213
mushroom 201
Musophagidae 291
Mustelidae 296
mycorrhizae 10
Myocricetodon irhoudi 294
parvus 294
Myohyrax oswaldi 291, 294
Myophiomys arambourgi 291
Myorycteropus chemeldoi 137, 239, 294
Myriapoda 290, 293

Nabha Series soil 215
Nachola 151, 241
Nagri Formation 158, 162, 166, 168-171, 173, 183, 184, 189, 192, 197, 202, 203, 206 207
sandstone tongue 171, 178, 188, 193, 211
village 159
Nairobi city 59, 235
Naisiusiu Beds 140
NaLag profile 133
Nalagarh village 201, 210, 245
Namibian Desert 241, 253
Nandi Hills 59
NaNo-A profile 133
Naranji Series paleosols 11, 13, 18, 27, 32, 39, 211, 217, 219, 220, 222, 223, 247, 248, 273
clay, type paleosol 11, 44, 177, 183, 211-217, 261, 271, 277, 281, 283, 286, 288
natric soil 132

natrocarbonatite 66, 67, 80
natrolite 46, 64, 107, 114
Natrustoll 125
Ndamathaia kubwa 291
Ndutu beds 140
needle fiber calcite 51
neocalcan 9, 10, 173, 198, 203, 213
neoformation 40
neomangan 26, 96
neomorphic spar 51
neosesquan 114, 126, 130, 192, 213
Nepal 153, 186, 203, 216, 259
nepheline 36, 64, 68, 73-75, 79, 81, 83, 114, 115
nephelinite 4, 12, 14, 16, 19-24, 26, 35, 36, 39, 41, 57, 62, 67, 71-73, 75-78, 81, 85, 90, 91, 95, 100, 112, 130, 133, 141, 144, 146
    clasts 82, 88, 96, 101, 102, 106, 107, 108, 111, 112, 113, 118, 119, 121, 126, 130, 131, 140, 141
    lahar 70, 80, 82
    lapilli 67,77
    lava 82, 90, 140, 142
    sandstone 34, 46, 78-80, 102, 107, 109, 121, 127, 129-132, 136, 140, 142, 143, 149, 234, 238, 240, 252
    tuff 77, 82, 90, 95, 106, 107, 113, 118, 121, 127, 134, 140, 149, 228, 229, 231
    volcano 74, 82, 120, 125, 129, 139, 144-146, 149, 150, 229
nephelinite-phonolite colluvium 238
nest building by human ancestors 259
nesting behavior, dung beetles 105, 182
New Mexico 107, 196, 206, 217
Newtonia 104, 231
Newtonia buchananii 93, 94
Ngeron village 70
Ngorongoro Crater 123, 125, 150
Ngorora Formation 151
    village 99
Nguruwe kijivium 110, 233, 292
Ni (nickel) 183, 285, 286
Nigeria 120
nilgai, antelope 221, 240, 246
Nitosol 116
nodular horizon 70
nodule, calcareous 11, 20, 39, 47, 71, 108, 110, 118, 122, 124, 125, 167, 173, 181, 183, 184, 187, 189, 190, 192-195, 198, 199, 202-204, 207-209, 211, 215, 222, 242, 250, 251, 254
    iron-manganese 26, 88, 101, 107, 118, 191, 206-209, 210, 213, 251
Non-calcic Brown Soil 123, 134, 200
North Africa 241, 259
North America 23, 107, 109, 134, 143, 181, 190, 195, 201, 205, 210, 249, 252
North Ruri hill 74
Notocricetodon petteri 291
nurturing explanation, human evolution 260
nutrient depleted soils 255
    rich soils 255
Nyakach Formation 63
    village 63
Nyando Valley 62
Nyanza Rift 54, 55-57
Nyanzapithecus 150
    vancouveringorum 99, 234, 291
Nyanzian System 59
Nycterididae 291
Nycticeboides simpsoni 296
nyerereite 47, 66, 71, 80, 91, 229
nyika vegetation 137
Nzuve-trench 17

Oakleaf limpopo profile 181
    makulek profile 92, 147, 190, 219
    mutale profile 181, 219
ocean floor lavas 169
Ochrept 88, 116, 119, 132
Odontotermes 176, 201
Oioceros 138, 237
    tanyceras 137, 138, 149, 239, 240, 294
okapi 294
old-growth forest 93, 195, 231, 246, 247
    woodland 120
Oldoinyo Lengai volcano 15, 23, 25, 47, 53, 57, 66-68, 70, 71, 77-80, 89, 90, 98, 100, 120, 125, 127-129, 139, 145, 150, 229
Oldowan culture 256
Olduvai Beds 79
    Gorge 23, 140
Olea 89, 104, 136, 231, 293
    africana 136, 293
    welwitschii 89
Oleaceae 293
oleander 293
Oligichnus 205
Oligocene 60, 94, 156, 176, 182, 241
olive tree 136, 293
olivine 64, 68, 74, 83
Ombo village 14, 151
Onitis 106
Ontario province 34
onthophagine 105, 106
Onthophagus 106
Onuria Series paleosols 11, 13, 17, 18, 25, 27, 30, 39, 41, 46, 47, 52, 77, 81, 120, 128, 139, 140, 147, 148, 206, 235, 237, 238
    clay, type paleosol 14, 34, 36-39, 49, 50, 79, 117, 126, 136, 137, 140-142, 144, 239, 240, 261, 266, 276, 280, 283, 285, 286, 288
opal phytoliths 143
opaque oxides 64, 70, 71, 83, 91, 112, 114, 115, 121, 122, 130, 131, 140, 141, 179, 184, 188, 192, 198, 203, 206, 207, 211-213, 217
Opeas 290
open country mammals 151, 152, 241
    grassland 3, 117, 135, 144
operculum of snail 15, 182, 195, 196, 246, 296
Ophidea 293, 296
Ophiomorpha 205
opportunistic theories of human origin 3, 258
oral pellets of termites 195, 201
orang-utan, 8, 154, 226, 240, 247, 249, 256, 259, 296
Ordovician 251
Oreopithecus 138, 240, 294
Organ I surface 217

organic carbon 26, 96, 100-103, 108, 109, 115, 119, 122, 123, 127, 132, 142, 174, 180, 181, 185, 187, 189, 193, 199, 203, 208, 214
Oriental-Asian mammal fauna 222, 247
Ormocarpum trichocarpum 104
Orthent 4, 88, 125, 127, 147, 150, 159
Orthic Luvisol 215, 219
  Solonetz 200
ortho-isotubule 91, 101, 184, 207
Orthox 4, 59, 123
Orycteropus browni 297
  minutus 292
ostrich 138, 294
otter 17, 71, 292, 296
Ouranopithecus 256
overthrusting 59, 154-158
oxidized 16, 38
Oxisol 116, 194
Oxysternon 182

paddyfield 182
Pakistan, Miocene paleoenvironment 217-226, 241-249
  Miocene stratigraphy 154-172
  studied localities 6-8, 154-172
  tectonic development 154-158
Palaeomerycidae 292, 294
Palaeotragus primaevus 137, 138, 239, 240, 294
paleochannel 53, 65, 71, 81, 107, 110, 129, 139, 153, 158, 162, 164, 167, 172, 175, 177, 181-183, 191, 196, 206, 210, 211, 214, 216, 222, 223, 227, 246, 257, 259
  abandoned 177
  bone assemblages 17, 227, 228, 257, 258
paleoclimate 88, 92, 98, 103, 109, 112, 116, 120, 124, 127, 134, 142, 154, 175, 181, 185, 190, 194, 200, 204, 209, 215, 224, 235
paleocurrent 79, 164, 168, 170
paleoecology 227, 228
paleoenvironmental heterogeneity 4, 152, 225, 226
paleogeography 253
paleogully 60, 67, 78, 79, 131, 135, 138, 149, 239, 240
paleokarst 12, 26, 27, 33, 34, 44, 65, 66, 85, 88, 90, 95, 149, 231, 250, 251
paleomagnetic dating 173, 183, 188, 192, 203, 207, 211, 242
  sedimentation rate 223
  studies 57
  transition 184
paleosol, chemical analyses 27-32, 282-286
  derivation of names 85, 91, 95, 100, 107, 111, 113, 118, 121, 126, 130, 140, 172, 177, 183, 187, 192, 196, 202
  descriptions 83-145, 172-217
  diagenesis 43-52
  features 10-42
  mineral composition 278-281
  molecular weathering ratios 287-289
  number 261
  recognition 10, 17, 34, 41-42
  series 83
  texture 274-277
paleotopographic setting of paleosols 90, 95, 99, 106, 110, 113, 117, 120, 125, 129, 139, 144, 177, 182, 187, 191, 196, 201, 206, 210, 216, 230, 236, 237, 243, 248
Paleozoic 228, 241
Palhyaena sivalensis 176, 221, 245, 246, 297
Pallichnus 105, 106, 108, 290
palm 186, 216, 259
Pamal Domeli, village 165, 171
Pan 256
Pandu Series paleosols 11, 15, 22, 31, 39, 44, 46, 49, 52, 142, 182, 184, 187, 197, 201, 202, 205, 206, 209, 211, 216, 217, 219, 220, 223-225, 243, 244, 246, 251
  clay loam 20, 197, 198, 203, 261, 270, 276, 280
  clay thin surface phase 197, 203, 261, 270, 277, 281
  clay, type paleosol 10, 11, 13, 37, 197, 202-204, 261, 270, 277, 281, 283, 286, 288
panicoid grass 14, 136, 143, 148, 238, 240, 252, 293
Papilionaceae 135, 148, 195, 240, 293
papule 188
Paradiceros 236
  mukirii 124, 137, 138, 149, 239, 240, 294
Parageogale aletris 291
Paramachaerodus 297
Paranomalurus 138, 232, 240, 294
  bishopi 99, 110, 149, 233, 291
  soniae 291
  walkeri 291
Paraphiomys hopwoodi 291
  pigotti 99, 138, 291, 294
Parapodemus 296
Parashorea 186, 201, 205
parasitic volcanic cone 68, 77, 80
Paratethys Ocean 228, 235
parent material of paleosols 18-20, 27, 35, 36, 39, 90, 100, 106, 111, 113, 118, 121, 125, 129, 139, 144, 177, 182, 187, 191, 196, 202, 206, 211, 217
Parinari 104, 231
Parkia filicoides 93
Paşalar 240
Passeriformes 94, 291
patella bone 210
patterns of root traces 12
pea 293
peat 136, 185, 187, 204
peccary 294
pedalfer 215
pedisediments 67
pedocal 215
peds 34, 35, 70, 85, 91, 108, 111, 113, 114, 119, 121-123, 173, 177, 179, 180, 183, 184, 192, 193, 199, 203, 211
Pelecypoda 296
Pelee's tears 85, 91, 101
Pellic Vertisol 104, 134
Pellustert 125
Pennisetum mezianum 120
  purpureum 98
  stramineum 120
Pentacme 186, 201, 205
Perciformes 296
Percrocuta 137, 294
  grandis 297
permafrost 250

permineralized wood 190, 201, 205, 210, 231
pervasively displacive fabric 24, 25, 71, 96, 97, 98, 100, 102, 112, 128, 229
petrocalcic horizon 141, 142, 145
Petroferric Haplustoll 116
pH 13, 15, 87, 92, 96, 102, 109, 110, 112, 115, 116, 119, 123, 127, 142, 175, 180, 185, 189, 194, 199, 204, 209, 214, 231
Phaeozem 72, 123, 200, 219
phalange, bone 176, 191, 210
Phalarborwa intrusion 104, 228
Phanaeus 182
Phasianidae 291
pheasant 296
phenocryst 64, 66, 74, 75, 81-83, 113-115, 121, 122, 125, 126, 130
phenol 250
Phillips-Norelco 12045 x-ray instrument 40
phillipsite 46, 78, 114, 115, 130
Phiomys andrewsi 291
Phoenicopteridae 291
Phoenix sylvestris 216
phonolite 18, 36, 46, 57, 62, 76, 78, 80, 81, 113-116, 118, 124, 133, 140, 144, 235
  clasts 77, 119, 121, 122, 123, 126, 130, 140, 141, 149
  colluvium 90, 121
  lava 118, 120, 139, 149, 150, 238
  plateau 125, 129, 238
  sandstone 121
Phonolite-nephelinite 62
Phosphate 53, 89, 231
Phragmites maximum 210
Phyllanthus 89
phytogeography 227, 254
phytokarst 251
phytolith 14, 49, 143, 144, 241, 253, 255
piedmont 80
pig 110, 137, 162, 191, 210, 221, 234, 241, 291, 292, 294, 297
pigeon 94
Pila prisca 15, 182, 187, 195, 247, 296
  globosa 182
Pila Series paleosols 11, 13, 18, 31, 35, 39, 182, 187, 195, 196, 199-201, 205, 206, 210, 211, 218-220, 223-225, 234, 242, 243, 244, 246, 251, 252, 269
  clay nodular variant paleosol 10, 27, 198, 199, 201, 202, 261, 269, 276, 280
  clay, type paleosol 21, 35, 197-199, 202, 261, 270, 276, 280, 283, 286, 288
pilotaxitic microfabric 75
pipe, volcanic 64
Pir Panjal mountains 158
pirssonite 80
Pisces 296
Pittosporum 104, 231
Placandept 146
placic horizon 26, 151
planar bedding 78, 173, 206
Planolites 89, 94, 290
Planosol 68, 134
plant-animal coevolution 253
plant fossils 9-15, 93, 135-137, 151, 152, 231, 234, 236, 237, 253, 290, 293, 296
  nutrition 105
Plantaginaceae 293
Plantago coronopus 136, 293
plantain 136, 293
plate tectonic theory 55
plateau phonolites 125, 129, 238
platy peds 111, 118, 119, 130, 179, 188, 207, 208, 214
Platybelodon 297
Pleistocene 60, 153, 172, 250, 252, 255
Plinian-style eruption 77
Plinthaquult 4
Plinthic Paleudalf 59
plinthite 116
Plio-Pleistocene 66, 67, 70, 71, 79, 80, 98, 106, 127, 158, 175, 176, 229, 241, 242, 254, 255, 260
pod 14, 99, 135, 293
podial, bone 176, 191, 210
podo 135
Podocarpaceae 293
Podocarpus 135, 136, 293
point bar 177, 221
point counting 19, 84, 172, 274-281
poison volcanic spring 80, 81
pokeweed 136, 293
pole trees 231
pollen 13, 54, 117, 134-136, 152, 225, 240, 293
Polygenetic Agglomerate 57
Polymict Agglomerate 57, 62, 82, 83
Polyscias fulva 98
Pomatiasidae 290
pondweed 136, 293
Pongo 256
  pygmaeus 8
pooid grass 136, 143
porosity dependent alteration during burial 27
Porphyrio 296
porphyroskelic microfabric 96, 100-102, 108, 111, 113, 114, 119, 121, 122, 130, 131, 141, 173, 178, 179, 183, 184, 188, 192, 193, 198, 203, 207, 211, 212
Portland cement 14
Post-Modesto II surface 217
postglacial 252
Potamogeton pectinatum 136, 293
Potamogetonaceae 293
Potamon emphysetum 205
potash 180, 185, 187, 194
  feldspar 49
Potwar Plateau 47, 158-162, 165, 166, 168-170, 172, 178, 183, 188, 192, 197, 202, 207, 211, 224, 242
  Silt 171, 172, 197
prairie 201
  Soil 123, 134, 147
Precambrian 59
  basement 55, 155, 242, 249, 255
  gneiss 46, 69, 150
  paleosols 9
  schist 145
  shield 45
predator 81, 137
  accumulation of bone 17, 176, 210
preferred orientation, bone 137

preservation of bones 8, 105, 154, 234, 256
primates 138, 291, 294, 295, 296
Primigulella 105, 290
    miocenica 94, 290
prismatic ped 35, 113, 114, 119, 122, 130, 132
Proboscidea 176, 191, 210, 292, 294, 297
Prochrysochloris miocaenicus 94, 233, 291
Proconsul 8, 150, 230, 240, 256, 259, 294
    africanus 54, 89, 94, 99, 138, 233, 234, 240, 291, 294
    major 54, 89, 94, 99, 110, 234, 291
    nyanzae 5, 54, 235
Prodeinotherium hobleyi 292
Progalago 232
    dorae 99, 291
    songhorensis 291
Progenetta 297
Progonomys debruijni 296
Proheliophobius leakeyi 291
Pronasilio ternanensis 294
Propalaeoryx 294
Propotamochoerus 243
    hysudricus 191, 221, 297
Propotto leakeyi 291
Protanancus macinnesi 294
Protarsomys macinnesi 291
Protenrec tricuspis 94, 233, 291
Proto-Gangetic River 40, 187, 191, 196, 202, 206, 210, 221, 241, 242, 244, 291
    braid plain 246
Protoryx 297
Protragelaphus skouzesii 297
provisioning home base, by human ancestors 259
proviverrine 292, 294
Prunus africanus 98
Psammentic Haplustalf 215, 219
Pseudoglessula 290
pseudomorph 36, 66, 71, 79, 113, 115, 121, 122, 131
Pseudopeas 290
Pteridophyta 135, 293
Pteropodae 291
Ptychotrema 290
puddling of soil 26, 142, 180, 182, 187, 191, 194, 196
pulmonate snail 15, 182
Punjab 103, 181, 190, 194, 200, 204, 209, 215, 247
Punjabi language 160
pupal cell 21, 48, 101, 105, 106, 179, 180, 290, 296
pupation 182
pyrite 81, 185, 187
pyroclastic base surge 78, 79
    flow 27, 73, 77, 78
pyroxene 18-20, 23-25, 36, 37, 44, 64, 71, 72, 77, 83, 85, 96, 101, 102, 107, 108, 111-115, 118, 119, 121, 122, 126, 130, 131, 140, 141, 145, 213, 217, 229
pyrrhotite 36
Python 296
python 94, 291, 296

Qadirpur village 169
quail 291
quartz 25, 37, 59, 67, 70-72, 76, 85, 90, 91, 96, 100-102, 107, 108, 111-114, 116, 118, 121, 122, 141, 150, 165, 173, 176, 177, 178, 179, 182-184, 188, 192, 193, 196, 198, 203, 206, 207, 211-214, 217, 242, 250
quartzite 59, 170, 217
quartzofeldspathic alluvium 8, 24, 39, 202, 221
    colluvium 107
    gneiss 133
quasimangan 185
quasisesquan 21, 114, 192, 213
Quaternary 13, 20, 45, 67, 92, 109, 129, 133

Rabuor Series paleosols 11, 13, 30, 35, 39, 41, 46, 52, 76, 121, 124, 139, 147, 148, 151, 236
    clay, type paleosol 50, 118, 120-123, 125, 127, 235, 238, 261, 264, 275, 279, 282, 285, 287
radial fiber calcite 67, 71
radiometric date 18, 60, 65, 68, 70, 76, 82, 118, 153, 160, 171
radius, bone 176, 191
rail 296
rainfall, Gorakhpur 222
    Kisozi 222
    Koru 59
    Messina 145
    Rawalpindi 159
    Roorkee 222
    Seronera 150
rain forest 12, 103, 104, 110, 135, 232, 253, 254
    shadow 225, 238, 240, 247
rainstorm lahar 79
Rairo Series paleosols 11, 13, 17, 19, 23, 35, 46, 52, 76, 118-121, 125, 147, 148, 151, 236, 238
    bouldery clay 113, 114, 119, 236, 261, 264, 275, 279
    clay, type paleosol 114, 118-120, 122, 124, 261, 264, 275, 279
Rangwa Banded Tuffs 80
Rangwapithecus gordoni 99, 110, 233, 234, 291, 295
rarity of ape fossils 258
rat snake 291, 296
rate of sediment accumulation 153, 168, 169, 171, 222, 223
ratel 296
rattan cane 186
Ravi River 158, 177, 183, 202, 217
Rawalpindi, city 159, 165, 171
    Group 165, 168
Recent 254
recognition of paleosols 9, 17, 42, 53
reconstruction of Miocene animals 227, 233, 239, 245
    Miocene landscapes 145-150, 217-225, 227, 229-230, 235-234, 241-243, 247, 248
    Miocene paleosols 87, 92, 96, 102, 109, 112, 115, 119, 122, 127, 132, 142, 174, 180, 185, 189, 194, 199, 204, 208, 214
    Miocene vegetation 145-152, 217-226, 229-232, 234, 236-238, 242-246, 247, 248, 252-256, 259
recrystallization 9, 27, 51, 52
Red Bed Member 63, 70, 72, 73, 100, 107, 112, 113
    beds 154
    Brown Earth 102, 109, 147
    Calcareous Soil 88, 147
    Clay, 219
    Earth 215, 219
    Sea 235

reddening 112, 115, 122, 132, 174, 180, 185, 189, 193, 199, 208, 214
reduction spot 120, 168
Regosol 128, 147, 159
relict bedding 19-21, 23, 111, 112, 118-121, 126, 131, 173-175, 177, 179, 180, 183, 184, 187-189, 191-193, 197, 198, 200, 203, 207, 209, 211-214, 217, 231, 238
  paleosol 249
Rendoll 142
Rensburg rensburg profile 185, 219
replacive microfabric 20-23, 46, 229
reptile 69, 94, 138, 149
Rhamnaceae 238, 293
rhinoceros 124, 137, 138, 149, 162, 176, 191, 210, 221, 234, 240, 241, 247, 249, 253, 292, 294, 297
rhizoconcretion 9, 10, 11, 21, 22, 174, 180, 181, 188-190, 204, 211, 213
rhizome 99, 252, 255
rhizomyid rat 296
Rhizomyoides 296
rhizosphere 12
rhyolite 59, 88
rice soils 120, 125
Ricinodendron heudelotii 93
Ries Crater 236
Rift Valleys of East Africa 55, 228, 235, 255
Rindkarren 33
riparian forest 220, 222, 231, 242, 247, 259
  woodland 135, 181, 222, 225
ripple drift cross lamination 96, 112
  mark 78, 79, 126, 207
river terrace 106, 196, 217
riverain forest 220, 247
rodent 99, 100, 110, 138, 247, 291, 292, 294, 296
Rondonin village 151
Roorkee city 222
root trace 9, 13, 19, 23, 24, 26, 42, 43, 46, 51, 60, 81, 85, 88, 91-93, 96, 101-103, 106-108, 111, 112, 114, 115, 117-124, 126-128, 130-132, 134, 136, 141, 143, 151, 167, 172-175, 177, 179, 181, 182, 184-186, 188, 192-196, 198-201, 203, 207, 209, 210, 218, 229, 238, 242, 244, 255, 290, 293, 296
  deeply penetrating 12
  tabular 12
Ruanda 26, 81, 109, 134, 146
Rubiaceae 136, 293
ruminant 205
Rusinga Island 14-17, 46, 48, 53, 80, 93, 95, 105, 106, 112, 129, 150, 228, 234, 235, 240, 252, 254
Russia 249

sabre-tooth cat 292, 297
Saccharum procerum 210
Sadhpur village 215
Sahara Desert 241, 253
Saharanpur town 222
Saka Series soils 123
Sakesar Limestone 162, 165
sal 186, 190, 195, 201, 205
saline soils 35, 180, 199, 209
salinization 88, 97, 109, 112, 123, 132, 133, 142, 150, 180, 185, 189, 199, 200, 204, 209, 214, 216
Salmalia malabarica 210
salt 23, 47, 127, 132
salt lake 249
Salt Range 156, 158, 162, 165, 169, 171
  Range Thrust 158
Samburu Hills 151
Samotherium africanum 137, 294
San Joaquin valley 107, 196, 217
sand wedge 250
Sangrut town 180
sanidine 75, 82, 114, 115
Sapindaceae 105
Sapium ellipticum 89, 98
saprolite 59, 82, 115
Sarang Series paleosols 11, 13, 19, 21, 39, 40, 46, 172, 173, 174, 176, 179, 181, 182, 187-193, 195, 201, 203, 206, 216, 218, 219, 220, 222, 223, 224, 225, 242, 243, 246-248, 257, 258, 266, 268-273
  clay cumulic surface phase paleosol 173, 178, 188, 261, 272, 277, 281, 283, 284, 286, 289
  clay eroded phase paleosol 173, 207, 208, 261, 271, 277, 281
  clay krotovina variant paleosol 261, 269, 276, 280
  clay loam paleosol 174, 207, 208, 261, 277, 281, 283, 286, 288
  clay rhizoconcretionary variant paleosol 174, 178, 261, 272, 277, 281, 284, 286, 289
  clay paleosol 173, 189, 261, 277, 281
  clay, type paleosol 10, 50, 172-174, 261, 266, 267, 276, 280
  silty clay paleosol 16
  silty clay massive surface phase paleosol 174, 211, 212, 261, 272, 277, 281
  silty clay rhizoconcretionary variant paleosol 174, 178, 187, 188, 261, 272, 277, 281, 283, 286, 288
Sargodha basement ridge 156, 158, 165, 242
  town 155
Satghara Series soils 26
Saturniidae 106
Saudi Arabia 259
savanna 3, 134
scale of development of soils 65, 75, 146, 166, 167, 218
  of calcareousness of soils 274-277
scanning electron microscope 13, 37, 143, 176
scaphoid, bone 205
scapula, bone 176, 191
Scarabaeidae 179, 180, 182
scavenger 81, 176
  accumulation of bone 17, 176, 210
scavenging by human ancestors 3, 258
schist 40, 51, 145, 196
  clasts 166, 169, 177, 179, 178, 182, 183, 188, 198, 203, 206, 207, 211-213, 217, 242
schistosity 27, 43
Schizochoerus gandakasensis 297
Scincidae 290
Scintag Pad V x-ray instrument 41
Sciuridae 294, 296
Sclerocarya 117
sclerophyll forest 216, 260
  woodland 103, 128, 201, 259

scoria 23, 65, 107
scour-and-fill 100, 140
scroll bar 187
scute, crocodile 210
sea level 235, 249
seasonal drainage 39
    flooding 9, 15
    puddling 125
    waterlogging 13, 14, 26
secondary porosity 38, 49
    grassland 59, 92, 146
    wooded grassland 68
sedge 99, 135, 144, 210, 252, 293
sedimentary facies model 227
seed 135
seepage 119, 120, 210
seive analysis for grain size 84
Selenoportax 246, 248
    lydekkeri 160, 162, 171, 177, 216, 221, 191, 245, 247, 297
    lydekkeri, zone of 171
    vexillarius 205, 221, 246, 297
semiarid climate 24, 40
semideciduous forest 93, 146
sepic microfabric 34, 38, 39
Serek village 76, 80, 82, 122
Serengeti National Park 116, 238
    Plains 18, 19, 36, 43, 46, 103, 107, 109, 120, 123-125, 128, 133-136, 140-143, 144, 145, 150, 181, 229, 235, 252
    woodlands 107
Seronera 150
Serpentes 291
sesquiargillan 91, 126, 188, 198, 203, 207, 213
sesquan 108, 114, 121, 122, 130, 131, 132, 141, 192, 184, 194, 196, 211, 213, 217
sesquioxides 36, 39
Sethi Nagri, village 169
sexual dimorphism of apes 256
    display by human ancestors 259
shaft of bone 176
Shahdara Series soils 177
shale clasts 177
shark teeth 162
sheep 205
sheetwash 121
shell 53, 55, 60, 67, 69, 85, 89, 91, 94, 195, 257
shield volcano 57
Shin Ghar Range 168, 169
Shorea 186, 190, 201, 205
    robusta 186, 190, 195, 201, 205, 244, 246
shrew 296
shrink-swell behavior 39, 120
shrub 136, 181, 234
shrubland 181, 190
Siberia 34
Sichuan 251
siderite 81, 136, 187
silan 35, 173, 183
silasepic microfabric 173, 141, 207
silica 90, 187, 196, 206, 217, 231, 282-284
silicification 81
sillimanite 158
Siluriformes 296
Simonimys genovefae 291
Sind Sea 156
sissu, plant 175
Sitala 290
Sivalhippus 297
Sivaonyx bathygnathus 296
Sivapithecus 8, 154, 225, 240, 243, 249, 256, 259, 260
    punjabicus 176, 221, 246, 247, 296
    sivalensis 5, 162, 175, 176, 221, 245-247, 296
Siwalik Chronofauna 249
    Group 14, 47, 153, 154, 158, 159, 165, 171, 172, 175, 181, 184, 189, 190, 193, 195, 199, 201, 203, 205, 210, 216, 224, 225, 227, 228, 242, 245, 257
    Hills 158
skeletan 212
skelinsepic microfabric 119, 122, 130
skelmasepic microfabric 39, 88, 96, 90, 121, 122, 198, 203, 212, 213
skelmosepic microfabric 39, 96, 102, 111, 112, 119, 121, 130, 131, 141, 178, 179, 183, 188, 198, 203, 212, 213
skelsepic microfabric 118, 121
skink 290
Skolithus 99, 106, 110, 113, 139, 174, 176, 177, 191, 201, 290, 293, 296
skull 16, 137, 162, 175-177, 235, 256
slickensides 26, 35, 113, 119-122, 123, 124, 173, 179, 183, 184, 186, 192, 193, 198, 200, 203, 207, 235
slit-faced bat 291
slow loris 294, 296
smectite 36, 40, 49, 92, 109, 116, 134, 177, 180, 185, 187, 190, 194, 229, 235, 242
Smilax 195
snail 15, 53, 55, 60, 67, 69, 85, 89, 91, 94, 99, 105, 107, 110, 117, 128, 129, 135, 137, 139, 149, 151, 152, 172, 182, 184, 186, 187, 195, 196, 228, 232-234, 240, 246, 253, 258, 290, 293, 295, 286
snake 94, 99, 110, 149, 291, 293, 296
Soan Formation 47, 158, 170
    Gorge 165
    industry 172
    River 158-171
    Syncline 47, 160, 172
socialization of human ancestors 259
soda 27, 180, 194
soda/potash ratio 46, 88, 92, 97, 102, 109, 123, 132, 133, 142, 180, 185, 189, 199, 202, 209, 214
sodic soils 35, 180, 199, 209
sodium 199
soft sediment deformation 102
    pan 136
softshell turtle 296
soil catena 124
    chronosequence 107
    clasts 37
    development 146
    geography 249
    horizons 17, 42
    map of the world 4
    structure 34, 42
soilscape 145, 146, 182, 217, 222

Solonetz 134, 200
Solonized Brown Soil 200
Somali-Masai *Acacia-Commiphora* wooded grassland 98, 135, 239
Somalia 92, 94, 104, 146
songbird 291
Songhor Hill 68, 69, 96, 100, 107, 111
  National Monument 43, 46, 48-50, 53, 63, 65, 67-69, 71, 79, 96, 97, 100, 101, 105, 107, 111, 112, 146, 148, 150, 234, 235, 240, 254, 258, 290-292, 295
  village 11, 12, 14, 16-18, 22, 25, 29, 36, 57, 60, 62, 63, 70, 74, 75, 100, 107, 111, 139, 145, 228, 229
Sonita Series paleosols 11-13, 22, 32, 35, 39, 40, 45, 52, 179, 180, 186, 187, 189-191, 193-195, 199, 202, 203, 211, 215, 217-219, 220, 222, 223, 224, 242, 246-248, 257, 268, 270-272
  clay ferric concretionary variant 183, 261, 267, 276, 280
  clay, type paleosol 44, 50, 178, 187-191, 261, 272, 277, 281, 283, 286, 288
  silty clay paleosol 50, 188, 189, 261, 277
Soricidae 296
Sossok Series soils 4
  clay loam soil 75
South Africa 102-104, 116, 124, 142, 145, 175, 181, 190, 194, 209, 210, 215, 228, 259
  America 182
  Dakota 241
Southeast Asia 154
sövite 20, 60, 64, 65, 70, 79, 85, 96, 101, 108, 126, 130, 140, 141
Spain 259
sparry calcite 9, 20, 21, 23, 48, 51, 66, 78, 96, 101, 102, 107, 118, 119, 126, 127, 130, 131, 141, 173, 176, 179, 180, 184, 188, 189, 198, 199, 211, 214, 229, 278-281
Spathodea campanulata 89
spherical micropeds 37, 190, 192, 194, 195, 201, 203, 250, 251
spheroidal weathering 118
Sphingidae 106
spider 16, 17, 53, 94, 99, 106, 110, 113, 129, 139, 149, 176, 191, 201, 205, 216, 221, 228, 290, 293, 296
spined turtle 296
spore 135, 225, 293
Sporobolus iocladus 144
  kentrophyllus 128
  marginatus 144
Spreiten 176
spring hare 99, 233, 257, 291, 294
spying predators by human ancestors 3, 258
squirrel 292, 294, 296
Sr (strontium) 71, 79, 183, 285, 286
stage, chronostratigraphic 159
staurolite 168
stegodont 297
Stegolophodon 176, 221, 246, 297
Steinheim Crater 236
stenogyrid snail 139
Stenoplesictis muhoroni 292
Stephania 105
Sterculia 94, 136, 293
  appendiculata 93
Sterculiaceae 293
stinging nettle 135, 293
stipular spine 135, 240, 241, 293
stomate 49, 144
stone line 19
  tool 80, 170, 172, 222, 256
  stripe 250
stork 296
stratigraphy Kenya 57, 58, 62
  Pakistan 161
streptaxid snail 94, 99, 290, 293
Streptostele 290
strike ridge 168
strongly developed paleosols 22, 24
strontium 71, 79, 183, 285, 286
Struthio 294
stump cast 21, 49, 60, 80, 81, 126-128, 133, 135, 139, 293
sub-Miocene erosion surface 59
subangular blocky peds 85, 108, 173, 192, 203
subhumid climate 24, 40
subsurface calcareous nodules 20
  clayey horizons 18
  horizons rich in iron manganese 26
Subulina 139
Subulinid snails 94, 99
Subulinidae 290, 293
Subulininae 105, 110, 290
succulent plants 128
Sudanian vegetation 117, 259
  woodlands 104, 231
Sudnatti town 201
sugarcane 59
Suidae 292, 294, 297
Sulcophanaeus 182
sulfur mineralization 81
Sultanpur Series soils 183, 217
surface water gley 191, 194, 196, 199, 208, 214
Surghar Range 165, 166
survey block, fossil collection 227
Sutlej River 158, 242
swale 221, 244
swamp 151, 153, 186, 201, 216, 222, 244, 246, 259
sweat bee 95
swelling clay 49
Swift Automatic Point Counter 84, 274-281
syenite-carbonatite intrusion 228
Syzygium claviflorum 186
  cumini 186, 216

tabular root system 103, 127, 204
talc 36
Tanzania 7, 15, 18, 19, 21, 23, 25-27, 34-36, 60, 66, 67, 70, 71, 78, 79, 90, 98, 100, 103, 106, 107, 109, 116, 120, 123, 125, 127-129, 133-136, 139, 140, 142, 144, 145, 150, 176, 181, 229, 235, 238
Taphozous incognita 291
Taucanamo 294
Tayassuidae 294
Tayloria 99, 105, 290, 293
Teclea 99
  nobilis 98
tectonic breccia 36

teeth 53, 91, 134, 138, 176, 191, 210, 246
Teleki Valley 90, 121
temperate climate 33
  soils 249, 250
temperature 45
  Gorakhpur 222
  Kisozi 146
  Koru 59
  Messina 145
  Roorkee
  Seronera 150
temporal completeness 223
Tenebrionidae 129, 293
tenrec 94, 291
tensional origin of rift valleys 55
tepee structures 35
tephrite 62
terai 170, 186, 200, 201, 218
Teratodon enigmae 89, 292
  spekei 89, 233, 292
Terminalia 94, 104, 105, 117, 128, 181, 190, 216, 231, 234
  arjuna 181, 216, 244, 246
  bellerica 205
  mannii 205
  tomentosa 190
  zambeziaca 93
termite 23, 89, 92, 94, 131, 135, 139, 173-176, 190, 191, 195, 198, 200, 201, 204, 206, 215, 221, 229, 246, 247, 250, 290, 296
  mound 37, 195
  pellets 36, 37
Termitichnus 94, 188, 173, 174, 176, 195, 206, 216, 290, 296
terrace, coastal 121
  fluvial 106, 196, 217
terrapin 296
tesselated pavement 111
Testudo 296
Tethyan evergreen laurel forests 254, 259
Tethys Ocean 156, 228, 235
Tetraconodon magnus 297
Thaganwali Dhok village 7
Thapsia 290, 293
Themeda arundina 201
theridomyid cane rat 138, 291, 294
thermal maturation of organic matter 38, 49, 52
thicket forest 53
thin section preparation 39
thorn scrub 158, 240
thorny twigs 14
threatening predators by human ancestors 259
three-toed horse 176, 182, 191, 205, 210, 221, 246, 247, 249, 297
throwing weapons by human ancestors 258
Thryonomyidae 249
Tibetan Plateau 156, 241
tibia, bone 176, 191
Tiliaceae 232
till 107, 252
Timboroa volcano 55, 57, 80, 83
time for formation of paleosols 90, 100, 107, 111, 113, 118, 121, 125, 129, 140, 145, 146, 177, 183, 187, 191, 196, 202, 206, 217
Tinderet basanite 64
  lavas 75
  melanephelinites 64
  Peak 59, 62, 72, 228
  volcano 6, 48, 55, 57, 59, 68, 74, 82, 83, 232
  Volcanics 61, 62, 68
titania 18, 27, 140, 211, 282-284
tomb bat 94, 291
tool wielding by human ancestors 260
tools, stone 80, 170, 172, 222, 256
  Acheulian 256
  Oldowan 256
  Soan 172

tooth 53, 91, 134, 138, 176, 191, 210, 246
  enamel 100
  microwear 138
Tororo village 34
tortoise 149, 181, 191, 205, 216, 221, 296
tourmaline 165
trace elements 183, 285-286
  fossils 9-13, 15-16, 89, 94, 128, 139, 140, 174-176, 179, 183, 195, 201, 205, 246, 247, 290, 292-294, 296
trachyphonolite 118
trachyte 57, 146
trachytoidal texture 20, 70, 71, 91, 114, 115
Tragulidae 249, 292, 294, 297
trampling damage to bone 81
transportation of bone 154
tree 134, 136, 143, 181, 186, 210, 234
  shrew 296
tremolite 36
Trewia nudiflora 186
Triassic 156, 224
Trichilia prieuriana 93
Triclisia 105
trimasepic microfabric 114, 119
Trinervitermes 176
Trionyx 296
Trochonanina 290, 293
Trochozonites 290
Tropept 88, 132
tropical climate 22, 23
  soils 6, 37, 227, 249, 251
Tropidophora miocenica 290
Tropudalf 4
true lizard 290
trunk cast 14, 78, 131
Tsavo West National Park 137
Tubiflorae 136, 293
Tubulidentata 292, 294, 297
tuff 27, 41, 47, 48, 53, 60, 62, 63, 65, 67, 71, 76, 77, 79, 82, 85, 88, 90-92, 94, 95, 99-102, 105-107, 109, 111, 113, 118, 120, 121, 127-129, 134, 140, 144, 145, 149, 150, 151, 228, 229, 231, 234, 238, 251, 257
  cone 100
Tugen Hills 151
Tunisia 134
Tunnel railway siding 62, 82
  Tuff Member 62, 82

Tupaiidae 296
turaco 94, 291
Turkana Lavas 57
Turkey 240, 256
Turoka System 59, 69
turtle 17, 71, 99, 110, 113, 149, 162, 290, 296
tusk 176, 191, 210
Tut Series paleosols 11, 13, 18, 19, 24, 27, 29, 39, 41, 45-47, 52, 70, 71, 96, 98, 100, 103, 105, 107, 139, 142, 147, 148, 151, 206, 230, 233, 234
  clay, type paleosol 22, 50, 97, 100, 101, 107-112, 113, 229, 261, 263, 274, 275, 278, 279, 282, 285, 287
twig casts 14, 49, 82, 290, 293
Typha 136, 293
Typhaceae 293
Typic Haplaquept 204, 219
  Tropudalf 75

U-sandstone 162, 170
Uasin Gishu Plateau 235
Udic Haplustalf 215
  Ustochrept 215
Uganda 14, 34, 53, 60, 89, 92, 93, 98, 99, 104, 109, 110, 123, 146, 231, 232, 252
Ulmaceae 232, 290, 293
ulna, bone 176, 191
Ultisol 116
umbrella tree 113
undulic microfabric 126
unionid 296
unistrial microfabric 173, 179, 188
United States 128, 134, 139, 140, 181, 186, 215, 237
upland drainage net 224
Upper Modesto surface 217
upright stance of human ancestors 258
upward-fining sequence 9, 27, 78, 140, 126, 193, 206
Urdu language 172, 177, 183, 187, 192, 196, 202, 207, 211, 219
Urocyclidae 290, 293
Urocyclinae 290
Urticaceae 135, 293
Ustochrept 180, 189, 190, 194, 215, 219
Ustorthent 128
Ustropept 92, 116, 119, 145, 147, 150, 180
Uttar Pradesh province 171, 180, 181, 185-187, 190, 195, 200, 201, 204, 205, 215, 216, 222, 242, 245, 247

Valsrivier lilydale profile 215, 219
  lindley profile 181
  marienthal profile 102, 109, 147, 194, 219
Varanidae 290
Varanus 296
variegated beds 154
vector magnitude 79
  mean azimuth 79
vegetation of Miocene time 12, 13, 54, 55, 92, 88, 98, 103, 109, 112, 117, 120, 124, 128, 134, 154, 143, 146, 149-152, 154, 175, 181, 186, 190, 195, 201, 204, 209, 215, 221-223, 225, 226, 229-232, 234, 236-238, 242-247, 252-256, 258-260
vermiculite 20, 22, 41, 67, 85, 90-92, 107, 108, 131
Vernonia 98
vertebra, bone 176
Vertic Argiustoll 123, 147
  Cambisol 116, 119, 120, 147
  Eutropept 116, 147
  Haplaquept 185, 186, 219
  soils 137
  Ustochrept 189, 219
  Ustropept 119, 147
Vertisol 35, 49, 116, 117, 123, 124, 134, 186, 210, 214, 235
vesicle 113, 114
vesicular soil structure 190
Vespertilionidae 291
Victoriapithecus 151
Vindhyan Supergroup 155
vine 181, 195, 216, 222, 225, 247
viper 291
Viperidae 291
Vitrandept 146
Vitrow Formation 158
Viverra chinjiensis 176, 221, 246, 296
Viverridae 176, 296
vlei grassland 252
volcanic apron facies 71, 80
  arena facies 65, 71
  breccia 36, 59, 60, 67, 70, 73, 76
  eruption 20, 23, 33
  glass 64
  mudflow 9, 73, 78
  soils 229
  spring 81
vomasepic microfabric 192, 193
Vulcanian-style eruptions 66, 80
Vulcanisciurus africanus 292
vulture 81, 138, 294

wading by human ancestors 258
Walangania 230, 233
  africanus 89, 94, 99, 105, 110, 137, 149, 233, 292, 294
Walkers Limestone 14, 63, 67
Walkley-Black determination of organic carbon 43, 84, 172, 282-284
Walther's facies rule 182
wart snake 296
wasp 95, 139, 250, 290, 293
water snake 291, 296
  table 117, 222, 224
waterlogging 12, 26, 39, 97, 120, 123, 139, 182, 186, 187, 204, 208-210, 218, 225, 226, 242, 250, 252
Watuni profile 4
wavy bedding 173
weasel 296
weathering rind 22, 33
Weaver-index 50
Weber-index 50
welded tuff 12, 26, 27, 33, 66, 149, 231, 251, 262
West Bengal province 180, 185
  Gangetic moist mixed deciduous forest 195
  Germany 236
Western Australia 156
  Ghats hills 156
  light plains alluvial sal 205
  Rift, Africa 26, 146

wet forest 191, 225, 257
wheat 256
whip scorpion 16, 129, 139, 149, 293
whole rock chemical analyses 84, 282-286
Wiesenboden 200, 209, 219
Willowbrook chinyika profile 209, 219
Winam Gulf 55
wolf 294
wollastonite 79, 107, 141, 145
women's liberation 259
wood 13, 60, 104, 151, 162, 190, 201, 203-205, 207, 209-211, 214, 225, 231, 250
wooded grassland 3, 13, 53, 54, 59, 72, 89, 93, 98, 104, 117, 120, 124, 128, 134-138, 143, 148, 150, 152, 159, 172, 181, 186, 190, 195, 201, 205, 210, 216, 218, 220, 222, 225, 234, 238, 239, 240, 246, 252-254, 256, 257, 259
  shrubland 137
woodland 3, 13, 53, 55, 104, 106, 135, 136, 138, 145, 175, 181, 186, 190, 195, 201, 204, 215, 216, 222, 225, 234, 240, 246, 254, 255, 258, 259, 260
  adapted fauna 241
worm lizard 291
wormwood 293
Wrightia tomentosa 201
Wyoming 139

x-ray diffractometry 40, 41, 51, 145
Xanthic Ferralsol 123
xenolith 64, 83, 118
Xerorthent 128

Yamuna River 158, 222
Yellow Tuffs and Agglomerates 66, 78, 90

Zaire 26, 81, 99, 146, 231
Zambezian deciduous woodland 103, 104, 110, 135, 150, 231, 234, 239
  miombo woodland 104, 231
  riparian woodland 105, 231, 240
  vegetation 105, 117, 146, 234, 259
  wooded grassland 150
zebra 241, 255
Zenkerella wintoni 291
zeolite 23, 46, 47, 52, 73, 77, 78, 80, 114, 115, 150, 229
zeolitization 46, 47
Zimbabwe 102, 103, 116, 120, 145
Zingis 290
zircon 165
zirconium 18, 27, 211, 285, 286
Zizyphus 105, 234, 240
  abyssinica 238
  rusingensis 135, 148, 293
  sivalicus 195
  zeyheriana 104
Zn (zinc) 183, 285, 286